AF595027

Remote Sensing of the Aquatic Environments

Remote Sensing of the Aquatic Environments

Editors

Giacomo De Carolis
Francesca De Santi

MDPI • Basel • Beijing • Wuhan • Barcelona • Belgrade • Manchester • Tokyo • Cluj • Tianjin

Editors
Giacomo De Carolis
Institute for Electromagnetic Sensing of the Environment (IREA)
National Research Council of Italy (CNR)
Milan
Italy

Francesca De Santi
Institute for Electromagnetic Sensing of the Environment (IREA)
National Research Council of Italy (CNR)
Milan
Italy

Editorial Office
MDPI
St. Alban-Anlage 66
4052 Basel, Switzerland

This is a reprint of articles from the Special Issue published online in the open access journal *Remote Sensing* (ISSN 2072-4292) (available at: www.mdpi.com/journal/remotesensing/special_issues/remote_sensing_aquatic_environments).

For citation purposes, cite each article independently as indicated on the article page online and as indicated below:

LastName, A.A.; LastName, B.B.; LastName, C.C. Article Title. *Journal Name* **Year**, *Volume Number*, Page Range.

ISBN 978-3-0365-1666-0 (Hbk)
ISBN 978-3-0365-1665-3 (PDF)

Contents

Preface to "Remote Sensing of the Aquatic Environments"

The observation of the aquatic environments represented by inland surface water, seas and oceans has been traditionally linked to the need for safe navigation and fishing locations. More recently, there has been a growing demand on monitoring capability due to increasing concerns about the contaminants produced by anthropogenic activities on the quality of inland and coastal waters.

Remote observations allow us to gather plenty of information about surface temperature, winds, currents, ocean color, coral reefs, sea and lake ice, suspended solid concentrations, algal blooms, and other bio-geophysical parameters related to the aquatic environment.

In this context, active and passive remote sensors offer suitable solutions for the synoptic monitoring of the water surface along with all the properties directly involved. Our aim is to develop methods and applications to extract detailed environmental information from multisensor observations.

This book—*Remote Sensing of the Aquatic Environments*—is focused on the relevant aspects related to the remote measurement of the bio-geophysical properties of bodies of water and the methodologies aimed at studying the relevant processes. It includes a collection of research efforts aimed at improving our capability to monitor inland waters such as lakes and lagoons, water reservoirs, and coastal areas including estuarine and river delta regions. These water districts represent the most sensitive regions of our planet where the delicate balance among all the ecological systems should be preserved.The book is aimed at a wide audience, ranging from graduate students, university faculty members and scientists to policy makers and managers.

Giacomo De Carolis, Francesca De Santi
Editors

Article

Empirical Estimation of Nutrient, Organic Matter and Algal Chlorophyll in a Drinking Water Reservoir Using Landsat 5 TM Data

Md Mamun [1], Jannatul Ferdous [2] and Kwang-Guk An [1,*]

[1] Department of Bioscience and Biotechnology, Chungnam National University, Daejeon 34134, Korea; mdmamun@o.cnu.ac.kr
[2] Climate Change Lab, Department of Civil Engineering, Military Institute of Science and Technology, Mirpur, Dhaka 1216, Bangladesh; jannatul@ce.mist.ac.bd
* Correspondence: kgan@cnu.ac.kr; Tel.: +82-42-821-6408; Fax: +82-42-822-9690

Abstract: The main objective of this study was to develop empirical models from Landsat 5 TM data to monitor nutrient (total phosphorus: TP), organic matter (biological oxygen demand: BOD), and algal chlorophyll (chlorophyll-a: CHL-a). Instead of traditional monitoring techniques, such models could be substituted for water quality assessment in aquatic systems. A set of models were generated relating surface reflectance values of four bands of Landsat 5 TM and in-situ data by multiple linear regression analysis. Radiometric and atmospheric corrections improved the satellite image quality. A total of 32 compositions of different bands of Landsat 5 TM images were considered to find the correlation coefficient (r) with in-situ measurement of TP, BOD, and CHL-a levels collected from five sampling sites in 2001, 2006, and 2010. The results showed that TP, BOD, and CHL-a correlate well with Landsat 5 TM band reflectance values. TP ($r = -0.79$) and CHL-a ($r = -0.79$) showed the strongest relations with B1 (Blue). In contrast, BOD showed the highest correlation with B1 (Blue) ($r = -0.75$) and B1*B3/B4 (Blue*Red/Near-infrared) ($r = -0.76$). Considering the r values, significant bands and their compositions were identified and used to generate linear equations. Such equations for Landsat 5 TM could detect TP, BOD, and CHL-a with accuracies of 67%, 65%, and 72%, respectively. The developed empirical models were then applied to all study sites on the Paldang Reservoir to monitor spatio-temporal distributions of TP, BOD, and CHL-a for the month of September using Landsat 5 TM images of the year 2001, 2006, and 2010. The results showed that TP, BOD, and CHL-a decreased from 2001 to 2006 and 2010. However, S3 and S4 still have water quality issues and are influenced by climatic and anthropogenic factors, which could significantly affect reservoir drinking water quality. Overall, the present study suggested that the Landsat 5 TM may be appropriate for estimating and monitoring water quality parameters in the reservoir.

Keywords: empirical models; multiple regression; Paldang Reservoir; water quality parameters

Citation: Mamun, M.; Ferdous, J.; An, K.-G. Empirical Estimation of Nutrient, Organic Matter and Algal Chlorophyll in a Drinking Water Reservoir Using Landsat 5 TM Data. *Remote Sens.* **2021**, *13*, 2256. https://doi.org/10.3390/rs13122256

Academic Editor: Giacomo De Carolis

Received: 26 April 2021
Accepted: 8 June 2021
Published: 9 June 2021

Publisher's Note: MDPI stays neutral with regard to jurisdictional claims in published maps and institutional affiliations.

1. Introduction

Freshwater reservoirs are significant natural resources within the biosphere that function as sources of drinking, irrigation and industrial water, tourism attractions, and aquatic organisms' habitats [1–3]. These reservoirs face a number of stressors, including land use change, pollution, intensive farming, climate change, and human activities, causing several water quality issues [1,4,5]. About fifty percent of the world's populations live near water resources, and human activities accelerate aquatic stressors like eutrophication and algal blooms [4]. Due to rapid urban population growth, industrialization, intensive agricultural farming, and global climate changes, reservoirs are facing significant challenges, most important of which are the rise in nutrients, algal blooms, and organic matter pollution [6–8]. This is a global environmental issue and a current research subject [3,9,10].

Paldang Reservoir is one of the main reservoirs in South Korea, formed by the construction of a hydroelectric dam in 1973 [11]. It has been used for various purposes such as irrigation, hydroelectric, fishing, recreation, and drinking water [12]. It has been declared a nationally protected resource and provides water for the Seoul metropolitan and surrounding areas [13]. Approximately half of the Korean population depends on the Paldang Reservoir for drinking water [14]. Simply put, the water quality of the reservoir is crucial to the Korean government. However, human activities have risen in the watershed, resulting in short-term algal blooms and organic pollution in the reservoir [15,16]. Urbanization, municipal pollutants, livestock farms, intensive farming practices, domestic and industrial wastewater, and inflowing rivers contribute to the water contamination of the reservoir [17,18]. Henceforth, monitoring the nutrients, organic matter, and algal chlorophyll concentrations and determining their spatial and temporal dynamics are essential to managing the reservoir water quality [3].

Traditional monitoring approaches, including in-situ measurements and laboratory analysis, allowed us to understand and categorize water-quality parameters [1,3,19]. Though this technique yields accurate measurements, it is time-intensive and laborious and may not provide an overview of water quality at a broad spatial scale [20]. Furthermore, current monitoring techniques can not cover the wider spectrum of spatial and temporal analysis which is necessary to resolve aquatic integrity and public health issues [4]. It is particularly true for large water bodies like Paldang Reservoir, one of Korea's largest freshwater sources.

Satellite remote sensing is currently one of the most powerful and most reliable methods for monitoring and managing water quality [3,21]. Readily accessible remote sensing data offers cost-effective and less time-intensive methods than in-situ methods by providing continuous spatial and temporal coverage of environmental processes [1]. This approach delivers a large-scale synoptic range of the systems [19,22]. Spectral satellite radiance measurement is interrelated to many water quality variables influencing an aquatic ecosystem's optical properties [23,24]. Several previous studies have shown that satellite systems' brightness data are closely associated with water quality variables [3,4,19–22,24,25].

Miller et al. [26] noted that the "*Landsat series provided an approximate annual economic benefit of 2.19 billion US dollars spread across several study areas for only the USA*". Since 1984, Landsat 5 has provided a steady stream of data with a moderate spatial resolution (30 m), multispectral imagery, and a sampling rate of 16 days [4,19,24]. Therefore, these pictures are appropriate for demonstrating the study of aquatic resources. The moderate spatial resolution of images allows us to study a small water body, about 8 ha [25]. It indicates that Landsat Thematic Mapper (TM) sensor can be used extensively to form the empirical relationships among water quality parameters and spectral reflectance values. The most common way to determine a relationship among spectral reflectance values and water quality parameters are through regression analysis. The most critical aspect of running a regression analysis is choosing a regression model with appropriate independent parameters (single bands, band ratios, and combinations of bands) that yield a high R^2 value. A high R^2 value reveals that the return equation is highly correlated with existing data and provides a relatively accurate model. However, previous studies demonstrated that the bands that best predict water quality parameters differ with water conditions and ecosystems. Therefore, empirical models must be individually developed for each variable at different systems. Researchers used Landsat 5 sensor to determine the spatial and temporal distribution of water quality parameters throughout the world, including chlorophyll-a (CHL-a), turbidity, Secchi depth (SD), total suspended solids (TSS), total phosphorus (TP), organic matter (BOD), electrical conductivity, etc. [4,19,21,22,25,27–30].

This study aimed to develop a method for using Landsat 5 TM data to determine TP, BOD, and CHL-a concentrations in Paldang Reservoir, Korea. This reservoir was selected for study due to its status as a nationally protected resource. This research's primary objectives were to: (i) determine the relationship among TP, BOD, and CHL-a with TM bands, band ratios, and combinations of bands and (ii) develop empirical models using regression analysis for monitoring TP, BOD, and CHL-a. The developed models were also used to evaluate the spatio-temporal variations in TP, BOD, and CHL-a among study sites during 2001, 2006, and 2010.

2. Materials and Methods

2.1. Study Area

The Paldang Reservoir is situated approximately 45 km northeast of Seoul and provides drinking water for 24 million people [14]. It has an area of 38.2 km^2 and a volume of 250 × 106 m^3 [11]. The mean and maximum depth of the reservoir is 6.5 m and 25 m, respectively [11]. Five reservoir sampling sites labelled S1–S5 were selected for this study. Sites S1 and S2 were located in the South Han River part of the reservoir. In contrast, S3, S4, and S5 were situated at the North Han River, Kyoungan Stream, and dam, respectively. The water intake tower for Paldang Reservoir is located at S5 (Figure 1). It receives water from three different sources, namely the Kyoungan Stream, South and North Han River, and directly affects the reservoir's hydrodynamics and water quality [2,11,12,16]. About 95% of the reservoir's water comes from the North and South Han Rivers, which have relatively good water quality [11]. In contrast with the two sources, Kyoungan Stream has a small flow rate and a lower water quality. The drinking water supply tower is located near Kyoungan Stream's confluence and this significantly impacts drinking water quality (Figure 1).

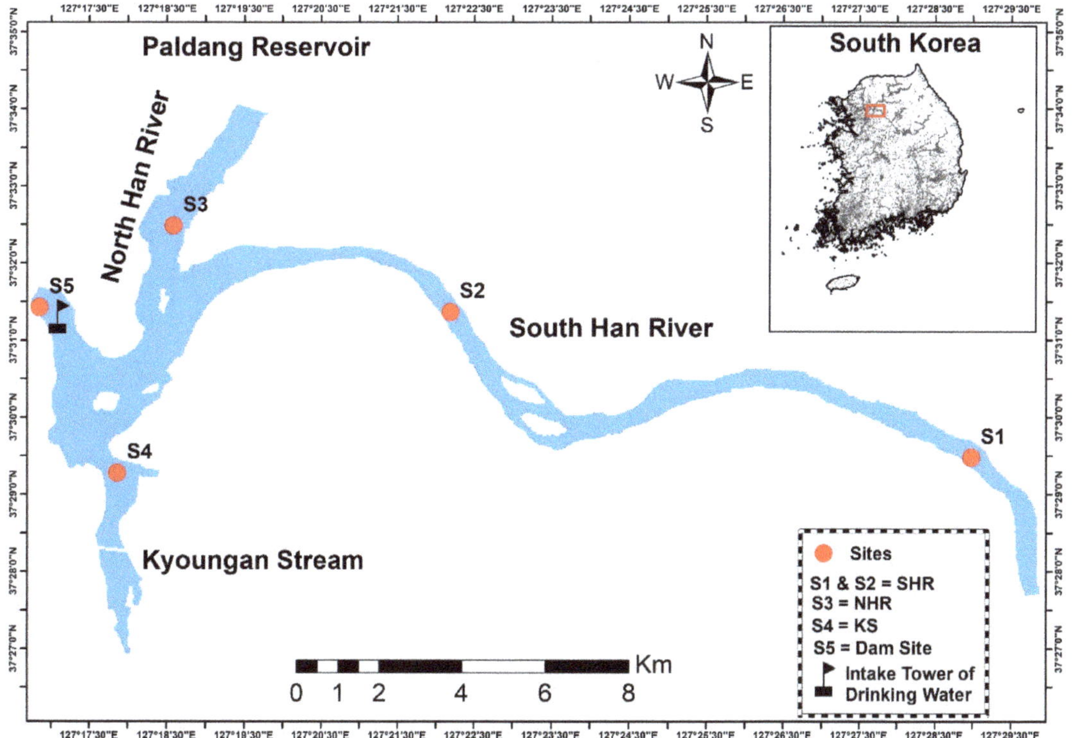

Figure 1. The map showing the study sites of Paldang Reservoir.

2.2. Methodological Approach

This study solely depends on secondary data. To monitor the water quality parameters (WQPs: TP, BOD, and CHL-a) of a reservoir, several Landsat 5 TM images with band values were acquired and processed. Finally, regression analysis was carried out to establish a relationship between band values of Landsat images and in-situ measurements of WQPs. Figure 2 illustrates the methodological approach of this study.

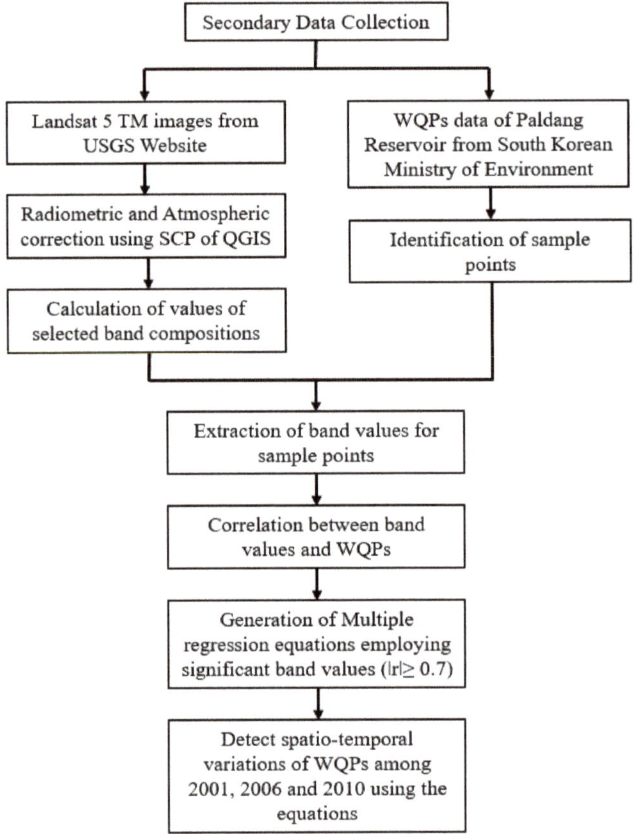

Figure 2. Methodological flow chart of the study.

2.2.1. Acquisition and Processing of Satellite Data

A total of 19 images of Landsat 5 TM (path/row: 116/34 and 115/34) were downloaded from the United States Geological Survey (USGS) Website (https://earthexplorer.usgs.gov/, accessed on 26 April 2021). Seven images from the year 2001 (dated 11 January, 16 March, 17 April, 19 May, 15 September, 27 November, and 13 December), six images from 2006 (dated 25 January, 17 February, 14 March, 5 August, 13 September, and 18 December), and six images from 2010 (dated 5 February, 16 March, 24 September, 19 October, 11 November, and 29 December) were selected due to availability of cloud-free images. To make these raw images more suitable to use, appropriate radiometric and atmospheric corrections were carried out using the semi-automatic classification plugin (SCP) of QGIS. To remove the effect of haze, this plugin employs dark object subtraction (DOS) method. SCP is a widely used plugin for preprocessing satellite images [31,32]. SCP uses the spectral radiance scaling method to convert the digital number (DN) to top of atmosphere (TOA) reflectance in two steps [33]. The procedure is described in the following sections. At first, the spectral radiance at the sensor's aperture L_λ ($Wm^{-2}\ sr^{-1}um^{-1}$) is measured from DN (Equation (1)) [34]:

$$L_\lambda = M_L \times Q_{cal} + A_L \quad (1)$$

where M_L = Band-specific multiplicative rescaling factor from Landsat metadata (RADIANCE_MULT_BAND_x, where x is the band number), A_L = Band-specific additive rescaling factor from Landsat metadata (RADIANCE_ADD_BAND_x, where x is the band number), Q_{cal} = Quantized and calibrated standard product pixel values (DN). After that DOS, the image-based atmospheric correction is carried out in SCP to calculate land surface reflectance (Equation (3)) for each pixel by calculating path radiance (Equation (2)):

$$L_p = M_L \times DN_{min} + A_L - 0.01 \times E_{SUN\lambda} \times cos\theta s/(\pi \times d^2) \quad (2)$$

where, L_p = path radiance, DN_{min} = minimum DN value of the scene, d = Earth-Sun distance in astronomical units, $E_{SUN\lambda}$ = mean solar exo-atmospheric irradiances, θs = solar zenith angle in degrees, which is equal to $\theta s = 90^\circ - \theta e$ where θe is the Sun elevation:

$$\rho = [\pi \times (L_\lambda - L_p) \times d^2]/(E_{SUN\lambda} \times cos\theta s) \quad (3)$$

where, ρ = land surface reflectance, L_λ = spectral radiance at the sensor's aperture, L_p = path radiance.

2.2.2. Assembling WQPs Data and Associated Band Values

The concentrations of different WQPs (TP, BOD, and CHL-a) of five sample collection points in Paldang Reservoir were collected from the South Korean Ministry of Environment for 2001, 2006, and 2010. These measurements are usually collected once a month. The dates of acquisition of satellite images were near the sampling days. A total of 95-pixel values for each band associated with these sample points were extracted from processed satellite images in the ArcGIS platform (Esri Inc., Redlands, CA, USA). For this analysis, four bands (blue, green, red and near infra-red) of Landsat 5 TM images were selected to extract, and a total of 32 band compositions were calculated in Microsoft Excel (Microsoft Office, Redmond, WA, USA).

2.2.3. Development of Multiple Regression Equation between WQPs and Landsat Band Values

After arranging the data, outliers of the dataset were identified by plotting box-whisker plots (Supplementary File Figures S1–S3). These box-whisker plots have identified one outlier for BOD (3.5 mg/L), six for TP (123, 138, 140, 142, 228, 236 µg/L), four for CHL-a (49.1, 56.3, 71.9, 132 µg/L). For developing the empirical models, these outliers were omitted. Pearson's coefficient of correlation (r) was calculated to find the strength of association among the band values and WQPs. To identify the significant band values, a threshold value of r was considered to be equal to or greater than 0.7, which represents strong correlation [21]. Multiple regression analysis was carried out in an online-based calculator to generate equations for each WQP. This analysis continued iteration until a significant p-value was obtained. This online-based calculator consideres all the assumptions of linear regression analysis (https://www.statskingdom.com/doc_linear_regression.html#multi, accessed on 26 April 2021). The assumptions are: (i) linearity—there is a linear relationship between the dependent variable, Y and the independent variables, Xi; (ii) residual normality; (iii) homoscedasticity (homogeneity of variance)—the variance of the residuals is constant and does not depend on the independent variables Xi; (iv) variables—the dependent variable, Y, should be continuous variable while the independent variables, Xi, should be continuous variables or ordinal variables; (v) multicollinearity—there is no perfect correlation among two or more independent variables, Xi.

To determine the efficiency of the generated models, root mean squared error (RMSE), root mean squared log error (RMSLE), mean relative error (MRE) and mean absolute error (MAE) were computed along with coefficient of determination (r^2) and p-values. The following are the equations of RMSE, RMSLE, MRE and MAE. These equations can be applied to radiometrically and atmospherically corrected Landsat 5 TM images to predict specific water properties (TP, BOD and CHL-a):

$$RMSE = \sqrt{\frac{\Sigma(P_i - O_i)^2}{n}} \tag{4}$$

$$RMSLE = \sqrt{\frac{\Sigma(\log(P_i + 1) - \log(O_i + 1))^2}{n}} \tag{5}$$

$$MRE = \frac{\Sigma\left|\frac{P_i}{O_i} - 1\right|}{n} \tag{6}$$

$$MAE = \frac{\Sigma|P_i - O_i|}{n} \tag{7}$$

where, P_i = predicted values of WQPs, O_i = observed values of WQPs, and n = sample size.

2.3. Spatio-Temporal Variation of WQPs

The spatio-temporal variation in WQPs of Paldang Reservoir for the years 2001, 2006, and 2010 were studied using the generated equations. Landsat 5 TM images of the month of September of these years were processed in SCP of QGIS, and the area of interest- Paldang Reservoir was extracted from the images. From their band values, values of WQPs were computed in Raster Calculator (Esri Inc., USA) and analyzed for the change detection study.

3. Results

3.1. Reservoir Conditions

The water quality parameters (TP, BOD, and CHL-a) of the Paldang Reservoir showed significant site variations (Table 1). The mean TP varied from 34.75–92.06 μgL^{-1} from sites S1–S5. Site S4 showed the highest TP (92.06 μgL^{-1}) value compared to all sites due to the reception of wastewater from industry and household activities. Moreover, Site S4 is highly impacted by the Kyoungan Stream. Nürnberg [35] proposed that TP concentrations > 30 μgL^{-1} indicate a eutrophic reservoir. Mean TP levels above 30 μgL^{-1} at all sites were observed in this study. High BOD values suggest that organic matter pollution is linked to wastewater effluents. The mean BOD ranged from 1.05 to 1.72 mgL^{-1} in the Paldang Reservoir. Like TP, the highest BOD had been observed in Site S4 (1.72 mgL^{-1}). It is well known that CHL-a is the primary indicator of eutrophication in the lentic system [5,36]. The mean CHL-a varied from 10.89 to 27.74 μgL^{-1}. Nürnberg [35] proposed that eutrophic reservoir should be indicated by CHL-a concentrations greater than 9 μgL^{-1}. Mean CHL-a concentrations at five sites were found above 9 μgL^{-1}. Like TP and BOD, the highest CHL-a was also observed at site S4. Industrial and household wastewater and the Kyoungan Stream highly affect the water quality of site S4. Eun and Seok [11] and Mamun et al. [2] found that the water quality of the Kyoungan Stream is in poor condition compared to the South Han River (sites S1 and S2) and North Han River (Site S3), and this could have a major effect on the reservoir's water quality.

Table 1. Summary statistics of Total Phosphorus (TP), Biological Oxygen Demand (BOD) and Chlorophyll-a (CHL-a) in Paldang Reservoir.

Sites	TP (μg/L) Mean ± SD (Min–Max)	BOD (mg/L) Mean ± SD (Min–Max)	CHL-a (μg/L) Mean ± SD (Min–Max)
S1	50.93 ± 29.51	1.15 ± 0.56	13.61 ± 6.16
	(28–142)	(0.4–2)	(1.1–49.1)
S2	52.81 ± 27.35	1.31 ± 0.61	15.37 ± 12.14
	(29–140)	(0.4–2.3)	(0.9–37.5
S3	34.75 ± 22.94	1.05 ± 0.35	10.89 ± 6.83
	(11–100)	(0.4–1.5)	(1.2–24.4)
S4	92.06 ± 66.66	1.72 ± 0.68	27.74 ± 14.37
	(11–236)	(0.8–3.5)	(3–132)
S5	43.25 ± 28.18	1.18 ± 0.37	16.71 ± 11.52
	(12–116)	(0.7–1.9)	(5.3–42.5)

A Pearson-based correlation analysis was used to identify the relationship among TP, BOD, and CHL-a ($p < 0.05$; Table 2). The BOD showed positive correlation with TP ($r = 0.249$) and CHL-a ($r = 0.627$). The positive correlation between BOD and TP suggests that nutrients (TP) flow into the Paldang Reservoir along with organic matter (BOD). The high positive correlation between BOD and CHL-a indicates that autochthonous organic matter production is primarily resulting from phytoplankton processes. CHL-a was positively related with TP ($r = 0.375$), which is the key factor regulating algal growth in the freshwater lentic system [5,9,10].

Table 2. Pearson correlation among total phosphorus (TP), biological oxygen demand (BOD) and chlorophyll-a (CHL-a).

Variables		*r* Value	*p*
BOD	TP	0.249	0.02
BOD	CHL-a	0.627	<0.001
TP	CHL-a	0.375	<0.001

3.2. Relations of Band Compositions with TP, BOD, and CHL-a

Values of 32 band compositions and associated TP, BOD, and CHL-a concentrations were employed to compute correlation (r) values for Landsat 5 TM sensors. The band compositions and their allied r values are shown in Table 3. Only four bands (blue, green, red, and near-infrared) provide the visibly displayed water quality parameter spectral reflectiveness (0.4–0.9 μm); that is why we used these four bands to determine TP, BOD, and CHL-a [20]. TP is a significant factor in deciding eutrophication in freshwater systems. The TM bands' correlation with TP ranged from −0.07 (B2/B4) to −0.79 (B1). Particularly, TP showed the strongest correlation with B1 (r = −0.79). The present findings have concurred with some previous studies [20]. BOD is the indicator of organic pollution in the aquatic systems. BOD and TM bands' correlation varied from −0.15 (B1/B3) to −0.76 (B1*B3/B4). Like TP, BOD showed the highest correlation with B1 (r = −0.75) and B1*B3/B4 (r = −0.76). CHL-a is a good indicator of overall algal biomass in the aquatic systems. The present results showed a dynamic relation between TM bands and CHL-a. The correlation among TM bands and CHL-a ranged from 0.12 (B2/B3) to −0.79 (B1). Like TP and BOD, CHL-a showed the highest correlation with B1 (r = −0.79). From the r values, influential bands and band compositions have been identified to generate empirical models for TP, BOD, and CHL-a (marked in bold; Table 3).

Table 3. Correlation matrix values for different band compositions of Landsat 5 TM with TP, BOD and CHL-a (B1: Blue, B2: Green, B3: Red, B4: NIR-Near-Infrared, TP: Total Phosphorus, BOD: Biological Oxygen Demand, CHL-a: Chlorophyll-a, influential bands and band compositions have been ($r \geq 0.7$) marked in bold.

Band Composition	*r*-Values		
	TP	BOD	CHL-a
B1	**−0.79**	**−0.75**	**−0.79**
B2	**−0.76**	**−0.74**	**−0.76**
B3	**−0.74**	**−0.71**	**−0.75**
B4	−0.68	−0.60	−0.53
B1*B2	**−0.70**	**−0.71**	**−0.72**
B1*B3	−0.68	−0.67	**−0.70**
B1*B4	−0.65	−0.63	−0.62
B2*B3	−0.66	−0.65	−0.68
B2*B4	−0.63	−0.62	−0.56
B3*B4	−0.61	−0.58	−0.58
B1*B2*B3	−0.58	−0.57	−0.61
B1*B2*B4	−0.57	−0.55	−0.55
B1*B3*B4	−0.55	−0.52	−0.53
B2*B3*B4	−0.54	−0.50	−0.51
B1/B2	−0.42	−0.25	−0.39
B1/B3	−0.20	−0.15	−0.28
B1/B4	−0.28	−0.36	−0.33
B2/B3	0.31	0.18	0.12
B2/B4	−0.07	−0.27	−0.18
B3/B4	−0.27	−0.41	−0.24
B1*B2/B3	**−0.76**	**−0.73**	**−0.75**
B1*B2/B4	**−0.72**	**−0.74**	**−0.72**
B1*B3/B2	**−0.78**	**−0.73**	**−0.78**
B1*B3/B4	**−0.78**	**−0.76**	**−0.75**
B1*B4/B2	**−0.72**	−0.61	−0.63
B1*B4/B3	**−0.73**	−0.63	−0.58
B2*B3/B1	−0.65	−0.64	−0.65
B2*B3/B4	**−0.75**	**−0.74**	**−0.72**
B1*B2*B3/B4	**−0.71**	**−0.71**	**−0.72**
B1*B2*B4/3	−0.67	−0.66	−0.59
B1*B3*B4/2	−0.63	−0.59	−0.60
B2*B3*B4/B1	−0.59	−0.55	−0.51

3.3. Empirical Model Development of TP, BOD, and CHL-a from Landsat 5 TM Data

Variables with high correlation values ($|r| \geq 0.70$) have only been used to generate the empirical model for TP, BOD, and CHL-a (Table 4). The analysis was performed in an online-based calculator until a significant relationship was indicated by the *p*-value ($p < 0.01$). Due to an online-based calculator's automatic iteration power, it is not easy to control the inclusion of any specific independent variables. The *p*-values for the model show that they have a significant relationship. The developed model using Landsat 5 TM images can detect TP 67% correctly while it was 65% and 72% for BOD and CHL-a, respectively (Table 4). The values of RMSE, RMSLE, MRE and MAE also depict the efficiency of developed models (Table 4). A more efficient TP, BOD, and CHL-a models from Landsat 5 TM can be developed using more sampling point data [21]. Considering our findings, further studies should be carried out with satellite sensors data to develop the empirical models of TP, BOD, and CHL-a. Scatter plots of the observed TP, BOD, and CHL-a data with predicted TP, BOD, and CHL-a values from the generated regression models are shown in Figure 3. For TP, the relationship between observed and predicted values displayed a correlation of 0.82 with $p < 0.01$. In contrast, it was 0.81 and 0.85 for BOD and CHL-a, respectively with $p < 0.01$.

Table 4. Linear equations for Landsat 5 TM to detect TP, BOD and CHL-a of Paldang Reservoir (B: Blue, G: Green, R: Red, NIR: Near-Infrared, TP: Total Phosphorus, BOD: Biological Oxygen Demand, CHL-a: Chlorophyll-a, WQPs: Water Quality Parameters).

Sensor	WQPs	Equations	R^2	p	RMSE	RMSLE	MRE	MAE
Landsat 5 TM	TP	=91.01 − 268.22*B*NIR/G − 347.50*G*R/NIR + 1194.55*B*G*R/NIR	0.67	<0.01	30.4	0.072	0.11	3.39
	BOD	=1.83 − 127.38*B*G + 13.39*B*G/NIR + 18.50*B*R/G − 36.74*B*R/NIR + 122.78*B*G*R/NIR	0.65	<0.01	0.08	0.058	0.25	0.23
	CHL-a	=39.40 + 548.80*G − 778.68*R + 1396.84*B*R − 243.21*B*G/R	0.72	<0.01	4.9	0.155	0.34	1.41

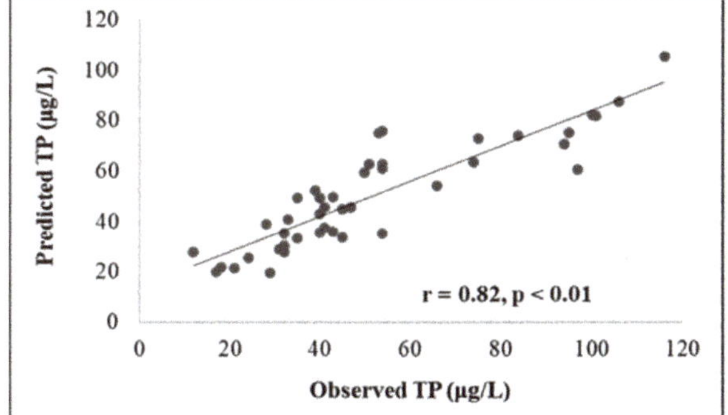

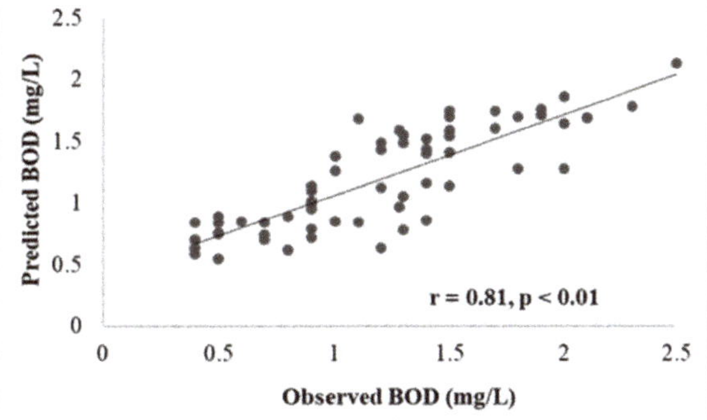

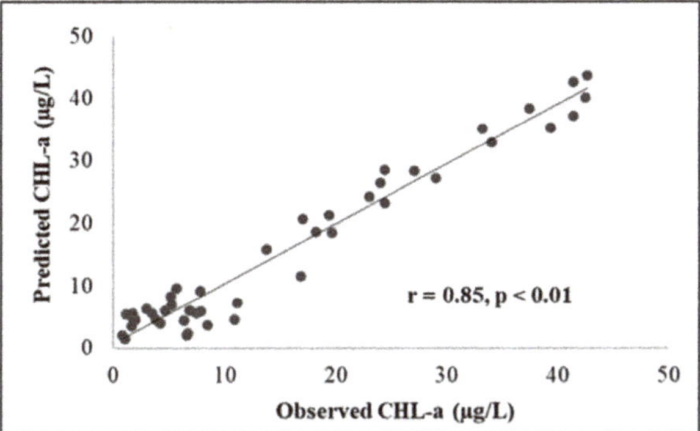

Figure 3. The relationship among observed and predicted TP, BOD and CHL-a for Landsat 5 TM (TP: Total Phosphorus, BOD: Biological Oxygen Demand, and CHL-a: Chlorophyll-a).

3.4. Spatial and Temporal Patterns of Water Quality Parameters

The developed empirical models were applied to all study sites on the Paldang Reservoir to monitor spatio-temporal distributions of WQPs on 15 September 2001; 13 September 2006, and 24 September 2010, using Landsat 5 TM images (Figures 4–6). Sites S1 and S2 are influenced by the South Han River, While S3 and S4 are affected by the North Han River and Kyoungan Stream. The TP concentration varied between 5.82 to 80.60 μgL^{-1} on 15 September 2001. The maximum TP concentrations gradually decreased from 2001 (80.60 μgL^{-1}) to 2006 (57.83 μgL^{-1}) and 2010 (55.32 μgL^{-1}) in the Paldang Reservoir due to new treatment facilities in the sewage treatment plants and the development of water quality management strategies in the reservoirs (Figure 4). However, the TP concentrations were still in a eutrophic state in site S3 and S4 during 2006 and 2010. Like TP, the maximum BOD concentration also showed decreasing pattern from 2001 (2.28 mgL^{-1}) to 2006 (2.10 mgL^{-1}) and 2010 (2.0 mgL^{-1}) (Figure 5). It indicates that the biological effluent treatment process may efficiently degrade the influent's degradable organic matter [16]. The highest BOD level was also observed in S4 during 2010 September. Like TP and BOD, the maximum level of CHL-a was also showed a declining trend from 2001 (35.46 μgL^{-1}) to 2006 (14.96 μgL^{-1}) and 2010 (14.31 μgL^{-1}) (Figure 6). During 2001 September, the entire reservoir showed a eutrophic state (>9 μgL^{-1}) based on CHL-a concentration. Although the water quality is getting better in terms of TP, BOD, and CHL-a from 2001 to 2010, Site S3 and S4 are still facing some problems. The authors suggest that sites S3 and S4 should be taken under special consideration as S4 highly influences the intake tower water quality.

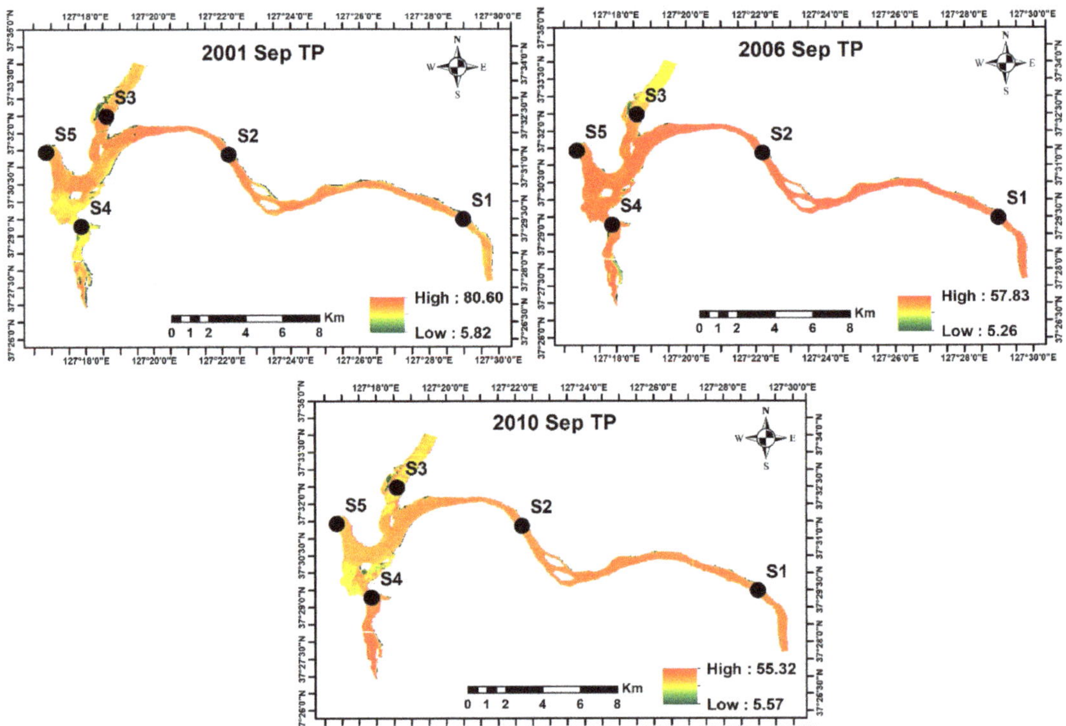

Figure 4. Spatial and temporal pattern of total phosphorus (TP) on September 2001, 2006 and 2010.

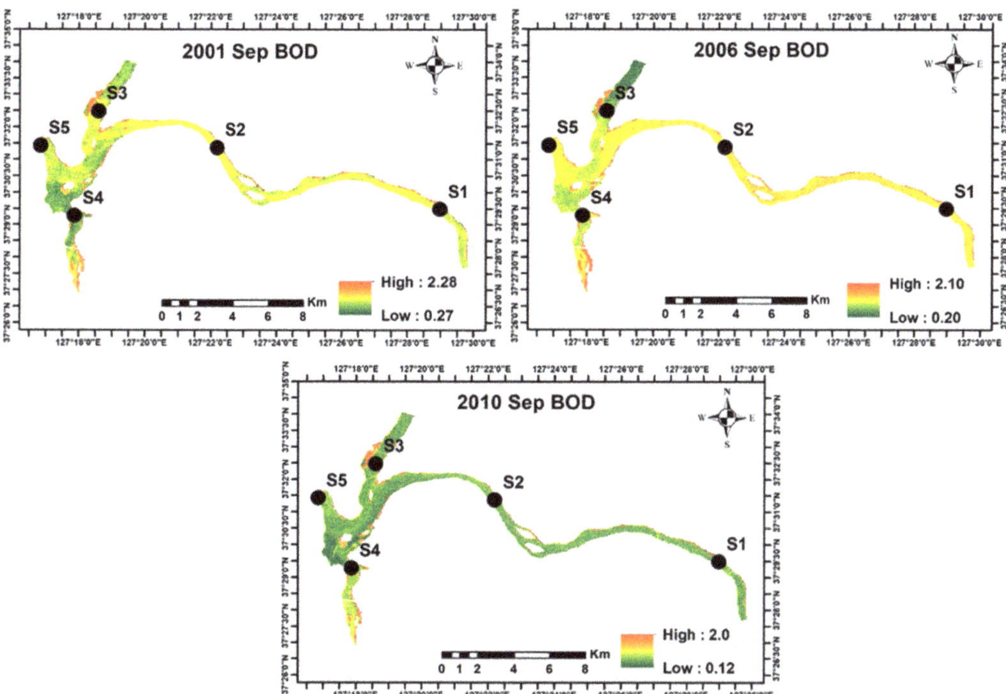

Figure 5. Spatial and temporal pattern of biological oxygen demand (BOD) on September 2001, 2006 and 2010.

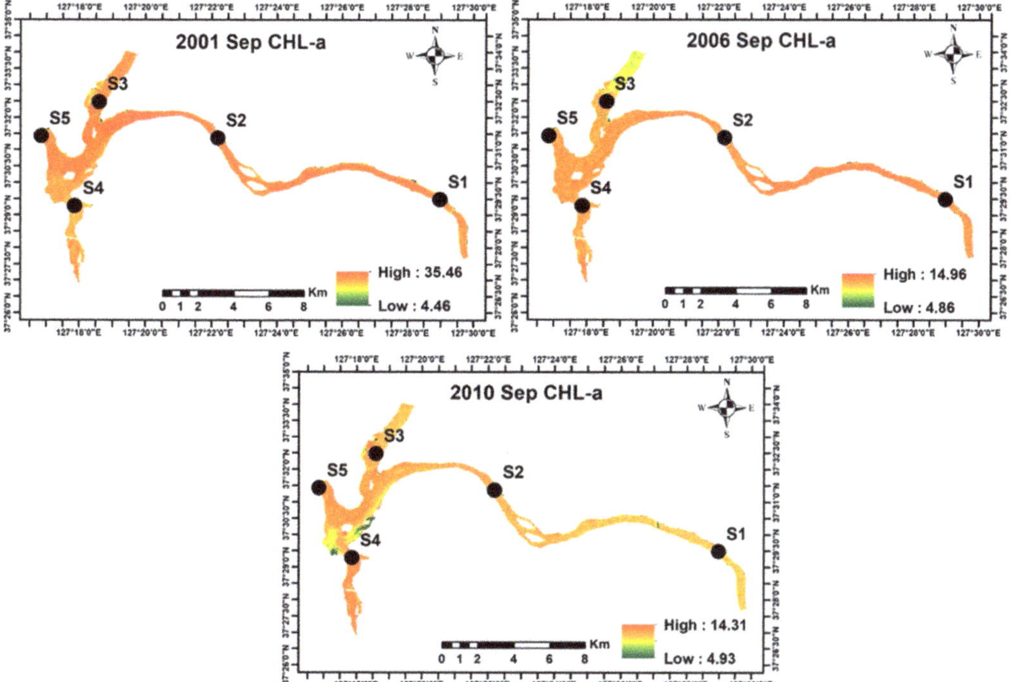

Figure 6. Spatial and temporal pattern of chlorophyll-a (CHL-a) on September 2001, 2006 and 2010.

4. Discussion

The present study shows that remote sensing technology can be a handy tool to detect water quality parameters. Landsat data series are useful for monitoring the water quality of freshwater bodies. Paldang Reservoir has experienced significant water quality changes due to urbanization, land use change, and intensive agricultural farming [14,15,17]. The observed mean TP and CHL-a concentration at all sites in the Paldang Reservoir showed eutrophic conditions. This indicates a moderate risk of cyanobacterial exposure in the reservoir [37]. Previous studies stated that blooms of cyanobacteria are associated with eutrophic conditions in water bodies [36]. CHL-a is a good predictor of total phytoplankton biomass and monitoring CHL-a is a direct tool for semiquantitative estimation of cyanobacterial biomass in aquatic environments [2]. Previous studies of Paldang Reservoir have suggested that cyanobacterial blooms occur during the spring season and identified the following genera: *Anabaena, Aphanocapsa, Chroococcus, Coelosphaerium, Dactylococcopsis, Microcystis, Merismopedia, Phormidium, Oscillatoria, and Pseudoanabaen* [2,16,18]. In addition, TP, BOD, and CHL-a levels at site S4 were constantly elevated. The water quality of site S4 is heavily impacted by industrial and domestic wastewater and the Kyoungan Stream. Eun and Seok [11] and Mamun et al. [2] found that the water quality of the Kyoungan Stream is in poor condition in comparison to the Southern Han River (Sites S1 and S2) and Northern Han River (Site S3) based on nutrients, organic matter and algal chlorphyll. It could have significant effects on the reservoir's water quality.

Variation in TP, BOD, and CHL-a concentrations of the Paldang Reservoir was prominent during the pre-monsoon, monsoon, and post-monsoon seasons [18]. TP concentrations were higher during the monsoon period due to intense precipitation, while BOD and CHL-a level at Paldang Reservoir was highest in the spring period [2,18]. The summer monsoon significantly influences the nutrient, organic matter, and algal chlorophyll in the Korean reservoirs [5,16,38]. Organic matter in aquatic systems may come from allochthonous or autochthonous sources. Allochthonous organic matter enters into the environment during precipitation events, while algae produce autochthonous organic matter by photosynthesis [18]. It was noticeable that 69% of the total organic matter was allochthonous in the Paldang Reservoir during monsoon season [18]. Inversely, during winter and spring, a high load of autochthonous organic matter had observed because of low flow rates and high water residence time [16]. Previous research on Paldang Reservoir indicated that 73% of autochthonous organic matter loading happens during the spring [16,18]. The high-level organic matter during spring corresponds to algae's maximum production [18]. It suggests that autochthonous production by algae (CHL-a) is dire to accumulate organic matter in the reservoir during spring; hence, the threat to the reservoir's water quality is highest in spring [18].

The water quality of the Paldang Reservoir varied from site to site and season to season due to climatic factors and anthropogenic activities. Since climatic conditions are uncontrollable, anthropogenic impacts should be kept to a minimum. For that reason, regular monitoring of water quality parameters is essentially mandatory. Traditional monitoring approaches are time-intensive, laborious, and cannot provide an overview of water quality at a broader scale [1,19].

On the other hand, satellite remote sensing is presently one of the most potent and reliable approaches for monitoring and managing water quality [4,20,21]. This study has confirmed the applicability of Landsat 5 TM to identify and map the water quality parameters in the reservoir. The developed empirical models by multiple linear regression analysis can identify TP 67%, BOD 65%, and CHL-a 72% accurately from Landsat 5 TM images. As shown in Table 4, blue*near-infrared/green (B1*B4/B2), green*red/near-infrared (B2*B3/B4), and blue*green*red/near-infrared (B1*B2*B3/B4) bands and band ratios are the significant predictors for TP concentrations in Paldang Reservoir. Previous studies also found that three visible bands (blue, green, and red) and NIR bands and their ratios are suitable for estimating TP concentrations in freshwater systems [4,20]. blue*green (b1*b2), blue*green/near-infrared(b1*b2/b4), blue*Red/Green (B1*B3/B2), blue*red/near-infrared

(B1*B3/B4), blue*green*red/near-infrared (B1*B2*B3/B4) bands and band ratios are the significant predictors for BOD concentrations in the reservoir. Quibell [39] reported that the NIR and red bands ratio were good predictors of CHL-a concentration in waters. Also, other bands and band ratios are good indicators of CHL-a [40]. The present result indicated that CHL-a was better explained by the green (B2), red (B3), blue*red (B1*B3) and blue*green/red (B1*B2/B3) bands and band ratios.

5. Conclusions

Nutrient and organic pollution and algal blooms regulate water quality in freshwater systems. For this reason, it is essential to develop a cost-effective remote sensing monitoring tool to estimate the water quality parameters for maintaining an effective water management system. The present study has successfully established Landsat 5 TM data's applicability to detect TP, BOD, and CHL-a for the surface water of Paldang Reservoir (Korea). The results showed that TP, BOD, and CHL-a are closely related to the Landsat 5 surface reflectance band values. TP ($r = -0.79$) and CHL-a ($r = -0.79$) showed the highest relations with B1 (blue) band. By contrast, BOD showed the highest negative correlation with B1 (blue) ($r = -0.75$) and B1*B3/B4 (blue*red/near-infrared) bands ($r = -0.76$). The developed empirical models of Landsat 5 TM data can estimate TP, BOD, and CHL-a correctly by around 67%, 65%, and 72%, respectively, for the reservoir. The results presented here revealed that the surface water quality of the reservoir varied from site to site. The water quality of sites S3 and S4 are affected by anthropogenic factors, which significantly impact reservoir's water quality. Considering the present findings, we should take a particular account for site S3 and S4 to maintain the water quality. The present developed models and methods could be applied to other Korean reservoirs for validation.

Supplementary Materials: The following are available online at https://www.mdpi.com/article/10.3390/rs13122256/s1. Figure S1 Outlier detection for BOD; Figure S2 Outlier detection for TP; Figure S3 Outlier detection for CHL-a.

Author Contributions: Conceptualization, M.M.; methodology, M.M. and J.F.; software, M.M. and J.F.; validation, M.M. and J.F.; formal analysis, M.M. and J.F.; resources, M.M. and J.F.; data curation, M.M.; writing—original draft preparation, M.M. and J.F.; writing—review and editing, M.M., J.F. and K.-G.A.; visualization, M.M. and J.F.; supervision, K.-G.A.; project administration, K.-G.A.; funding acquisition, K.-G.A. All authors have read and agreed to the published version of the manuscript.

Funding: This research was funded by the Korea Environmental Industry & Technology Institute (KEITI) project (Project:2020-1008-01) "Development of Longitudinal Connectivity Evaluation System, Based on the Fish Movement".

Institutional Review Board Statement: Not applicable.

Informed Consent Statement: Not applicable.

Data Availability Statement: The datasets presented in this study are available on reasonable request from the corresponding author.

Acknowledgments: The authors would like to acknowledge the Korea Environmental Industry & Technology Institute (KEITI) for their assistance.

Conflicts of Interest: The authors declare no conflict of interest.

References

1. Mushtaq, F.; Nee Lala, M.G. Remote estimation of water quality parameters of Himalayan lake (Kashmir) using Landsat 8 OLI imagery. *Geocarto Int.* **2017**, *32*, 274–285. [CrossRef]
2. Mamun, M.; Kim, J.Y.; An, K.-G. Multivariate Statistical Analysis of Water Quality and Trophic State in an Artificial Dam Reservoir. *Water* **2021**, *13*, 186. [CrossRef]
3. Li, Y.; Zhang, Y.; Shi, K.; Zhu, G.; Zhou, Y.; Zhang, Y.; Guo, Y. Monitoring spatiotemporal variations in nutrients in a large drinking water reservoir and their relationships with hydrological and meteorological conditions based on Landsat 8 imagery. *Sci. Total Environ.* **2017**, *599*, 1705–1717. [CrossRef]

4. Torbick, N.; Hession, S.; Hagen, S.; Wiangwang, N.; Becker, B.; Qi, J. Mapping inland lake water quality across the Lower Peninsula of Michigan using Landsat TM imagery. *Int. J. Remote Sens.* **2013**, *34*, 7607–7624. [CrossRef]
5. Mamun, M.; Kwon, S.; Kim, J.E.; An, K.G. Evaluation of algal chlorophyll and nutrient relations and the N:P ratios along with trophic status and light regime in 60 Korea reservoirs. *Sci. Total Environ.* **2020**, *741*, 140451. [CrossRef] [PubMed]
6. Smith, V.H. Responses of estuarine and coastal marine phytoplankton to nitrogen and phosphorus enrichment. *Limnol. Oceanogr.* **2006**, *51*, 377–384. [CrossRef]
7. Rabalais, N.N.; Turner, R.E.; Díaz, R.J.; Justić, D. Global change and eutrophication of coastal waters. *ICES J. Mar. Sci.* **2009**, *66*, 1528–1537. [CrossRef]
8. Li, Y.; Cao, W.; Su, C.; Hong, H. Nutrient sources and composition of recent algal blooms and eutrophication in the northern Jiulong River, Southeast China. *Mar. Pollut. Bull.* **2011**, *63*, 249–254. [CrossRef]
9. Jones, J.R.; Obrecht, D.V.; Thorpe, A.P. Chlorophyll maxima and chlorophyll: Total phosphorus ratios in Missouri reservoirs. *Lake Reserv. Manag.* **2011**, *27*, 321–328. [CrossRef]
10. Atique, U.; An, K.G. Landscape heterogeneity impacts water chemistry, nutrient regime, organic matter and chlorophyll dynamics in agricultural reservoirs. *Ecol. Indic.* **2020**, *110*, 105813. [CrossRef]
11. Eun, H.N.; Seok, S.P. A hydrodynamic modeling study to determine the optimum water intake location in Lake Paldang, Korea. *J. Am. Water Resour. Assoc.* **2005**, *41*, 1315–1332. [CrossRef]
12. Boopathi, T.; Wang, H.; Lee, M.-D.; Ki, J.-S. Seasonal Changes in Cyanobacterial Diversity of a Temperate Freshwater Paldang Reservoir (Korea) Explored by using Pyrosequencing. *Environ. Biol. Res.* **2018**, *36*, 424–437. [CrossRef]
13. MOE. *(ECOREA) Environmental Review 2015*; Ministry of Environment: Sejong City, Korea, 2015; Volume 1.
14. Lee, J.E.; Youn, S.J.; Byeon, M.; Yu, S.J. Occurrence of cyanobacteria, actinomycetes, and geosmin in drinking water reservoir in Korea: A case study from an algal bloom in 2012. *Water Sci. Technol. Water Supply* **2020**, *20*, 1862–1870. [CrossRef]
15. Kim, D.W.; Min, J.H.; Yoo, M.; Kang, M.; Kim, K. Long-term effects of hydrometeorological and water quality conditions on algal dynamics in the Paldang dam watershed, Korea. *Water Sci. Technol. Water Supply* **2014**, *14*, 601–608. [CrossRef]
16. Park, H.K.; Cho, K.H.; Won, D.H.; Lee, J.; Kong, D.S.; Jung, D. Il Ecosystem responses to climate change in a large on-river reservoir, Lake Paldang, Korea. *Clim. Chang.* **2013**, *120*, 477–489. [CrossRef]
17. Youn, S.J.; Kim, H.N.; Yu, S.J.; Byeon, M.S. Cyanobacterial occurrence and geosmin dynamics in Paldang Lake watershed, South Korea. *Water Environ. J.* **2020**, 1–10. [CrossRef]
18. Park, H.K.; Byeon, M.S.; Shin, Y.N.; Jung, D. Il Sources and spatial and temporal characteristics of organic carbon in two large reservoirs with contrasting hydrologic characteristics. *Water Resour. Res.* **2009**, *45*, 1–12. [CrossRef]
19. Khattab, M.F.O.; Merkel, B.J. Application of landsat 5 and landsat 7 images data for water quality mapping in Mosul Dam Lake, Northern Iraq. *Arab. J. Geosci.* **2014**, *7*, 3557–3573. [CrossRef]
20. Lim, J.; Choi, M. Assessment of water quality based on Landsat 8 operational land imager associated with human activities in Korea. *Environ. Monit. Assess.* **2015**, *187*, 1–17. [CrossRef]
21. Ferdous, J.; Rahman, M.T.U. Developing an empirical model from Landsat data series for monitoring water salinity in coastal Bangladesh. *J. Environ. Manag.* **2020**, *255*, 109861. [CrossRef]
22. Kumar, V.; Sharma, A.; Chawla, A.; Bhardwaj, R.; Thukral, A.K. Water quality assessment of river Beas, India, using multivariate and remote sensing techniques. *Environ. Monit. Assess.* **2016**, *188*, 1–10. [CrossRef]
23. Zhang, Y.; Zhang, Y.; Shi, K.; Zhou, Y.; Li, N. Remote sensing estimation of water clarity for various lakes in China. *Water Res.* **2021**, *192*. [CrossRef]
24. Zhou, W.; Wang, S.; Zhou, Y.; Troy, A. Mapping the concentrations of total suspended matter in Lake Taihu, China, using Landsat-5 TM data. *Int. J. Remote Sens.* **2006**, *27*, 1177–1191. [CrossRef]
25. Brezonik, P.; Menken, K.D.; Bauer, M. Landsat-based remote sensing of lake water quality characteristics, including chlorophyll and colored dissolved organic matter (CDOM). *Lake Reserv. Manag.* **2005**, *21*, 373–382. [CrossRef]
26. Miller, H.; Sexton, N.; Koontz, L.; Loomis, J.; Koontz, S.; Hermans, C. *The Users, Uses, and Value of Landsat and Other Moderate-Resolution Satellite Imagery in the United States—Executive Report*; U.S. Geological Survey: Reston, VA, USA, 2011.
27. Wu, C.; Wu, J.; Qi, J.; Zhang, L.; Huang, H.; Lou, L.; Chen, Y. Empirical estimation of total phosphorus concentration in the mainstream of the Qiantang River in China using Landsat TM data. *Int. J. Remote Sens.* **2010**, *31*, 2309–2324. [CrossRef]
28. Gholizadeh, M.H.; Melesse, A.M.; Reddi, L. A comprehensive review on water quality parameters estimation using remote sensing techniques. *Sensors* **2016**, *16*, 1298. [CrossRef] [PubMed]
29. Gholizadeh, M.H. Water Quality Modelling Using Multivariate Statistical Analysis and Remote Sensing in South Florida. Ph.D. Thesis, Florida International University, Miami, FL, USA, 2016.
30. Nas, B.; Ekercin, S.; Karabörk, H.; Berktay, A.; Mulla, D.J. An application of landsat-5TM image data for water quality mapping in Lake Beysehir, Turkey. *Water Air Soil Pollut.* **2010**, *212*, 183–197. [CrossRef]
31. Ghorai, D.; Mahapatra, M. Correction to: Extracting Shoreline from Satellite Imagery for GIS Analysis. *Remote Sens. Earth Syst. Sci.* **2020**, *3*, 23. [CrossRef]
32. Lehmann, J.R.K.; Prinz, T.; Ziller, S.R.; Thiele, J.; Heringer, G.; Meira-Neto, J.A.A.; Buttschardt, T.K. Open-source processing and analysis of aerial imagery acquired with a low-cost Unmanned Aerial System to support invasive plant management. *Front. Environ. Sci.* **2017**, *5*, 1–16. [CrossRef]

33. Rahman, M.T.U.; Ferdous, J. Detection of Environmental Degradation of Satkhira District, Bangladesh Through Remote Sensing Indices. In *GCEC 2017*; Pradhan, B., Ed.; Lecture Notes in Civil Engineering; Springer: Singapore, 2019; Volume 9, pp. 1053–1066. ISBN 9789811080166.
34. Congedo, L. *Semi-Automatic Classification Plugin User Manual Release 5.3.6.1*; RoMEO: Paterson, NJ, USA, 2017.
35. Nürnberg, G.K. Trophic state of clear and colored, soft- and hardwater lakes with special consideration of nutrients, anoxia, phytoplankton and fish. *Lake Reserv. Manag.* **1996**, *12*, 432–447. [CrossRef]
36. Carlson, R.E.; Havens, K.E. Simple graphical methods for the interpretation of relationships between trophic state variables. *Lake Reserv. Manag.* **2005**, *21*, 107–118. [CrossRef]
37. WHO. *Guidelines for Drinking Water Quality: Management of Cyanobacteria in Drinking Water Suppliers Information for Regulators and Water Suppliers*, 4th ed.; World Health Organization: Geneva, Switzerland, 2011.
38. Jung, S.; Shin, M.; Kim, J.; Eum, J.; Lee, Y.; Lee, J.; Choi, Y.; You, K.; Owen, J.; Kim, B. The effects of Asian summer monsoons on algal blooms in reservoirs. *Inland Waters* **2016**, *6*, 406–413. [CrossRef]
39. Quibell, G. The effect of suspended sediment on reflectance from freshwater algae. *Int. J. Remote Sens.* **1991**, *12*, 177–182. [CrossRef]
40. Kloiber, S.M.; Brezonik, P.L.; Olmanson, L.G.; Bauer, M.E. A procedure for regional lake water clarity assessment using Landsat multispectral data. *Remote Sens. Environ.* **2002**, *82*, 38–47. [CrossRef]

Article

A Deep Learning Model Using Satellite Ocean Color and Hydrodynamic Model to Estimate Chlorophyll-*a* Concentration

Daeyong Jin [1], Eojin Lee [1], Kyonghwan Kwon [2] and Taeyun Kim [1,*]

[1] Environment Data Strategy Center & Environmental Assessment Group, Korea Environment Institute, Sejong 30147, Korea; dyjin@kei.re.kr (D.J.); eojinlee@kei.re.kr (E.L.)
[2] Ocean Environment Group, Oceanic, Seoul 07207, Korea; khkwon@oceaniccnt.com
* Correspondence: kimty@kei.re.kr; Tel.: +82-44-415-7415

Abstract: In this study, we used convolutional neural networks (CNNs)—which are well-known deep learning models suitable for image data processing—to estimate the temporal and spatial distribution of chlorophyll-*a* in a bay. The training data required the construction of a deep learning model acquired from the satellite ocean color and hydrodynamic model. Chlorophyll-*a*, total suspended sediment (TSS), visibility, and colored dissolved organic matter (CDOM) were extracted from the satellite ocean color data, and water level, currents, temperature, and salinity were generated from the hydrodynamic model. We developed CNN Model I—which estimates the concentration of chlorophyll-*a* using a 48 × 27 sized overall image—and CNN Model II—which uses a 7 × 7 segmented image. Because the CNN Model II conducts estimation using only data around the points of interest, the quantity of training data is more than 300 times larger than that of CNN Model I. Consequently, it was possible to extract and analyze the inherent patterns in the training data, improving the predictive ability of the deep learning model. The average root mean square error (RMSE), calculated by applying CNN Model II, was 0.191, and when the prediction was good, the coefficient of determination (R^2) exceeded 0.91. Finally, we performed a sensitivity analysis, which revealed that CDOM is the most influential variable in estimating the spatiotemporal distribution of chlorophyll-*a*.

Keywords: deep learning; convolutional neural network; chlorophyll-*a*; satellite; hydrodynamic model

Citation: Jin, D.; Lee, E.; Kwon, K.; Kim, T. A Deep Learning Model Using Satellite Ocean Color and Hydrodynamic Model to Estimate Chlorophyll-*a* Concentration. *Remote Sens.* **2021**, *13*, 2003. https://doi.org/10.3390/rs13102003

Academic Editor: Giacomo De Carolis

Received: 6 April 2021
Accepted: 17 May 2021
Published: 20 May 2021

1. Introduction

Marine environments experience continuous deterioration owing to the influx of pollutants from rivers and various infrastructure projects including breakwater construction, dredging, and reclamation. To restore marine environments, numerous mitigation plans have been established using various prediction and evaluation techniques. Nevertheless, several limitations still remain: first, the ocean is a complex three-dimensional system that is difficult to model accurately; second, sea water constituents exhibit dynamic movements due to external forces such as wind, tides, currents, density, etc.; third, a significant amount of time and effort is required to observe oceanic trends; and finally, despite significant developments in marine environment prediction technology, several assumptions and additional research area information are still required [1–4].

The water quality model has been widely employed in marine environment prediction, although professional knowledge and experience, various input data, and model validation procedures are required to utilize it. However, owing to the complex and interconnected nature of marine environments, major problems such as eutrophication, harmful algal blooms (HABs), and hypoxia, are difficult to identify and solve. Consequently, considerable research has been conducted on the development of efficient and reliable prediction techniques. Since 2015, deep learning technology that makes predictions using big data has been widely used in various atmospheric, financial, medical, and scientific fields [5–8].

Marine research using deep learning technology can be divided into prediction-related research, classification-related research, and research on methods to correct missing values. Prediction-related research has been applied to various topics, such as the El Niño Index, chlorophyll-*a* time series, and sea surface temperature [9–11]. Classification-related research has been conducted to classify marine life using image data. For example, studies have been conducted to identify the harmful algae that adversely affect marine ecosystems and to classify coral reefs and monitor aquatic ecosystems [12–14]. However, observations using sensors can contain a significant amount of missing data. Consequently, various methods have been developed to estimate the missing data using deep learning techniques [15].

In addition to water quality modeling and deep learning studies, significant research has also been conducted to evaluate the status of plankton and other environmental factors related to marine environments using remote sensing. Ocean color sensors have been used in remote sensing satellites for decades. Those currently in operation include the Chinese Ocean Color and Temperature Scanner (COCTS) onboard HY-1D; Geostationary Ocean Color Imager (GOCI) onboard the Communication, Ocean, and Meteorological Satellite (COMS); Moderate Resolution Imaging Spectroradiometer (MODIS) onboard Aqua; Multi-Spectral Instrument (MSI) onboard Sentinel-2A and Sentinel-2B; Ocean and Land Color Instrument (OLCI) onboard Sentinel-3A and Sentinel-3B; Visible Infrared Imaging Radiometer Suite (VIIRS) onboard Suomi NPP; and Second-Generation Global Imager (SGLI) onboard GCOM-C [16,17]. Ocean color sensors provide vast amounts of spatial data that cannot be obtained from in situ measurements, and consequently, various analyses of spatiotemporal trends are possible. Therefore, extensive research has been conducted to retrieve marine inherent optical properties from ocean color remote sensing and verify ocean color data [18–21]. The data obtained from ocean color sensors are calibrated and verified by comparing them with in situ measurements and the results of existing ocean color sensors [22,23]. Recently, the measurement of ocean color data products such as colored dissolved organic matter (CDOM), chlorophyll-*a*, and total suspended sediment (TSS) has been improved using various neural network methods [24–26].

Another significant problem is the occurrence of HABs, which induce hypoxia and kills fish in marine environments. An HAB is caused by complex external environmental processes and factors such as eutrophication, currents, and salinity gradients [27,28]. Monitoring and predicting the spatiotemporal distribution of chlorophyll-*a* are vital to minimize the damage of HABs [29]. A variety of spatial information is required to predict the spatiotemporal distribution of chlorophyll-*a*, owing to the complex interaction of various physical, chemical, and biological factors. Although CDOM, TSS, and chlorophyll-*a* data can be obtained using ocean color sensors, the extraction of physical information such as currents, velocity, and salinity is limited, and in situ measurements can only provide some information. The continued development of hydrodynamic models has significantly improved their prediction ability, providing physical information with a root mean square error (RMSE) of $\pm 10\%$, $\pm 10\%$ to $\pm 20\%$, ± 0.5 °C, and ± 1 psu for water level, velocity, temperature, and salinity, respectively [30].

In this study, we aim to develop a tool that can estimate the spatial distribution of chlorophyll-*a* using deep learning technology. Satellite ocean color and hydrodynamic model data are used as the training data for the deep learning model. The CDOM, TSS, visibility, and chlorophyll-*a* data recorded on an hourly basis were extracted from a geostationary satellite. The hydrodynamic model data include temperature, salinity, water level, and velocity. The developed tool estimates the spatial distribution of chlorophyll-*a* using the spatial information of CDOM, TSS, visibility, water level, velocity, temperature, and salinity. The accuracy and applicability of the developed prediction tool is demonstrated by comparing the predicted results against the satellite data. As the variables applied to the prediction of chlorophyll-*a* contribute both individually and collectively, the contribution of each variable to the estimation of chlorophyll-*a* is examined as well.

2. Material and Methods

2.1. Study Area

The study area is a semi-closed maritime region surrounded by Hadong-gun, Sacheon, and Namhae-gun in South Korea, and is connected to the sea through the Daebang channel to the east, the Noryang channel to the west, and the Changsun channel to the south, as shown in Figure 1. The study area is approximately 19 km long along the north–south direction, and 13 km long along the east–west direction. The length of the coastline is approximately 136 km and the bounded area is approximately 180 km^2. The average depth is approximately 3.6 m, the depth of the central area is approximately 10 m, and the deepest area—in the channels—is approximately 30–40 m. In summer, a large volume of river water flows into the study area through the channels due to high rainfall. Consequently, although it is a semi-closed sea area, seawater exchange occurs. Sprayed shellfish farming is actively carried out in the region, gradually increasing from 230 tons in 2000, to 730 tons in 2010, and 2410 tons in 2014 [31]. Consequently, sustainable water quality management is vital in such semi-closed marine environments with active aquaculture.

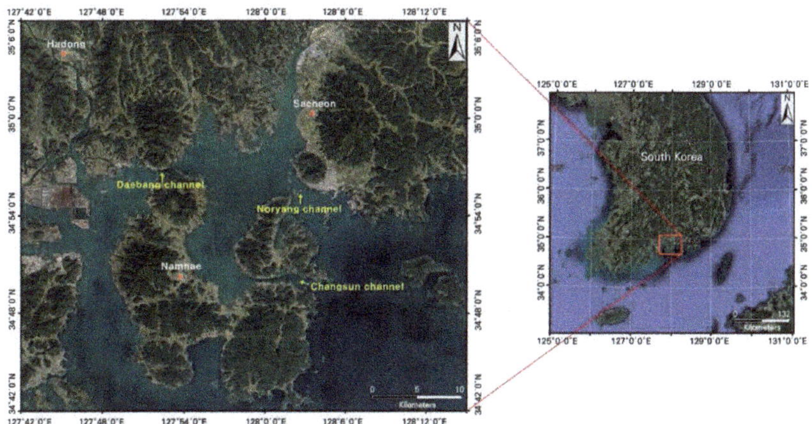

Figure 1. Study area in South Korea (Source: Google Earth).

2.2. Satellite Ocean Color

Various satellites with ocean color sensors have been launched from around the world, and Korea launched COMS in 2010 for ocean observation [32,33]. COMS performs meteorological and ocean observations and provides communication services. Ocean color observations are made using the GOCI. The GOCI observes an area of 2500 km × 2500 km, centered on the Korean Peninsula. The resolution of each grid is 500 m, both in width and height, as shown in Figure 2. As COMS is a geostationary satellite, the GOCI records data eight times a day (from 9:00 to 16:00), with images recorded for 30 min every hour. The primary role of the GOCI is to monitor the marine ecosystems around the Korean Peninsula, including long- and short-term marine environmental and climatic changes, coastal and marine environmental monitoring, coastal and marine resource management, and the generation of marine and fishery information [34,35].

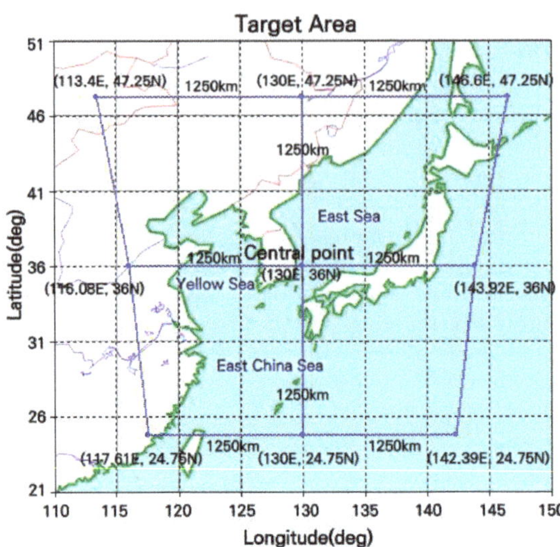

Figure 2. Spatial information observed by the GOCI http://kosc.kiost.ac.kr/p20/kosc_p21.html (accessed on 26 May 2020).

The GOCI has six visible bands with band centers of 412 nm (B1), 443 nm (B2), 490 nm (B3), 555 nm (B4), 660 nm (B5), and 680 nm (B6), and two near-infrared bands with band centers of 745 nm (B7) and 865 nm (B8). Bands B1–B5 are used to record the water quality parameters. The main applications of each band are B1 for yellow substances and turbidity; B2 for chlorophyll absorption maximum; B3 for chlorophyll and other pigments; B4 for turbidity and suspended sediment; and B5 for baseline of fluorescence signal, chlorophyll, and suspended sediment [36]. The amount of light recorded by the optical sensor onboard the satellite is converted to an electronic value and stored in the satellite image. Radiometric calibration is used to precisely define the relationship between the amount of light and the electronic value, and geometric correction is performed to correct the positional information of each pixel in the image. Subsequently, first-order outputs, such as the top-of-atmosphere radiance, and secondary outputs, such as the remote sensing reflectance, chlorophyll-*a*, TSS, and CDOM concentrations, are verified. Various calibration and validation studies have been performed on the GOCI data to improve its accuracy [35,37–39]. The ocean data products used herein were obtained from the GOCI using a software GDPS including atmospheric correction and ocean environment analysis algorithms. The GDPS enables real-time data processing using a Windows-based GUI. The data products obtained from the GDPS include the water leaving radiance (Lw), normalized water leaving radiance (nLw), chlorophyll-*a*, TSS, and CDOM [40].

2.3. Hydrodynamic Model

A hydrodynamic model was used to generate marine physical factors, such as the currents, water level, salinity, and temperature, in the study area. The Delft 3D model, which has been applied in several research areas, was used to simulate three-dimensional hydrodynamics [41–44]. The model domain extended for 58 km along the north–south direction and 53 km along the east–west direction, to sufficiently cover the study area. The model grid contained 155 $\times$ 245 horizontal cells and, to optimize the computational time, fine and coarse grids were formed in the study area and open sea area, respectively. A total of five vertical layers were modeled to replicate the interaction between the vertical layers and the vertical distribution of salinity and water temperature. Bathymetry for the study area was obtained from the latest navigational charts and the survey data of the Korea Hydrographic and Oceanographic Agency (KHOA). As shown in the bathymetry chart in Figure 3, the bay has a relatively shallow depth and the channels are relatively deep.

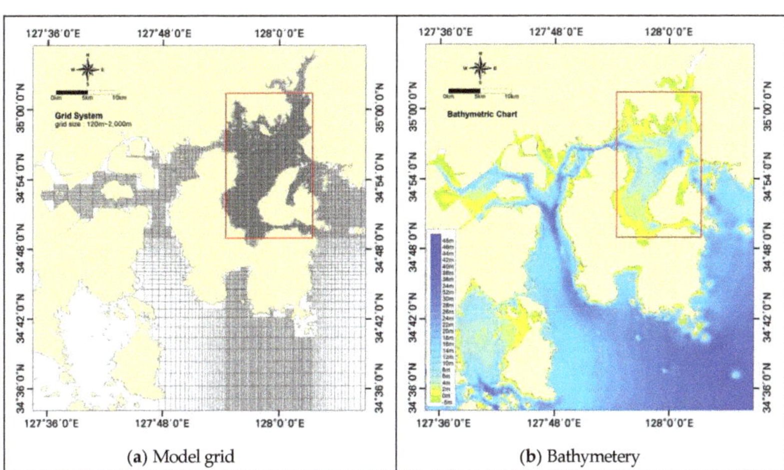

Figure 3. (**a**) Model grid in the study area; (**b**) Bathymetry in the study area.

The boundary conditions of the study area must be defined to execute the hydrodynamic model. The water levels, salinity, and temperatures observed at different measurement sites (GoSung-JaRan, TongYong3, NamHae3) by the Korea Marine Environment Management Corporation (KOEM) were set as the sea boundary conditions, and the monthly average flow rates at GwanGok, BakRyeon, MukGok, GaWa, and SaCheon were set as the river boundary conditions. Meteorological data, such as the wind direction, wind speed, air temperature, and relative humidity, measured at the NamHae site of the Korea Meteorological Administration (KMA), were also used as model input data. The initial conditions of the water level and velocity were set to zero, and the initial conditions of temperature and salinity were derived from the measured data at the five KOEM stations shown in Figure 4. The hydrodynamic model was simulated for a total of five years from 1 January 2015 to 31 December 2019. As the data used in the deep learning model include the water level, current, salinity and temperature, these data were verified. The water level was verified using the data observed at the T1 site operated by KHOA, which is located inside the bay. The current was validated against the data recorded at the PC1 site operated by KHOA, between 24 July 2015 and 26 August 2015. The salinity and water temperature were validated against the data measured at the JinJuMan 1 and JinJuMan 2 sites, operated by KOEM, and the SamCheonPo site, operated by KHOA, as shown in Figure 4.

The water levels in the study area fluctuated by approximately 3 m and were primarily affected by the tides. The average difference in the water level between the hydrodynamic model and the observed values was approximately 10 cm, and the absolute error was within 8–10%, with slight differences every year. The currents observed between 24 July 2015 and 26 August 2015 were classified into a U-component—moving east–west—and a V-component—moving north–south. As shown, the U-component was the dominant current in the study area. The U-component current flowed as fast as 0.5 m/s and fluctuated based on the tidal cycle. Although the hydrodynamic model results appear to underestimate the current patterns, the results are reproduced well. The temperature was below 10 °C during winter and almost 30 °C during summer, with clearly noticeable seasonal variations. The water temperature varied between 13 °C and 20 °C during spring and autumn, with the lowest temperature in February and the highest temperature in August. Considering the predicted daily temperatures, the hydrodynamic model adequately reproduced the annual temperature-change pattern, and the average RMSE of the temperature was 0.862 °C. The salinity was highly influenced by the river flow, i.e., during spells of high rainfall, the salinity temporarily decreased before increasing to approximately 32–33 psu. The average RMSE of the salinity was 0.6 psu, as shown in Figure 5.

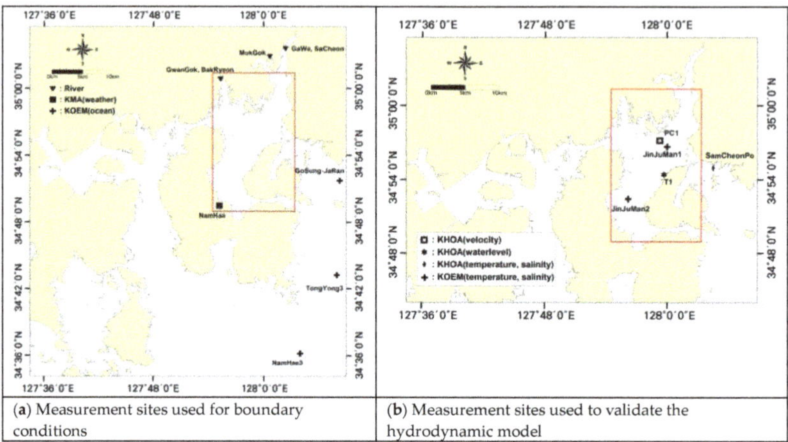

(**a**) Measurement sites used for boundary conditions

(**b**) Measurement sites used to validate the hydrodynamic model

Figure 4. Locations of the KMA, KOEM, KHOA, and river monitoring stations in the study area. (**a**) Measurement sites used for boundary conditions. (**b**) Measurement sites used to validate the hydrodynamic model.

2.4. Data Structure for Deep Learning Model

The satellite data of the study area, which was required to construct the deep learning model, was provided by the Korea Ocean Satellite Center (KOSC) in the Korea Institute of Ocean Science and Technology. The data were recorded eight times per day between 9:00 and 16:00, from January 2015 to December 2019. The data obtained included the entire Korean Peninsula, and the total size of the data was approximately 14 TB. No satellite data could be extracted when the study area was covered by clouds. The total number of extracted data was 391 in 2015, 276 in 2016, 266 in 2017, 271 in 2018, and 128 in 2019. Generally, a large amount of data were recorded during winter, when the weather was good, and a small amount of data were recorded during summer, owing to the increased rainfall and typhoons.

The hydrodynamic model results were extracted for the same area as the satellite measurements, as shown in Figure 6. The hourly salinity, temperature, currents, and water levels between 2015 and 2019 were converted into a grid format. As the resolution of the satellite data was 500 m, the data from the area adjacent to the coastline could not be obtained. Therefore, only the data pertaining to the sea area 500 m away from the coastline were used to train the deep learning model. Accordingly, the hydrodynamic model results of the area adjacent to the coastline were also neglected.

2.5. Deep Learning Model Structure

As the satellite and hydrodynamic model data were in the form of a 48 $\times$ 27 grid, they could be treated as image data. Consequently, an image-based deep learning method was applied herein. Each 48 $\times$ 27 grid was referred to as an 'image,' and each point in the image was referred to as the 'data' or 'point'. The satellite chlorophyll-*a* data were treated as ground-truth data, as several studies have shown a high correlation between the ground-truth chlorophyll-*a* data and satellite chlorophyll-*a* data. Accordingly, we constructed a deep learning model to estimate the temporal and spatial distribution of chlorophyll-*a* using both the satellite and the hydrodynamic model data. Specifically, the deep learning model estimated the temporal and spatial distribution of chlorophyll-*a* at a given time (t) by integrating the satellite data, such as the CDOM, TSS, and visibility, and the hydrodynamic model data, such as the currents, water level, temperature, and salinity, at the same time (t), as illustrated in Figure 7.

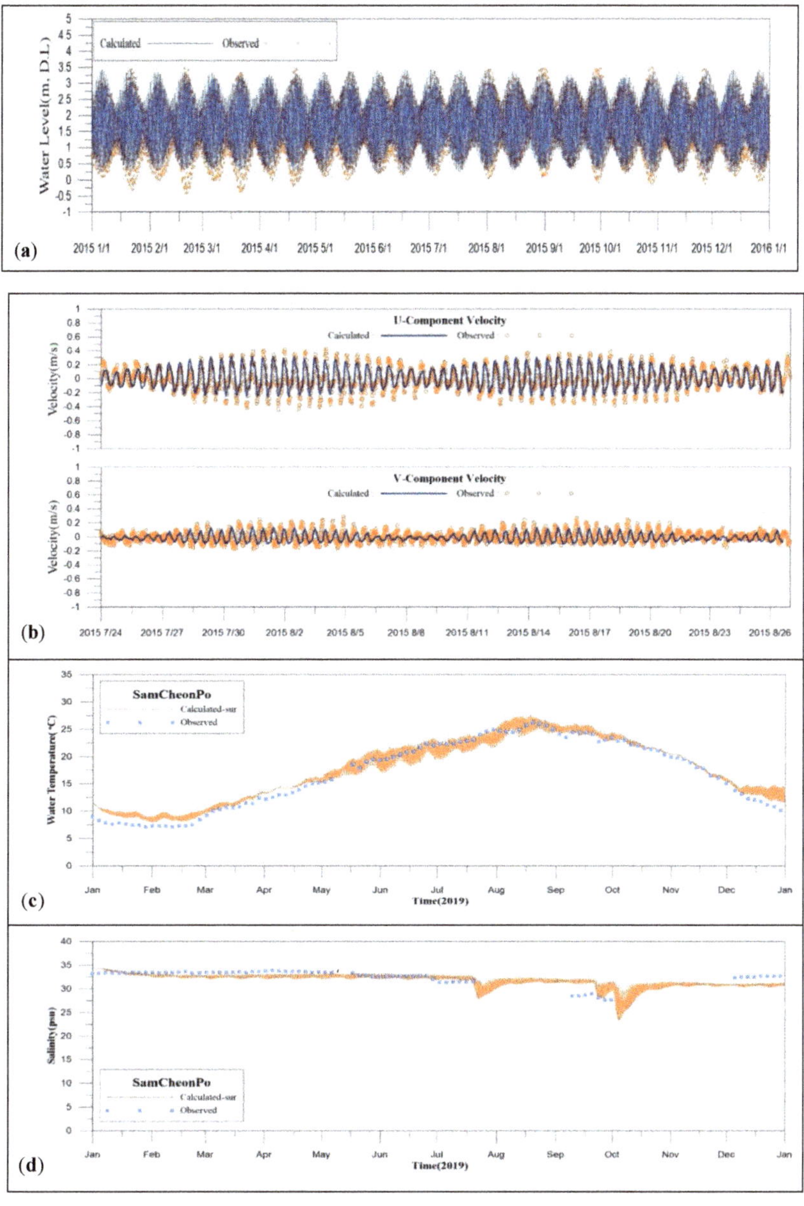

Figure 5. (**a**) Temporal variations of water level; (**b**) Temporal variation of currents; (**c**) Temporal variation of salinity; (**d**) Temporal variation of temperature (points are observations and lines are model results).

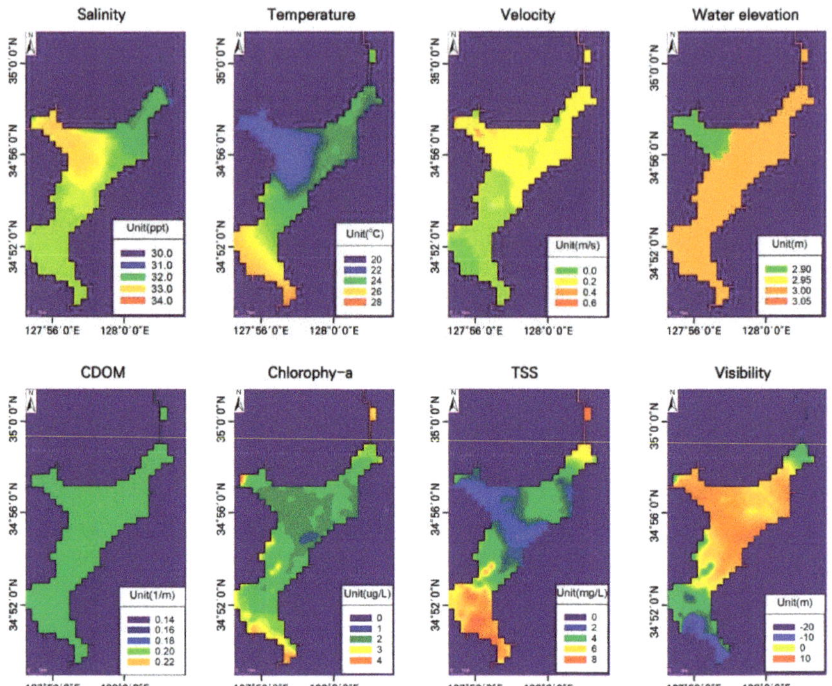

Figure 6. Spatial distribution of training data in the study area: salinity, temperature, currents, and water levels from the hydrodynamic model, and CDOM, chlorophyll-*a*, TSS, and visibility from the satellite ocean color data.

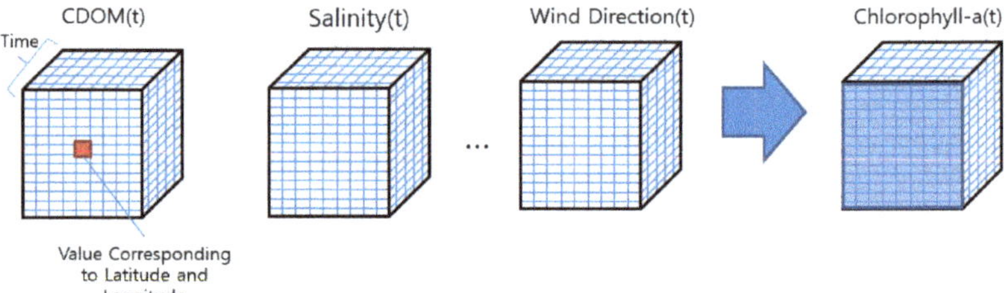

Figure 7. Construction of the deep learning model for estimating the temporal and spatial distribution of chlorophyll-*a*. To utilize spatial information, the input data were organized in a matrix accumulated over time. The value corresponding to each row and column corresponds to the latitude and longitude of each data.

A convolutional neural network (CNN) is a well-known deep learning model that is suitable for image data processing. A CNN model consists of multiple convolutional layers that extract features from an image and pool the layers through subsampling, leaving only the important patterns behind. Classification and estimation are performed through iterative convolutional and pooling operations. We designed two approaches to estimate chlorophyll-*a* based on a CNN. The first CNN model, called 'CNN Model I', estimates the chlorophyll-*a* concentration from an image in a 48 × 27 grid format by integrating a total of seven images—three images from the satellite data, such as the CDOM, TSS, and visibility, and four images from the hydrodynamic model data, such as the currents, water level, temperature, and salinity—as shown in Figure 8. Notably, as the image size was small, there was no pooling layer. Consequently, the pooling layer for information compression was ineffective. The second CNN model, called 'CNN Model II', predicted the chlorophyll-*a* concentration using segmented images.

Additional preprocessing is required to use segmented images as the model input. For example, in the case of 7 × 7 segmented images, the chlorophyll-*a* value is estimated by using segmented images of seven individual input variables. The difference between CNN Model I and CNN Model II is that the former estimates one chlorophyll-*a* image by integrating the images of seven individual input variable changes, whereas the latter estimates the chlorophyll-*a* value by integrating segmented images of seven individual input variables, as shown in Figure 9. As CNN Model II estimates the chlorophyll-*a* value using the data around a point of interest, we believe that it also reflects the local characteristics well.

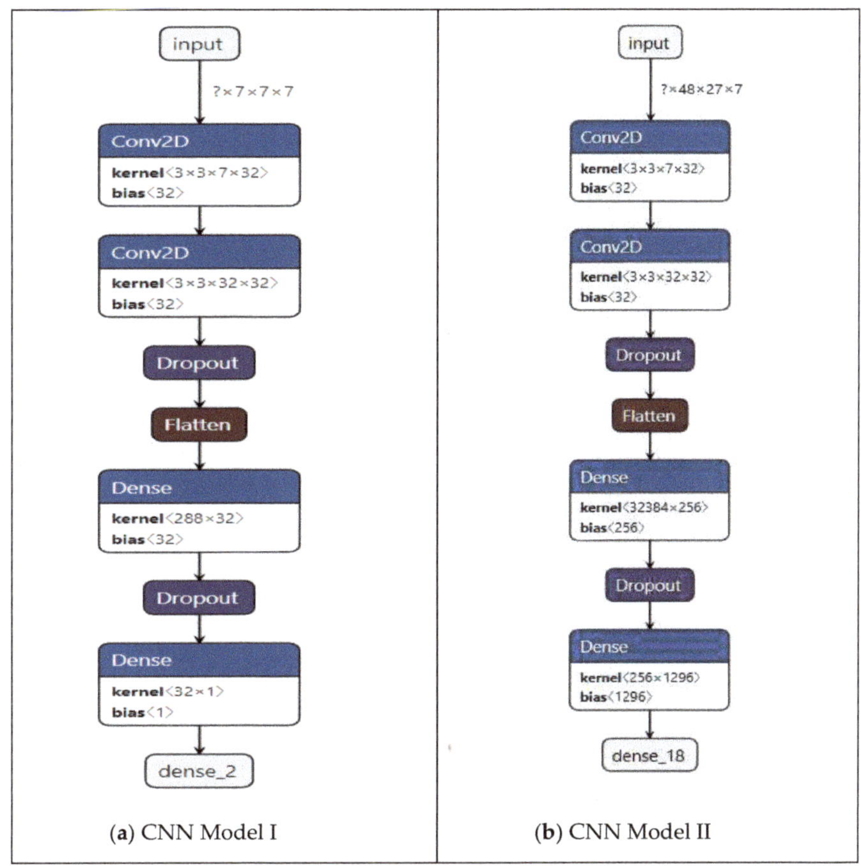

Figure 8. (**a**) Algorithm of CNN Model I and (**b**) CNN Model II. CNN Model I uses seven images of 48 × 27 grid size and estimates the chlorophyll-*a* value in a 48 × 27 grid format. CNN Model II uses segmented images in a 7 × 7 grid format and estimates the chlorophyll-*a* value.

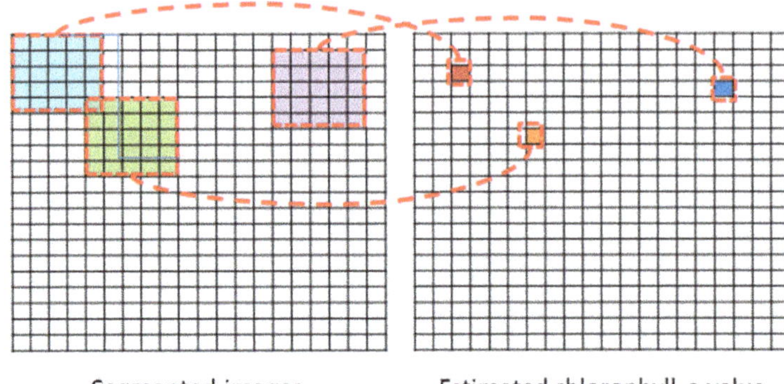

Figure 9. Schematic diagram of the application of segmented images in the CNN Model II; segmented images are generated by iteratively moving the window cell-by-cell. The CNN Model II estimates a chlorophyll-*a* value integrating segmented images of seven individual input variables.

To verify the reliability of the deep learning model, the data were divided into training data, validation data, and test data, considering the seasonal characteristics over an entire year. For CNN Model I, 932 images were used for training, 271 images for validation, and 128 images for testing. For CNN Model II, the images in a 48 × 27 grid format were divided into segmented images with a 7 × 7 grid format. Consequently, the number of images used for training, validation, and testing increased to 293,580, 85,365, and 40,320, respectively. As CNN Model II did not have the segmented images required to estimate the values of three columns and three rows at the edge of each image, the values related to these regions were not predicted. The quantity of available data varied from one year to another as the satellite measurements could not be obtained on days with poor weather. In particular, the quantity of data obtained during summer was relatively small compared to that obtained during the other seasons owing to increased rainfall and typhoons, as shown in Table 1.

Table 1. Information of training data, validation data, and test data in the CNN Model I and CNN Model II.

Category	Training Data	Validation Data	Test Data
Period (year)	2015–2017	2018	2019
CNN Model I (# of images)	932	271	128
CNN Model II (# of segmented images (7 × 7))	293,580	85,365	40,320

3. Results

3.1. CNN Model I

The RMSE, which is the difference between the predicted chlorophyll-*a* and the satellite chlorophyll-*a* values, was used to evaluate the accuracy of the CNN models designed herein. The RMSE was calculated as:

$$\mathrm{RMSE} = \sqrt{\frac{1}{n}\sum\nolimits_{1}^{n}(pred(i) - target(i))^2} \quad (1)$$

where *pred(i)* represents the predicted chlorophyll-*a* pixel value for of the *i*th point and *target(i)* represents the satellite chlorophyll-*a* pixel value for the *i*th point in each image.

CNN Model I was used to estimate the chlorophyll-*a* value of 128 images recorded in 2019. In most cases, the RMSE was approximately 0.2–0.6 and the average RMSE was 0.436, as shown in Figure 10. The minimum RMSE was 0.106 and the maximum RMSE was 1.242, which is a significant gap. Therefore, specific analyses were performed for the cases with RMSE = 0.106, RMSE = 0.506, and RMSE = 1.209, as shown in Figure 11.

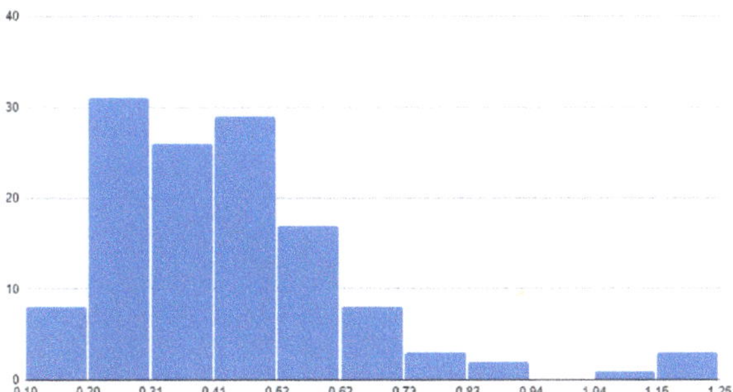

Figure 10. RMSE distribution for 128 images using CNN Model I: histogram with the range of RMSE values on the *X*-axis and the number of images on the *Y*-axis.

In the case with the lowest RMSE (RMSE = 0.106), the model results showed that there was a slight predictive error in the image, but the overall trend was well estimated. In the case with the RMSE close to the average value (RMSE = 0.506), the overall change in chlorophyll-*a* in the entire image was clearly estimated, but the accuracy of the estimation of the local changes in chlorophyll-*a* was limited. In the case with the high RMSE (RMSE = 1.209), the model was unable to estimate the satellite chlorophyll-*a* value. The measured values clearly indicate a change in the spatial chlorophyll-*a* values, whereas the estimated values tend to converge to the average value at most points. Thus, the model appeared to have a tendency to approximate the average value as the estimated value when the training data were insufficient, as shown in Figure 11. Consequently, the coefficient of determination (R^2), which represents how well the model results fit the satellite data, was applied herein. R2 is represented by a value of 0.0–1.0, where a value of 1.0 indicates a perfect fit. When the RMSE was relatively low, the R^2 was around 0.673, and when the RMSE was high, $R^2 < 0.5$. When $R^2 < 0.5$, the higher the chlorophyll-*a* value of the satellite data, the lower the predictive ability, as shown in Figure 12.

The results of CNN Model I tended to be averaged by assimilating the surrounding values instead of estimating local changes. As deep learning models such as a CNN estimate values by analyzing patterns from training data, the prediction patterns could not be determined from insufficient training data. Therefore, CNN Model I, which was trained using only 1203 training and validation images, could predict the overall trends but failed to predict local changes. Notably, if additional training data is provided, the prediction accuracy of CNN Model I can be improved.

3.2. CNN Model II

Chlorophyll-*a* estimation was also performed using CNN Model II, which utilized 300 times more training and validation data than CNN Model I, owing to the use of segmented images. The RMSE values of CNN Model II were around 0.05–0.8. Most of the RMSE values were less than or equal to 0.2, with an average of 0.167. Compared to the results of CNN Model I, the RMSE values of CNN Model II were significantly lower, confirming the excellent predictive ability of the latter. Notably, RMSE was less than or equal to 0.12 in almost half the total number of predictions. A detailed analysis was performed by classifying the RMSE values of CNN Model II into good, average, and bad cases, as shown in Figure 13.

In the case of a low RMSE value (RMSE = 0.055), the predicted chlorophyll-*a* values were almost the same as those of the satellite chlorophyll-*a* values. Furthermore, the spatial variations of chlorophyll-*a* concentration were properly estimated. The case with an RMSE value close to the average value (RMSE = 0.204) also demonstrated similar results to the observed values. In particular, the changes in the spatial concentration were estimated accurately. In the case of a high RMSE value (RMSE = 0.775), the model accurately reproduced the spatial concentration pattern but tended to underestimate the concentration at some points. The satellite data exhibited large variations in the concentration between adjacent points, whereas the deep learning model corrected this drastic change and estimated it smoothly in space, as shown in Figure 14.

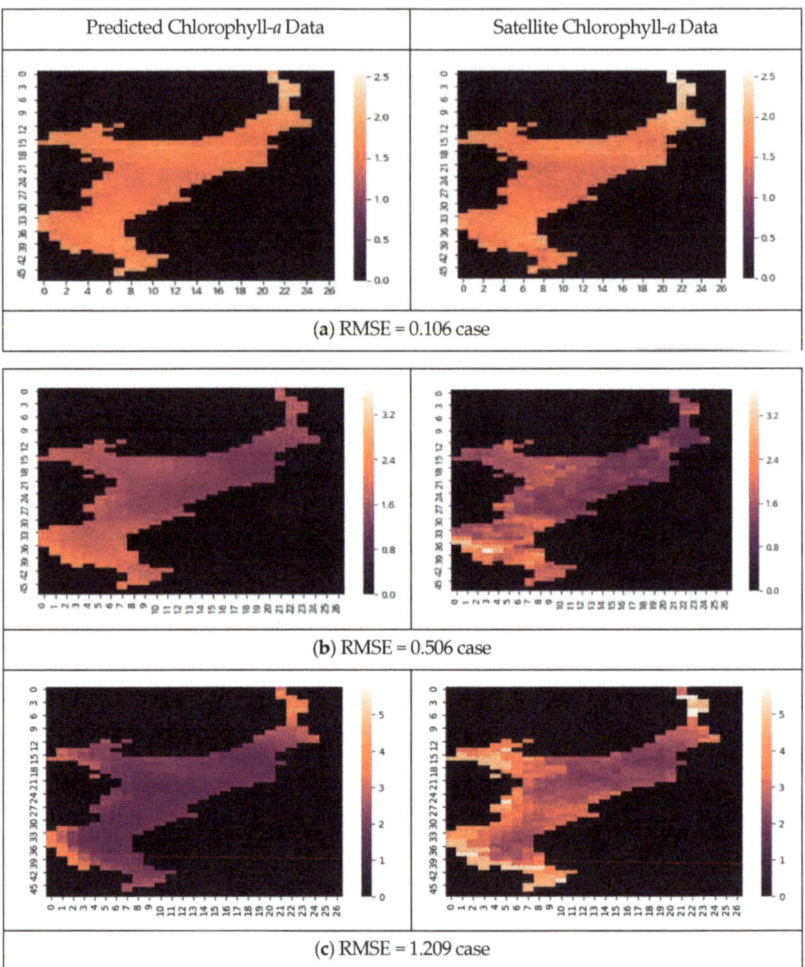

Figure 11. Chlorophyll-*a* results estimated using the CNN Model I: The left section shows the predicted chlorophyll-*a* values and the right section shows the satellite chlorophyll-*a* values corresponding to the left section. The RMSE values for the three cases are (**a**) 0.106, (**b**) 0.506, and (**c**) 1.209, respectively.

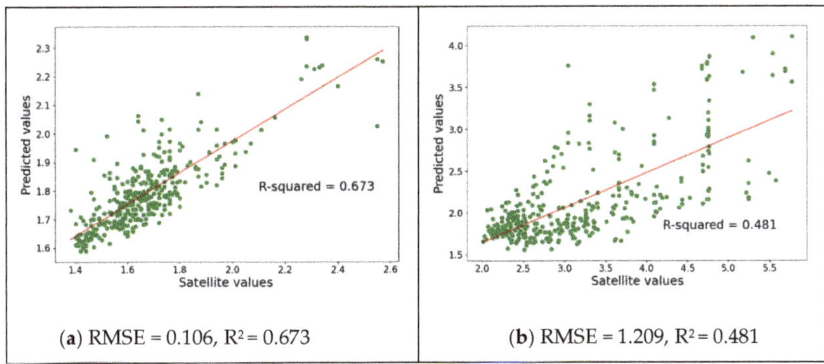

(**a**) RMSE = 0.106, R^2 = 0.673 (**b**) RMSE = 1.209, R^2 = 0.481

Figure 12. Examples of (**a**) good R^2 and (**b**) bad R^2 values among the results of the CNN Model I.

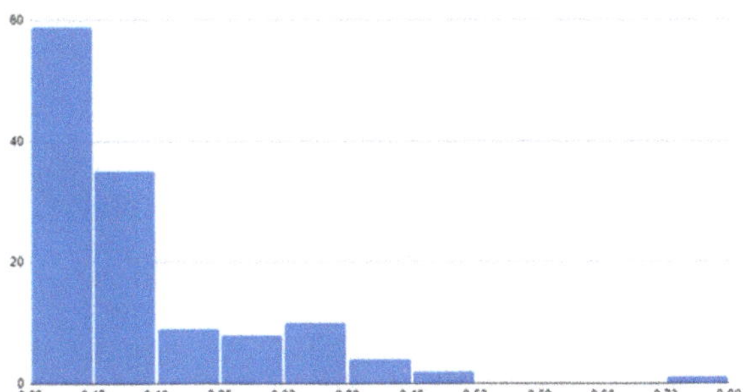

Figure 13. RMSE distribution for 128 images using the CNN Model II: histogram with the range of RMSE values on the *X*-axis and the number of images on the *Y*-axis.

Compared to CNN Model I, CNN Model II has significantly better chlorophyll-*a* estimation ability, and the spatial change pattern of chlorophyll-*a* was successfully estimated in all the model results. Furthermore, the coefficient of determination (R^2) improved significantly. When RMSE = 0.055, R^2 = 0.91, and when RMSE = 0.775, which suggests a high degree of error, the overall trend was reproduced well and R^2 = 0.661, as shown in Figure 15. Although both models used the same CNN technique, the difference in their estimation abilities is likely due to the large difference in their respective training data volumes.

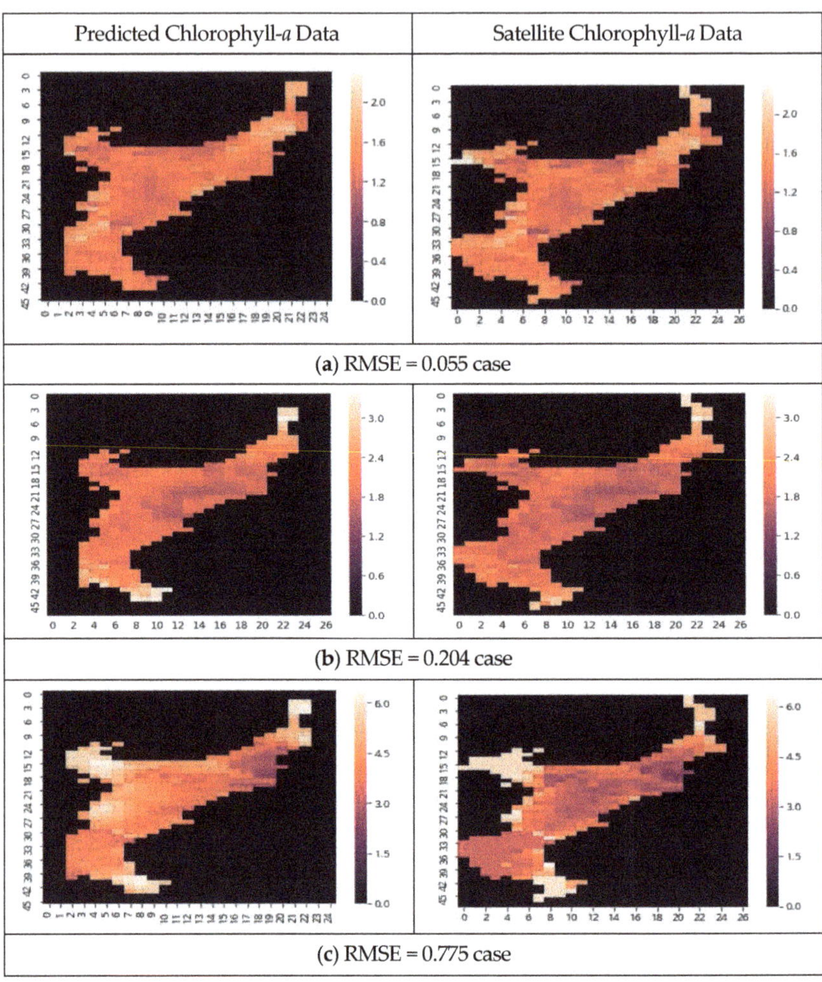

Figure 14. Chlorophyll-*a* results estimated using CNN Model II. The left section shows the predicted chlorophyll-*a* values and the right section shows the corresponding satellite chlorophyll-*a* image values. The corresponding RMSE values are (**a**) 0.055, (**b**) 0.204, and (**c**) 0.775, respectively.

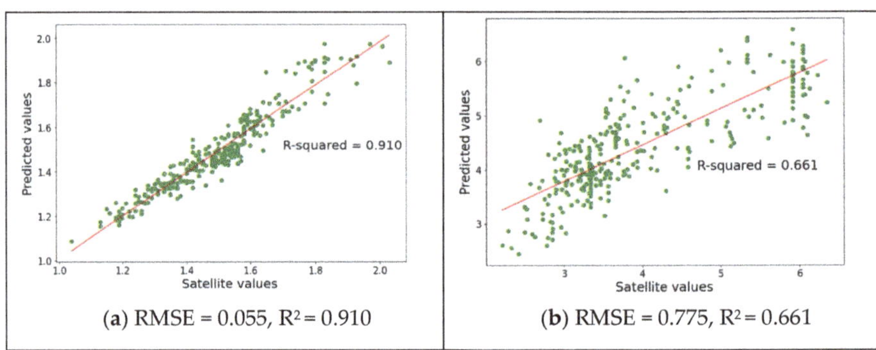

Figure 15. Examples of (**a**) good R^2 and (**b**) bad R^2 values among the results of the CNN Model II.

4. Discussion

Plankton growth is affected by various factors such as water flow, water temperature, nutrients, and light. The concentration of plankton is relatively high in shallow water coastal areas and upwelling regions, as they have a rich supply of nutrients. The surface salinity and temperature of the study area change significantly as high salinity and low temperature seawater flows through the Daebang channel, located in the northeast. The satellite data reveals that the seawater flowing in from the Daebang channel contains low concentrations of chlorophyll-*a*, resulting in a relatively low chlorophyll-*a* concentration in the center of the study area. Moreover, the study area is connected to a river, and large amounts of river water flow into the study area during the rainy summer season, affecting the growth of plankton. As the growth of each type of plankton depends on the water temperature, it is important to predict the seasonal changes in plankton concentration.

The monthly averaged satellite data and model data were compared to determine whether the prediction model developed herein can adequately estimate the seasonal changes in plankton concentration. In 2016 and 2018, the chlorophyll-*a* concentration was low in January—the winter season—but high during spring and summer. The concentration decreased again in November, which clearly demonstrates the seasonal fluctuations in plankton concentration in the study area. The developed model successfully estimated the seasonal fluctuations in plankton concentration in 2016 and 2018. Notably, although the seasonal fluctuations in 2019 were relatively small compared to those in 2016 and 2018, the developed model accurately estimated the small seasonal and local concentration changes, as shown in Figure 16.

We performed a sensitivity analysis to determine the influence of each input variable in the model results. To do so, the performance of the model was investigated by only using individual input variables as training data for the deep learning model. The results of the sensitivity analysis (Table 2) indicated that CDOM contributes significantly to the estimation of chlorophyll-*a*, with an RMSE of 0.231. The visibility, TSS, and temperature are also relatively important variables, whereas the remaining input variables have a relatively low contribution to the improvements in model performance. Notably, when all the input variables, except for CDOM, were integrated, the RMSE increased to 0.330. Thus, although the individual input variables have a negligible effect on the model performance, the integration of the input variables has a complementary effect and improves model prediction. When all the input variables were used, the RMSE was 0.191, which represents the best model performance.

Predictive studies on plankton concentrations have been conducted for decades using various water quality models. However, there are numerous challenges and limitations owing to the complex interactions between water quality parameters, uncertainty of hydrodynamic information, and lack of boundary nutrient loadings and validation data. For example, the results of studies that predicted the level of chlorophyll-*a* in Chesapeake Bay by employing a 3D water quality model had a correlation coefficient of less than 0.5 [45,46]. The main objective of this study was to develop a prediction tool that can be used in combination with existing water quality models, wherein the currents, water level, salinity, and temperature calculated from the hydrodynamic model were used to predict chlorophyll-*a* concentration. As the hydrodynamic model results have an error of only 10–20%, they can be used as training data for deep learning models [30]. Accordingly, satellite data such as CDOM, TSS, and visibility, which were validated through various studies, were used as training data to develop a chlorophyll-*a* prediction tool. The prediction model developed herein—CNN Model II—has good accuracy in the estimation of chlorophyll-*a* concentration, as evidenced by an R^2 of 0.66–0.91 and an RMSE of 0.055–0.775. Although the data used in the model are not in situ measurements, satellite data and hydrodynamic model data have continuously improved in recent years, and provide spatiotemporal data that cannot be obtained from in situ measurements. In addition, the developed model can predict the spatiotemporal chlorophyll-*a* concentration based on changes in individual parameters such as an increase in water temperature due to climate change, an increase in

CDOM due to land development, and an increase in TSS as a result of poor flushing due to the presence of coastal structures, etc.

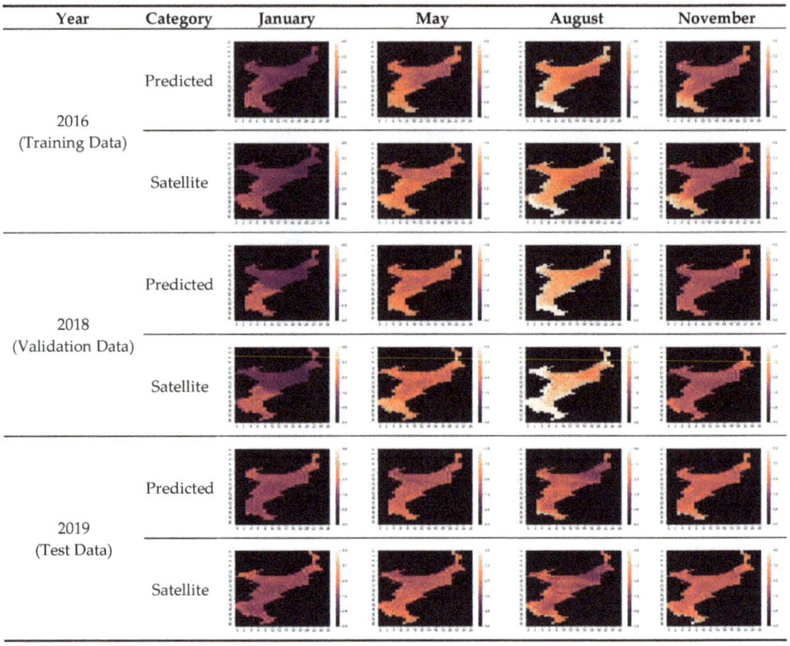

Year	Category	January	May	August	November
2016 (Training Data)	Predicted				
	Satellite				
2018 (Validation Data)	Predicted				
	Satellite				
2019 (Test Data)	Predicted				
	Satellite				

Figure 16. Monthly averaged spatial distribution of model results and satellite chlorophyll-*a* images (CNN Model II).

Table 2. Sensitivity analysis results showing RMSE values corresponding to input variables.

Input Variables	RMSE
CDOM	0.231
TSS	0.526
Visibility	0.492
Currents	0.651
Salinity	0.648
Temperature	0.545
Water level	0.653
All except CDOM	0.330
All	0.191

The model results must be compared to real-world measurement data to validate the performance of the model. However, spatiotemporal chlorophyll-*a* data cannot be obtained through in situ measurements. The performance of the chlorophyll algorithms used for the GOCI radiometric data were evaluated using in situ measurements collected at 491 stations [47]. The evaluation results of the coincident in situ pairs of Rrs and chlorophyll measurements demonstrated that the mean uncertainty was <35%, with a correlation of around 0.8. Therefore, assuming that the data from GOCI are close to the real-world values, the model results were validated by comparing them against the satellite data. To improve the developed model, it is necessary to conduct a validation study with the measurement data of the study area and a comparative study with the state-of-the-art methods.

5. Conclusions

In this study, we developed a deep learning model using a CNN to predict the spatiotemporal changes in chlorophyll-*a* in a bay in Korea. The data used to train the deep learning model were the spatial data of chlorophyll-*a*, total suspended sediment (TSS), visibility, and colored dissolved organic matter (CDOM) obtained from the Geostationary Ocean Color Imager (GOCI) on board COMS, and the water level, currents, temperature, and salinity calculated by a verified hydrodynamic model. CNN MODEL I, which estimates chlorophyll-*a* images in a 48×27 grid format, was developed using the same 48×27 grid size of the CDOM, TSS, visibility, water level, currents, temperature, and salinity data. The RMSE between the satellite image and the predicted image from the model was calculated, and was between 0.2 and 0.6 in most cases. Although CNN Model I was able to estimate the overall trend, there were significant differences between the predicted results and the satellite data in some cases. As the deep learning model improves the predictive ability of the model by extracting and analyzing the inherent patterns in the training data, if the training data is insufficient, the predictive ability of the model decreases significantly.

To solve the problem of insufficient data, we designed another deep learning model—CNN Model II—using segmented images in a 7×7 grid format. CNN Model II estimates target values only using the data around the point of interest and, consequently, the volume of training data used in CNN Model II is around 300 times more than that of CNN Model I. Therefore, CNN Model II can extract and analyze inherent patterns in the training data more accurately. The average RMSE of CNN Model II was 0.191, which is significantly lower than that of CNN Model I, which was 0.463. Moreover, the spatial concentration of chlorophyll-*a* was well estimated by CNN Model II, thereby proving the efficacy of the deep learning model.

A sensitivity analysis was performed to determine the influence of each input variable on the model performance, and CDOM was found to have the most influence on the prediction of chlorophyll-*a*. The visibility, TSS, and temperature were also relatively important variables. The input variables with a strong influence on the model performance have a direct relationship with nutrients, photosynthesis, and temperature, which influence plankton growth. Therefore, the data-based deep learning model considers the major factors related to the growth of plankton and makes predictions. Additionally, the predictive accuracy of the deep learning model was improved if the training data also included the currents, velocity, and salinity.

Author Contributions: Conceptualization, D.J. and T.K.; methodology, E.L. and T.K.; software, D.J., T.K. and K.K.; validation, D.J. and K.K.; data curation, E.L.; writing—original draft preparation, D.J. and T.K.; writing—review and editing, D.J. and T.K; visualization, D.J. and K.K. All authors have read and agreed to the published version of the manuscript.

Funding: This research received no external funding.

Institutional Review Board Statement: Not applicable.

Informed Consent Statement: Not applicable.

Data Availability Statement: Data sharing is no applicable to this article.

Acknowledgments: This paper was written following the research work "A Study on Marine Pollution Using Deep Learning and its Application to Environmental Impact Assessment (II)" (RE2021-08), funded by the Korea Environment Institute (KEI).

Conflicts of Interest: The authors declare that they have no conflict of interest.

References

1. Beck, M.B. Water quality modeling: A review of the analysis of uncertainty. *Water Resour. Res.* **1987**, *23*, 1393–1442. [CrossRef]
2. Zheng, L.; Chen, C.; Zhang, F.Y. Development of water quality model in the Satilla River Estuary, Georgia. *Ecol. Model.* **2004**, *178*, 457–482. [CrossRef]
3. Jia, H.; Xu, T.; Liang, S.; Zhao, P.; Xu, C. Bayesian framework of parameter sensitivity, uncertainty, and identifiability analysis in complex water quality models. *Envrion. Model. Softw.* **2018**, *104*, 13–26. [CrossRef]

4. Yan, J.; Xu, Z.; Yu, Y.; Xu, H.; Gao, K. Application of a hybrid optimized BP network model to estimate water quality parameters of Beihai Lake in Beijing. *Appl. Sci.* **2019**, *9*, 1863. [CrossRef]
5. Vargas, M.R.; de Lima, B.S.L.P.; Evsukoff, A.G. Deep learning for stock market prediction from financial news articles. In Proceedings of the 2017 IEEE International Conference on Computational Intelligence and Virtual Environments for Measurement Systems and Applications (CIVEMSA), Annecy, France, 26–28 June 2017. [CrossRef]
6. Razzak, M.I.; Naz, S.; Zaib, A. Deep learning for medical image processing: Overview, challenges and the future. In *Classification in BioApps*; Dey, N., Ashour, A., Borra, S., Eds.; Springer: Cham, Switzerland, 2018. [CrossRef]
7. Matsuoka, D.; Watanabe, S.; Sato, K.; Kawazoe, S.; Yu, W.; Easterbrook, S. Application of deep learning to estimate atmospheric gravity wave parameters in reanalysis data sets. *Geophys. Res. Lett.* **2020**, *47*, e2020GL089436. [CrossRef]
8. Singh, R.; Agarwal, A.; Anthony, B.W. Mapping the design space of photonic topological states via deep learning. *Opt. Express* **2020**, *28*, 27893. [CrossRef] [PubMed]
9. Xiao, C.; Chen, N.; Hu, C.; Wang, K.; Xu, Z.; Cai, Y.; Xu, L.; Chen, Z.; Gong, J. A spatiotemporal deep learning model for sea surface temperature field prediction using time-series satellite data. *Envrion. Model. Softw.* **2019**, *120*, 104502. [CrossRef]
10. Guo, Y.; Cao, X.; Liu, B.; Peng, K. El Niño index prediction using deep learning with ensemble empirical mode decomposition. *Symmetry* **2020**, *12*, 893. [CrossRef]
11. Shin, Y.; Kim, T.; Hong, S.; Lee, S.; Lee, E.; Hong, S.; Lee, C.; Kim, T.; Park, M.; Park, J.; et al. Prediction of chlorophyll-*a* concentrations in the Nakdong River using machine learning methods. *Water* **2020**, *12*, 1822. [CrossRef]
12. Park, S.; Kim, J. Red tide algae image classification using deep learning based open source. *Smart MediaJ.* **2018**, *7*, 34–39.
13. Lumini, A.; Nanni, L.; Maguolo, G. Deep learning for plankton and coral classification. *Appl. Comput. Inform.* **2019**. [CrossRef]
14. Raphael, A.; Dubinsky, Z.; Iluz, D.; Benichou, J.I.C.; Netanyahu, N.S. Deep neural network recognition of shallow water corals in the Gulf of Eilat(Aqaba). *Sci. Rep.* **2020**. [CrossRef] [PubMed]
15. Velasco-Gallego, C.; Lazakis, I. Real-time data-driven missing data imputation for short-time sensor data of marine systems: A comparative study. *Ocean Eng.* **2020**, *218*, 108261. [CrossRef]
16. ICCG. Current Ocean-Colour Sensors. Available online: https://ioccg.org/resources/missions-instruments/current-ocean-colour-sensors/ (accessed on 16 April 2021).
17. McKinna, L.I.W. Three decades of ocean-color remote-sensing *Trichodesmium* spp. In the World's oceans: A review. *Prog. Oceanogr.* **2015**, *131*, 177–199. [CrossRef]
18. Hu, S.B.; Cao, W.X.; Wang, G.F.; Xu, Z.T.; Lin, J.F.; Zhao, W.J.; Yang, Y.Z.; Zhou, W.; Sun, Z.H.; Yao, L.J. Comparison of MERIS, MODIS, SeaWiFS-derived particulate organic carbon, and in situ measurements in the South China Sea. *Int. J. Remote Sens.* **2016**, *37*, 1585–1600. [CrossRef]
19. Werdell, P.J.; McKinna, L.I.W.; Boss, E.; Ackleson, S.G.; Craig, S.E.; Gregg, W.W.; Lee, Z.; Maritorena, S.; Roesler, C.S.; Rousseaus, C.S.; et al. An overview of approaches and challenges for retrieving marine inherent optical properties from ocean color remote sensing. *Prog. Oceanogr.* **2018**, *160*, 186–212. [CrossRef] [PubMed]
20. Scott, J.P.; Werdell, P.J. Comparing level-2 and level-3 satellite ocean color retrieval validation methodologies. *Opt. Express* **2019**, *27*, 30140–30157. [CrossRef]
21. Niroumand-Jadidi, M.; Bovolo, F.; Bruzzone, L. Novel spectra-derived features for empirical retrieval of water quality parameters: Demonstrations for OLI, MSI, and OLCI sensors. *IEEE Trans. Geosci. Remote Sens.* **2019**, *57*, 10285–10300. [CrossRef]
22. Crout, R.L.; Ladner, S.; Lawson, A.; Martinolich, P.; Bowers, J. Calibration and validation of multiple ocean color sensors. In Proceedings of the OCEANS 2018 MTS/IEEE Conference, Charleston, SC, USA, 22–25 October 2018; pp. 1–6. [CrossRef]
23. Wang, M.; Ahn, J.H.; Jiang, L.; Shi, W.; Son, S.H.; Park, Y.J.; Ryu, J.H. Ocean color products from the Korean Geostationary Ocean Color Imager (GOCI). *Opt. Express* **2013**, *21*, 3835–3849. [CrossRef]
24. Hieronymi, M.; Muller, D.; Doerffer, R. The OLCI neural network swarm(ONNS): A Bio-Geo-Optical algorithm for open ocean and costal waters. *Front. Mar. Sci.* **2017**, *4*, 140. [CrossRef]
25. Brockmann, C.; Doerffer, R.; Peters, M.; Stelzer, K.; Embacher, S.; Ruescas, A. Evolution of the C2RCC neural network for SENTINEL 1 and 3 for the retrieval of ocean colour products in normal and extreme optically complex waters. In Proceedings of the Living Planet Symposium 2016, Prague, Czech Republic, 9–13 May 2016.
26. Xie, F.; Tao, Z.; Zhou, X.; Lv, T.; Wang, J.; Li, R. A prediction model of water in situ data change under the influence of environment variables in remote sensing validation. *Remote Sens.* **2021**, *13*, 70. [CrossRef]
27. Cui, A.; Xu, Q.; Gibson, K.; Liu, S.; Chen, N. Metabarcoding analysis of harmful algal bloom species in the Changjiang Estuary, China. *Sci. Total Environ.* **2021**, *782*, 146823. [CrossRef]
28. Breitburg, D. Effects of hypoxia, and the balance between hypoxia and enrichment, on coastal fishes and fisheries. *Estuaries* **2002**, *26*, 767–781. [CrossRef]
29. Zhao, N.; Zhang, G.; Zhang, S.; Bai, Y.; Ali, S.; Zhang, J. Temporal-Spatial Distribution of Chlorophyll-a and Impacts of Environmental Factors in the Bohai Sea and Yellow Sea. *IEEE Access* **2019**, *7*, 160947–160960. [CrossRef]
30. Williams, J.J.; Esteves, L.S. Guidance on setup, calibration, and validation of hydrodynamic, wave, and sediment models for shelf seas and estuaries. *Adv. Civ. Eng.* **2017**, *2017*, 5251902. [CrossRef]
31. Lee, G.; Hwang, H.J.; Kim, J.B.; Hwang, D.W. Pollution status of surface sediment in Jinju bay, a spraying shellfish farming area, Korea. *J. Korean Soc. Mar. Env. Saf.* **2020**, *26*, 392–402. [CrossRef]

32. Groom, S.; Sathyendranath, S.; Ban, Y.; Bernard, S.; Brewin, R.; Brotas, V.; Brockmann, C.; Chauhan, P.; Choi, J.; Chuprin, A.; et al. Satellite ocean colour: Current status and future perspective. *Front. Mar. Sci.* **2019**, *6*, 485. [CrossRef]
33. Minnett, P.J.; Alvera-Azcárate, A.; Chin, T.M.; Corlett, G.K.; Gentemann, C.L.; Karagali, I.; Li, X.; Marsouin, A.; Marullo, S.; Maturi, E.; et al. Half a century of satellite remote sensing of sea-surface temperature. *Remote Sens. Environ.* **2019**, *233*, 111366. [CrossRef]
34. Kim, D.K.; Yoo, H.H. Analysis of temporal and spatial red tide change in the south sea of Korea using the GOCI Images of COMS. *J. Korean Assoc. Geogr. Inf. Stud.* **2014**, *22*, 129–136.
35. KIOST. Korea Ocean Satellite Center. Available online: https://www.kiost.ac.kr/eng.do (accessed on 30 November 2020).
36. Choi, J.K.; Park, Y.J.; Ahn, J.H.; Lim, H.S.; Eom, J.; Ryu, J.H. GOCI, the world's first geostationary ocean color observation satellite, for the monitoring of temporal variability in coastal water turbidity. *J. Geophys. Res.* **2012**, *117*, C09004. [CrossRef]
37. Lee, K.H.; Lee, S.H. Monitoring of floating green algae using ocean color satellite remote sensing. *J. Korean Assoc. Geogr. Inf. Stud.* **2012**, *15*, 137–147. [CrossRef]
38. Huang, C.; Yang, H.; Zhu, A.; Zhang, M.; Lu, H.; Huan, T.; Zou, J.; Li, Y. Evaluation of the geostationary ocean color imager (GOCI) to monitor the dynamic characteristics of suspension sediment in Taihu Lake. *Int. J. Remote Sens.* **2015**, *36*, 3859–3874. [CrossRef]
39. Concha, J.; Mannino, A.; Franz, B.; Bailey, S.; Kim, T. Vicarious calibration of GOCI for the SeaDAS ocean color retrieval. *Int. J. Remote Sens.* **2019**, *40*, 3984–4001. [CrossRef]
40. Ryu, J.H.; Han, H.J.; Cho, S.; Park, Y.J.; Ahn, Y.H. Overview of geostationary ocean color imager(GOCI) and GOCI Data Processing System(GDPS). *Ocean Sci. J.* **2012**, *47*, 223–233. [CrossRef]
41. Lesser, G.R.; Roelvink, J.A.; van Kester, J.A.T.M.; Stelling, G.S. Development and validation of a three-dimensional morphological model. *Coast. Eng.* **2004**, *51*, 883–915. [CrossRef]
42. Hu, K.; Ding, P.; Wang, Z.; Yang, S. A 2D/3D hydrodynamic and sediment transport model for the Yangtze estuary, China. *J. Mar. Syst.* **2009**, *77*, 114–136. [CrossRef]
43. Dissanayake, P.; Hofmann, H.; Peeters, F. Comparison of results from two 3D hydrodynamic models with field data: Internal seiches and horizontal currents. *Inland Waters* **2019**, *9*, 239–260. [CrossRef]
44. Ramos, V.; Carballo, R.; Ringwood, J.V. Application of the actuator disc theory of Delft3D-FLOW to model far-field hydrodynamic impacts of tidal turbines. *Renew. Energy* **2019**, *139*, 1320–1335. [CrossRef]
45. Xia, M.; Jiang, L. Application of an unstructured grid-based water quality model to Chaeapeake Bay and its adjacent coastal ocean. *J. Mar. Sci. Eng.* **2016**, *4*, 52. [CrossRef]
46. Hartnett, M.; Nash, S. An integrated measurement and modelling methodoloby for estuarine water quality management. *Water Sci. Eng.* **2015**, *8*, 9–19. [CrossRef]
47. Kim, W.K.; Moon, J.E.; Park, Y.J.; Ishizaka, J. Evaluation of chlorophyll retrievals form Geostationary Ocean Imager(GOCI) for the North-East Asian region. *Remote Sens. Environ.* **2016**, *184*, 482–495. [CrossRef]

remote sensing

MDPI

Article

RCSANet: A Full Convolutional Network for Extracting Inland Aquaculture Ponds from High-Spatial-Resolution Images

Zhe Zeng [1], Di Wang [2], Wenxia Tan [3], Gongliang Yu [4,*], Jiacheng You [1], Botao Lv [1] and Zhongheng Wu [5]

1. College of Oceanography and Space Informatics, China University of Petroleum, Qingdao 266580, China; zengzhe@upc.edu.cn (Z.Z.); s19160017@s.upc.edu.cn (J.Y.); z20160108@s.upc.edu.cn (B.L.)
2. State Key Laboratory of Information Engineering in Surveying, Mapping and Remote Sensing, Wuhan University, Wuhan 430079, China; d_wang@whu.edu.cn
3. Key Laboratory for Geographical Process Analysis & Simulation of Hubei Province, College of Urban and Environmental Sciences, Central China Normal University, Wuhan 430079, China; tanwenxia@mail.ccnu.edu.cn
4. Key Laboratory of Algal Biology, Institute of Hydrobiology, Chinese Academy of Sciences, Wuhan 430072, China
5. NavInfo Co., Ltd., Beijing 100094, China; keykeywu@hotmail.com

* Correspondence: yugl@ihb.ac.cn

Abstract: Numerous aquaculture ponds are intensively distributed around inland natural lakes and mixed with cropland, especially in areas with high population density in Asia. Information about the distribution of aquaculture ponds is essential for monitoring the impact of human activities on inland lakes. Accurate and efficient mapping of inland aquaculture ponds using high-spatial-resolution remote-sensing images is a challenging task because aquaculture ponds are mingled with other land cover types. Considering that aquaculture ponds have intertwining regular embankments and that these salient features are prominent at different scales, a Row-wise and Column-wise Self-Attention (RCSA) mechanism that adaptively exploits the identical directional dependency among pixels is proposed. Then a fully convolutional network (FCN) combined with the RCSA mechanism (RCSANet) is proposed for large-scale extraction of aquaculture ponds from high-spatial-resolution remote-sensing imagery. In addition, a fusion strategy is implemented using a water index and the RCSANet prediction to further improve extraction quality. Experiments on high-spatial-resolution images using pansharpened multispectral and 2 m panchromatic images show that the proposed methods gain at least 2–4% overall accuracy over other state-of-the-art methods regardless of regions and achieve an overall accuracy of 85% at Lake Hong region and 83% at Lake Liangzi region in aquaculture pond extraction.

Keywords: aquaculture ponds; extraction; inland lake; self-attention

Citation: Zeng, Z.; Wang, D.; Tan, W.; Yu, G.; You, J.; Lv, B.; Wu, Z. RCSANet: A Full Convolutional Network for Extracting Inland Aquaculture Ponds from High-Spatial-Resolution Images. *Remote Sens.* **2021**, *13*, 92. https://doi.org/10.3390/rs13010092

Received: 10 December 2020
Accepted: 26 December 2020
Published: 30 December 2020

1. Introduction

Aquaculture has become one of the main sources of animal protein and increasingly contributes to food security for many inland cities with large populations in Asia. Freshwater aquaculture products such as fish, crustaceans, and molluscs are supplied from aquaculture ponds built around natural lakes. Aquaculture in China already accounts for 60% of global production [1]. Aquaculture foods provided by inland aquaculture ponds have become predominant contributors of aquatic foods in Chinese banquets [2]. Provinces in the middle and lower reaches of the Yangtze River basin account for more than half the country's total freshwater production. In recent years, pond aquaculture has become predominant and has contributed on average 71 percent to total freshwater production (China Fishery Statistical Yearbook 2004–2016), maintaining an average growth rate of 5.8 percent per year. The area under pond aquaculture has greatly increased. However, intensive aquaculture has a severely destructive effect on the environment, including high levels of water use, local environmental pollution, and the loss of services provided by the freshwater ecosystems of natural lakes [3,4].

Remotely sensed imagery has been used as an effective means for global monitoring of aquaculture ponds in coastal areas [5–7] and nearby inland lakes [8]. An object-based image analysis (OBIA) method was used on Landsat TM images to extract aquaculture ponds in coastal areas of south-eastern China [9]. Tran et al. used maximum likelihood classification on Landsat and SPOT5 images to obtain long-term land-cover and land-use changes in a delta in Vietnam, where aquaculture ponds were one of the classes [10]. Ottinger et al. used the geometric features of aquaculture ponds for image segmentation on Sentinel-1 Synthetic Aperture Radar (SAR) images to extract fish ponds in several delta areas in Asia [5,11]. Zeng et al. used Landsat and Gaofen-1 satellite images to extract aquaculture ponds around inland lakes using boundary curve features and a Support Vector Machine (SVM) classifier [8]. In state-of-the-art methods for aquaculture pond extraction, object-oriented classification is usually integrated with hand-crafted features, and the spatial resolution of the satellite images commonly used is generally 10 meters or coarser. However, aquaculture ponds close to inland lakes are mixed with water bodies that are approximately the same size as these ponds. Accurately mapping aquaculture ponds using finer spatial resolution (up to a few meters) remote-sensing images and applying a more generalized approach, rather than manual feature engineering, remains a technical challenge for inland lake mapping.

Because semantic segmentation can understand images at the pixel level, statistics- and geometry-based image segmentation methods have been replaced by methods that depend on Deep Convolutional Neural Networks (DCNNs) [12]. DCNNs have been recognized by industry and have become widely used, advancing from LeNet-5's success in zip encoding recognition in the 1980s to AlexNet's victory in the 2012 ImageNet competition [13]. Subsequently, a deep CNN architecture proposed by Visual Geometry Group of Oxford University (VGG) [14], a residual network architecture proposed by He (ResNet) [15] and other DCNN structures have become the basic learning framework for advanced feature extraction from visual images. The fully convolutional network (FCN) constitutes a breakthrough in semantic image segmentation by converting the fully connected layer in traditional DCNNs, such as VGG, into a fully convolutional layer, thereby successfully achieving end-to-end labelling [16]. Badrinarayanan et al. [17] proposed Segnet to achieve pixel level classification through a deep convolutional encoder-decoder architecture in which the decoder upsamples the lower-resolution feature maps. Chen et al. proposed the Deeplab architecture and its revised versions, which introduced atrous convolution and atrous spatial pyramid pooling (ASPP) models into the deep encoder-decoder architecture for semantic segmentation [18–20]. Deep learning techniques for semantic segmentation have been developed for various computer vision tasks such as autonomous vehicles, medical research and many other applications in recent years [21]. However, implementing semantic segmentation of deep neural networks on remote-sensing images must overcome specific problems, including different data sources and scales [22]. For example, SegNet and ResNet have been efficiently implemented on multi-modal remote-sensing data using the FuseNet principle [23]. FCN has been used for slum mapping by transfer learning [24]. FCN was re-designed and used for automatic raft labelling in offshore waters by a dual-scale structure [25] or a U-Net [26].

Aquaculture ponds are shallow artificial water bodies that commonly have distinctly man-made shapes for efficient aquaculture production [10]. The ponds around inland lakes are formed gradually by embankment, partition, and regularization of other land cover types, such as cropland or natural lake water bodies. Because the shoreline of a natural lake winds along the surrounding terrain, its boundary shape is generally extremely irregular. On the other hand, the borders of aquaculture ponds are constructed on the principle of cost-saving, and straight lines are often used to delimit the boundary in a local area. Hence, the boundaries of aquaculture ponds have more regular shapes overall. Furthermore, when the human eye perceives satellite images where aquaculture ponds are densely distributed, the aquaculture ponds with their intertwining regular boundaries will be visual attention areas because human perception commonly pays attention to parts of visual space where patterns can be acquired, according to neuroscience and cognitive science literature [27].

Attention mechanisms have been extensively used for various visual tasks. The recurrent attention model is used for object recognition through a recurrent neural network (RNN) integrated with reinforcement learning to mimic the process of the human visual system as it recurrently determines the attention region. The attention mechanism on top of the RNN proposed by the neural machine translation community [28,29], was also adopted to perform image captioning by assigning different weights to image representations [30]. The self-attention mechanism without the RNN model is exploited in a super-resolution image generator [31], which is a variant of the TRANSFORMER [32], a cutting-edge deep neural network for language translation. Furthermore, self-attention mechanisms have been introduced into scene segmentation for modelling feature dependencies from spatial and channel dimensions [33]. In remote sensing, attention models have also been used for object classification in various satellite images. For instance, attention mechanisms are integrated into multi-scale and feedback strategies of deep neural networks for pixel-wise classification of very-high-resolution satellite images [34]. The attention model is combined with a learning layer to capture class-specific feature dependencies [35].

When human beings visually identify densely distributed aquaculture ponds on remote-sensing images, the intertwining regular embankments around these ponds are prominent visual attention features. This paper is inspired by this visual attention mechanism used for human interpretation of satellite images. Moreover, the intertwining regular embankments are a salient feature that is available at different scales. The two motivations of this study are first to develop a novel attention mechanism that can mimic the process of the human visual system to recurrently determine the attention region, which is the intertwining regular embankments of aquaculture ponds, and to evolve multi-scale visual attention through the encoder-decoder, fully convolutional network architecture that integrates the attention mechanism with atrous convolutions to better extract aquaculture ponds.

Therefore, the main contributions of the paper can be summarized as follows:

(1) Propose the Row-wise and Column-wise Self-Attention (RCSA) mechanism, which can work in parallel to capture visual emphasis on salient pixels in the context of rows and columns from a remote-sensing image.
(2) Propose an improved fully convolutional network based on the RCSA mechanism that is combined with an ASPP structure for multi-scale attention.
(3) Evaluate the validity of the proposed method on a developed dataset that contains abundant aquaculture ponds around inland lakes.

2. Materials

2.1. Study Area

Hubei Province, known as the province of thousands of lakes, lies in the middle reaches of the Yangtze River and has densely distributed lakes. Hubei has a mature freshwater aquaculture industry with large numbers of aquaculture ponds developed surrounding natural lakes. As shown in Figure 1, six regions with densely distributed aquaculture ponds were selected as study areas from three large lakes (Lake Liangzi, Lake Futou, and Lake Hong) along the Yangtze River because these are typical inland aquaculture areas in China. Among them, Lake Hong and Lake Liangzi are the two largest freshwater lakes in Hubei Province. The population in this part of China is dense, and aquaculture is very developed. Lake Liangzi, and its surroundings, however, have been relatively well protected since the 1980s. The six selected regions were divided into two categories: type I and type II. The type I regions, including regions A and B, are used for testing, whereas type II regions are used for training. Region A is an area of 73.76 km^2 close to eastern Lake Hong which is an artificial lake, and region B is an area of 33.92 km^2 close to eastern Lake Liangzi, which has been preserved in a state more like a natural lake.

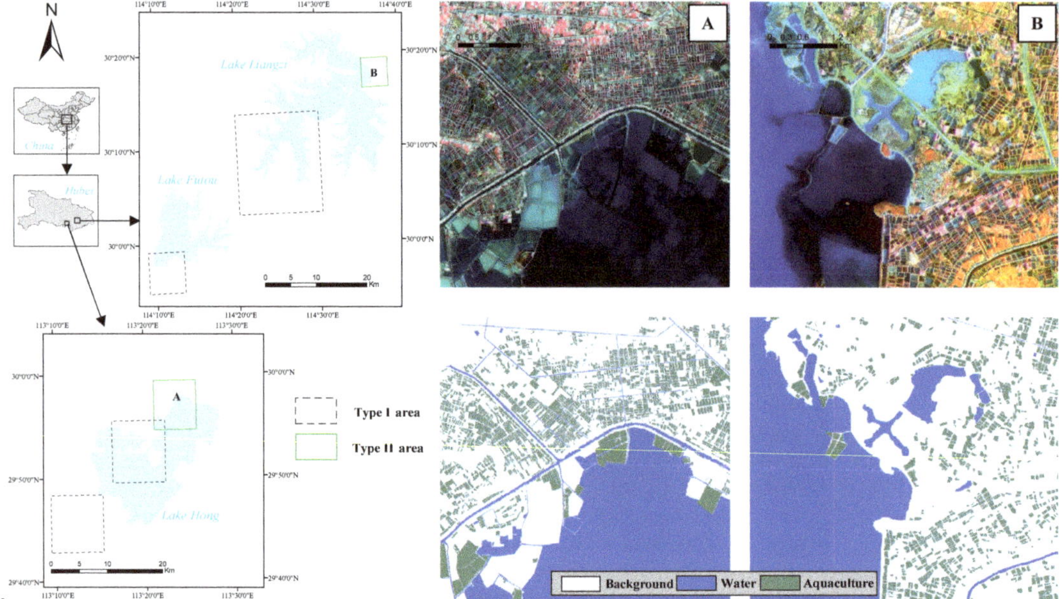

Figure 1. Location of the study area. The pseudo-colour images (**A**,**B**) are pansharpening images using the near infrared, the red and the green band as red, green and blue. The corresponding labelling image for each is given below.

2.2. Dataset

The Landsat multispectral images were selected because of their long history. The bands such as the near infrared can be beneficial for extracting water bodies. However, the spatial resolution of Landsat multispectral data is only 30 m. Panchromatic images with 2–2.5 m spatial resolutions from the panchromatic and multispectral (PMS) camera of the GaoFen-1 (GF-1) satellite [36], the panchromatic remote-sensing instrument for stereo mapping (PRISM) of the ALOS satellite, and the NAD panchromatic sensor of the ZiYuan-3 (ZY-3) satellite [37] were also used to improve recognition and extraction of aquaculture ponds and natural water bodies. Table 1 lists the images used for the selected study regions. The Landsat multi-spectral images used in this study were captured in the winter of 2010–2011 and 2013–2014 and the spring of 2015. High-resolution panchromatic images were used for fusion with multi-spectral images. The panchromatic images were mainly selected from the GF-1 satellite and had acquisition dates close to the corresponding OLI images from Landsat satellite, whereas panchromatic images from the ALOS satellite were used instead for 2010 TM images from the Landsat satellite. However, when ALOS or GF-1 panchromatic images with similar acquisition dates were still not found, panchromatic images from the ZY-3 satellite captured in same season as Landsat images from a nearby year were selected because the ZY-3 satellite was launched in 2012.

Three classes: aquaculture ponds (artificial water surfaces), natural water surfaces and background (non-water surfaces) were included in the reference dataset (Figure 1), which was mainly generated by human visual interpretation. Field investigations were also conducted on some difficult-to-identify features, in cases where aquaculture ponds were mixed with small natural water surfaces (Figure 2).

Table 1. Satellite images information for various regions.

Regions	Multispectral Images			Panchromatic Images		
	Sensors	Spatial Resolution (m)	Date	Sensors	Spatial Resolution (m)	Date
Lake Hong (west, type I region)	TM (Landsa 5)	30	2011.01.15	PAN-NAD(ZY-3)	2.1	2013.01.27
Lake Hong (west, type I region)	OLI (Landsat 8)	30	2014.01.23	PMS2(GF-1)	2	2014.01.23
Lake Hong (middle, type I region)	TM (Landsat 5)	30	2011.01.15	PAN-NAD(ZY-3)	2.1	2013.01.27
Lake Hong (middle, type I region)	OLI (Landsat 8)	30	2014.01.23	PMS2(GF-1)	2	2014.01.23
Lake Futou (south, type I region)	OLI (Landsat 8)	30	2015.03.31	PAN-NAD(ZY-3)	2.1	2017.01.22
Lake Liangzi (west, type I region)	TM (Landsat 5)	30	2010.11.12	PRISM(ALOS)	2.5	2010.11.06
Lake Liangzi (west, type I region)	OLI (Landsat 8)	30	2014.02.01	PMS2(GF-1)	2	2014.01.31
Lake Hong (east, type II region A)	TM (Landsat 5)	30	2011.01.15	PAN-NAD(ZY-3)	2.1	2013.01.27
Lake Hong (east, type II region A)	OLI (Landsat 8)	30	2014.01.23	PMS2(GF-1)	2	2014.01.23
Lake Liangzi (east, type II region B)	TM (Landsat 5)	30	2010.11.12	PRISM(ALOS)	2.5	2010.11.06
Lake Liangzi (east, type II region B)	OLI (Landsat 8)	30	2014.02.01	PMS2(GF-1)	2	2014.01.31

Figure 2. Field photos of inland aquaculture ponds in Hubei Province, China. Aquaculture ponds are usually equipped with air pumps. (**A**) A branch of a natural lake. (**B**) An aquaculture pond equipped with oxygen pumps. (**C**) Below is a small river (natural water body) and above are several aquaculture ponds.

3. Methodology

To better understand the effectiveness of the proposed method for aquaculture pond segmentation, the methodology will be introduced in three parts: data pre-processing, the basic model, and a fusion strategy designed to further improve accuracy. In the preprocessing stage, the multi-spectral image and the corresponding 2 m panchromatic image were pansharpened. The pansharpened image was then fed into the proposed network, i.e., RCSANet, for semantic segmentation. The result generated from the network was finally fused with a water surface extraction image using the water index to further improve segmentation quality.

3.1. Preprocessing

Multi-spectral satellite images contain more spectral information, especially in the infrared spectral bands, which is beneficial for aquaculture pond identification, whereas panchromatic satellite images have higher spatial resolution, which helps to better distinguish the shape of the aquaculture pond. To use both together, the multi-spectral and high-spatial-resolution panchromatic images must be pansharpened to obtain images with both spectral information and higher spatial resolution. First, multi-spectral images were synthesized by selecting the three bands (green, red, NIR) that are useful for water body identification. The pixel values were normalized and then mapped to the range (0, 255). Similarly, the gray values of panchromatic images were also normalized and mapped to the range of (0, 255). The multi-spectral images were re-projected into the coordinate system of the corresponding panchromatic images to ensure consistent coordinates. The multi-spectral and panchromatic images were fused by the GRAM-SCHMIDT method [38], which is a widely used high-quality pansharpening method providing a fusion of panchromatic image and multi-spectral images with any number bands through orthogonalization of different multi-spectral bands [39].

3.2. Basic Model

3.2.1. Network Architecture

The deep neural network architecture, depicted in Figure 3a, for semantic segmentation of aquaculture ponds in the proposed method is based on an FCN framework, that uses ResNet-101 [15] as the encoder to generate multiple semantic features. The encoding part produces the feature maps through five convolution layers, including the first convolution layer (Conv1) followed by a pooling layer, and the other four convolution layers (Res-1 to Res-4) are all residual subnetworks. The feature maps are abstract representations of the input image at different levels. Semantic segmentation by the FCN framework is a dense prediction procedure in that the coarse outputs of the convolution layers are connected by upsampling to produce pixel-level prediction. In the proposed method, the RCSA mechanism (introduced in Section 3.2.2) was developed on the coarse outputs at different levels of abstract representation (detailed in Section 3.2.3). Next, channel attention blocks (CAB), which were designed to assign different weights to features at different stages for consistency [40], were used to connect the coarse abstract representations from the encoder with the upsampling feature at the decoder in the whole dense prediction procedure. The spatial size of the coarse outputs derived from the different convolution layers were kept consistent by the upsampling blocks (Figure 3c) to achieve end-to-end learning through backward propagation. Specifically, to accurately capture aquaculture ponds and their context information at multiple scales, the ASPP module combined with the RCSA mechanism (ASPP-RC) forms a branch from Conv1 to the end of the decoder before a 1×1 convolution layer and is integrated with the corresponding feature as a skip connection. To extract spatial context information at different scales, atrous convolutions with different rates, followed by the RCSA mechanism, were performed in parallel on the low-level feature map in the ASPP-RC module. These branches for capturing features at different scales are connected by weighting each branch in terms of its own importance (Figure 3b, introduced in Section 3.2.4).

3.2.2. RCSA Mechanism

When human beings use visual perception to understand remote-sensing images containing inland lakes with densely distributed aquaculture ponds, the ponds as a group will be eye-catching. The attention focuses on the spatial dependencies of aquaculture ponds and their surroundings. To mimic this human visual mechanism, the proposed model first establishes inter-pixel contextual dependencies through bidirectional gated recurrent units (GRUs) [41], which are a powerful variant of RNN, and then the self-attention modules are used on top of the bidirectional GRUs to establish this visual attention.

The self-attention mechanism is essentially a special case of the attention model. The unified attention model contains three types of inputs: key, value, and query [42], as depicted in Figure 4. The key and the value are a pair of data representations. Assume that there are T pairs $< k_i, v_i >$ $(i \in 1, \dots, T)$. By evaluating the similarity between a query q and each key, the model essentially captures the weight coefficient of each key and then weights the corresponding values to derive their final attention values. The attention mechanism first scores the similarity between a query and a key pair by the f function:

$$e_i = f(k_i, q) \tag{1}$$

Then the original scores e_i are normalized by a Softmax function to obtain the weight coefficients:

$$\begin{aligned} a_i &= g(e_i) \\ &= softmax(e_i) \\ &= \frac{\exp(e_i)}{\sum_{j=1}^{T} \exp(e_j)} \end{aligned} \tag{2}$$

Finally, the context vector c_t is evaluated by a weighted sum of the values:

$$c_t = \sum_i a_i v_i \tag{3}$$

The attention model can be presented in a unified form

$$c_t = Attention(K, Q, V) = Softmax(f(K, q))V \tag{4}$$

The attention model becomes a self-attention mechanism when all inputs, including the query, the key, and the value, have the same value.

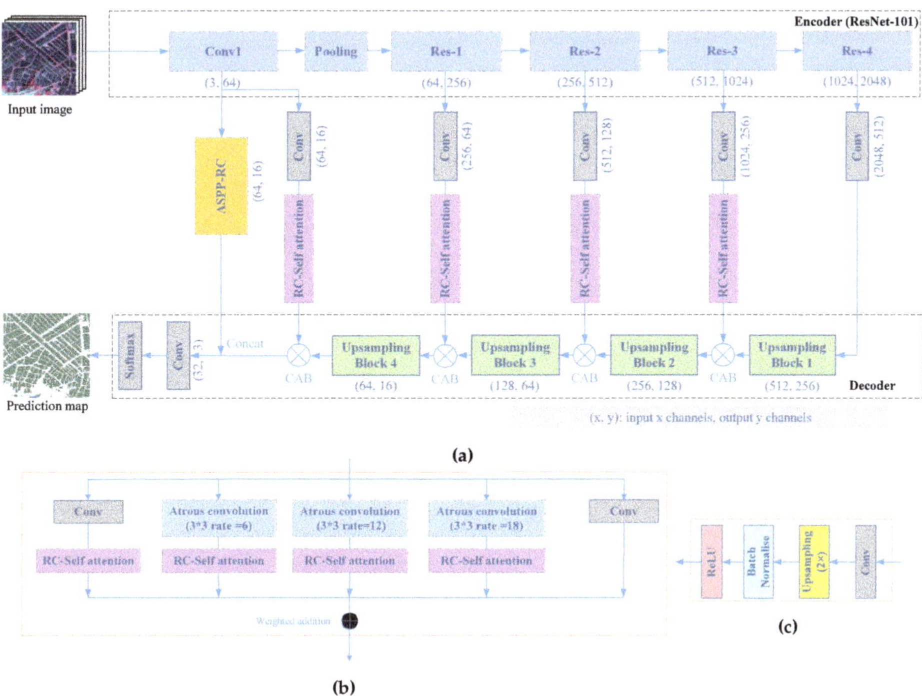

Figure 3. RCSANet: FCN architecture combined with RCSA mechanism for semantic segmentation of aquaculture ponds: (**a**) Network architecture; (**b**) ASPP-RC module; (**c**) Upsampling block. The input image of the entire deep neural network is a 256 × 256 pansharpening patch with three spectral channels. Through encoding and decoding, a three-channel matrix for classification was output through a 1 × 1 convolution layer at the end, and finally a Softmax layer gave a prediction map with the same size as the input image.

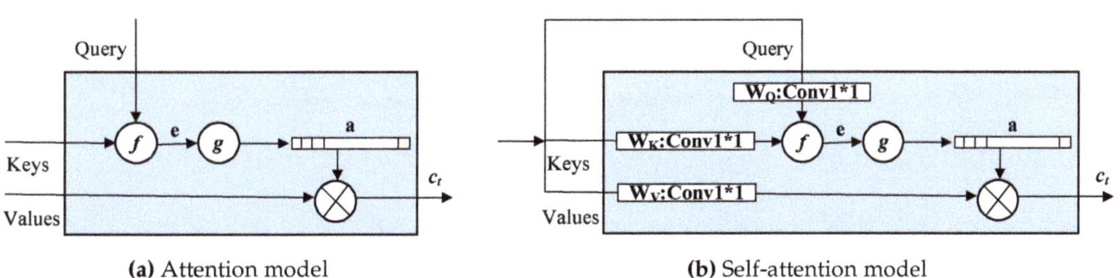

Figure 4. Attention models.

The RCSA mechanism takes a feature map, which is the convolutional result from the previous layer or the input image, as an input $x \in \mathbb{R}^{h \times w \times C}$, where h, w, and C are the number of rows, columns, and channels respectively. The feature map can be spatially divided into h rows $r_i \in \mathbb{R}^{1 \times w \times C} (i \in 1 \dots h)$ or w columns $c_j \in \mathbb{R}^{h \times 1 \times C} (j \in 1 \dots w)$. RCSA enables the construction of spatial dependencies between pixels within a row or a column by the self-attention mechanism. Hence, the RCSA mechanism consists of two parallel branches, column-wise and row-wise self-attention, which are subsequently concatenated by summation, as shown in detail in Figure 5. In the upper branch, the row-wise self-attention mechanism first uses the bidirectional GRU model to depict the dependencies between the pixels in a row of the feature map

$$r_i' = BiGRU(r_i) \tag{5}$$

Then the outcome from the GRUs r_i' is fed into the self-attention model by which the importance of the dependencies between pixels in the row is evaluated. The self-attention model is a specific variant of the attention model, in which the input query, key, and value have the same value, as shown in Figure 4b. The r_i' are respectively conducted by three 1×1 convolution kernels, W_Q, W_K,and W_V, so that the query, the key, and the value can be obtained by $Q = W_Q * r_i', K = W_K * r_i', V = W_V * r_i'$, where "*" is the convolution operation. Then they are substituted into the following Equation (4):

$$\begin{aligned} c_t =& Attention(W_K r_i', W_Q r_i', W_V r_i') \\ =& Softmax(f(W_K r_i', W_Q r_i')) W_V r_i' \end{aligned} \tag{6}$$

where the similarity function $f(K, Q) = \frac{QK^T}{\sqrt{d_k}}$ and d_k is the dimension of the key. The computation for one row can traverse to each row of the feature map. Equivalently, in the bottom branch, the same operations are performed in parallel on each column of the feature map. Eventually, the two branches are combined with equal weights.

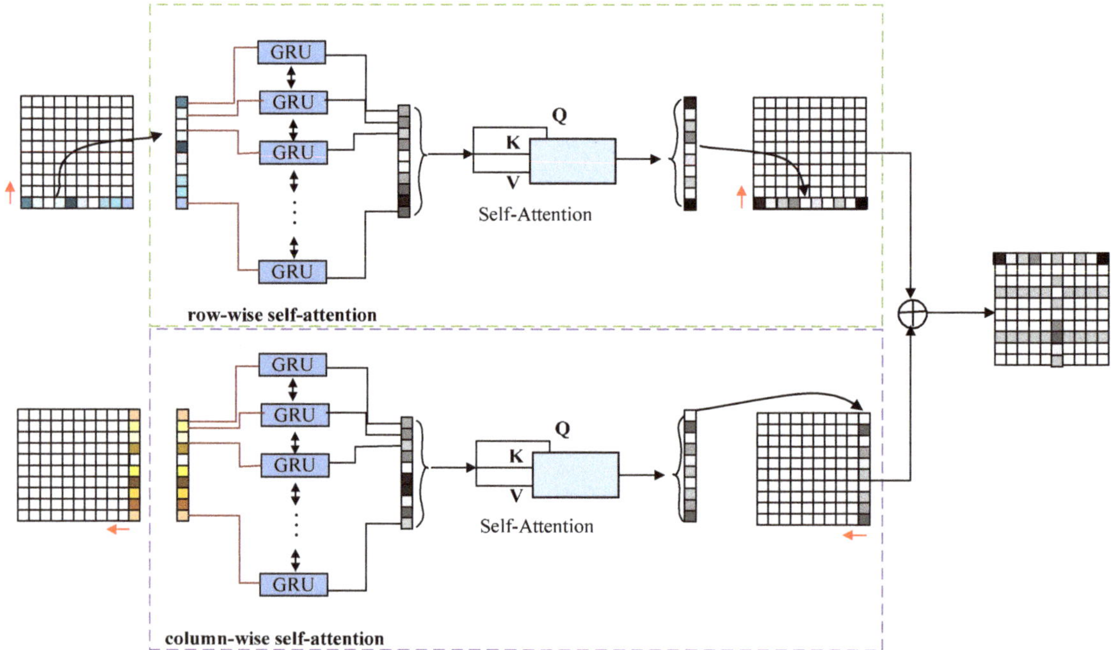

Figure 5. Attention layer consisting of column-wise and row-wise self-attention models.

3.2.3. RCSA for Dense Prediction

In semantic segmentation of remote-sensing images, dense prediction must fuse abstract representations of different levels from the encoder to improve pixel-level prediction. Visual attention on densely distributed aquaculture ponds could be involved in the dense prediction procedure. Consequently, the outputs of different convolution blocks in the encoding part are conducted by RCSA and then participate in dense prediction. These RCSA modules in the lateral connection enhance the features pixel-wise by assigning different weights to achieve a reasonable optimization of visual attention. In fact, this optimization takes place in a two-dimensional space made up of row and column vectors. However, the importance of different band channels must also be emphasized. The CAB module is directly used to fuse encoder and decoder features through assigning different weights to channels.

3.2.4. ASPP-RC Module

Atrous convolutions at different rates can enlarge the field of view so that spatial information at different scales can be extracted. Aquaculture ponds, which are water bodies surrounded by dikes with regular shapes, are densely distributed close to inland lakes. These features show visual salience in remote-sensing images. Hence, the RCSA block is arranged next to atrous convolution to selectively focus attention. After the first convolution blocks of the encoder, in the ASPP-RC module, the low-level feature map is executed in parallel by atrous convolutions with different rates combined with RCSA. Eventually, the branches are connected by:

$$I = \sum_{i=1}^{5} w_i \cdot b_i \tag{7}$$

where b_i is the feature map produced by the ith branch in which the atrous convolution and RCSA are conducted in sequence and w_i is the weight of the ith branch that evaluates the importance of different scales. This is unlike the original ASPP structure in which each branch has the same importance. The importance of each branch is adjusted adaptively in the proposed ASPP-RC module. All weight parameters are initially defined by a random vector w_i^0, which can be optimized during backpropagation when training the whole network. Finally, these weights are normalized using a Softmax function:

$$w_i = softmax(w_i^0) \tag{8}$$

3.3. *Fusion Strategy*

To further improve the segmentation quality of aquaculture ponds, the normalized difference water index (NDWI) maps from pansharpening images are fused with the prediction probability matrices from the proposed network to produce the final classification result (Figure 6). This implementation is called "RCSANet-NDWI". The classification probability matrices are produced from the three classes (aquaculture ponds, natural water surfaces, and background) probability maps after the Softmax layer. Both aquaculture ponds and natural water surfaces are water bodies surrounding inland lakes. Hence, the water extraction index, which is a typical representation of the spectral characteristics of a water body used to distinguish ground features, has been extensively used. The NDWI maps were used to provide prior knowledge for aquaculture pond extraction. Through OTSU threshold binary segmentation [43], NDWI maps were divided into water and non-water parts. The water parts in the NDWI maps were used to refine the three-class probability matrix described earlier. Assume that the original probability matrix P_0 and the refined matrix P are both $h \times w \times c$ in size, whereas the NDWI map S is $h \times w$ in size. c is the channel number, k is the channel ID, and the kth channel represents the kth class. Hence, $k = 1, 2, 3$ represent background, water, and aquaculture ponds, respectively. The fusion operation can be defined as:

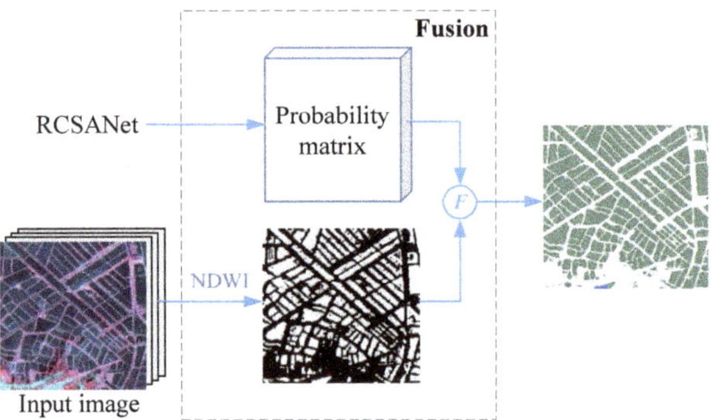

Figure 6. Fusion strategy.

$$p^{ijk} = \begin{cases} P_0^{ijk}, k \neq 1 \\ y^{ij} \cdot P_0^{ijk}, k = 1 \end{cases} \tag{9}$$

where y is the indicator variable

$$y^{ij} = \begin{cases} 0, S^{ij} = 1 \\ 1, S^{ij} = 0 \end{cases} \tag{10}$$

For the pixel in the i-th row and j-th column of S, if its value is 1 (representing water), the corresponding background probability ($k = 1$) in P_0 is set to 0, and the fused matrix P is generated. The final classification maps can be obtained using the maximum probability judgment. The maximum probability judgment is the usual method for mapping the probability matrix to the final label image: the classification label of this pixel is determined with maximum probability: $l_{ij} = \arg\max_k(p_{ij}^k)$, where probability p_{ij}^k is the probability of a pixel in the i-th row and j-th column from k different sources. With the NDWI, the interference from the background of the water body extraction is eliminated because the probability of the non-water part is set to 0.

4. Experiments

This section describes a series of qualitative and quantitative comprehensive evaluations that were conducted using the proposed methods with the dataset introduced in Section 2.

4.1. Experimental Set-Up

The inputs of the proposed network were 256 × 256 pansharpening patches with three spectral channels. Table 2 lists the parameters of the convolution kernels, which are basic operators of different modules in the entire process of the proposed RCSANet. Parameter rate means that the convolution kernels in different atrous convolution branches of the ASPP-RC module have different padding and dilation configurations, which are set to 6, 12, and 18, respectively, according to Figure 3b. Validation consisted of two parts:

(1) Evaluating the performance of the proposed methods. The pansharpening images of the six regions (both type I and type II in Figure 1) were segmented into image patches 256 × 256 pixels in size. These image slices were randomly divided into training and test sets, of which 80% (4488 images) made up the training set and 20% (1122 images) made up the test set. The overall accuracy, user's accuracy, producer's accuracy, and kappa coefficients were used as the main evoluation metrics.

(2) To assess the quality of aquaculture pond extraction and evaluate the generalization and migration capabilities of RCSANet, four regions (type I) were used as training data, and the other two regions (type II) were used as test areas. The overall accuracy,

user's accuracy, producer's accuracy, and kappa coefficients were calculated to assess aquaculture pond extraction accuracy on the 2 m spatial resolution pansharpened images.

Table 2. Parameters used for the convolution kernels in the various modules in RCSANet.

Module	Kernel Size	Stride	Padding	Dilation
Conv	1×1	1	0	1
RCSA	1×1	1	0	1
Upsampling block	1×1	1	0	1
ASPP-RC(Conv)	1×1	1	0	1
ASPP-RC(Atrous conv)	3×3	1	*rate*	*rate*

In addition, the proposed methods were divided into two versions: RCSANet (without NDWI fusion) and RCSANet-NDWI (with NDWI fusion) to verify the role of NDWI fusion. Three state-of-the-art segmentation methods, including DeeplabV3+ [20], Reseg [44], and Homogeneous Convolutional Neural Network (HCN) [25] were selected for comparison. In addition, the performance of SVM was also assessed as a representative of traditional machine learning methods that directly use each pixel as a feature. DeeplabV3+ is an FCN method for semantic segmentation with the help of an ASPP module. Reseg is a hybrid deep network for semantic segmentation. Except for CNN, the bidirectional GRU is also used in Reseg to capture contextual dependencies. HCN was originally proposed for automatic raft labelling and is now considered to have potential for aquaculture pond extraction. HCN was implemented following the settings in [25], and Resnet-101 was simultaneously used as the encoder in DeeplabV3+, Reseg, and the proposed methods.

In the present experiments, the parameters of the proposed methods were optimized by minibatch stochastic gradient descent using a momentum algorithm with a batch size of 2. The learning rate was set to 10^{-2} and decayed with training epoch according to the "polynomial" strategy. The number of training epochs was configured as 40. The SVM was implemented with the help of the LIBSVM package[45], and two important factors, C and γ, were determined through a five-fold cross validation grid search. Except for the HCN, which was operated using TensorFlow 1.9.0, the other deep learning-based algorithms were implemented in Pytorch 1.1.0. All deep learning methods were implemented on a single NVIDIA GeForce GTX 1080 GPU.

4.2. Results

The performance of the various semantic segmentation methods in Part 1 of the experiments is depicted in Table 3. Clearly, the deep learning-based methods perform better than the traditional SVM algorithm because the latter cannot perceive spatial semantic information in the image. DeeplabV3+ is a state-of-the-art FCN method that has been widely used. Resnet-101 was also chosen as the backbone for DeeplabV3+. HCN is a deep convolutional neural network for automatic raft labelling, and Reseg is a deep recurrent neural network for semantic segmentation. The classification accuracy of the proposed methods for natural water surfaces and aquaculture ponds was consistently better than the other methods. Meanwhile, compared with DeeplabV3+, the overall accuracy in the two versions of the proposed methods led to an improvement of more than 7% and the Kappa coefficients of the proposed methods were greater than 0.72, indicating that the proposed method is significantly better than DeeplabV3+. Moreover, the results also demonstrated the effect of the proposed fusion strategy because RCSANet-NDWI further surpassed RCSANet on most metrics.

Table 3. Performance comparison of different methods for semantic segmentation of aquaculture ponds in Part 1 of the experiments (%).

Methods	Overall Accuracy (%)	Kappa	Producer's Accuracy Natural Water (%)	Producer's Accuracy Aqua-Culture (%)	User's Accuracy Natural Water (%)	User's Accuracy Aqua-Culture (%)
SVM	26.90	9.71	54.80	15.96	52.43	76.60
Deeplabv3+	79.16	59.23	90.32	55.26	97.90	93.14
Reseg	84.52	68.23	90.74	71.18	97.44	90.31
HCN	74.53	49.86	86.83	48.21	92.71	85.74
RCSANet	86.95	72.83	92.83	74.36	98.13	93.99
RCSANet-NDWI	89.31	77.28	93.28	80.81	98.07	93.57

Figure 7 gives a detailed display of the classification results in Part 1 of the experiment. Inland water areas contain various natural water bodies as well as aquaculture ponds. These natural water bodies greatly interfere with the segmentation result for aquaculture ponds, making pixel-scale classification intricate. Figure 7 shows that the SVM classification results misclassified many aquaculture ponds as natural water bodies and many natural water bodies as aquaculture ponds, indicating that the traditional pixel-based method cannot efficiently distinguish natural water bodies from aquaculture ponds. The segmentation maps created by DeeplabV3+ look significantly better than those from SVM, but in some difficult zones where natural water bodies look similar to aquaculture ponds, they are also trapped by their own performance limitations and misclassified natural water bodies as aquaculture ponds (area in the 7th row) or aquaculture ponds as natural water bodies (districts in the 5th row). HCN, which has good performance for raft-culture extraction in offshore waters, performed poorly on semantic segmentation of inland aquaculture ponds and serious misclassifications also happened with HCN. Reseg, which combines CNN and bidirectional GRU, can perform semantic segmentation for aquaculture ponds. However, the identification of natural water bodies that closely resemble aquaculture ponds around inland lakes is not as good as with the proposed methods. In Table 3, the overall accuracy of the Reseg method can reach greater than 80% but its Kappa coefficient is less than 0.7. This shows that Reseg has established a spatial relationship through the construction of GRU, which has a certain effect on the segmentation of aquaculture ponds around inland lakes, but it is not good enough. In the Reseg segmentation map, many objects are stuck together, and the edges of aquaculture ponds are not well displayed. Among these result maps, the two versions of the proposed method separated natural water bodies and aquaculture ponds more satisfactorily than the other methods. The ASPP-RC module of the proposed method feeds back the details at different scales into the low-level feature map, which can draw visual attention to the decoding part. This facilitates identification of the thin edges surrounding the aquaculture ponds in semantic segmentation. Hence, the edges of aquaculture ponds were clearly identified in most cases, as shown in the results from RCSANet and RCSANet-NDWI. Finally, note that RCSANet-NDWI further improved the quality of aquaculture pond extraction compared with RCSANet.

Table 4 provides assessment results for the various algorithms in Part 2 of the experiment and shows the corresponding extraction accuracies of the aquaculture pond and natural water surface classes in the two experimental areas (regions A and B in Figure 1) by different sensors. The overall accuracy and Kappa coefficient show that the two versions of the proposed method (RCSANet and RCSANet-NDWI) both performed better than the other methods, regardless of sensor or area. Moreover, compared with RCSANet, the accuracy of RCSANet-NDWI was further improved with the aid of NDWI fusion. In region A, their overall accuracies in pansharpening images from different sensors were greater than 85 percent, and the Kappa coefficients were definitely greater than 0.7. These results were better than those of other deep learning-based methods, not to mention SVM. In region B, the proposed methods still performed the best. Unlike region A, where the lake is greatly influenced by residents living nearby, causing the aquaculture ponds to be neatly and regularly distributed, the aquaculture ponds in region B have a sparser distribution.

Region B is relatively well protected, and some small natural water bodies, which are easily confused with aquaculture ponds and interfere with network identification, were produced when the lake was split for artificial development. Hence, the situations in the two regions are completely different, which shows the stability of the proposed methods under various scenarios. The overall accuracies of the proposed methods in pansharpening images from different regions were close to or greater than 80 percent. In addition, it should be noted that user accuracy in identifying natural water bodies in almost all methods is relatively high. This is because natural water bodies tend to be extensive, homogeneous, self-contained, and distributed in aggregates, a situation that is easier to recognize for the classifier. Compared with the proposed methods, Reseg and DeeplabV3+ may also obtain higher user accuracy in some cases. However, because of their limited recognition ability, they cannot explicitly judge the difference between aquaculture ponds and natural water bodies (Figure 8).

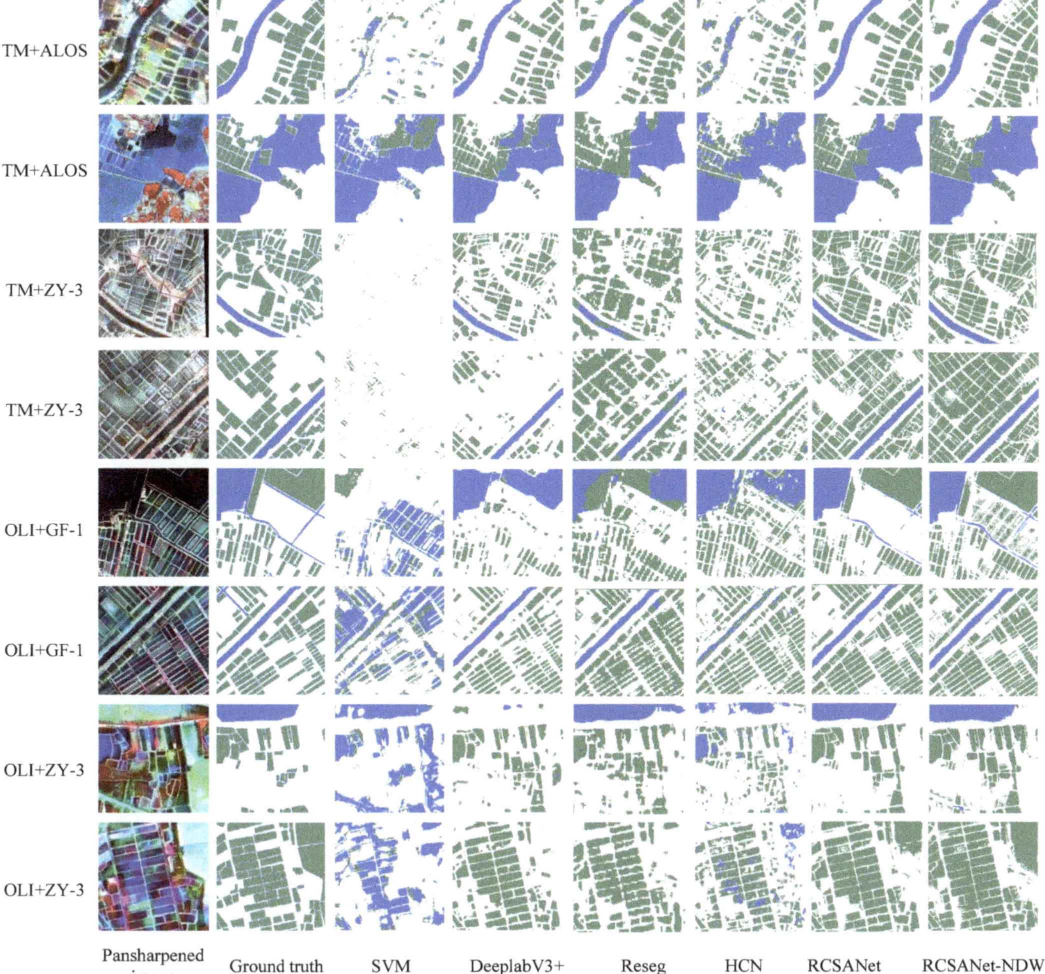

Figure 7. Semantic segmentation results for 256 × 256 pixel image patches from test set in Part 1 of the experiment. The leftmost column gives the sensors or satellites to which the multispectral and panchromatic data of the pansharpened images belong, and the bottom row lists the different methods by which the semantic segmentation images in the same column were obtained.

Table 4. Accuracy evaluation for the two classes, aquaculture ponds and natural water surfaces, in each experimental area (%).

Regions	Sensors	Methods	Overall Accuracy (%)	Kappa	Producer's Accuracy		User's Accuracy	
					Natural Water (%)	Aqua-Culture (%)	Natural Water (%)	Aqua-Culture (%)
Lake Hong (East, type II region A)	TM+ZY-3	SVM	24.44	5.09	27.55	23.80	53.83	86.72
		Deeplabv3+	81.30	60.93	87.87	64.71	98.25	83.30
		Reseg	84.74	66.92	88.97	74.06	97.85	82.52
		HCN	77.37	52.85	88.83	48.42	96.73	79.79
		RCSANet	86.79	70.83	90.79	76.70	98.25	84.47
		RCSANet-NDWI	88.77	74.78	91.08	82.94	98.21	84.42
	OLI+GF-1	SVM	67.60	25.99	38.82	78.75	47.09	82.16
		Deeplabv3+	84.96	69.73	87.60	79.57	96.82	90.01
		Reseg	76.47	55.10	78.30	72.75	92.44	83.38
		HCN	73.76	48.23	82.90	55.07	88.02	86.79
		RCSANet	85.36	69.14	90.57	74.70	93.59	90.38
		RCSANet-NDWI	86.61	71.43	91.07	77.50	93.61	90.14
Lake Liangzi (East, type II region B)	TM+ALOS	SVM	39.26	14.51	67.37	5.32	86.62	19.22
		Deeplabv3+	74.60	53.48	86.73	50.69	99.25	82.83
		Reseg	75.27	54.33	84.94	56.20	97.86	83.28
		HCN	67.68	40.23	90.05	23.60	92.51	86.44
		RCSANet	79.95	62.01	89.31	61.50	99.56	90.03
		RCSANet-NDWI	83.85	68.42	89.63	72.45	99.51	89.03
	OLI+GF-1	SVM	39.43	3.29	48.19	7.58	94.43	5.90
		Deeplabv3+	82.31	55.98	91.45	49.06	98.91	77.99
		Reseg	87.97	67.84	93.74	66.96	97.85	85.59
		HCN	77.25	43.83	91.80	24.31	97.19	81.96
		RCSANet	90.90	75.86	93.00	83.26	99.21	83.97
		RCSANet-NDWI	91.71	77.83	93.19	86.31	99.20	83.69

Figure 8 shows the classification results in the two study regions. Extracting aquaculture ponds in region B is more difficult than in region A because region B contains more natural water bodies that are hard to distinguish from aquaculture ponds. The two versions of the proposed method performed significantly better than the other methods for aquaculture pond extraction. The proposed methods were predominantly successful in predicting aquaculture ponds that are divided into regular shapes by embankments, as well as the natural water bodies in the two regions. In region A, the proposed methods generally extracted almost all aquaculture ponds compared with the ground truth, whereas other methods failed, especially in the upper part of the scene. In region B, compared with Reseg, the proposed methods had lower misclassification rates, and the natural rivers located at the bottom, which could not be identified by Reseg, were not misclassified as aquaculture ponds by the proposed methods. Moreover, the shapes of the ponds are best retained, as shown in the results of the proposed method. The advantage of the proposed method is the proposed RCSA mechanism for determining salient pixels in a row or column, which is essentially a description of the pixel-level context. This enables the proposed method to identify detailed features of the 2 m spatial resolution image, where the dikes around aquaculture ponds are such pixel-level details. Hence, the aquaculture ponds in region B were more fully extracted by the proposed RCSANet than by other state-of-the-art methods, such as DeeplabV3+ and Reseg. On the other hand, fusion using NDWI can better distinguish water surfaces, including natural water bodies and aquaculture ponds, from background. In effect, the proposed method with NDWI re-segments the leaked water surface from the background, which improves the producer's accuracy of the aquaculture pond. However, this also entails a phenomenon whereby a small part of the background is mistakenly classified as water surface.

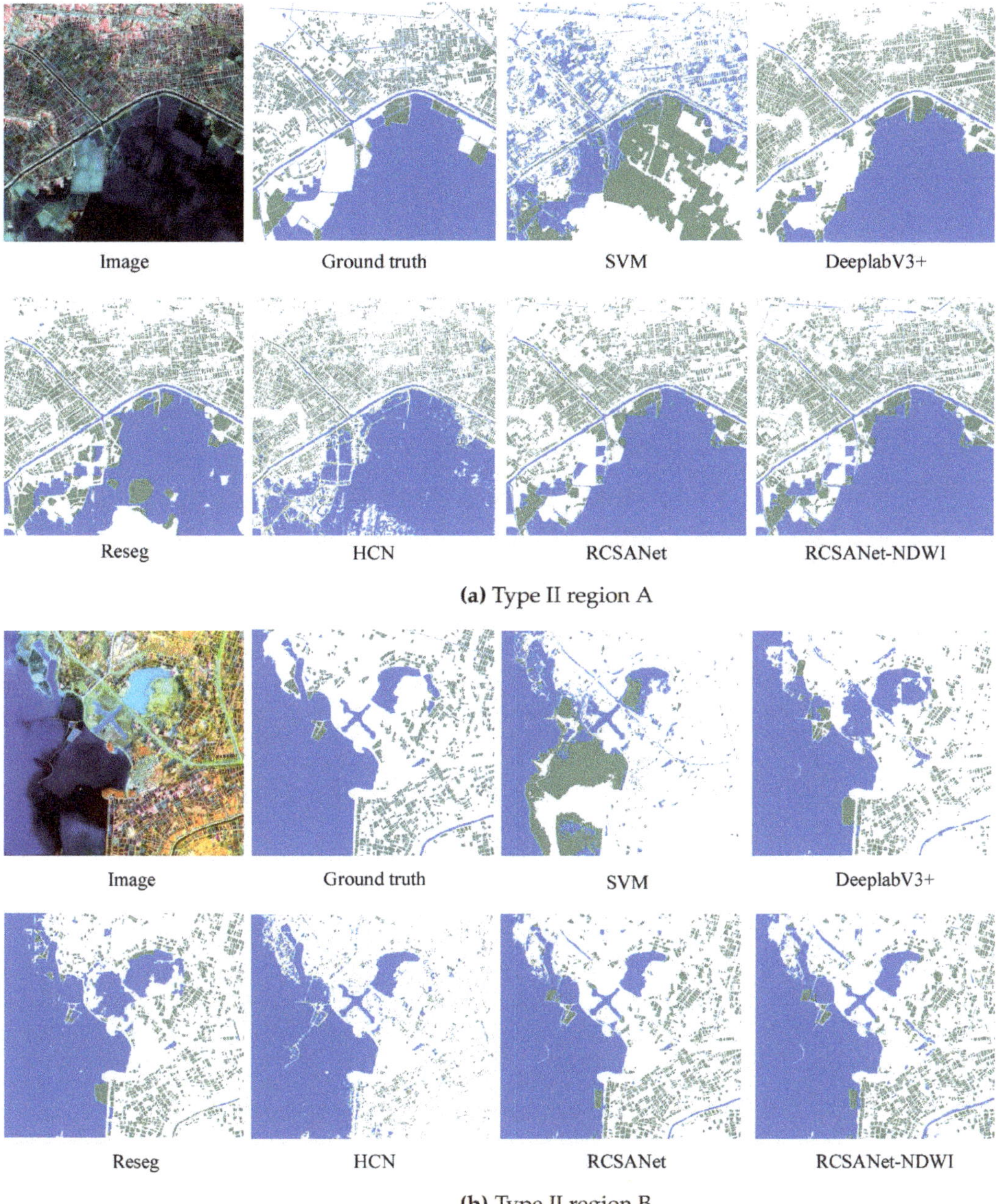

Figure 8. Semantic segmentation results for aquaculture ponds and natural water bodies by various methods: (**a**) in region A, using the pansharpened image with the TM multispectral image captured in January 2011 and the ZY-3 panchromatic image in January 2013; and (**b**) in region B, using the pansharpened image with the OLI multispectral image captured in February 2014 and the GF-1 panchromatic image in January 2014.

5. Discussion

This study has used a fully convolutional network architecture with row- and column-wise self-attention to semantically segment aquaculture ponds around inland lakes. Artificial aquaculture ponds around inland lakes are small, and the dikes between these ponds are only about 2 m wide. On medium-resolution multispectral images, water pixels are firstly separated from land, and then water objects are formed based on connectivity. After that, these water objects are classified as natural water bodies and aquaculture ponds using geometric characteristics [8]. However, for inland lake area where aquaculture ponds are intensively distributed with narrow dikes (e.g., Lake Hong), the 15–30 m spatial resolution of the image limits the capability of the object based method to accurately extract aquaculture ponds. Hence, finer-spatial-resolution images are considered for pond extraction. By fusing multi-spectral information into panchromatic images from the GF-1, ZY-3 or ALOS satellites, the spatial resolution of the resultant satellite images can achieve up to 2 meters, enabling the identification of thin narrow dams. Meanwhile, the multi-spectral capability is utilized to recognize water. From the segmentation results, the proposed network structure was shown to be capable of extracting these regular pond boundaries, mainly because semantic segmentation of the aquaculture ponds benefits from establishing a spatial relationship between pixels in the same direction by the self-attention model. Although HCN was also an FCN-based method used to automatic raft labeling [25], nevertheless, its performance for extracting aquaculture ponds around inland lakes are not as effective as that for labeling raft-culture. Because the spatial context of raft-culture in coastal area is much simple than that of the inland lake area. In general, through high-spatial-resolution images that incorporate multi-spectral and panchromatic data, the proposed RCSANet enables the extraction of large-scale aquaculture ponds around inland lakes where complex spatial contexts of water surfaces exist. However, it is still challenging for the recognition of small water bodies in such complex spatial context. The experimental region B was in the process of recovering aquaculture ponds and farmland as lake area from 2011 to 2014. Therefore, various aquaculture ponds and natural water bodies are spatially mixed on the images of pansharpening multispectral and panchromatic data from 2011 and 2014, which poses great challenges for semantic segmentation of aquaculture ponds. For example, Figure 9c,d are images of the same area, which changed significantly between 2011 and 2014. Several small reservoirs were apparent in Figure 9c, but the profiles of these reservoirs had changed significantly in Figure 9d, and the left side of this area had been recovered into a large lake. The segmentation results in Figure 9g,f show that the restored large lake has been well segmented, but the small reservoirs are easily classified as aquaculture ponds or missed segmentations.

In the paper, extracting aquaculture ponds is performed on images that pansharpen multi-spectral data from Landsat satellites and panchromatic data from other satellites in the same period, and therefore the semantic segmentation might also be affected by the spectral range of the panchromatic image. Table 5 gives the results of an accuracy analysis that divided the training data of Part 2 of the experiment into two portions: pansharpened TM images and pansharpened OLI images. The predicted results of pansharpened TM images from Region B are based on RSCANet, which was trained by fusing TM images with panchromatic images from ZY-3 or ALOS satellites. The predicting results of pansharpened OLI images from Region B is based on RSCANet, which was trained by fusing OLI images with panchromatic images from GF-1 satellites. Table 5 shows that the results of pansharpened OLI images with panchromatic images from GF-1 satellites are significantly better than the results of pansharpened TM images with panchromatic images from ZY-3 or ALOS satellites. The spectrum of panchromatic images from GF-1 satellites ranges from 0.45 to 0.90 μm, which can completely cover the three NIR, red, and green bands of Landsat OLI data. However, the spectrum of panchromatic images from ZY-3 or ALOS satellites can only partly cover the NIR band of the TM sensor. The acquisition time of the TM images was earlier than 2012, and therefore it is difficult to use a GF-1 panchromatic image for pansharpened TM images.

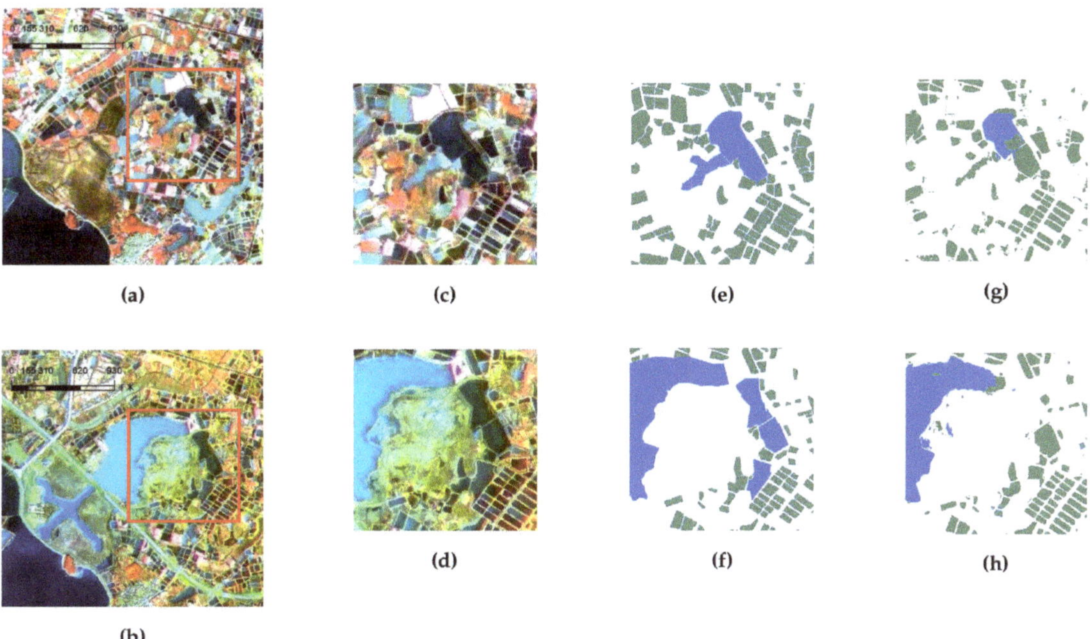

Figure 9. Influence of multi-year changes on semantic segmentation results for aquaculture ponds. (**a**,**b**) are pansharpened images of the significantly changed area from Region B in 2010 and 2014 respectively. (**c**,**d**) are magnified images. (**e**,**f**) are labelling images for (**c**,**d**). (**g**,**h**) are semantic segmentation results for (**c**,**d**) using the proposed method.

The RCSANet can extract aquaculture ponds around inland lakes on 2 m satellite images more accurately than other methods because the involvement of two connection groups from the encoder to the decoder. The first connection group is the combination of the RCSA module and the ASPP-RC module, which links Conv1 of encoder part to decoder part. The second is the RCSA modules, linking Res-1, Res-2, and Res-3 of encoder part to decoder part. Table 6 indicates that the first connection group of RCSANet achieves an additional 2.32% overall accuracy gains over $RCSANet_1$, and the second connection group brings 3.59% overall accuracy gains over $RCSANet_2$, i.e., a plain FCN architecture based on ResNet-101 model. Nevertheless, the connections expend more computing resources because they involve the non-local self-attention mechanism, which contains many inner-product operations. Moreover, the RCSANet is an encoder-decoder architecture in which the gradual upsampling are conducted, requiring more memory and calculation time. Table 7 shows that the RCSANet consumes more memory and training and prediction time than Deeplabv3+ and Reseg methods. It is feasible to sacrifice some computing resources to achieve higher accuracy of aquaculture pond extraction, especially the GPU performance will increase gradually.

Table 5. Accuracy comparation of semantic segmentation results for region B by dividing the training data of Part 2 of the experiment into pansharpened TM images and pansharpened OLI images (%).

Regions	Sensors	Methods	Training Data	Overall Accuracy (%)	Producer's Accuracy		User's Accuracy		
					Kappa	Natural Water (%)	Aqua-Culture (%)	Natural Water (%)	Aqua-Culture (%)
Lake Liangzi (East, type II region B)	TM+ALOS	RCSANet	All pansharpened images from type I regions	79.95	62.01	89.31	61.50	99.56	90.03
		RCSANet-NDWI		83.85	68.42	89.63	72.45	99.51	89.03
		RCSANet	Pansharpened images of fusing TM images with panchromatic images from ZY-3 or ALOS satellites, from type I regions	77.44	57.54	90.98	50.75	98.71	91.26
		RCSANet-NDWI		81.84	64.68	91.23	63.33	98.65	91.00
	OLI+GF-1	RCSANet	All pansharpened images from type I regions	90.90	75.86	93.00	83.26	99.21	83.97
		RCSANet-NDWI		91.71	77.83	93.19	86.31	99.20	83.69
		RCSANet	Pansharpened images of fusing OLI images with panchromatic images from GF-1 satellites, from type I regions	88.01	68.78	92.92	70.12	98.96	85.70
		RCSANet-NDWI		89.41	72.09	93.14	75.85	98.89	85.93

Table 6. Accuracy evaluation for RCSANet and its two varants $RCSANet_1$ and $RCSANet_2$ in Part 1 of the experiments (%). $RCSANet_1$ is a RCSANet variant ablating the first connection group and $RCSANet_2$ is the other variant ablating both connection groups.

Methods	Overall Accuracy	Kappa Coefficient
$RCSANet_2$	81.04	66.15
$RCSANet_1$	84.63	71.59
RCSANet	86.95	72.83

Table 7. Performance evaluation of different deep-learning based methods for semantic segmentation of aquaculture ponds in Part 2 of the experiments.

Methods	Training Time (seconds)	Occupied Memory of GPU for Training (MB)	Prediction Time for Region B (seconds)	Occupied Memory of GPU for Prediction (MB)
RCSA	60,280	7563	35	1543
HCN	66,000	7709	64	7843
Deeplabv3+	16,760	3113	12	1417
Reseg	10,640	2343	16	1083

6. Conclusions

This study has implemented a semantic segmentation network on high-spatial-resolution satellite images for aquaculture pond extraction. A row- and column-wise self-attention (RCSA) mechanism has been proposed to capture the intertwining regular embankments of aquaculture ponds in feature maps, and then a fully convolutional network framework combined with the RCSA mechanism is proposed for semantic segmentation of aquaculture ponds. The proposed methods have been evaluated on high-spatial-resolution pansharpened images obtained by fusing multi-spectral and panchromatic images in typical regions with inland lakes and densely distributed aquaculture ponds. Experiments on satellite images of both a highly developed lake and a reserved lake show that the overall accuracy of the proposed method is significantly better than those of other methods (3–8% overall accuracy gains at Lake Liangzi and 1–2% overall accuracy gains at Lake Hong over the best of other methods). Specifically, from the experimental semantic segmentation results for large regions, detailed information, such as the embankments of aquaculture ponds, can be more accurately identified by the proposed method. It can be concluded that the proposed method is effective for large-scale extraction of aquaculture ponds. In addition, RCSANet-NDWI further improves the accuracy of the proposed method compared with RCSANet, indicating the significance of the proposed NDWI fusion strategy. For future study, the proposed methods can be extended to raft-culture extraction in offshore waters.

Author Contributions: Conceptualization, Z.Z., W.T. and G.Y.; methodology, Z.Z. and D.W.; software, D.W. and J.Y.; validation, D.W., J.Y., B.L. and Z.W.; investigation, Z.Z., W.T. and G.Y.; writing—original draft preparation, Z.Z.; writing—review and editing, W.T., D.W. and G.Y.; visualization, D.W. and J.Y. All authors have read and agreed to the published version of the manuscript.

Funding: This work was supported partially by the National Key Research and Development Program of China under Grant 2017YFC1405600, partially by the Fundamental Research Funds for the Central Universities under Grant 18CX02060A and CCNU19TD002, and partially by the National Natural Science Foundation of China under grant 41506208.

Institutional Review Board Statement: Not applicable.

Informed Consent Statement: Not applicable.

Data Availability Statement: The data presented in this study are available on request from the corresponding author.

Acknowledgments: We are particularly grateful for help in the field and with high-resolution imagery from Jianhua Huang, National & Local Joint Engineering Research Center of Satellite Navigation and Location Service, Guilin University of Electronic Technology. We would like to acknowledge the invaluable input and assistance from the rest of the College of Oceanography and

Space Informatics, China University of Petroleum, including Xuxu Kong, Weipeng Lu, Yachao Chen and Sai Zhang.

Conflicts of Interest: The authors declare no conflict of interest.

References

1. Cao, L.; Naylor, R.; Henriksson, P.; Leadbitter, D.; Metian, M.; Troell, M.; Zhang, W. China's aquaculture and the world's wild fisheries. *Science* **2015**, *347*, 133–135. [CrossRef] [PubMed]
2. Cai, C.; Gu, X.; Ye, Y.; Yang, C.; Dai, X.; Chen, D.; Yang, C. Assessment of pollutant loads discharged from aquaculture ponds around Taihu Lake, China. *Aquac. Res.* **2013**, *44*, 795–806. [CrossRef]
3. Luo, J.; Pu, R.; Ma, R.; Wang, X.; Lai, X.; Mao, Z.; Zhang, L.; Peng, Z.; Sun, Z. Mapping long-term spatiotemporal dynamics of pen aquaculture in a shallow lake: Less aquaculture coming along better water quality. *Remote Sens.* **2020**, *12*, 1866. [CrossRef]
4. Zhang, H.; Kang, M.; Shen, L.; Wu, J.; Li, J.; Du, H.; Wang, C.; Yang, H.; Zhou, Q.; Liu, Z.; et al. Rapid change in Yangtze fisheries and its implications for global freshwater ecosystem management. *Fish Fish.* **2020**, *21*, 601–620. [CrossRef]
5. Ottinger, M.; Clauss, K.; Kuenzer, C. Large-scale assessment of coastal aquaculture ponds with Sentinel-1 time series data. *Remote Sens.* **2017**, *9*, 440. [CrossRef]
6. Ren, C.; Wang, Z.; Zhang, Y.; Zhang, B.; Chen, L.; Xi, Y.; Xiao, X.; Doughty, R.B.; Liu, M.; Jia, M.; et al. Rapid expansion of coastal aquaculture ponds in China from Landsat observations during 1984–2016. *Int. J. Appl. Earth Obs. Geoinf.* **2019**, *82*, 101902. [CrossRef]
7. Stiller, D.; Ottinger, M.; Leinenkugel, P. Spatio-temporal patterns of coastal aquaculture derived from Sentinel-1 time series data and the full Landsat archive. *Remote Sens.* **2019**, *11*, 1707. [CrossRef]
8. Zeng, Z.; Wang, D.; Tan, W.; Huang, J. Extracting aquaculture ponds from natural water surfaces around inland lakes on medium resolution multispectral images. *Int. J. Appl. Earth Obs. Geoinf.* **2019**, *80*, 13–25. [CrossRef]
9. Zhang, T.; Li, Q.; Yang, X.; Zhou, C.; Su, F. Automatic mapping aquaculture in coastal zone from TM imagery with OBIA approach. In Proceedings of the 2010 18th International Conference on Geoinformatics, Geoinformatics 2010, Beijing, China, 18–20 June 2010. [CrossRef]
10. Tran, H.; Tran, T.; Kervyn, M. Dynamics of land cover/land use changes in the Mekong Delta, 1973–2011: A Remote sensing analysis of the Tran Van Thoi District, Ca Mau Province, Vietnam. *Remote Sens.* **2015**, *7*, 2899–2925. [CrossRef]
11. Prasad, K.A.; Ottinger, M.; Wei, C.; Leinenkugel, P. Assessment of coastal aquaculture for India from Sentinel-1 SAR time series. *Remote Sens.* **2019**, *11*, 357. [CrossRef]
12. Geng, Q.; Zhou, Z.; Cao, X. Survey of recent progress in semantic image segmentation with CNNs. *Sci. China Inf. Sci.* **2018**, *61*, 051101. [CrossRef]
13. Lecun, Y.; Bengio, Y.; Hinton, G. Deep learning. *Nature* **2015**, *521*, 436–444. [CrossRef] [PubMed]
14. Simonyan, K.; Zisserman, A. Very Deep Convolutional Networks for Large-Scale Image Recognition. In Proceedings of the International Conference on Learning Representations, San Diego, CA, USA, 7–9 May 2015.
15. He, K.; Zhang, X.; Ren, S.; Sun, J. Deep residual learning for image recognition. In Proceedings of the IEEE Computer Society Conference on Computer Vision and Pattern Recognition, Las Vegas, NV, USA, 27–30 June 2016; pp. 770–778. [CrossRef]
16. Shelhamer, E.; Long, J.; Darrell, T. Fully Convolutional Networks for Semantic Segmentation. *IEEE Trans. Pattern Anal. Mach. Intell.* **2017**, *39*, 640–651. [CrossRef] [PubMed]
17. Badrinarayanan, V.; Kendall, A.; Cipolla, R. SegNet: A Deep Convolutional Encoder-Decoder Architecture for Image Segmentation. *IEEE Trans. Pattern Anal. Mach. Intell.* **2017**, *39*, 2481–2495. [CrossRef] [PubMed]
18. Chen, L.C.; Papandreou, G.; Kokkinos, I.; Murphy, K.; Yuille, A.L. DeepLab: Semantic Image Segmentation with Deep Convolutional Nets, Atrous Convolution, and Fully Connected CRFs. *IEEE Trans. Pattern Anal. Mach. Intell.* **2018**, *40*, 834–848. [CrossRef] [PubMed]
19. Chen, L.; Papandreou, G.; Schroff, F.; Adam, H. Rethinking Atrous Convolution for Semantic Image Segmentation. *arXiv* **2017**, arXiv:1706.05587.
20. Chen, L.; Zhu, Y.; Papandreou, G.; Schroff, F.; Adam, H. Encoder-Decoder with Atrous Separable Convolution for Semantic Image Segmentation. *arXiv* **2018**, arXiv:1802.02611.
21. Lateef, F.; Ruichek, Y. Survey on semantic segmentation using deep learning techniques. *Neurocomputing* **2019**, *338*, 321–348. [CrossRef]
22. Du, P.; Bai, X.; Tan, K.; Xue, Z.; Samat, A.; Xia, J.; Li, E.; Su, H.; Liu, W. Advances of Four Machine Learning Methods for Spatial Data Handling: A Review. *J. Geovisualization Spat. Anal.* **2020**, *4*, 13. [CrossRef]
23. Audebert, N.; Le Saux, B.; Lefèvre, S. Beyond RGB: Very high resolution urban remote sensing with multimodal deep networks. *ISPRS J. Photogramm. Remote Sens.* **2018**, *140*, 20–32. [CrossRef]
24. Wurm, M.; Stark, T.; Zhu, X.X.; Weigand, M.; Taubenböck, H. Semantic segmentation of slums in satellite images using transfer learning on fully convolutional neural networks. *ISPRS J. Photogramm. Remote Sens.* **2019**, *150*, 59–69. [CrossRef]
25. Shi, T.; Xu, Q.; Zou, Z.; Shi, Z. Automatic Raft Labeling for Remote Sensing Images via Dual-Scale Homogeneous Convolutional Neural Network. *Remote Sens.* **2018**, *10*, 1130. [CrossRef]
26. Cui, B.; Fei, D.; Shao, G.; Lu, Y.; Chu, J. Extracting raft aquaculture areas from remote sensing images via an improved U-net with a PSE structure. *Remote Sens.* **2019**, *11*, 2053. [CrossRef]

27. Mnih, V.; Heess, N.; Graves, A.; Kavukcuoglu, K. Recurrent Models of Visual Attention. *Adv Neural Inf Process Syst.* **2014**, *2*, 2204–2212.
28. Bahdanau, D.; Cho, K.; Bengio, Y. Neural Machine Translation by Jointly Learning to Align and Translate. *arXiv* **2015**, arXiv:1409.0473.
29. Luong, T.; Pham, H.; Manning, C.D. Effective Approaches to Attention-based Neural Machine Translation. In Proceedings of the 2015 Conference on Empirical Methods in Natural Language Processing, Lisbon, Portugal, 17–21 September 2015; pp. 1412–1421. [CrossRef]
30. Xu, K.; Ba, J.L.; Kiros, R.; Cho, K.; Courville, A.; Salakhutdinov, R.; Zemel, R.S.; Bengio, Y. Show, attend and tell: Neural image caption generation with visual attention. In Proceedings of the 32nd International Conference on Machine Learning, ICML 2015, Lille, France, 6–11 July 2015; Volume 3, pp. 2048–2057.
31. Parmar, N.; Vaswani, A.; Uszkoreit, J.; Kaiser, Ł.; Shazeer, N.; Ku, A.; Tran, D. Image Transformer. In Proceedings of the 35th International Conference on Machine Learning, Stockholmsmässan, Stockholm, Sweden, 10–15 July 2018.
32. Vaswani, A.; Shazeer, N.; Parmar, N.; Uszkoreit, J.; Jones, L.; Gomez, A.N.; Kaiser, L.; Polosukhin, I. Attention Is All You Need. In Proceedings of the 31st Conference on Neural Information Processing Systems (NIPS 2017), Long Beach, CA, USA, 4–9 December 2017.
33. Fu, J.; Liu, J.; Tian, H.; Li, Y.; Bao, Y.; Fang, Z.; Lu, H. Dual Attention Network for Scene Segmentation. *arXiv* **2019**, arXiv:1809.02983.
34. Xu, X.; Huang, X.; Zhang, Y.; Yu, D.; Xu, X.; Huang, X.; Zhang, Y.; Yu, D. Long-Term Changes in Water Clarity in Lake Liangzi Determined by Remote Sensing. *Remote Sens.* **2018**, *10*, 1441. [CrossRef]
35. Hua, Y.; Mou, L.; Zhu, X.X. Recurrently exploring class-wise attention in a hybrid convolutional and bidirectional LSTM network for multi-label aerial image classification. *ISPRS J. Photogramm. Remote Sens.* **2019**, *149*, 188–199. [CrossRef]
36. Gao, H.; Gu, X.; Yu, T.; Liu, L.; Sun, Y.; Xie, Y.; Liu, Q. Validation of the calibration coefficient of the GaoFen-1 PMS sensor using the landsat 8 OLI. *Remote Sens.* **2016**, *8*, 132. [CrossRef]
37. Jiang, Y.H.; Zhang, G.; Tang, X.M.; Li, D.; Huang, W.C.; Pan, H.B. Geometric calibration and accuracy assessment of ZiYuan-3 multispectral images. *IEEE Trans. Geosci. Remote Sens.* **2014**, *52*, 4161–4172. [CrossRef]
38. Maurer, T. How to pan-sharpen images using the Gram-Schmidt pan-sharpen method—A recipe. In Proceedings of the ISPRS Hannover Workshop 2013, Hannover, Germany, 21–24 May 2013; Volume XL-1/W1, pp. 239–244. [CrossRef]
39. Sekrecka, A.; Kedzierski, M.; Wierzbicki, D. Pre-processing of panchromatic images to improve object detection in pansharpened images. *Sensors* **2019**, *19*, 5146. [CrossRef] [PubMed]
40. Yu, C.; Wang, J.; Peng, C.; Gao, C.; Yu, G.; Sang, N. Learning a Discriminative Feature Network for Semantic Segmentation. In Proceedings of the IEEE Computer Society Conference on Computer Vision and Pattern Recognition, Salt Lake City, UT, USA, 18–22 June 2018; Volume 1, pp. 1857–1866. [CrossRef]
41. Zhao, R.; Wang, D.; Yan, R.; Mao, K.; Shen, F.; Wang, J. Machine Health Monitoring Using Local Feature-Based Gated Recurrent Unit Networks. *IEEE Trans. Ind. Electron.* **2018**, *65*, 1539–1548. [CrossRef]
42. Galassi, A.; Lippi, M.; Torroni, P. Attention in Natural Language Processing. *arXiv* **2020**, arXiv:1902.02181.
43. Otsu, N. A Threshold Selection Method from Gray-Level Histograms. *IEEE Trans. Syst. Man Cybern.* **1979**, *9*, 62–66. [CrossRef]
44. Visin, F.; Romero, A.; Cho, K.; Matteucci, M.; Ciccone, M.; Kastner, K.; Bengio, Y.; Courville, A. ReSeg: A Recurrent Neural Network-Based Model for Semantic Segmentation. In Proceedings of the 2016 IEEE Conference on Computer Vision and Pattern Recognition Workshops (CVPRW), Las Vegas, NV, USA, 28 June–1 July 2016; pp. 426–433. [CrossRef]
45. Chang, C.C.; Lin, C.J. LIBSVM: A Library for Support Vector Machines. *ACM Trans. Intell. Syst. Technol. TIST* **2013**, *2*, 1–39. [CrossRef]

Article

Monitoring the Seasonal Hydrology of Alpine Wetlands in Response to Snow Cover Dynamics and Summer Climate: A Novel Approach with Sentinel-2

Bradley Z. Carlson [1,*], Marie Hébert [2], Colin Van Reeth [1], Marjorie Bison [1], Idaline Laigle [1] and Anne Delestrade [1]

[1] Centre de Recherches sur les Ecosystèmes d'Altitude (CREA), Observatoire du Mont-Blanc, 74400 Chamonix, France; cvanreeth@creamontblanc.org (C.V.R.); marjorie.bison@gmail.com (M.B.); ilaigle@creamontblanc.org (I.L.); adelestrade@creamontblanc.org (A.D.)

[2] France Nature Environnement Haute-Savoie (FNE), 74730 Pringy, France; marie.hebert@fne-aura.org

* Correspondence: bcarlson@creamontblanc.org

Received: 7 May 2020; Accepted: 15 June 2020; Published: 17 June 2020

Abstract: Climate change in the European Alps during recent years has led to decreased snow cover duration as well as increases in the frequency and intensity of summer heat waves. The risk of drought for alpine wetlands and temporary pools, which rely on water from snowmelt and provide habitat for specialist plant and amphibian biodiversity, is largely unknown and understudied in this context. Here, we test and validate a novel application of Sentinel-2 imagery aimed at quantifying seasonal variation in water surface area in the context of 95 small (median surface area <100 m^2) and shallow (median depth of 20 cm) alpine wetlands in the French Alps, using a linear spectral unmixing approach. For three study years (2016–2018), we used path-analysis to correlate mid-summer water surface area to annual metrics of snowpack (depth and duration) and spring and summer climate (temperature and precipitation). We further sought to evaluate potential biotic responses to drought for study years by monitoring the survival of common frog (*Rana temporaria*) tadpoles and wetland plant biomass production quantified using peak Normalized Difference Vegetation Index (NDVI). We found strong agreement between citizen science-based observations of water surface area and Sentinel-2 based estimates (R^2 = 0.8–0.9). Mid-summer watershed snow cover duration and summer temperatures emerged as the most important factors regulating alpine wetland hydrology, while the effects of summer precipitation, and local and watershed snow melt-out timing were not significant. We found that a lack of summer snowfields in 2017 combined with a summer heat wave resulted in a significant decrease in mid-summer water surface area, and led to the drying up of certain wetlands as well as the observed mortality of tadpoles. We did not observe a negative effect of the 2017 summer on the biomass production of wetland vegetation, suggesting that wetlands that maintain soil moisture may act as favorable microhabitats for above treeline vegetation during dry years. Our work introduces a remote sensing-based protocol for monitoring the surface hydrology of alpine wetland habitats at the regional scale. Given that climate models predict continued reduction of snow cover in the Alps during the coming years, as well as particularly intense warming during the summer months, our conclusions underscore the vulnerability of alpine wetlands in the face of ongoing climate change.

Keywords: French Alps; optical remote sensing; multitemporal; linear spectral unmixing; NDVI; drought; *Rana temporaria*; ecohydrology; mountain temporary pools

1. Introduction

Recent climate warming in the European Alps is currently reshaping alpine landscapes and ecosystems. Increases in mean temperature in the Alps are amplified with respect to the global average [1] and warming has accelerated markedly since the 1980s [2]. Rising mean temperatures have been accompanied by an increase in the frequency and intensity of extreme events, such as summer heat waves, during the last twenty years [3]. Shifts in air temperature have led to significant reductions in glacier mass and extent [4,5], as well as a 4–5 week reduction in snow cover duration since the 1970s in the Swiss Alps [6]. Mountain plant species are moving upslope and increasing biomass production in response to climate warming [7–9], and vegetation belts within the Alps are expected to continue to shift upward in response to 21st century climate change [10].

Alpine wetlands are situated at the confluence of the aforementioned recent changes in climate, cryosphere, hydrology, and vegetation. Located between the treeline and snowfields and glaciers, alpine semipermanent pools and ponds (hereafter referred to more generally as wetlands) are present throughout most of world's mountains and are defined as small (1 m^2 to a few square hectares) and shallow water bodies characterized by at least the seasonal presence of surface water [11]. The hydrology of alpine wetlands is understood to be tightly linked to watershed runoff from rain and snowmelt [11]. Predicted decreases in snow cover duration combined with continued glacier retreat are expected to diminish summer water runoff, particularly during the second half of the 21st century [12], which could decrease available surface water for alpine wetlands. In the Swiss Alps, recent warming has been associated with increases in the abundance of generalist thermophilous plant species at the expense of specialized wetland species [13]. Amphibian populations are known to be declining at the global scale due to climate change, disease, and habitat degradation [14,15], and amphibian habitat loss has been documented in mountainous regions throughout Australia, North America, and Central America [16]. Notably, the drying up of wetland pools in Australia has been linked to the local mortality of an endangered Australian frog species, *Pseudophyrne pengilleyi* [17]. In light of these examples, studies linking climate, wetland hydrology, and biodiversity responses in the Alps remain lacking. Improving our knowledge of the ecohydrological functioning of alpine wetlands is of particular importance in order to inform wetland biodiversity conservation measures and also from the standpoint of ecosystem services, given that wetland habitats are known to provide important downstream regulatory services such as aquifer recharge, flood mitigation, and denitrification [18].

Recent improvements in widely available optical satellite imagery are enabling unprecedented opportunities for tracking the responses of mountain ecosystems to climate variability and change. Specifically, the Sentinel-2 satellite mission, launched in 2015, includes an unprecedented combination of 5 day temporal revisit, 10 m spatial resolution, ten spectral bands ranging from visible to short-wave infrared, and a free and open access data policy. In mountainous regions, Sentinel-2 has already been utilized for a number of applications, including for example generating high-resolution snow cover maps across the Alps and Pyrenees [19], quantifying the effects of snow cover duration on alpine plant community habitat [20], and improving land cover maps of dwarf Ericaceae shrubs above treeline [21]. We propose that the relatively high spatial, temporal, and spectral resolution of Sentinel-2 could be utilized to enhance monitoring of the seasonal hydrology of alpine wetlands.

In this study, we focus on alpine pools and ponds characterized by seasonal snow cover and fluctuating amounts of surface water over the course of the summer season. In contrast to alpine lakes, biological communities in small alpine water bodies are strongly driven by water availability and hydroperiod [11,22]. In this context, binary classification approaches based on spectral indices [23], or object-oriented classification that have previously been used to map large wetlands and water bodies [24], may be insufficient to map small (e.g., <100 m^2) and constantly fluctuating mountain pools. Furthermore, high spatial resolution images with pixel size below 25 m^2, obtained, for example, using aerial photographs, specialized satellite platforms such as PLEIADES or SPOT 6/7, or drone imagery, tend to have highly limited temporal and spectral resolution and can require expensive and time consuming acquisition and preprocessing protocols [25].

To respond to these methodological challenges, we propose using linear spectral unmixing, which was originally developed as a method for mapping desert shrublands using Landsat imagery [26]. This approach allows for mapping sub-pixel fractions that correspond to objects that are smaller than the spatial resolution of available imagery, which, in this case, means smaller than a 100 m^2 Sentinel-2 pixel. Based on the assumption that the reflectance values of image pixels are the result of varying mixtures of components, or endmembers, this technique has been used for numerous applications in plant cover mapping and forestry (see, e.g., in [27,28]), soil and erosion mapping (see, e.g., in [29]) and more recently for snow algae detection in Antarctica [30]. We propose multi-temporal spectral unmixing as a potential approach for quantifying the surface dynamics of wetlands, in terms of snow cover and surface water.

In this paper, we test the use of linear spectral unmixing as a means of quantifying seasonal variation in the water surface area of small alpine wetlands located in the northern French Alps with respect to field observations of water surface area. We further utilize available Sentinel-2 imagery to quantify snow cover duration at the watershed and local wetland scales. For three study years (2016 to 2018), we correlate interannual variability in meteorological and snow cover parameters to mid-summer water surface area, in order to identify the most important parameters linked to drought risk. Finally, we assess the potential of our method for quantifying biotic responses of wetland communities to drought, by monitoring the development and survival of common frog (*Rana temporaria*) tadpoles in select sites and by quantifying wetland vegetation biomass production.

2. Materials and Methods

2.1. Study Area and Fieldwork

This work was carried out in the Chamonix valley in the Mont-Blanc massif and located in the northern French Alps (Figure 1). Studied sites (N = 95) were identified over the course of two field seasons carried out between mid-July and mid-August: 20 wetlands were identified in 2010, while an additional 75 wetlands were visited during the summers of 2017 and 2018. We focused on ponds and pools located in open areas without forest cover above 1800 m a.s.l., and that contained surface water at the time of the field survey (lakes, rivers, streams, peat bogs, and wetlands characterized by damp soil were excluded from the analysis). Observed water surface area of target wetlands varied from 3 to 5000 m^2 with a mean of 511 m^2, with water depth ranging from 3 cm to 3 m with a mean depth of 44 cm. The elevation of studied wetlands ranged from 1820 to 2600 m a.s.l., with a mean elevation of 2100 m a.s.l.. Sites were distributed across 22 alpine watersheds (Figure 1C). Bedrock varied from limestone in the western portion of the study area, to schist and limestone for central and northeastern wetlands and granite in the case of the southeastern most watershed. Many of the watersheds are characterized by persistent snowfields during the summer months (Figure 1C).

During the summers of 2016, 2017, and 2018 we carried out weekly visits to four wetlands at the Loriaz site (Figure 1D) to monitor the development of common frog (*Rana temporaria*) tadpoles. During each visit, we recorded the presence of frog eggs as well as the number of egg clusters, followed by the phenological stage attained by the tadpoles (see Figure S1). We also noted local mortality of the tadpole population, in the case of drought.

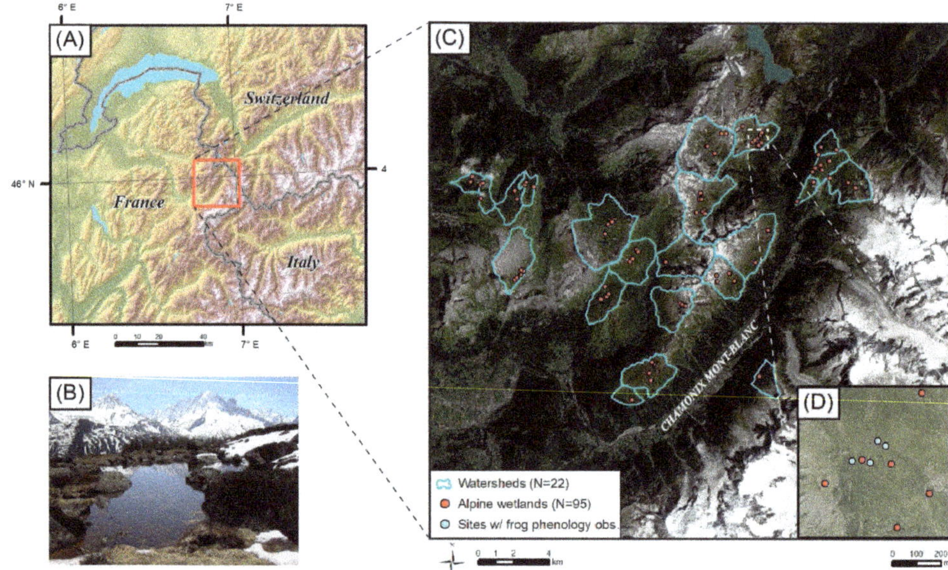

Figure 1. (**A**) Locator map displaying the study area in the Chamonix valley, France. (**B**) Photo of a typical alpine pond habitat in the Mont-Blanc massif. (**C**) Map of the 22 alpine watersheds (above 1800 m a.s.l.) and the 95 alpine wetlands considered in this analysis. (**D**) Inset map showing the locating of four wetlands visited weekly during the summer months to monitor the phenology of common frog (*Rana temporaria*) populations and periodically to survey wetland surface water extent.

2.2. Sentinel-2 Imagery

For this analysis, we relied on multi-spectral images acquired by the Sentinel-2 satellites (2A and 2B), which jointly provide a 5 day revisit time since March 2017. Available spectral bands range from visible to shortwave infrared, with 4 bands at 10 m spatial resolution (B2: 490 nm, B3: 560 nm, B4: 665 nm, B8: 842 nm) and 6 bands at 20 m spatial resolution (B5: 705 nm, B6: 740 nm, B7: 783 nm, B8a: 865 nm, B11: 1610 nm, B12: 2190 nm). We downloaded 76 Sentinel-2A and B (hereafter referred to as Sentinel-2) scenes for the T32TLS tile and for the years 2016, 2017, and 2018 from the French THEIA platform.

Acquired images were preprocessed by the THEIA to level 2A (i.e., orthorectified product in surface reflectance). Images were corrected for both atmospheric and topographic effects [31,32], which are particularly important in mountainous areas where slope angle and aspect affects pixel illumination [33]. Level-2A products were provided with cloud and shadow masks, which we applied to all spectral bands. In total, we analyzed 23 dates in 2016, 25 dates in 2017, and 28 dates in 2018 between the months of February and November of each year.

2.3. Endmember Selection and Spectral Unmixing

Spectral endmembers were identified using a 50 cm resolution aerial photograph acquired in 2009 covering the study area. We hypothesized that all alpine wetland pixels could be composed of a mixture of the following endmembers, water, vegetation, rock and bareground, and snow. Given that all sites were located above the treeline with low-stature vegetation, we considered it unnecessary to include shadow as a spectral endmember. Using the aerial photographs for reference, we identified between three and four visually pure Sentinel-2 pixels for each target endmember class, representing an area of homogenous cover for at least a 50 × 50 m zone in order to avoid edge effects. We then extracted spectral values for identified endmember pixels from a Sentinel-2A image acquired on August 3, 2016. We averaged spectral values for each endmember class and carried out a principal component analysis (PCA) using the *ade4* package in R.

Linear spectral unmixing allows for the estimation of sub-pixel fractions of target endmembers and is suitable for mapping phenomena that vary at finer scales than the spatial resolution of available imagery [26]. By resolving an ordinary least squares equation, the algorithm estimates the fractional cover of endmembers resulting in observed spectral values [34]. In Equation (1), ρ_j represents the observed spectral band values (in this case from the Sentinel-2 reflectance value for a given band and pixel), F_i represents endmember fractions that are equivalent to slope values in the linear model, $\rho_{j,\,i}$ represents the known reflectance values of target endmembers, and E_j represents an error term estimated for each band. The sum of estimated fractions ($\sum F_i$) is constrained to be equal to 1 (Equation (2)). In this configuration, F_i is the only unknown value and can be solved for using the following equations:

$$\rho_j = \sum_{i=1}^{m} F_i \times \rho_j + E_j \tag{1}$$

$$\sum_{i=1}^{m} F_i = 1 \tag{2}$$

where j corresponds to the number spectral bands, i corresponds to a given endmember, and m represents the the number of endmembers.

For each acquired Sentinel-2 scene, we resampled 20 m bands to 10 m using bilinear interpolation, and created a stack covering the study area. We extracted Sentinel-2 spectral values for the GPS coordinates of each wetland using two approaches: (i) a simple method extracting spectral values for the pixel overlapping the coordinates, and (ii) using a 20 m buffer to extract spectral values in the vicinity of the target wetland. The resulting tables were defined as spectral libraries, with each column representing one of the ten Sentinel-2 bands and each row representing a target pixel. We then used the "unmix" function in the *hsdar* R package to estimate the fractional cover of endmembers for each pixel and for each date. Error values and fraction estimates resulting from the simple extract were stored directly for each wetland. For results of the 20 m buffer, we summed the surface area of each endmember across all pixels and stored values for each Sentinel-2 scene date. In order to quantify water levels during the critical midsummer period when drought is most likely, we calculated the mean water surface area for available dates between July 15 and August 15 (*Water surface area*; Figure S2).

2.4. Field Validation

In order to validate Sentinel-2 based estimates of water surface area, we carried out field observations with different volunteer groups for four summer dates between 2016 and 2019. Volunteers included local elementary school students accompanied by teachers, local adults, visiting university students, and tourists. Group size ranged from 6 to 15 participants, which were split into at least two groups. In the field, groups were asked to delimit a 30 × 30 m grid centered on the target wetland and GPS coordinate, composed of nine smaller 10 m sub-squares (see Figure S3). For each 10 × 10 m square, volunteers conducted visual estimates of the percent cover of different endmembers (rock and bareground, vegetation, water, and snow). For each plot, two separate groups observed the central grid cell in order to quantify observer error. Based on values for the nine grid cells, we calculated mean and total water surface area for each site as noted by volunteers.

For target wetlands, we extracted water surface area estimates for the nine nearest Sentinel-2 pixels centered on the wetland GPS coordinate. Field visits were coordinated so as to occur either the same day or within 1 day of the passage of one of the Sentinel-2 satellites. We calculated mean and total water surface area from Sentinel-2 fractional estimates for the nine pixels overlapping the target water body, and correlated these values with field observations. Given that both field and satellite observations were subject to errors, we utilized standardized major axis regression in the *smatr* R-package to assess agreement between Sentinel-2 and ground-based observations.

2.5. Mapping Snow Cover and Wetland Plant Biomass

For each Sentinel-2 scene, we calculated the Normalized Difference Snow Index (NDSI) and the Normalized Difference Vegetation Index (NDVI) and the using Equations (3) and (4):

$$NDSI = \frac{(Rgreen - Rswir)}{(Rgreen + Rswir)} \tag{3}$$

$$NDVI = \frac{(Rnir - Rred)}{(Rnir + Rred)} \tag{4}$$

where NIR is the reflectance in the near-infrared band (Band 8), red corresponds reflectance in Band 4, green corresponds to Band 3, and SWIR is shortwave infrared (Band 11). We applied a 0.4 threshold to NDSI values to estimate the presence or absence of snow cover for 10 m pixels for each Sentinel-2 scene [35]. For each date, we also calculated the percentage of watersheds covered by snow. We also estimated local snow melt out date for wetlands based on the following criteria, NDSI < 0.4 and a snow fraction estimate of 0. Finally, for each year and each wetland, we extracted the maximum value of NDVI between July 15 and August 15 as a proxy of peak biomass and plant photosynthetic activity.

2.6. Meteorological Data and Structural Equation Modeling

We obtained daily meteorological and snowpack data for the last thirty years for 300 m elevation bands within the Mont-Blanc massif from the Safran-CROCUS atmosphere-snowpack reanalysis, provided by Météo-France and the Centre d'Etudes de la Neige [36,37]. Data were downloaded from the open access Aeris portal [38]. Based on our hypotheses, we extracted the following monthly parameters for each year (1988 to 2018): mean snowpack height for the month of March (March snowpack depth), sum of growing degree days (>0 °C) for 2 m air temperature in March and April (March–April GDD), sum of growing degree days (>0 °C) for June and July (June–July GDD), and the sum of rainfall in June and July (June–July precip.). For study years (2016–2018), we assigned values from the 300 m elevation class corresponding to each study site. In order to contextualize study years within the last thirty years and for the 1950 to 2250 m a.s.l. elevation band, we calculated average values and annual anomalies for all variables over the 1988 to 2018 period.

We derived the following spatial snow cover variables from Sentinel-2 imagery: the date that watershed snow cover fell below 30%, based on NDSI-based binary snow cover maps (Watershed snowmelt), the date of local snow melt-out for wetlands (defined by fractional snow cover below 10% for wetland pixels, Wetland snowmelt) and the average percent snow cover of watersheds between July 15 and August 15 based on NDSI snow cover maps (Summer snowfields).

We utilized a structural equation modeling (SEM) approach to test linear relationships between meteorological and snowpack parameters and observed midsummer water surface area. As an extension of path analysis, SEM represents an appropriate statistical framework for modeling multivariate and hierarchical effects of predictors on target response variables [39,40]. We first standardized values using the "decostand" function in the vegan R package, and subsequently verified that pairwise relationships between variables could be modeled using ordinary least squares linear regressions. We then tested multiple model pathways, based on the following hypotheses; in all models, we expected midsummer Water surface area to be correlated with June-July GDD, June–July precip., and with snowpack. In order to identify which snowpack parameter was the most significant, we tested models using Watershed snowmelt, Wetland snowmelt, and Summer snowfields as predictors. We also systematically tested models with and without summer precipitation as a predictor of Water surface area. To test the effects of snowpack depth on spatial snow cover parameters, we correlated each snowpack parameter with March snowpack depth and an air temperature variable (April–May GDD for Wetland and Watershed snowmelt, and June–July GDD in the case of Summer snowfields). We compared models using the following criteria: goodness-of-fit requiring a χ^2 chi-square p-value greater than 0.05 [41], the Akaike Information Criterion (AIC) value, and explained variance (R^2) of Water surface area.

3. Results

3.1. Spectral Endmember Differentiation and Method Validation

Selected endmembers exhibited distinct spectral signals with respect to Sentinel-2 bands, both in terms of mean reflectance values (Figure 2A) and multivariate PCA coordinates (Figure 2B). Vegetation and bareground were the most closely related spectral endmembers, while snow and water exhibited highly distinctive spectral signatures. Error resulting from spectral unmixing was low (around 0.01%) across scene dates, with slightly higher errors at the beginning (June) and end (September) of the summer season, potentially due to persistent snow cover and the variable state of plant canopies during the fall season compared to the early-August reference used for endmember definition (Figure S2).

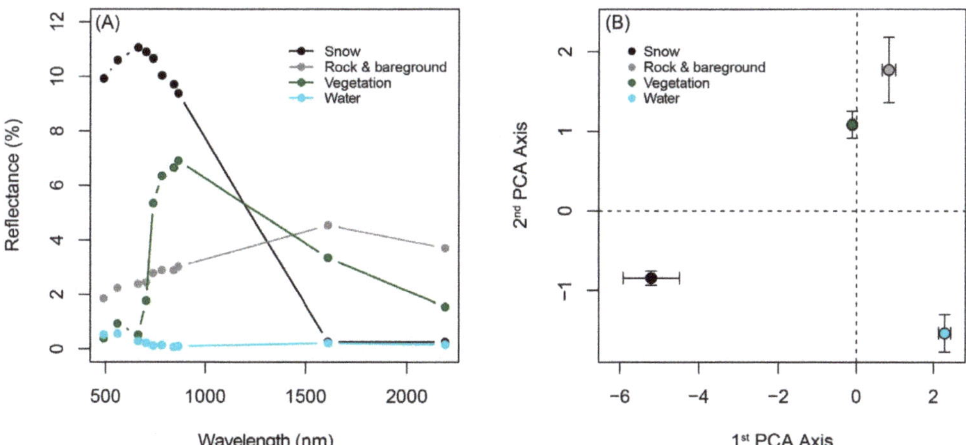

Figure 2. Spectral signatures of target endmembers, identified using 50 cm resolution aerial photographs and extracted from a Sentinel-2A image on August 3, 2016. (**A**) Mean reflectance values for spectral endmembers for the 10 utilized Sentinel-2 bands. (**B**) Mean PCA coordinates of spectral endmembers, showing 95% confidence intervals.

Wetlands underwent a characteristic transition from full snow cover during the winter months to a mixture of water, bareground, and vegetation between snowmelt-out in the spring and snow onset in the fall (Figure 3). Water surface area varied over the course of the summer season, which was corroborated by field observations and visually confirmed using drone imagery (Figure 3). Although localized drought was apparent immediately surrounding a target wetland in July 2017 (Figure 3A), the utilization of the 20 m buffer enabled the detection of water within a broader vicinity adjacent to the target wetland (Figure 3).

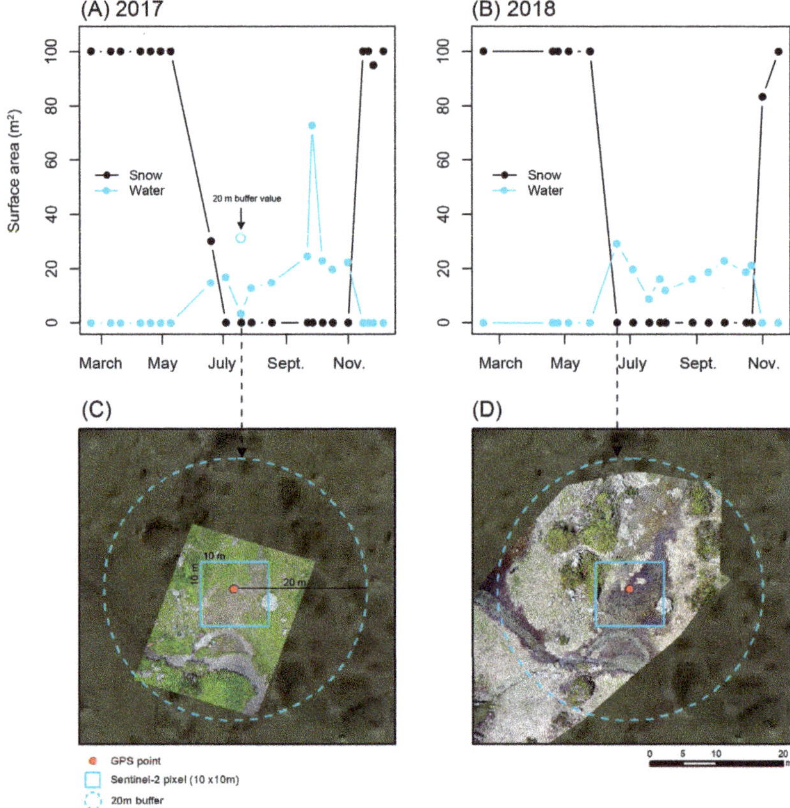

Figure 3. Example time series of fractional snow and water cover for a select wetland extracted from Sentinel-2 imagery in 2017 (**A**) and 2018 (**B**). Solid point values are result of fractional estimates resulting from a simple extract of spectral values for the GPS point. The hollow point in panel (**A**) indicates the fractional estimate for a 20 m buffer around the GPS point, which was the method used for all subsequent analysis. Panels (**C**,**D**) show drone images of the targeted wetland in July, 2017, and June, 2018. The red point indicates the GPS point used for field monitoring and image analysis, the blue rectangle shows the overlapping Sentinel-2 pixel, and the diameter of the 20 m buffer used to estimate surface water in the vicinity of target wetlands.

We observed strong agreement between field observations of water surface area and estimates derived from Sentinel-2 images (Figure 4). We found positive linear relationships with respect to mean water surface area (Figure 4A, $R^2 = 0.88$) and total water surface area (Figure 4B, $R^2 = 0.79$) for target wetland areas. Some systematic error was apparent in the case of mean water surface area, and Sentinel-2 tended to slightly underestimate water surface area for low quantities of water (<10 m^2) with respect to field observations (Figure 4A). Although the linear relationship was not as strong compared to mean water surface area, in the case of total water surface area, the relationship exhibited no systematic error and the linear trend closely followed the 1:1 reference line (Figure 4B). We observed a mean absolute error (MAE) of ±27 m^2 for total water surface area over a 900 m^2 area, which corresponded to a 3% error rate. Different volunteer groups demonstrated strong agreement between visual estimates of endmember ground cover, including for water (R^2 = 0.76), vegetation (R^2 = 0.76), and rock and bare ground (R^2 = 0.80; Figure S4).

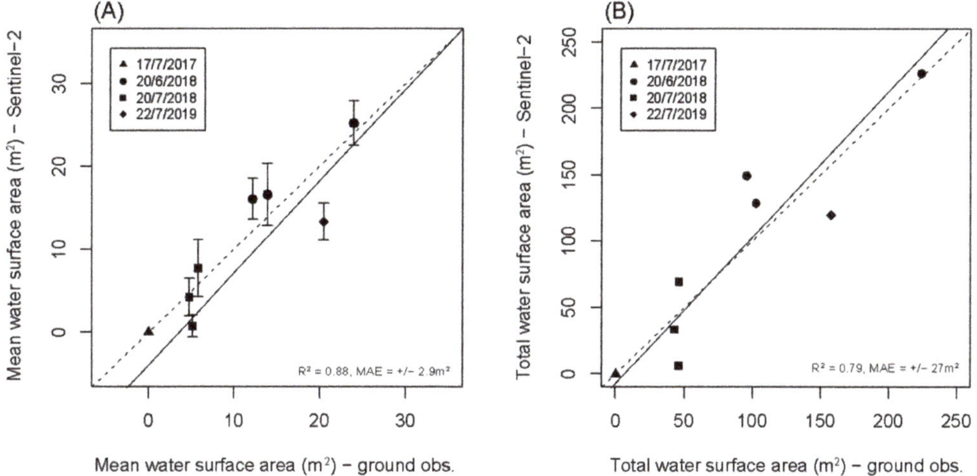

Figure 4. Validation of Sentinel-2 based estimates of water surface area relative to visual ground-based estimates carried out by volunteers. Values are based on a grid of nine 10 × 10 m pixels centered on the GPS point of a target wetland (see Figure S1). (**A**) Mean water surface area observed on the ground compared to mean water surface area estimated using Sentinel-2, with error bars representing 95% confidence intervals. (**B**) Total water surface area observed on the ground compared to total water surface area estimated using Sentinel-2. The dashed line represents a 1:1 relationship while the solid line represents a line of best fit using linear standardized major axis regression. Point symbols correspond to do different visit dates in the field, which were coordinated with the passage of the satellite. MAE = mean absolute error.

3.2. *Characterization of Wetland Seasonal Hydrology (2016–2018)*

Regardless of the threshold used to identify drought risk, 2017 stood out as the driest summer of the three-year study period (Figure 5). Summer 2016 was slightly drier than 2018 for thresholds below 60 m^2. Based on the error and uncertainty presented in Figure 4, we identified 25 m^2 as a potential drought risk threshold, meaning that we considered wetlands with less than 25 m^2 of estimated surface water (within the 20 m buffer zone corresponding to approximately 1300 m^2) to be at risk of drought. Based on this threshold, of the 95 target wetlands, eight were dry in 2017, five were dry in 2016, and only one was dry in 2018 (Figure 5).

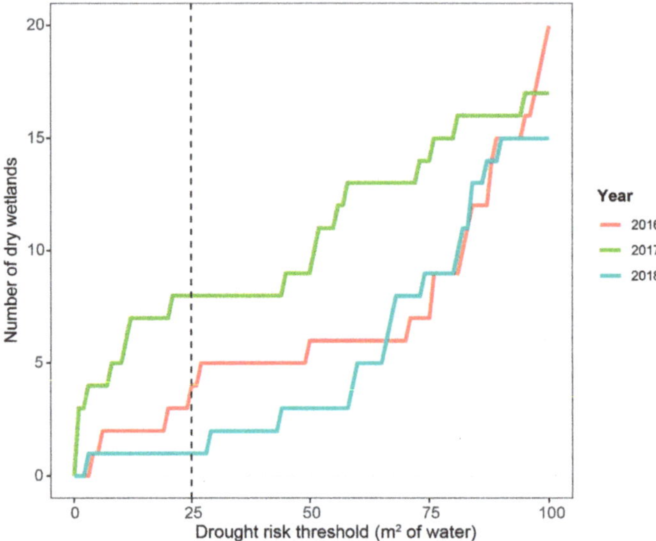

Figure 5. Number of dry wetlands ($N = 95$) depending on the threshold used to identify drought risk. Values are based on sum of water surface area within a 20 m buffer around wetlands or approximately a 1300 m^2. For monitoring purposes, we propose utilizing a threshold of 25 m^2 given the uncertainty of the method presented in Figure 4B.

Regarding the persistence of mid-summer snowfields at the uppermost elevations of watersheds, in late-July and early-August snow covered 5 to 6% of watersheds in 2016 and 2018 and less than 1% of watersheds in 2017 (Figure 6). On average, watersheds became snow-free approximately three to four weeks earlier in 2017 compared to 2016 and 2018 (Figure 7A). In 2017, watersheds dropped below 20% snow cover on average by late-May, whereas in both 2016 and 2018, watershed snow cover did not decline rapidly until late-June (Figure 7B).

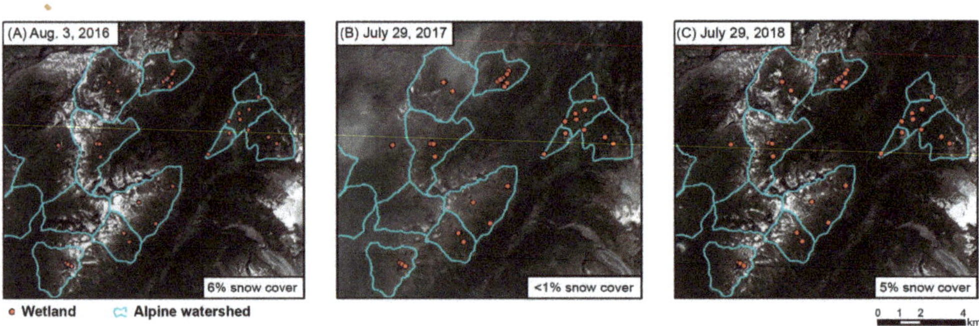

Figure 6. Mid-summer images for (**A**) 2016, (**B**) 2017, and (**C**) 2018 showing the spatial distribution of persistent snowfields, watershed boundaries, and wetlands for a portion of the study area. Although cirrus clouds are present in panel (**B**), masked pixels did not cover snowfield areas.

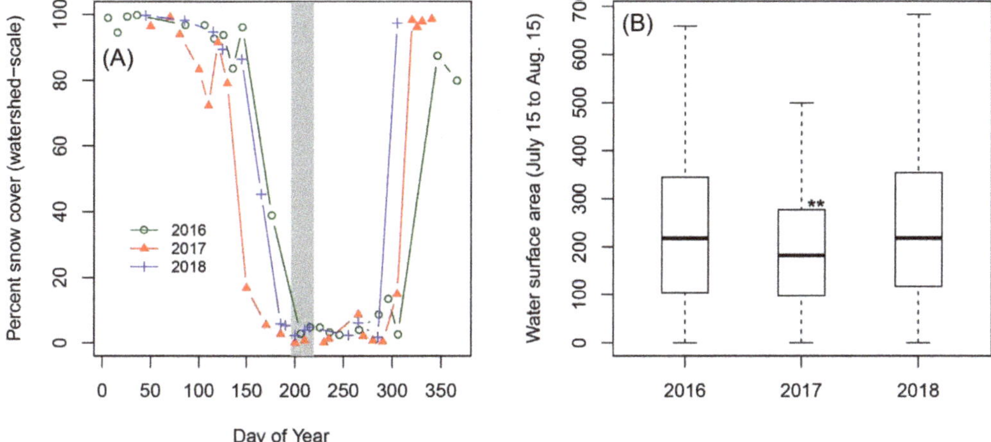

Figure 7. (**A**) Mean percent snow cover for 22 watersheds >1800 m a.s.l. for 2016, 2017, and 2018; the gray rectangle indicates the July 15 to August 15 period considered for drought-risk. (**B**) Average water surface area, applying the 20 m buffer, for alpine wetlands during the July 15 to August 15 period in 2016, 2017, and 2018; ** denotes a significant difference in water surface area values between 2017 and the other years according to a Pairwise Wilcoxon signed rank test: 2017 vs. 2016, p-value = 0.001; 2017 vs. 2018, p-value = 0.004, 2016 vs. 2018, not significant).

The three study years were highly contrasted in terms of March snowpack height and summer temperature and precipitation for the 1950–2250 m a.s.l. elevation band (Figure S5). With respect to the 30 year reference period, 2016 was close to average in terms of March snowpack depth, June–July GDD and June–July precipitation (Figure S5). The year 2017, however, was characterized by low snowpack high summer temperatures and average precipitation, with a 50 cm deficit in snowpack depth in March and a more than 100 °C surplus in June–July GDD. Last, snowpack depth in 2018 was 1 m above average, summer precipitation was well below average (−43 mm), and June–July GDD was well above average (+94 °C; Figure S5). Values extracted for wetlands exhibited the same inter-annual variability in snowpack and meteorological parameters (Figure S6).

For the July 15 to August 15 critical period, water surface area of target wetlands varied between 0 and 500–700 m^2 (Figure 6B). The mid-summer water surface area of wetlands was significantly lower in 2017 compared to 2016 and 2018 (Figure 6B), with a median decrease of 40–50 m^2 of surface water in 2017 (Figure 7B).

3.3. Structural Equation Modeling Results

Only one of the SEM models we tested was statistically acceptable (χ^2 p-value = 0.68, Table S1 Model 1A). This model also exhibited the lowest AIC value (16.76) compared to all other models (Table S1). Model 1A demonstrated a significant positive effect of March snowpack depth and a significant negative effect of June–July GDD on the persistence of Summer snowfields, followed by a significant positive effect of Summer snowfields and a negative effect of June–July GDD on Water surface area (Table S1, Figure 8). Interestingly, summer precipitation (June–July precip.) was not significantly related to Water surface area for any of the tested models, with the exception of Model 2B which exhibited a nearly significant negative relationship (p-value = 0.06, Path coefficient estimate of −0.16). Furthermore, Summer snowfields was the only measured snowpack parameter to have a significant effect on Water surface area, and local snow melt-out date of wetlands and at watershed snowmelt-out were not significant. We also tested for a direct effect of March snowpack height on Water surface area (results not shown), but did not observe a significant relationship or an improvement in model performance. Tested models explained between 4% and 11% of variance in Water surface area. Summary statistics for tested path models are provided in Table S1.

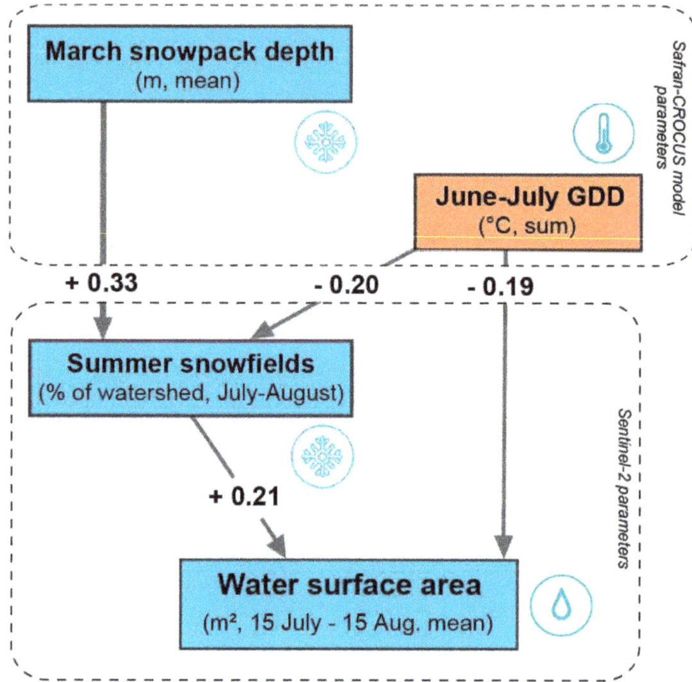

Figure 8. Structural equation model results for the retained model (Table S1, Model 1A), linking hydrological (blue) and meteorological (orange) parameters to interannual (2016–2018) variability in the mid-summer water surface area of wetlands. Line thickness is proportional to the values of path coefficient estimates. Boxes with dashed outlines indicate variables measured using the Safran-CROCUS atmosphere-snowpack model, and spatial variables measured using Sentinel-2 satellite imagery.

3.4. Observed Effects of Drought on Tadpole Development and Plant Biomass

In 2017, we observed the drying up of two monitored wetlands at the Loriaz site (Figure 1D) during a field visit on July 17. A subsequent field visit on August 3, 2017 confirmed the local mortality of frog tadpoles in these sites. Sentinel-2 estimates confirmed a lack of surface water for the July 17 date (Figure 4). We did not observe mid-summer drought and associated tadpole mortality for Loriaz sites in either 2016 or 2018. The 2017 drought occurred when tadpoles were in the process of developing from stage 3 to 4, which typically occurs during the critical drought-risk period from July 15 to August 15 for the Loriaz wetlands (Figure S7).

We did not observe a significant difference in peak NDVI (NDVI max) for wetland vegetation during the three years of the study (Figure 9A). Field observations and drone imagery indicated that even dry wetlands were surrounded by productive and lush vegetation, despite the lack of surface water (Figure 9B).

We did not observe vegetation with a high degree of senescence during our mid-summer field observations of Loriaz wetlands in 2017.

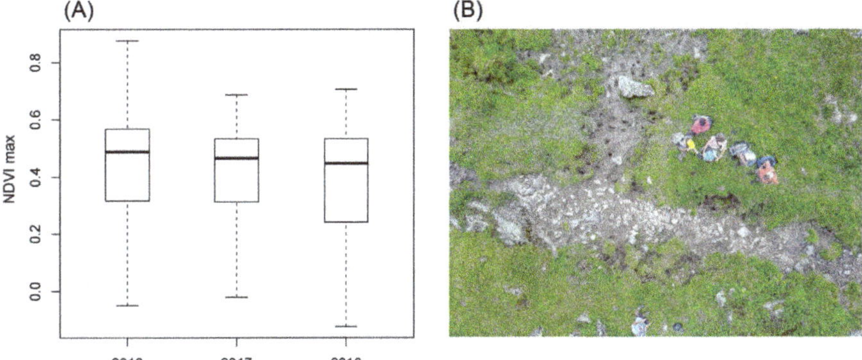

Figure 9. (**A**) Boxplots of maximum Normalized Difference Vegetation Index (NDVI) values observed for wetlands during the July 15 to August 15 period for the three study years. (**B**) Drone photograph of a dried out wetland in 2017, showing lush vegetation despite the dry conditions.

4. Discussion

Our proposed use of Sentinel-2 imagery enables a novel approach for monitoring the seasonal hydrology of alpine wetlands and temporary pools, which constitute a ubiquitous and understudied feature of mountain landscapes with important implications for biodiversity and ecosystem services. We advocate for use of satellite imagery for automated monitoring of water surface area combined with field-based observations to quantify the consequences of drought on wetland plant and amphibian communities. In this paper, we provide examples of biotic responses to drought in terms of tadpole survival and plant biomass; however, further studies are needed in order to assess the population-level the effects of more frequent droughts on long-lived amphibian and plant species in a climate change context.

4.1. Methodological Limits & Perspectives

Our application of linear spectral unmixing provides a solution for mapping small and seasonally variable water bodies in alpine habitats, characterized by surface water extents of less than 100 m^2. Validation of spectral unmixing algorithm results demonstrated strong agreement with ground observations (Figure 4); however, with a degree of error that highlights certain methodological limitations. First, target wetlands were generally shallow (with a median water depth of 20 cm) and therefore exhibited spectral signatures that likely reflected a blend of surface water and underlying rocks, mud, and vegetation. This phenomenon is visually apparent in Figure 3D. Accordingly, we expect that the slight underestimation of water surface area for low water levels shown in Figure 4A results from blurred spectral signals in the case of extremely shallow water. One possible methodological improvement would be to include derived spectral indices such as the Bare Soil Index [42] in the spectral library used for endmember differentiation and spectral unmixing. This approach, i.e., including spectral index information in addition to band reflectance values, proved to be effective for multitemporal spectral unmixing and forest species mapping in the northeastern United States using NDVI in addition to Landsat 7 and 8 bands [28].

The difficulty of detecting extremely shallow surface water is an important consideration from an ecohydrological standpoint in the context of temporary alpine pools, given that field observations and previous studies indicate that small patches of shallow water between 10 and 15 cm of depth are sufficient for the growth and development of the common frog [43] as well as considerable zooplankton diversity [22]. Due to this uncertainty, for monitoring purposes we propose using a threshold of 25 m^2 (quantified within the 20 m buffer) below which drought is possible but difficult to detect with a high degree of certainty using our method. Accordingly, we propose that our method is sufficiently sensitive to quantify interannual variation in water surface area as well as drought risk for alpine wetlands distributed at the regional scale; however, field observations, potentially combined with higher resolution imagery, remain necessary in order to confirm the occurrence of drought for target ponds and pools.

We did not test the sensitivity of our results to spectral endmember selection, which is known to affect fractional cover estimates over time and space [44]. Given that our main target for this analysis was mapping fractional water cover, we assumed stationarity in the spectral properties of water over the course of the snow-free season from late spring to early fall. This assumption was supported by the consistently low residual error resulting from fraction estimates, particularly during the mid-summer season (Figure S2). Water spectral endmembers were defined in our case by selecting central pixels within lakes throughout the study area with low sediment concentration, which we consider to be a reproducible and consistent approach that could be utilized in other temperate mountain study areas. Should our multitemporal spectral unmixing approach be used for other applications such as mapping plant canopies or cover, which are influenced by species composition and are known to show highly variable spectral signals over course of the growing season [21], we recommend careful initial endmember selection using field spectroscopy as well as iteratively redefining spectral endmembers for each scene date [45].

4.2. Influence of Snowpack and Summer Climate on Wetland Hydrology

We relied on interannual variability in meteorological parameters during three highly contrasted study years as a means to assess the potential implications of climate change on alpine wetlands in our study area. The year 2017, which stood out as the driest of the three years considered, exhibited some of the most important climate changes currently underway in the Alps, including early snowmelt followed by the occurrence of summer heat waves (Figure S5). Our findings, which highlight an important regulatory role of snow cover parameters for the hydrology of alpine wetlands, align with results from the Canadian Rockies reporting that while alpine peat bogs are more resilient to drought due to increased soil water storage, mineral alpine wetlands and shallow pools with thin soils are likely to be more sensitive to interannual variability in climate [46]. It is important to point out that occasional drought occurrence is part of the usual functioning of semi-temporary basins in mountain environments, with wetland environments ranging along a continuum from permanent basins to temporary basins that dry shortly after snowmelt-out [11]. Our findings demonstrate that this continuum is sensitive to interannual variability in snow cover duration and summer temperature, with the implication that climate change could lead to a general shift toward increasingly dry and temporary alpine wetlands in the years ahead.

While we did not detect a positive effect of summer precipitation on midsummer water surface area, previous work has established the relevance of this parameter for the hydrology of alpine ponds and pools [11]. Given that the Chamonix valley is characterized by especially high average precipitation levels with respect to other alpine regions, it is also possible that summer precipitation has a stronger direct effect on water levels in drier systems such as the nearby southern French Alps. There may also be threshold effects, for example a minimum amount of received summer rainfall below which the risk of wetland drought dramatically increases, as was observed in the case of Australian wetlands during particularly hot and dry summers [16]. We also expect the effect of summer precipitation to depend in part on other climate parameters, and hypothesize that summer precipitation may become a critical hydrological input below certain snowpack levels or above certain temperatures. Last, while we did not test for the effect of geologic substrate on wetland hydrology or drought risk due to insufficient sample size across different bedrock types, we expect that the most important climate parameters may differ for example in limestone vs. granite watersheds. To address these questions, we recommend extending the remote sensing-based approach presented here, as well as field monitoring protocols of drought occurrence and amphibian phenology, at broader spatial scales in order to account for spatial non-stationarity in the relationships between meteorological parameters and wetland hydrology. Last, we propose that our method could be extended to estimate not only drought occurrence but also the duration and frequency of dry spells throughout the summer season.

4.3. Implications of Ongoing Climate Change for Alpine Wetland Habitat and Flora and Fauna

Twenty-first century scenarios of climate change for the European Alps predict ongoing and accelerating warming in future climate scenarios, including increasingly frequent and intense summer heat waves [47]. Summer precipitation is also expected to decrease by between 10 and 20% between now and 2050 in the Chamonix region depending on the scenario [48], suggesting that rainfall may become an increasingly important limiting factor in our study area in the years ahead. Furthermore, at 1500 m a.s.l., snow cover duration in the northern French Alps is expected to decrease with respect to the 1986–2005 reference period by approximately 3 weeks in 2030 and 4–5 weeks in 2050 [49]. Collectively, and in light of our findings, these predictions highlight increased drought risk for alpine wetlands in the coming years with strong implications for associated specialist flora and fauna. In terms of ecosystem services, reduced surface water and runoff linked to decreased snowpack, reduced summer rainfall and increased evapotranspiration, particularly in deglaciated watersheds, are projected to reduce summer stream flow and downstream water availability in alpine watersheds [50].

Our findings confirm the mortality of common frog tadpoles as a biotic response to drought, suggesting that consecutive years of dry conditions could affect the local viability of amphibian populations dependent on seasonally present water [22]. While, on the one hand, earlier snowmelt allows more time for tadpole development and growth in systems that are strongly limited by snow cover duration [51], on the other hand, reduced water availability associated with a thinner spring snowpack and smaller midsummer snow patches increases the risk of wetland drought during the summer (Figures 8 and 9). A study conducted in the Rocky Mountains suggests that wetland habitat diversity and connectivity increases the resilience of frog populations to interannual climate variability, with shallow ephemeral sites being more favorable in wet and cold years and deeper and more permanent wetlands being more favorable to in warm and dry years [52]. Indeed the low R-squared values of our structural equation models highlight the remarkable habitat heterogeneity and complexity of alpine wetland pools, with highly variable characteristics despite similar climate conditions. Local characteristics including microtopography, bedrock, and soil heterogeneity and water runoff networks undoubtedly account for a large portion of unexplained variance in our model, and this habitat diversity may allow for certain pools to remain viable despite an overall decrease in water availability. Finally, while we quantified drought risk for wetlands considered as independent spatial entities, further work should seek to assess the potential effects of habitat configuration and spatial proximity of wetlands on amphibian survival and population dynamics in response to climate variability.

Our use of NDVI to quantify interannual responses of wetland plant biomass did not show any significant differences in plant productivity during the three study years, including for 2017 (Figure 9). Although additional fieldwork would be necessary to test this hypothesis, our initial observations (Figure 9B) suggest that soils adjacent to wetlands retain moisture and remain saturated even during periods of drought, and therefore can provide locally favorable conditions for plant growth when elsewhere plant growth might be hampered by moisture availability. Accordingly, wetlands could provide important refugia for above-treeline vegetation in the warmer and drier years ahead, particularly for subalpine to low-alpine vegetation, whose growth has been shown to be negatively affected by heat wave summers [53]. More frequent and sustained droughts could also lead to increases in the local abundance of generalist plant species at the expense of specialist hygrophile vegetation, which was observed in the Swiss Alps following 10 years of monitoring wetland plant community composition [13]. Given the combined threat to specialist wetland plant species posed by climate change and competitive exclusion by generalist plants, we recommend using Sentinel-2 not only for hydrological monitoring of snow cover and surface water, but also to detect potential changes in both plant biomass and shrub cover [21], in combination with repeated field surveys of wetland plant diversity.

5. Conclusions

Our work introduces and validates a novel application of multitemporal spectral unmixing, used to quantify the seasonal hydrology of alpine wetlands in the northern French Alps. We demonstrate that decreased snowpack and hot summers increase drought risk for alpine wetlands, with strong implications for the habitat requirements of specialist wetland flora and fauna. Our approach has the potential to be readily upscaled to systematically monitor above-treeline mountain pools at the regional scale for the duration of the Sentinel-2 satellite mission. We recommend combining remote sensing-based monitoring with systematic field observations to further enhance our method and to improve our understanding of ground-level responses of wetland plant communities and amphibian populations to ongoing climate changes.

Supplementary Materials: The following are available online at http://www.mdpi.com/2072-4292/12/12/1959/s1, Figure S1. Photos of tadpole development and different phenological stages observed in the field at the Loriaz site (Figure 1D).Figure S2. Boxplots of error resulting from linear spectral unmixing for pixels overlapping alpine wetlands for (A) 2016, (B) 2017, and (C) 2018. Figure S3. Illustration of the method used to validate Sentinel-2 based estimates of water surface area with respect to ground observations. (A) Example 900 m^2 grid centered on a target wetland, composed of nine 10×10 m pixels, within which volunteers made visual estimates water surface area on the ground (B) Photo of ground observations underway, using a simple grid constructed using measuring tapes and 30 m lengths of string. Figure S4. Comparison of visual estimates of the surface area of target endmembers: water, snow, vegetation and rock. Figure S5. Contextualization of the three study years (2016, 2017, and 2018, shown in red) with respect to 30-year (1988–2018) anomalies of mean March snowpack height and the sum of growing degree days (>0 °C) in June and July. Figure S6. Boxplots of seasonal climate and snowpack variables for 2016, 2017, and 2018 extracted for elevation bands pertaining to studied alpine wetlands in the Mont-Blanc massif (1650–2850 m a.s.l.). Figure S7. Observed dates of phenological stages of the common frog tadpoles for a survey wetland in 2017 and 2018 (see Figure 1D for location). Table S1. Structural equation model results for all tested models.

Author Contributions: Conceptualization, B.Z.C. and A.D.; methodology, B.Z.C., M.H., M.B., I.L., and C.V.R.; software, B.Z.C. and I.L.; data curation, M.H. and A.D.; validation, B.Z.C. and C.V.R.; formal analysis, B.Z.C.; writing—original draft preparation, B.Z.C.; writing—review and editing, M.H., C.V.R., M.B., I.L., and A.D.; supervision, A.D. All authors have read and agreed to the published version of the manuscript.

Funding: This work was funded by the Agence de l'eau Rhône-Méditerranée Corse, with additional support from the Fondation Terre d'Initiatives Solidaires, the Fondation Caisse d'Epargne, the European Union, the Auvergne-Rhône-Alpes Region, the Haute-Savoie Department and the Communauté de Communes de la Vallée de Chamonix Mont-Blanc.

Acknowledgments: We are particularly grateful for help in the field and with high-resolution imagery from Guillaume Belissent. We would like to acknowledge the invaluable input and assistance from the rest of the CREA Mont-Blanc staff, including Irene Alvarez, Hillary Gerardi, and Julie Gavillet. Last, we would like to warmly thank the volunteers and interns who participated in identifying wetland sites in the field, as well as volunteers from the University of Colorado Boulder, participants from the 2019 Arcteryx Academy and local residents who carried out field observations of wetland water surface area. This research was conducted within the Long-Term Socio-Ecological Research (LTSER) platform Zone Atelier Alpes, a member of the eLTER network.

Conflicts of Interest: The authors declare no conflicts of interest. The funders had no role in the design of the study; in the collection, analyses, or interpretation of data; in the writing of the manuscript; or in the decision to publish the results.

References

1. Beniston, M. Mountain climates and climatic change: An overview of processes focusing on the European Alps. *Pure Appl. Geophys.* **2005**, *162*, 1587–1606. [CrossRef]
2. Gobiet, A.; Kotlarski, S.; Beniston, M.; Heinrich, G.; Rajczak, J.; Stoffel, M. 21st century climate change in the European Alps—A review. *Sci. Total Environ.* **2014**, *493*, 1138–1151. [CrossRef] [PubMed]
3. Corona-Lozada, M.C.; Morin, S.; Choler, P. Drought offsets the positive effect of summer heat waves on the canopy greenness of mountain grasslands. *Agric. For. Meteorol.* **2019**, *276*, 107617. [CrossRef]
4. Gardent, M.; Rabatel, A.; Dedieu, J.P.; Deline, P. Multitemporal glacier inventory of the French Alps from the late 1960s to the late 2000s. *Glob. Planet. Chang.* **2014**, *120*, 24–37. [CrossRef]
5. Huss, M. Extrapolating glacier mass balance to the mountain range scale: The European Alps 1900–2100. *Cryosphere Discuss.* **2012**, *6*, 1117–1156. [CrossRef]
6. Klein, G.; Vitasse, Y.; Rixen, C.; Marty, C.; Rebetez, M. Shorter snow cover duration since 1970 in the Swiss Alps due to earlier snowmelt more than to later snow onset. *Clim. Chang.* **2016**, *139*, 637–649. [CrossRef]
7. Steinbauer, M.J.; Grytnes, J.A.; Jurasinski, G.; Kulonen, A.; Lenoir, J.; Pauli, H.; Bjorkman, A.D. Accelerated increase in plant species richness on mountain summits is linked to warming. *Nature* **2018**, *556*, 231–234. [CrossRef]
8. Carlson, B.Z.; Corona, M.C.; Dentant, C.; Bonet, R.; Thuiller, W.; Choler, P. Observed long-term greening of alpine vegetation—A case study in the French Alps. *Environ. Res. Lett.* **2017**, *12*, 114006. [CrossRef]
9. Filippa, G.; Cremonese, E.; Galvagno, M.; Isabellon, M.; Bayle, A.; Choler, P.; Carlson, B.Z.; Gabellani, S.; Morra di Cella, U.; Migliavacca, M. Climatic Drivers of Greening Trends in the Alps. *Remote Sens.* **2019**, *11*, 2527. [CrossRef]
10. Rubel, F.; Brugger, K.; Haslinger, K.; Auer, I. The climate of the European Alps: Shift of very high-resolution Köppen-Geiger climate zones 1800–2100. *Meteorol. Z.* **2017**, *26*, 115–125. [CrossRef]
11. Wissinger, S.A.; Oertli, B.; Rosset, V. Invertebrate communities of alpine ponds. In *Invertebrates in Freshwater Wetlands*; Springer: Berlin/Heidelberg, Germany, 2016; pp. 55–103.
12. Hanzer, F.; Förster, K.; Nemec, J.; Strasser, U. Projected cryospheric and hydrological impacts of 21st century climate change in the Ötztal Alps (Austria) simulated using a physically based approach. *Hydrol. Earth Syst. Sci.* **2018**, *22*, 1593–1614. [CrossRef]
13. Moradi, H.; Fakheran, S.; Peintinger, M.; Bergamini, A.; Schmid, B.; Joshi, J. Profiteers of environmental change in the Swiss Alps: Increase of thermophilous and generalist plants in wetland ecosystems within the last 10 years. *Alp. Bot.* **2012**, *122*, 45–56. [CrossRef]
14. Blaustein, A.R.; Kiesecker, J.M. Complexity in conservation: Lessons from the global decline of amphibian populations. *Ecol. Lett.* **2002**, *5*, 597–608. [CrossRef]
15. Collins, J.P. Amphibian decline and extinction: What we know and what we need to learn. *Dis. Aquat. Org.* **2010**, *92*, 93–99. [CrossRef] [PubMed]
16. Stuart, S.N.; Chanson, J.S.; Cox, N.A.; Young, B.E.; Rodrigues, A.S.; Fischman, D.L.; Waller, R.W. Status and trends of amphibian declines and extinctions worldwide. *Science* **2004**, *306*, 1783–1786. [CrossRef] [PubMed]
17. Scheele, B.C.; Driscoll, D.A.; Fischer, J.; Hunter, D.A. Decline of an endangered amphibian during an extreme climatic event. *Ecosphere* **2012**, *3*, 1–15. [CrossRef]
18. Maltby, E.; Acreman, M.C. Ecosystem services of wetlands: Pathfinder for a new paradigm. *Hydrol. Sci. J.* **2011**, *56*, 1341–1359. [CrossRef]
19. Gascoin, S.; Grizonnet, M.; Bouchet, M.; Salgues, G.; Hagolle, O. Theia Snow collection: High-resolution operational snow cover maps from Sentinel-2 and Landsat-8 data. *Earth Syst. Sci. Data* **2019**, *11*, 493. [CrossRef]
20. Dedieu, J.P.; Carlson, B.Z.; Bigot, S.; Sirguey, P.; Vionnet, V.; Choler, P. On the importance of high-resolution time series of optical imagery for quantifying the effects of snow cover duration on alpine plant habitat. *Remote Sens.* **2016**, *8*, 481. [CrossRef]
21. Bayle, A.; Carlson, B.Z.; Thierion, V.; Isenmann, M.; Choler, P. Improved Mapping of Mountain Shrublands Using the Sentinel-2 Red-Edge Band. *Remote Sens.* **2019**, *11*, 2807. [CrossRef]

22. Tavernini, S.; Mura, G.; Rossetti, G. Factors influencing the seasonal phenology and composition of zooplankton communities in mountain temporary pools. *Int. Rev. Hydrobiol. A J. Cover. All Asp. Limnol. Mar. Biol.* **2005**, *90*, 358–375. [CrossRef]
23. Gao, B.C. NDWI—A normalized difference water index for remote sensing of vegetation liquid water from space. *Remote Sens. Environ.* **1996**, *58*, 257–266. [CrossRef]
24. Ozesmi, S.L.; Bauer, M.E. Satellite remote sensing of wetlands. *Wetl. Ecol. Manag.* **2002**, *10*, 381–402. [CrossRef]
25. Duffy, J.P.; Cunliffe, A.M.; DeBell, L.; Sandbrook, C.; Wich, S.A.; Shutler, J.D.; Myers-Smith, I.H.; Varela, M.R.; Anderson, K. Location, location, location: Considerations when using lightweight drones in challenging environments. *Remote Sens. Ecol. Conserv.* **2018**, *4*, 7–19. [CrossRef]
26. Sohn, Y.S.; McCoy, R.M. Mapping desert shrub rangeland using spectral unmixing and modeling spectral mixtures with TM data. *Photogramm. Eng. Remote Sens.* **1997**, *63*, 707–716.
27. Hostert, P.; Röder, A.; Hill, J. Coupling spectral unmixing and trend analysis for monitoring of long-term vegetation dynamics in Mediterranean rangelands. *Remote Sens. Environ.* **2003**, *87*, 183–197. [CrossRef]
28. Gudex-Cross, D.; Pontius, J.; Adams, A. Enhanced forest cover mapping using spectral unmixing and object-based classification of multi-temporal Landsat imagery. *Remote Sens. Environ.* **2017**, *196*, 193–204. [CrossRef]
29. De Asis, A.M.; Omasa, K.; Oki, K.; Shimizu, Y. Accuracy and applicability of linear spectral unmixing in delineating potential erosion areas in tropical watersheds. *Int. J. Remote Sens.* **2008**, *29*, 4151–4171. [CrossRef]
30. Huovinen, P.; Ramírez, J.; Gómez, I. Remote sensing of albedo-reducing snow algae and impurities in the Maritime Antarctica. *ISPRS J. Photogramm. Remote Sens.* **2018**, *146*, 507–517. [CrossRef]
31. Hagolle, O.; Huc, M.; Villa Pascual, D.; Dedieu, G. A multi-temporal and multi-spectral method to estimate aerosol optical thickness over land, for the atmospheric correction of FormoSat-2, LandSat, VENμS and Sentinel-2 images. *Remote Sens.* **2015**, *7*, 2668–2691. [CrossRef]
32. Hagolle, O.; Huc, M.; Desjardins, C.; Auer, S.; Richter, R. MAJA Algorithm Theoretical Basis Document. 2017. Available online: https://www.theia-land.fr/wp-content-theia/uploads/sites/2/2018/12/atbd_maja_071217.pdf (accessed on 4 April 2020). [CrossRef]
33. Dymond, J.R.; Shepherd, J.D. Correction of the topographic effect in remote sensing. *IEEE Trans. Geosci. Remote Sens.* **1999**, *37*, 2618–2619. [CrossRef]
34. Smith, M.O.; Ustin, S.L.; Adams, J.B.; Gillespie, A.R. Vegetation in deserts: I. A regional measure of abundance from multispectral images. *Remote Sens. Environ.* **1990**, *31*, 1–26. [CrossRef]
35. Dozier, J. Spectral signature of alpine snow cover from the Landsat Thematic Mapper. *Remote Sens. Environ.* **1989**, *28*, 9–22. [CrossRef]
36. Durand, Y.; Laternser, M.; Giraud, G.; Etchevers, P.; Lesaffre, B.; Mérindol, L. Reanalysis of 44 yr of climate in the French Alps (1958–2002): Methodology, model validation, climatology, and trends for air temperature and precipitation. *J. Appl. Meteorol. Climatol.* **2009**, *48*, 429–449. [CrossRef]
37. Durand, Y.; Giraud, G.; Laternser, M.; Etchevers, P.; Mérindol, L.; Lesaffre, B. Reanalysis of 47 years of climate in the French Alps (1958–2005): Climatology and trends for snow cover. *J. Appl. Meteorol. Climatol.* **2009**, *48*, 2487–2512. [CrossRef]
38. Vernay, M.; Lafaysse, M.; Hagenmuller, P.; Nheili, R.; Verfaillie, D.; Morin, S. The S2M meteorological and snow cover reanalysis in the French mountainous areas (1958–present). *AERIS* **2019**. [CrossRef]
39. Grace, J.B.; Bollen, K.A. Interpreting the results from multiple regression and structural equation models. *Bull. Ecol. Soc. Am.* **2005**, *86*, 283–295. [CrossRef]
40. Shipley, B. *Cause and Correlation in Biology: A User's Guide to Path Analysis, Structural Equations and Causal Inference with R*; Cambridge University Press: Cambridge, UK, 2016.
41. Grace, J.B. *Structural Equation Modeling and Natural Systems*; Cambridge University Press: Cambridge, UK, 2006.
42. Li, S.; Chen, X. A new bare-soil index for rapid mapping developing areas using Landsat 8 data. *Int. Arch. Photogramm. Remote Sens. Spat. Inf. Sci.* **2014**, *40*, 139. [CrossRef]
43. Gibbons, M.M.; McCarthy, T.K. Growth, maturation and survival of frogs *Rana temporaria* L. *Ecography* **1984**, *7*, 419–427. [CrossRef]

44. Dennison, P.E.; Roberts, D.A. Endmember selection for multiple endmember spectral mixture analysis using endmember average RMSE. *Remote Sens. Environ.* **2003**, *87*, 123–135. [CrossRef]
45. Liu, S.; Bruzzone, L.; Bovolo, F.; Du, P. Unsupervised multitemporal spectral unmixing for detecting multiple changes in hyperspectral images. *IEEE Trans. Geosci. Remote Sens.* **2016**, *54*, 2733–2748. [CrossRef]
46. Mercer, J.J. Insights into Mountain Wetland Resilience to Climate Change: An Evaluation of the Hydrological Processes Contributing to the Hydrodynamics of Alpine Wetlands in the Canadian Rocky Mountains. Ph.D. Thesis, University of Saskatchewan, Saskatoon, SK, Canada, 2018.
47. Jacob, D.; Petersen, J.; Eggert, B.; Alias, A.; Christensen, O.B.; Bouwer, L.M.; Georgopoulou, E. EURO-CORDEX: New high-resolution climate change projections for European impact research. *Reg. Environ. Chang.* **2014**, *14*, 563–578. [CrossRef]
48. Cremonese, E.; Carlson, B.; Filippa, G.; Pogliotti, P.; Alvarez, I.; Fosson, J.P.; Ravanel, L.; Delestrade, A. AdaPT Mont-Blanc Rapport Climat: Changements Climatiques Dans le Massif du Mont-Blanc et Impacts Sur Les Activités Humaines. Report Written in the Context of the AdaPT Mont-Blanc Projet and Financed by the European Union (Alcotra France-Italy 2014–2020). November 2019, p. 101. Available online: http://espace-mont-blanc.com/asset/rapportclimat.pdf (accessed on 4 April 2020).
49. Verfaillie, D.; Lafaysse, M.; Déqué, M.; Eckert, N.; Lejeune, Y.; Morin, S. Multi-component ensembles of future meteorological and natural snow conditions for 1500 m altitude in the Chartreuse mountain range, Northern French Alps. *Cryosphere* **2018**, *12*, 1249–1271. [CrossRef]
50. Huss, M.; Bookhagen, B.; Huggel, C.; Jacobsen, D.; Bradley, R.S.; Clague, J.J.; Vuille, M.; Buytaert, W.; Cayan, D.R.; Greenwood, G.; et al. Toward mountains without permanent snow and ice. *Earth's Future* **2017**, *5*, 418–435. [CrossRef]
51. Corn, P.S. Amphibian breeding and climate change: Importance of snow in the mountains. *Conserv. Biol.* **2003**, *17*, 622–625. [CrossRef]
52. McCaffery, R.M.; Maxell, B.A. Decreased winter severity increases viability of a montane frog population. *Proc. Natl. Acad. Sci. USA* **2010**, *107*, 8644–8649. [CrossRef]
53. Cremonese, E.; Filippa, G.; Galvagno, M.; Siniscalco, C.; Oddi, L.; di Cella, U.M.; Migliavacca, M. Heat wave hinders green wave: The impact of climate extreme on the phenology of a mountain grassland. *Agric. For. Meteorol.* **2017**, *247*, 320–330. [CrossRef]

remote sensing

MDPI

Article

Spatio-Temporal Variations and Driving Forces of Harmful Algal Blooms in Chaohu Lake: A Multi-Source Remote Sensing Approach

Jieying Ma [1], Shuanggen Jin [1,2], Jian Li [1,*], Yang He [2] and Wei Shang [2]

[1] School of Remote Sensing and Surveying Engineering, Nanjing University of Information Science and Technology, Nanjing 210044, China; mjy@nuist.edu.cn (J.M.); sgjin@shao.ac.cn (S.J.)
[2] Shanghai Astronomical Observatory, Chinese Academy of Sciences, Shanghai 200030, China; heyang2014@whu.edu.cn (Y.H.); shangwei@shao.ac.cn (W.S.)
* Correspondence: lijian@nuist.edu.cn; Tel.: +86-25-58235371

Abstract: Harmful algal blooms (hereafter HABs) pose significant threats to aquatic health and environmental safety. Although satellite remote sensing can monitor HABs at a large-scale, it is always a challenge to achieve both high spatial and high temporal resolution simultaneously with a single earth observation system (EOS) sensor, which is much needed for aquatic environment monitoring of inland lakes. This study proposes a multi-source remote sensing-based approach for HAB monitoring in Chaohu Lake, China, which integrates Terra/Aqua MODIS, Landsat 8 OLI, and Sentinel-2A/B MSI to attain high temporal and spatial resolution observations. According to the absorption characteristics and fluorescence peaks of HABs on remote sensing reflectance, the normalized difference vegetation index (NDVI) algorithm for MODIS, the floating algae index (FAI) and NDVI combined algorithm for Landsat 8, and the NDVI and chlorophyll reflection peak intensity index (ρ_{chl}) algorithm for Sentinel-2A/B MSI are used to extract HAB. The accuracies of the normalized difference vegetation index (NDVI), floating algae index (FAI), and chlorophyll reflection peak intensity index (ρ_{chl}) are 96.1%, 95.6%, and 93.8% with the RMSE values of 4.52, 2.43, 2.58 km^2, respectively. The combination of NDVI and ρ_{chl} can effectively avoid misidentification of water and algae mixed pixels. Results revealed that the HAB in Chaohu Lake breaks out from May to November; peaks in June, July, and August; and more frequently occurs in the western region. Analysis of the HAB's potential driving forces, including environmental and meteorological factors of temperature, rainfall, sunshine hours, and wind, indicated that higher temperatures and light rain favored this HAB. Wind is the primary factor in boosting the HAB's growth, and the variation of a HAB's surface in two days can reach up to 24.61%. Multi-source remote sensing provides higher observation frequency and more detailed spatial information on a HAB, particularly the HAB's long-short term changes in their area.

Keywords: HAB; multi-source remote sensing; MODIS; Landsat; sentinel; Chaohu Lake

Citation: Ma, J.; Jin, S.; Li, J.; He, Y.; Shang, W. Spatio-Temporal Variations and Driving Forces of Harmful Algal Blooms in Chaohu Lake: A Multi-Source Remote Sensing Approach. *Remote Sens.* **2021**, *13*, 427. https://doi.org/10.3390/rs13030427

Academic Editors: Giacomo De Carolis and SeungHyun Son

Received: 4 December 2020
Accepted: 19 January 2021
Published: 26 January 2021

1. Introduction

As a vital freshwater resource, lakes provide essential and diverse habitats and ecosystem functions, and play vital roles in climate regulation, global carbon, nutrient cycles, thereby contributing to the industrial, agricultural, and food industries around the lakes [1]. However, the aquatic environment has been put at risk by both climate change and anthropogenic factors [2,3]. Wastewater discharge, farmland drainage, soil erosion, and agricultural fertilization are also primary nutrient sources leading to lake eutrophication. Besides, nitrogen and phosphorus pollution from inefficient sewage treatment systems and agricultural practices threaten to increase pollution and cause inland lakes' eutrophication [4]. Lake eutrophication may cause a harmful algal bloom (HAB), which is widely distributed, adaptable, and destructive [5]. A HAB increases oxygen consumption in the water, releases toxins, degrades the water quality, and critically affects drinking water safety [4]. Comprehensive monitoring of HAB is vital in governing and repairing the lake environment [6], which has recently attracted more attention from both governments and the academic community.

Since both the environmental and meteorological factors may influence the breakout and spread of a HAB, it is important to study how these driving factors affect the HAB for effective management. Environmental factors, including the nutrients in water from fertilizer, agricultural nitrogen fixation, grain nitrogen, and feed nitrogen, are the primary sources of lake eutrophication [7–12], which have certain effects on HAB growth. Iron is an important component of the nitrate and nitrite reductase system, and its effect on enhancing the reduction efficiency and transfer rate of nitrate substances by algae is very observable [13]. Meteorological factors, including temperature, wind speed, precipitation, sunshine hours, etc., are also vital in HAB breakout. Previous research proved that the growth of cyanobacteria was directly proportional to the water temperature when greater than 18 °C, and that the activity of microcystis decreased when the temperature was greater than 30 °C [2], and HABs mostly occur in summer with proper temperature and sunshine hours. Variations in rainfall lead to a significant increase in nitrogen, which may lead to a HAB [14]. However, the influences of these factors on HAB are varied in different lakes, which requires further research in the region of interest.

It is challenging to capture the HAB dynamics using a conventional field sampling method due to the significant spatial-temporal variations of HAB [15]. Satellite remote sensing has been extensively used for monitoring the spatial coverage and temporal trends of HAB [16]. Many HAB detecting methods, including visual interpretation, supervised classification [17], single-band threshold [18], the spectral index method [19], and the water quality inversion method [16] have been developed. The visual interpretation delineates the HAB distribution using false-color composite satellite images [20], which is high-precision but low efficiency and is prone to personal misjudgment. The single threshold or spectral index methods, such as the normalized cyanobacteria index (NDI_CB) for Landsat-7 ETM+ [21] and *FAI* for Terra/Aqua MODIS, apply a single threshold for single or multiple bands data for HAB detection, which is simple and easy to implement [20,22]. Moreover, some research uses algal or chlorophyll concentration derived from remote sensing images to monitor HAB [23]. For example, the HABs were identified using chlorophyll inversion models on SeaWiFS from 1988 to 2002, on the Korean coast [24]. However, the uncertainties of these methods depend on regional water properties, sensor selection, and a threshold determination, which thus requires comprehensive assessments for method selection and implementation.

Among existing satellite images, Terra/Aqua MODIS imagery has been preferred for HAB monitoring due to its high temporal and spectral resolution [25]. However, the capabilities of Terra/Aqua MODIS are still limited by the low spatial resolution (250/500/1000 m), making it different to identify HABs in small and medium inland lakes [26]. For example, the optimal spatial resolution to monitor HAB in the Great Lakes is at most 50 m [27]. The Landsat TM/ETM+/OLI provides a higher spatial resolution (30 m), but its low revisit period (16 days) cannot track HAB's variations over time [22]. Sentinel-2A/B satellites launched on 23 June 2015 by the European Space Agency have wider spatial coverage and higher temporal resolution for monitoring of HABs [28]. Therefore, there is a pressing need for an effective and practical approach to capturing spatio-temporal variability of inland lake HAB integrating multi-source remote sensing techniques, which involves determining the appropriate algorithm and threshold for varied satellite sensors, and integration of HAB results.

Given this background, in this paper, multi-satellite images, including Sentinel-2A, Landsat 8 OLI, and Terra/Aqua MODIS, are used to monitor the spatial and temporal variations of HAB in Chaohu Lake, mostly its short-term variations. The proper algorithm was evaluated and adopted for different satellite sensors, and the accuracy and uncertainty were analyzed. Based on HAB results from multi-source data, the variations and driving forces of HAB in Chaohu Lake for environmental management are discussed.

2. Study Area and Data

2.1. Study Area

Chaohu Lake, located in Hefei City, Anhui Province, is the fifth largest freshwater lake in China (Figure 1, projection: Gauss—Kruger projection, geographic coordinate system: World Geodetic System 1984). The tributaries of Chaohu Lake mainly include the Nanfei River, Shiwuli River, Pai River, Hangbu River, Baishitian River, Zhao River, Yuxi River, and Shuangqiao River. Chaohu Lake has an inflow of 344.2 million m^3 and an outflow of 23 million m^3. The center of Chaohu Lake is located at 29°47′–31°16′ north and 115°45′–117°44′ east, with an average water depth of 2.89 m and an average annual lake temperature of about 20 °C [29]. The terrain around the lake is mostly mountains and hills, and the Chaohu Lake basin is cultivated mainly by rice, wheat, rape, cotton, and corn. The agricultural land around the lake makes it easily accumulate nutritive salt in the water, causing severe non-point source pollution, which caused the lake's external pollution load, mainly originating from the northwestern part of the basin [30,31]. Nutrients in farmland are mainly composed of phosphorus and nitrogen, and the inflow of total phosphorus and total nitrogen is one of the main reasons for the eutrophication of Chaohu Lake. Chaohu Lake has become one of the most eutrophic lakes in China [32]. The total phosphorus concentration was one of the main driving factors affecting Anabaena and microcystins' spatial and temporal distribution [33,34]. The farming period is from June to November. The average annual rainfall in Chaohu Lake is 224 mm, which drives the farmland nutrients to the lake during the farming period [35]. Moreover, the rain stirs up the mud at the bottom of Chaohu Lake, and large amounts of nutrient salts in the mud turn up, increasing the concentration of nutrient salts in Chaohu Lake. The total phosphorus content in Chaohu Lake is 0.131mg/L, and the total nitrogen content is 2.04 mg/L. The nitrogen and phosphorus ratio of optimum reproduction of the dominant species of HAB in Chaohu Lake was about 11.8:1 [36]. According to the monitoring data over the years, the ratio of nitrogen to phosphorus in Chaohu Lake is between 10:1 and 15:1, resulting in an outbreak situation of non-point source HAB [37]. When algae proliferate and die, they accelerate the consumption of dissolved oxygen in water, resulting in the death of many aquatic animals and plants, weakening the purification capacity of water, and causing severe harm to human health [5]. Therefore, it is essential to monitor the water environment with joint multi-source remote sensors.

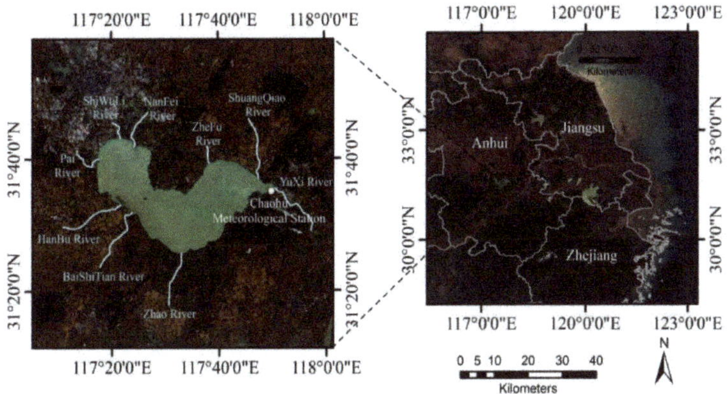

Figure 1. Location of Chaohu Lake.

2.2. Remote Sensing Data

A total of 420 images of Terra/Aqua MODIS L-1B data (MOD02) in 2019 were selected and downloaded from Earthdata's website (https://search.earthdata.nasa.gov/). Two images of Landsat 8 OLI (Level 1) were downloaded from the USGS official website of shared data (https://earthexplorer.usgs.gov/). A total of 16 images of Sentinel-2 MSI satellite data (L1C) were downloaded from the official website of ESA (https://scihub.copernicus.eu/). Clear and cloudless images were picked out (see Table 1) and preprocessed, including re-projection and geometric correction. Figure 2 shows the different cloudless products distributed in the space in 2019 so one can picture the time lag between the different satellite acquisitions.

Table 1. Multi-sensor data of the cloudless images of Chaohu Lake in 2019.

2019	Resolution	Revisit Period	May	June	July	August	September	October	November
Terra/MODIS	250 m	1 day	4	13	10	16	12	12	12
Aqua/MODIS	250 m	1 day	2	3	3	2	3	5	5
Landsat8 OLI	30 m	16 days	0	0	0	2	0	0	0
Sentinel-2A MSI	20 m	10 days	2	1	2	0	2	3	1
Sentinel-2B MSI	20 m	10 days	0	1	0	0	0	2	2
Total	-	-	8	17	13	20	17	19	18

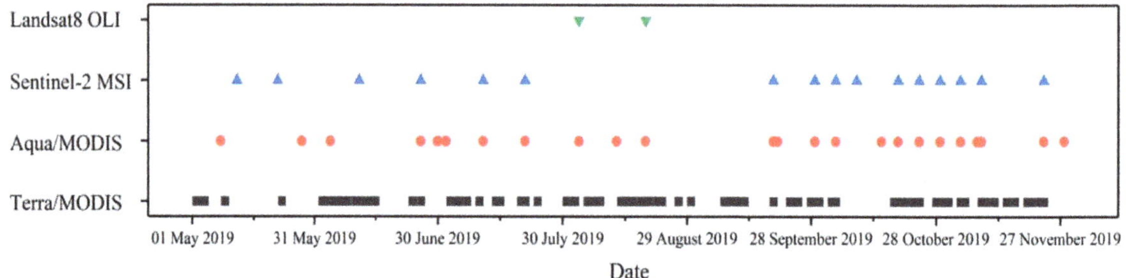

Figure 2. Annual distribution of cloudless images from multi-sensor data.

2.3. Environmental and Meteorological Data

The meteorological analysis data were obtained from the Meteorological Center of the National Meteorological Administration (http://www.cma.gov.cn/) (Figure 3). In 2019, Chaohu Meteorological Station's maximum sunshine hours, maximum temperature, and maximum wind speed occurred in May, July, and August, respectively. The variation range of wind speed was 0.5–6.4 m/s, the maximum number of sunshine hours was 12.9 h, and the time of direct sunlight was half a day. The average rainfall was 224 mm. The average maximum temperature was 33.9 °C.

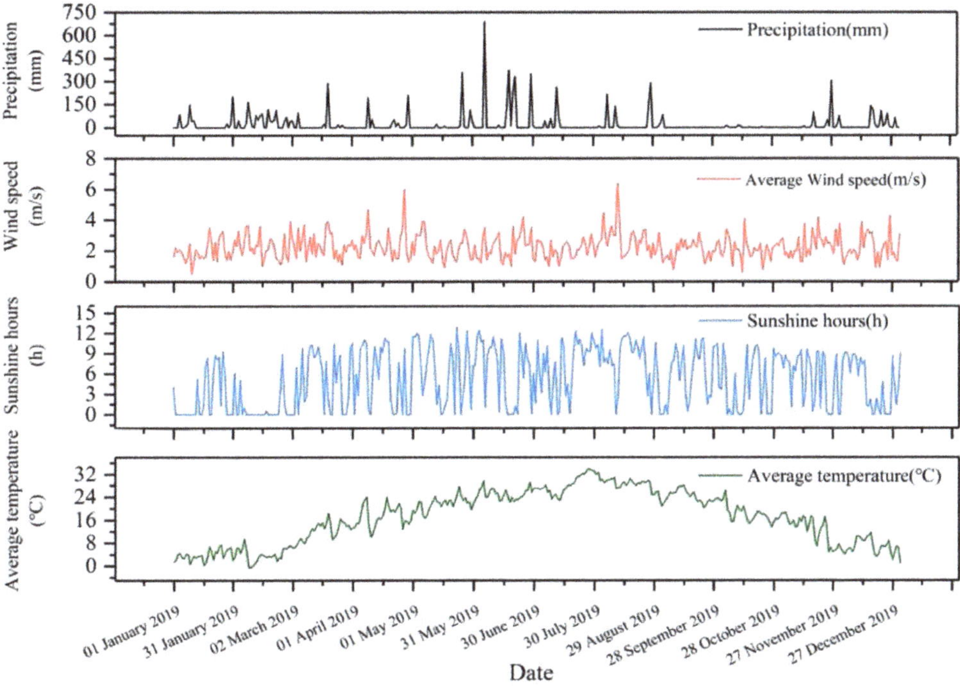

Figure 3. The variation diagram of rainfall, sunshine hours, average temperature, and precipitation in Chaohu Station in 2019.

3. Methods

Figure 4 is the technical flow chart of this paper, using which the original satellite data were obtained and preprocessed. The most appropriate algorithms were selected respectively for Sentinel-2 MSI, Terra/Aqua MODIS, and Landsat 8 OLI to obtain the distribution map of HAB in Chaohu Lake, and we checked the accuracy of the algorithms with visual interpretation results. Finally, the formation and distribution of HAB were analyzed by combining various meteorological factors.

3.1. Data Preprocessing

The preprocessing steps mainly included geometric correction, radiometric calibration, and atmospheric correction. Landsat-8 OLI and Terra/Aqua MODIS data were preprocessed using ENVI software (ENVI 5.3) to convert DN (digital number) values into TOA (top of atmosphere reflectance) radiance or reflectance after radiometric calibration, and then different atmospheric correction models were selected according to different data sources. The FLAASH atmospheric correction module (Fast Line-of-sight Atmospheric Analysis of Spectral Hypercubes) was adopted for Landsat 8 OLI, which was based on the MODTRAN-4 (Moderate Spectral Resolution Atmospheric Transmittance Algorithm and Computer Model) radiation transmission model, with high accuracy. It can maximally eliminate the influences of water vapor and aerosol scattering over case II waters, and has been successfully used in previous studies from Landsat 8 OLI [38,39]. MODIS images were atmospherically corrected using the dark-objects method [40–42]. The procedure was to select the relatively clean area as a region of interest in the eastern part of Chaohu Lake, and statistically analyze the pixel brightness value of each band, while using a non-zero pixel with a suddenly increased brightness value as the dark pixel value. The selected dark pixel was used as the distance luminance value for atmospheric correction. Sentinel-2A/B original L-1C images were mainly processed using SEN2COR (version: Sen2Cor-02.08.00-win64) for radiometric calibration and atmospheric correction. SEN2COR is a plug-in released by the European Space Agency (ESA) specifically for Sentinel-2 atmospheric calibration. The spectral curve of the image by SEN2COR with atmospheric correction of Sentinel-2 images is consistent with the trend of the actual spectral curve on the ground [43]. The reflectance after atmospheric correction was compared with the field spectra of 39 ground objects; R^2 was 0.82 and the root mean square error was 0.04 [44], indicating high accuracy. All the images selected in the experiment were mostly cloudless. Before determining the HAB, cloud-covered regions of the remote sensing images were made into a cloud mask product by the single-band threshold method to eliminate the influence of clouds [45].

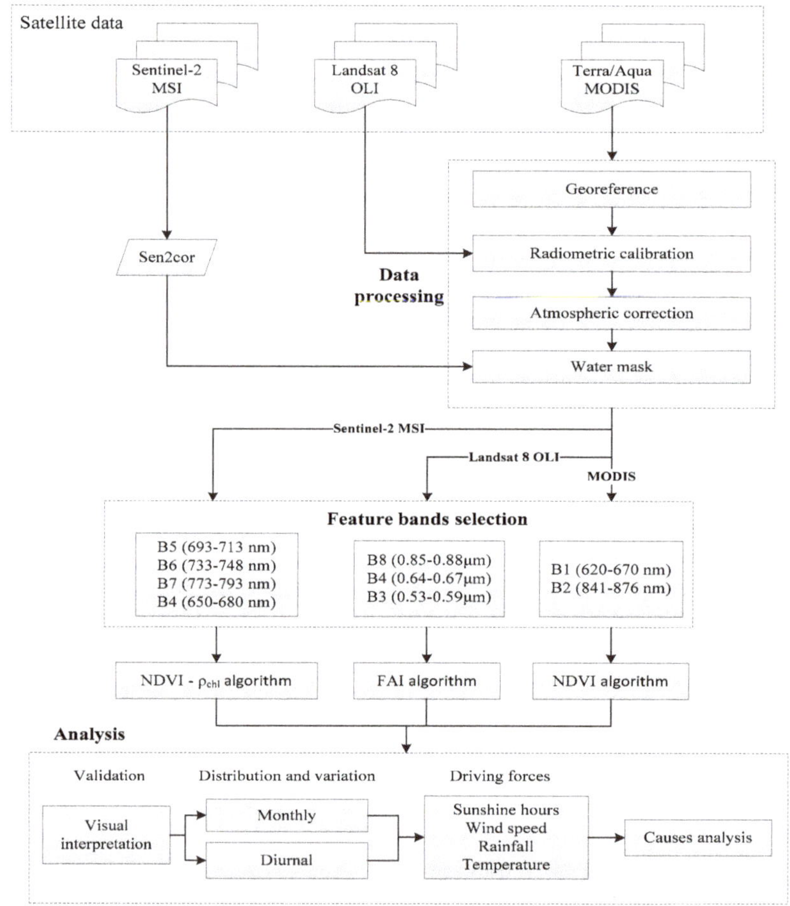

Figure 4. Technical route flow chart.

3.2. Extraction Algorithm of HAB

Algae in water would cause an absorption peak near the wavelength of 620–630 nm and a reflection peak at 650 nm, with a sharp increase in reflectance at around 700 nm [46]. High absorption in the red band by vegetation pigments and high reflection in the near-infrared band have been used for a long time to detect vegetation coverage, and eliminate some radiation errors. NDVI can reflect the background influence of the vegetation canopy. Therefore, the NDVI algorithm of MODIS was used for monitoring HAB in Chaohu Lake [47]. RGB band synthesis of Landsat/OLI B8 (0.85–0.88 μm), B4 (0.64–0.67 μm), and B3 (0.53–0.59 μm) renders HABs in a reddish color, in strong contrast with the bloom-free dark water, making it easy do distinguish bloom and non-bloom areas. Due to the influences of lake currents and wind, HAB areas generally present as elongated strips [48,49]. The *FAI* algorithm can eliminate the impact of the atmosphere by using the combination of these three bands. Compared with NDVI algorithm easily influenced by the observation environment, *FAI* would be suitable for the Landsat images. Unlike MODIS and Landsat 8, Sentinel-2 MSI was equipped with multiple spectral bands and 20 m ground resolution. Three special bands, B5 (693–713 nm), B6 (733–748 nm), and B7 (773–793 nm), are set for vegetation monitoring, which is also sensitive for HABs [50,51]. Therefore, the ρ_{chl}-NDVI algorithm is used for improving the accuracy of acquiring HAB in Chaohu Lake by fusing these 5 characteristic bands. Detailed descriptions of these algorithms are included in Figure 5.

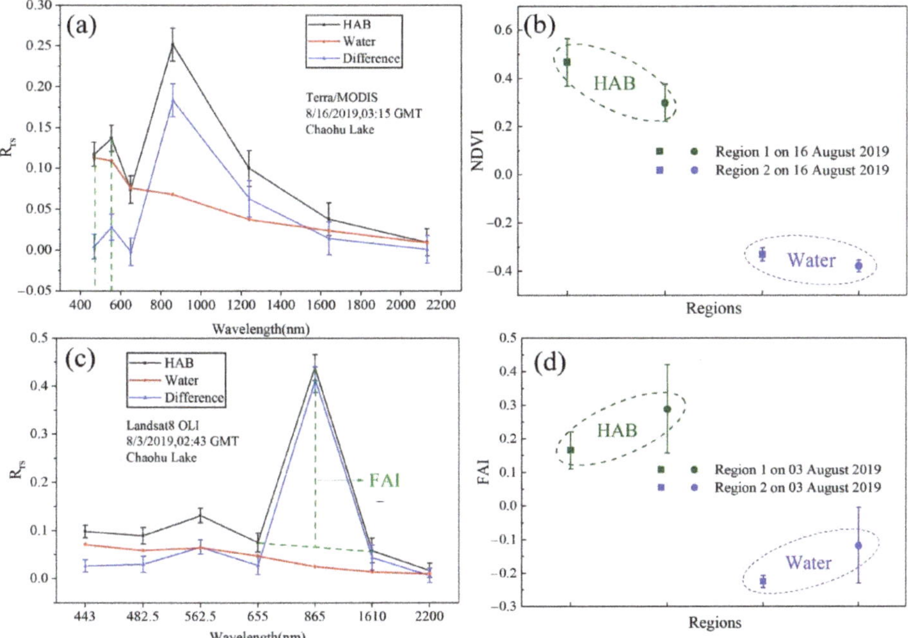

Figure 5. The interpretation of the algorithm and the results of each algorithm's reflectivity diagram (**a**,**c**,**e**) are the reflectances of a harmful algal bloom (HAB) and a nearby non-HAB lake, and (**b**,**d**,**f**) are means and standard deviations of HAB and non-HAB water lake reflectance. For the two regions of MODIS, Landsat8, and Sentinel-2 150 × 150 image pixels with 9359 × 9459, 7651 × 7791, and 0980 × 10,980 HAB classification pixels respectively.

3.2.1. Normalized Vegetation Index (NDVI)

Rouse [52] first used Landsat-1 MSS data to propose a NDVI based on the characteristic that the reflectivity of all vegetation increases dramatically near 700 nm. NDVI can reflect surface vegetation coverage [53]. Therefore, as the most common method, NDVI has been widely used in the study of algal extraction [54–56], which can eliminate the influences of terrain, shadow, and solar elevation angle [57]:

$$\text{NDVI} = \frac{\rho NIR - \rho RED}{\rho NIR + \rho RED} \quad (1)$$

where ρ_{RED} and ρ_{NIR} represent the reflectances of the red band and near-infrared band.

3.2.2. Floating Algae Index (FAI)

The floating algae index was first proposed by Hu [58]. *FAI* is defined as a linear spread of reflectivity in the near-infrared, red, and short-wave infrared regions, and can be applied to monitor proliferating algae, such as Ulva or Sargassum spp [59]. The observation results of this algorithm provide strong robustness. *FAI* is less affected by atmospheric environment, observation conditions, and water reflectivity absorption in the near-infrared band [60]. *FAI* is often used to identify dense HABs in marine and inland waters [61]. Therefore, the spectral information of the red band, near-infrared band, and short-wave infrared band can be used to correct the atmospheric effects [35]. The algorithm is as follows:

$$FAI = R_{NIR} - R'_{NIR} \tag{2}$$

$$R'_{NIR} = R_{RED} + (R_{SWIR} - R_{RED}) \times \frac{\lambda_{NIR} - \lambda_{RED}}{\lambda_{SWIR} - \lambda_{RED}} \tag{3}$$

where R_{RED}, R_{NIR}, and R_{SWIR} represent the reflectances of red, near-infrared, and short-wave infrared bands respectively; λ_{RED}, λ_{NIR}, and λ_{SWIR} represent the central wavelengths; and R'_{NIR} is the interpolating reflectance—namely, the reflectivity information of the infrared band can be obtained by linear interpolation of the red band and the short-wave infrared band.

The gradient contrast method was used for *FAI* algorithm to determine the threshold of HAB. The experimental results showed that $FAI < -0.01$ and $FAI > 0.02$ were non-bloom regions [19]. According to the average threshold value of the gradient diagram, $FAI > -0.002$ was finally determined to be the region of HAB.

3.2.3. Chlorophyll Reflection Peak Intensity Algorithm

Algae also contain chlorophyll, like land plants, so when the algae aggregates, the spectrum shows a vegetation-like characteristic [62,63]. Chlorophyll shows troughs at 420–500 nm (blue and violet light band) and 625 nm, and a small peak value is found at the central wavelength of the green band [36]. Based on the correlation between algae and chlorophyll concentration, the following model was constructed to identify the concentration of HAB [37,64]:

$$\rho_{\text{chl}} = \rho(560) - \frac{\rho(490) + \rho(665)}{2} \tag{4}$$

where $\rho(490)$, $\rho(560)$, and $\rho(665)$ correspond to the reflectivity of the blue, green, and vegetation red edge bands of the Sentinel-2A satellite.

3.3. Accuracy Assessment

To obtain the reference data or "truth data" for accuracy assessment of HAB detection from different satellite data, the visual interpretation method was used on false-color images. The verification data of the spatial distribution and area statistics of HAB were also obtained from the Department of the Ecological Environment of Anhui Province (http://sthjt.ah.gov.cn/), which have been checked through ground monitoring points, field investigations, and validation. The root mean square error (RMSE) and relative error (RE) were used to evaluate the accuracy of the HAB extractions using the NDVI algorithm. Additionally, the accuracies of different HAB detection methods were assessed using following indexes [17]:

Correct extraction rate (R) is the percentage of the extracted HAB area over the true data:

$$R = \frac{A_{\text{r}}}{A_{\text{truth}}} \times 100\% \tag{5}$$

Over-extraction rate (W) is the percentage of mixed extracted HAB area over the true data:

$$W = \frac{A_{\text{w}}}{A_{\text{truth}}} \times 100\% \tag{6}$$

Omitted extraction rate (M) is the percentage of the unextracted HAB area over the truth data:

$$M = \frac{A_{\mathrm{m}}}{A_{\mathrm{truth}}} \times 100\% \quad (7)$$

The reference data of HAB were denoted as A_{truth}. The area statistic of HAB extracted by various extraction methods was designated as A. The overlapping part of A and A_{truth} was regarded as the correct extracted part, which was denoted as A_{r}. The disjoint part of A is considered to be the extracted by mistake, which was denoted as A_{w}. In A_{truth}, the disjoint part was regarded as the missing part, which was denoted as A_{m}.

4. Results

Visual interpretation was analyzed based on 86 MODIS images and 2 Landsat images; 16 Sentinel-2 images were used to be the verification data to compare the accuracy of each extraction algorithm (Figure 6).

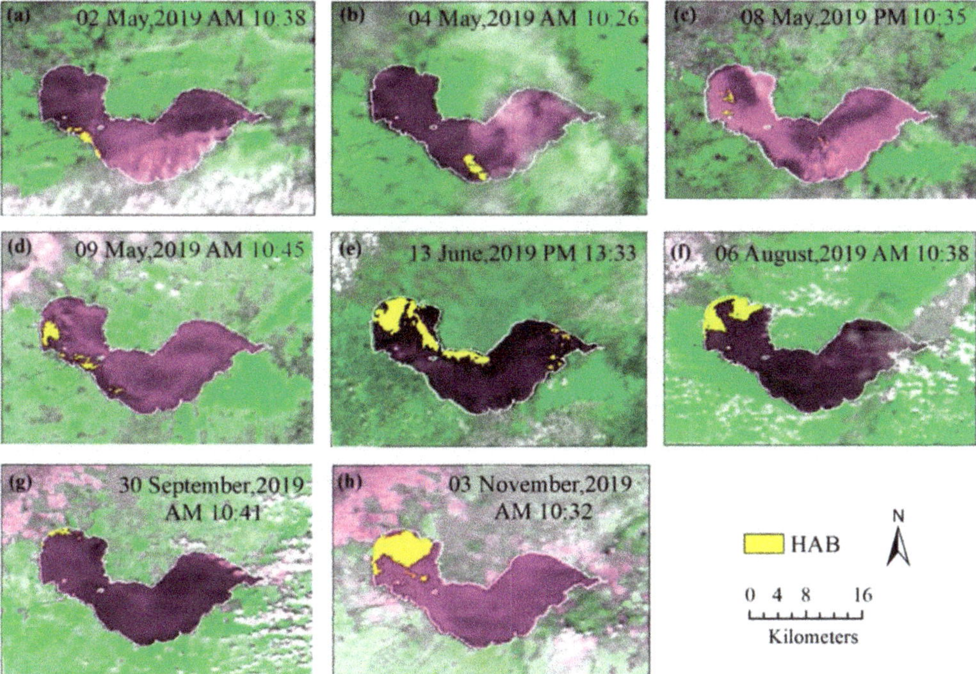

Figure 6. Spatial distribution of HAB in Chaohu Lake from visual interpretation.: (**a**) HAB distribution at 10:38 on 2 May, (**b**) HAB distribution at 10:26 on 4 May, (**c**) HAB distribution at 10:35 on 8 May, (**d**) HAB distribution at 10:45 on 9 May, (**e**) HAB distribution at 13:33 on 13 June, (**f**) HAB distribution at 10:38 on 6 August, (**g**) HAB distribution at 10:41 on 30 September, (**h**) HAB distribution at 10:32 on 3 November.

4.1. Accuracy of HAB Algorithms

Depending on the algorithm selection and analysis in Section 3.2, NDVI was used for MODIS to extract HAB. The comparison of NDVI and ρ_{chl} values showed that for a low concentration of HAB, the threshold for ρ_{chl} was 0.05, and the NDVI threshold was 0.24. For a moderate or high algae concentration, the threshold for ρ_{chl} was 0.09, and NDVI was larger than 0.68. Therefore, a pixel with an NDVI > 0 was first classified as a vegetation pixel, and then combined with $\rho_{\mathrm{chl}} > 0.05$ was judged as belonging to a HAB. NDVI < 0 and $\rho_{\mathrm{chl}} > 0.03$ was an "algal-water" suspension region and also judged as HAB.

The RMSE was 4.27 km^2 and RE was 15.9% when compared to HAB products reached by visual interpretation (Figure 7). For the significance test, $p < 0.05$, the results showed that the HAB region observed by satellite was consistent with the visual interpretation. Residual normal distribution of HAB areas extracted by MODIS and Sentinel-2 was showed on Figure 8, R^2 was 0.98 and 0.99 between MODIS, Sentinel-2 and visual interpretation, respectively. The Sentinel-2 MSI, MODIS, and Landsat 8 OLI randomly selected the day of the HAB outbreak, and a confusion matrix was used to evaluate the classification accuracy between the monitoring results and the visual interpretation (Table 2).

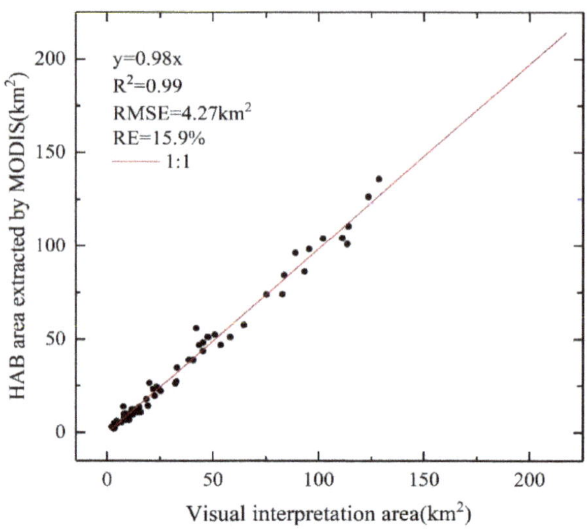

Figure 7. Comparison of HAB extracted by MODIS and visual interpretation.

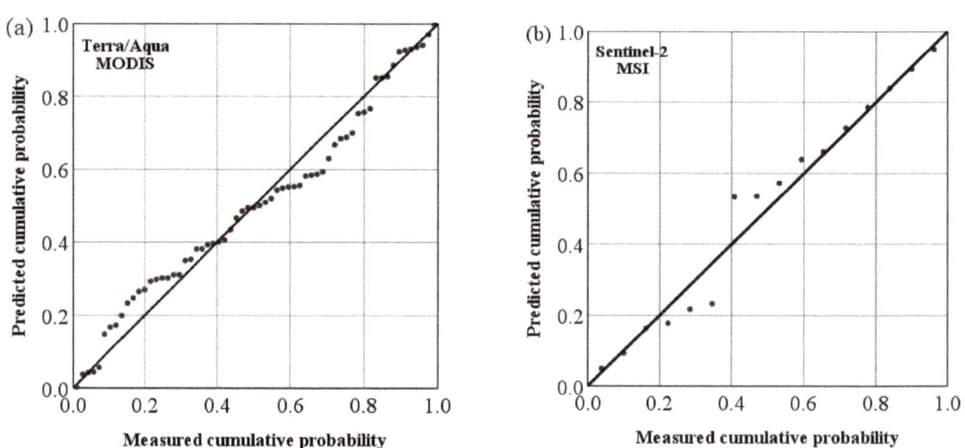

Figure 8. Residual normal distribution of HAB areas extracted by MODIS and Sentinel-2: (**a**) the coefficient of determination (R^2) was 0.98 between MODIS and visual interpretation, and (**b**) R^2 was 0.99 between Sentinel-2 and visual interpretation.

Table 2. The confusion matrix between the extraction results and visual interpretation.

	Class	HAB	Water	Cloud	Total	Accuracy
Sentinel-2 MSI 22 May 2019	HAB	78.47	0.02	0.00	0.92	Overall Accuracy = (17,105,160/17,189,550) 99.5091% Kappa Coefficient = 0.9002
	Water	20.57	99.76	0.26	97.49	
	Cloud	0.96	0.23	99.73	1.58	
	Total	100.00	100.00	100.00	100.00	
Landsat 8 OLI 19 August 2019	HAB	95.93	0.01	0.53	1.62	Overall Accuracy = (1,907,160/1,909,950) = 99.8539% Kappa Coefficient = 0.9972
	Water	4.07	99.99	8.07	97.51	
	Cloud	0.00	0.00	91.40	0.77	
	Total	100.00	100.00	100.00	100.00	
Terra/MODIS 1 August 2019	HAB	93.86	0.00	18.29	2.71	Overall Accuracy = (6124/6237) 98.1882% Kappa Coefficient = 0.8605
	Water	6.14	100.00	12.98	93.55	
	Cloud	0.00	0.00	68.73	3.74	
	Total	100.00	100.00	100.00	100.00	

NDVI and *FAI* were combined to detect HAB using Landsat 8 OLI images; NDVI and ρ_{chl} were combined for Sentinel-2 MSI data. Table 3 shows the accuracy evaluation results when compared to visual interpretation products, which demonstrated that HAB extracted by NDVI and *FAI* has a relatively correct extraction rate of about 95%. The RMSE of HAB from *FAI* algorithm was 0.56 km^2 and RE was 3.9%. However, the NDVI extraction method was affected by thin cloud or fog, and the cloud shadow was misidentified as a HAB. Moreover, NDVI method may miss pixels with lower algae concentrations, when compared with FAI. By comparing the extraction results on 3 August 2019 and 19 August 2019, the over-extraction areas of the NDVI method due to the mixed pixels and clouds were found to be 1.46 and 0.18 km^2, respectively. A comprehensive comparison shows that the extraction of HAB by the two methods was consistent, but the *FAI* method was better than NDVI at the details. Better results were obtained by combining NDVI with the chlorophyll reflection peak ρ_{chl}, especially for regions with lower concentrations of HAB. According to this method, the correct extraction rate of the Sentinel-2 data reached 96.01%, while RMSE and RE were 2.4 km^2 and 6.2%, respectively.

Table 3. Accuracy for HAB extraction of Landsat 8 OLI, Sentnel-2 MSI, and Terra/Aqua MODIS data.

	Extraction Method	Extracted Area (km^2)	Omission Area (km^2)	Overestimated Area (km^2)	Correct Area (km^2)	Missing Rate (%)	Over-Extraction Rate (%)	Correct Rate (%)
3 August 2019	*FAI*	16.30	0.02	3.31	12.98	0.12%	25.49%	99.88%
	NDVI	16.98	0.52	4.49	12.48	3.97%	34.57%	96.03%
	Visual interpretation	13.00						
4 October 2019	Sentinel		0.55	3.75	13.27	3.99%	27.12%	96.01%
	Visual interpretation	13.82						
3 November 2019	MODIS		1.84	18.88	10.21	3.92%	40.16%	96.08%
	Visual interpretation	47.02						

4.2. Monthly Variations of HAB

MODIS images were mainly used to track monthly HAB changes in Chaohu Lake in 2019 with the advantage of its high temporal resolution. The HAB in Chaohu Lake occurs between May and November (Figure 9). The northwestern part of Chaohu Lake is more seriously polluted by algae than the eastern, and the area of HAB reaches its maximum in July. The monthly frequency map is the ratio of the number of outbreaks in each region and month to the total numbers of the whole lake. The distribution frequency map indicates the probability of a HAB outbreak in each region of Chaohu Lake. Although HAB breaks out sometimes in a small region, they often occur in the west of the lake. According to the frequency distribution of inter-month HAB, it is increased in June and remains high from

June to November. The highest outbreak frequency occurs in the northwestern part of the lake in October, and the peak of distribution frequency of Chaohu Lake in the eastern lake appears in June.

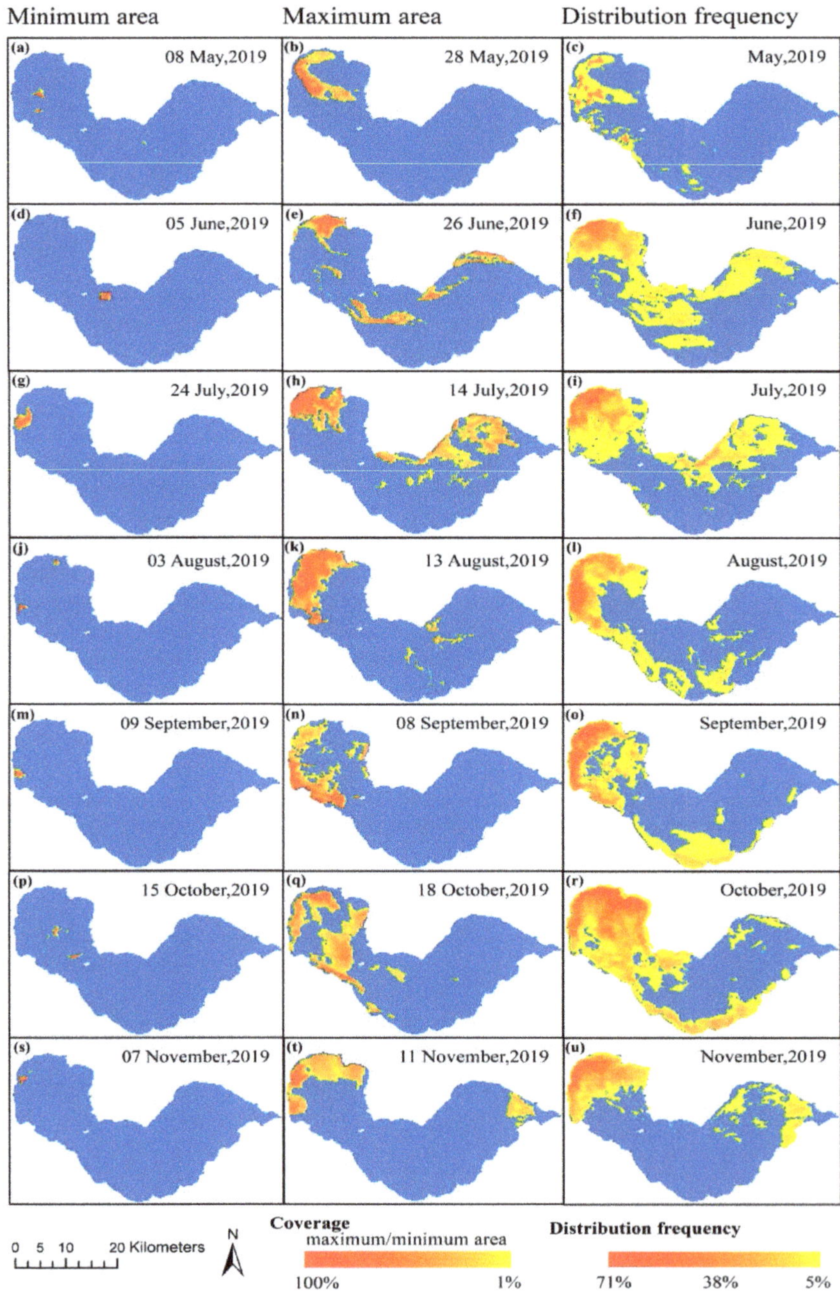

Figure 9. Spatial distribution of the minimum area, the maximum area, and the frequency of the monthly HAB in Chaohu Lake: (**a**,**d**,**g**,**j**,**m**,**p**,**s**) The minimum HAB area of each month from May to November, (**b**,**e**,**h**,**k**,**n**,**q**,**t**) The maximum HAB area of each month from May to November, (**c**,**f**,**i**,**l**,**o**,**r**,**u**) The HAB distribution frequency of each month from May to November.

The monthly coverage rates for the maximum, minimum, and average HAB area are shown in Figure 10. Adding up the maximum and the minimum area accounts for up to 50% of the total monthly HAB area in May, but the maximum HAB area was only 53.69 km^2. The average monthly coverage area was less than 20 km^2, which was the lowest in 2019. This indicates that the level of HAB in May was not serious. In contrast, from June to November, the maximum HAB area accounted for less than 25% to the total HAB area, and a HAB area exceeding 100 km^2 was always found in the mid-month. In July, the maximum area of HAB reached 217 km^2, accounting for 28.6% of the Chaohu Lake area, covering the northwestern and central parts of the lake. In 2019, the minimum HAB area was 1.625 km^2, which occurred on November 7, accounting for 0.2% of the total lake area. The average monthly coverage was lower than that in the period of HAB in Chaohu Lake (June to October). It indicated that the activity of HAB in Chaohu Lake began to decrease in November.

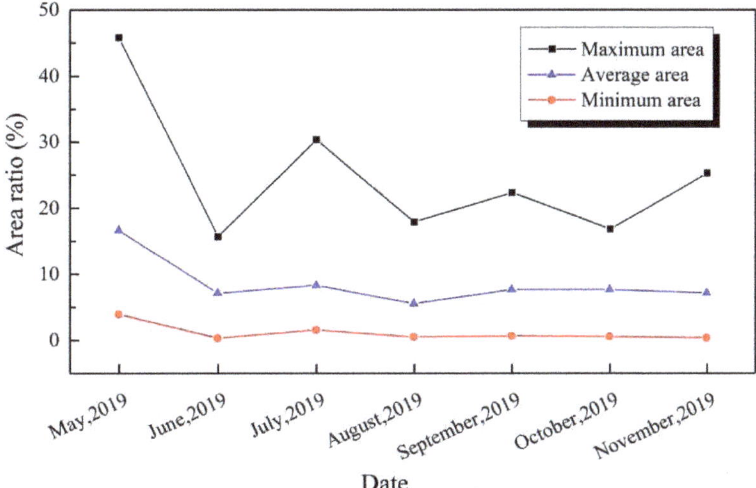

Figure 10. The ratios of the monthly maximum, minimum, and average HAB area to the total HAB area per month (total HAB area: monthly statistics of the area where HAB occurs each time).

4.3. Diurnal Variation of HAB

The spatial-temporal patterns of HABs can be easily affected by hydrology and meteorological factors, and thus induce dramatic variation in a short time, which requires high-frequency monitoring by the integration of a multi-satellite sensor. To reveal the diurnal variations of HAB in Chaohu Lake, multi-source satellite, including Sentinel-2 MSI, Landsat 8 OLI, and Terra/Aqua MODIS are integrated, as shown in Figure 11. While HAB is concentrated and stable, such as on 4 October 2019, the difference of extraction regions between Sentinel-2 MSI and Terra/MODIS is the smallest. Significant differences were observed due to the scattered distribution of HAB on June 26. In the surrounding areas with low algal density, MODIS had a lower spatial resolution; the result may be biased due to the mixed pixels. Since Terra/MODIS is the morning satellite, it passes through the equator from north to south at about 10:30 local time, and Aqua/MODIS is the afternoon satellite and passes through the equator from south to north at about 13:30 local time. According to the common influence of all factors, the monitored HAB area and distribution were different at different times of passing the territory. Besides, there will also be weather effects, such as the possibility of cloud cover in the afternoon compared with the morning in the study area, which will also have impacts on the extraction and identification of HAB.

The HAB diurnal changes from Landsat 8 and MODIS images on 19 August 2019 have no significant differences in the area and distribution. The morphology of HAB monitored by Terra (Figure 12a) was different from that of Landsat8 (Figure 12b), which may be due to the low quality (cloud coverage) of Terra/MODIS images on 3 August 2019. HAB region was disturbed by thin clouds, which could not represent the real distribution pattern at that time. The reliability of this result was also verified by the distribution diagram of bloom morphology in an Aqua image (Figure 12a). Compared with the result of Landsat 8 image (Figure 12d), the Aqua image (Figure 12e) result on August 3 showed a decrease in the distribution of HAB and a concentration increase in the coverage center. As the Terra image on August 3 was covered by clouds and fog, Figure 12 does not show the HAB distribution in the morning.

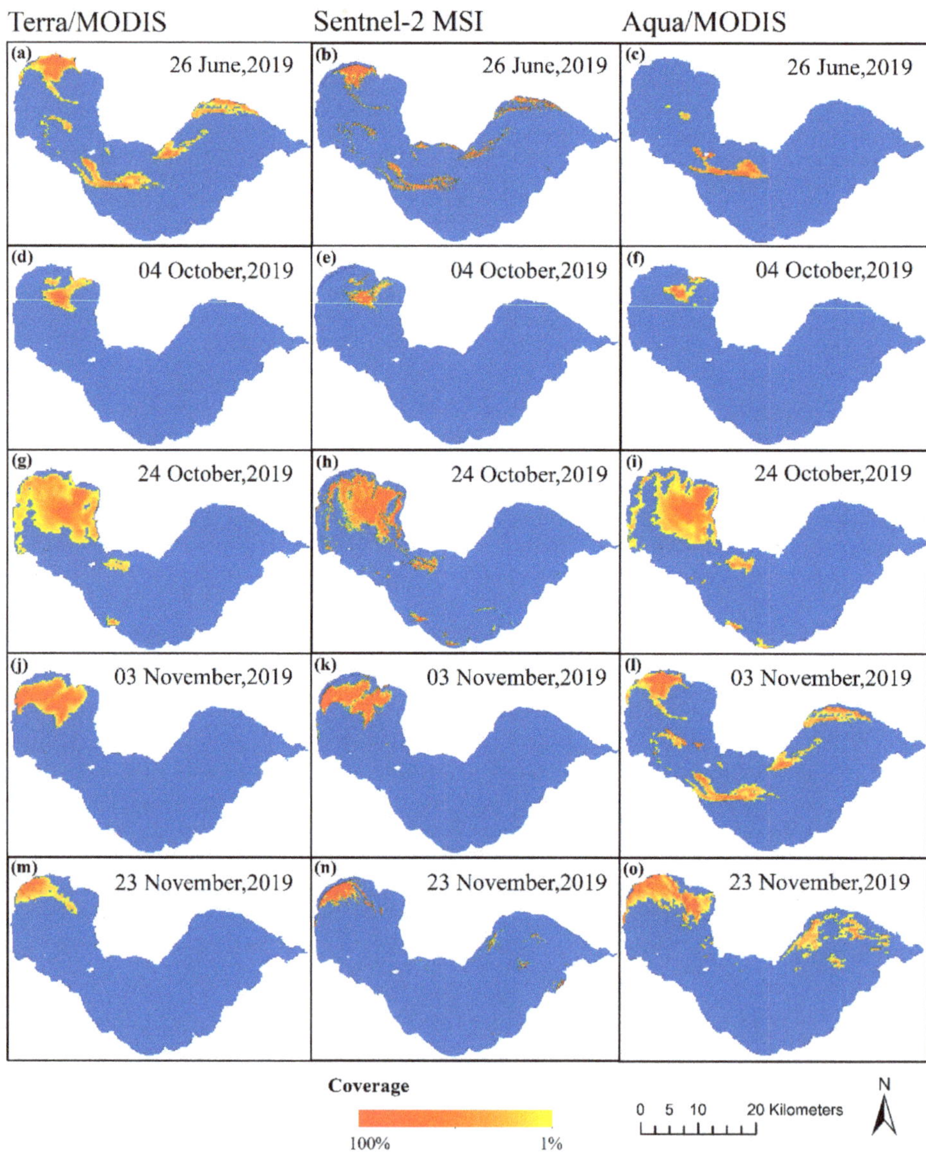

Figure 11. Variations of HAB from multi-satellite data: (**a**,**d**,**g**,**j**,**m**) results from Terra/MODIS, (**b**,**e**,**h**,**k**,**n**) results from Sentinel-2 MSI, (**c**,**f**,**i**,**l**,**o**) results from Aqua/MODIS.

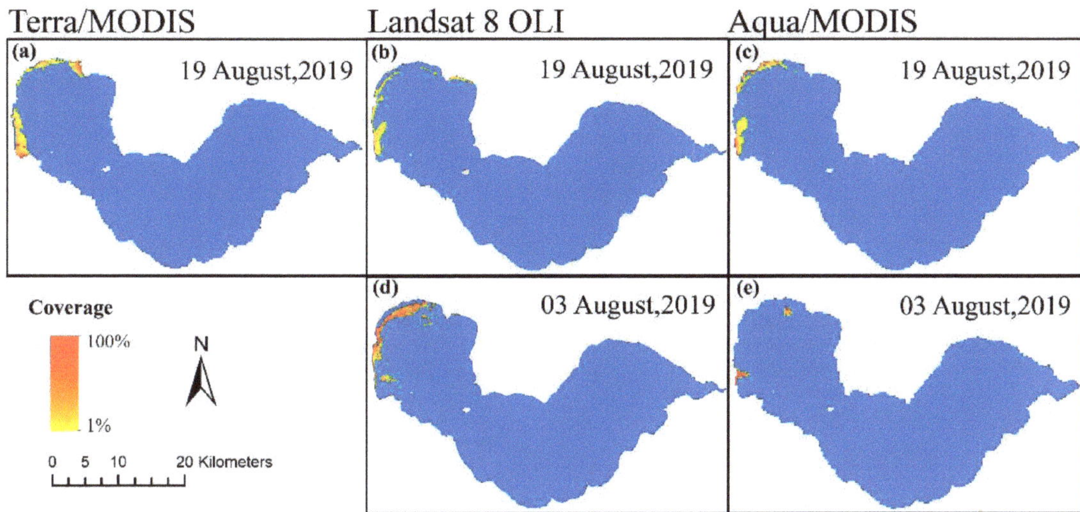

Figure 12. Harmful algal blooms from Terra/MODIS (**a**), Landsat 8 OLI (**b**,**d**), and Aqua/MODIS (**c**,**e**).

5. Discussion

5.1. Driving Forces of HAB

The driving forces for the breakout of HAB are of great concern for HAB control and management. Among many factors, the temperature, rainfall, sunshine hours, wind, radiation, etc., have drawn great attention [1]. Some previous research demonstrated that the degree of HAB is positively correlated with temperature, sunshine hours, and global radiation changes, and negatively correlated with precipitation and wind speed [65]. Our results showed similar results on the correlation between the HAB areas and both temperature and sunshine hours, but the R^2 was quite low (<0.05). However, increased temperature promotes the growth of HABs, and colder months may delay the occurrence of HAB [66]. It can be seen that the maximum and minimum areas of Chaohu Lake HAB in July were higher than in other months (Figure 13). The maximum, average, and minimum values of the HAB area in August and September were close. However, the number of hours of sunshine in September was 77.5 h lower than that in August. The low sunshine hours made it difficult for algae to reproduce and grow through photosynthesis, which inhibited the accumulation and explosion of large areas of HAB. However, too much sunshine will make algae inactive and also inhibit HAB growth. This is consistent with the conclusions from Zhang's research demonstrating that under high temperatures and with many sunshine hours, there will be no large-scale HAB [67,68]. Therefore, appropriate sunshine hours and temperature can promote the photosynthesis of algae.

The effect of precipitation showed a weak negative correlation with the HAB. The HAB on rainy days of August 3 and 7 was decreased by 79.3% and 61.3%, respectively, when compared with the previous days. This may indicate that the rainfall may dilute or inhibit the occurrence of HABs. HAB was often found on days after scattered rain, such as on 27 May, 26 August, and 17 October. In contrast, the total precipitation in September was half of that of August, and the scattered rain provided favorable conditions for the growth and reproduction of algae. Therefore, the rainfall was the main driving force of the monthly variations of the HAB from July to September. However, May–June rainfall is the highest and most frequent, which reduces the temperature of the water surface, and also reduces the density of algae and the concentrations of nutrients, making the probability of the occurrence of HABs only slightly increased in June compared with May. Rainfall decreased in July, the temperature increased, and the occurrence of HAB increased sharply. Therefore, the low occurrence of HABs in June was caused by precipitation. Based on the analysis of previous data, it was found that the period of highest temperature is inconsistent with the month with the highest probability of HAB, and atmospheric temperature is the main meteorological factor affecting HAB [69,70]. From mid-July to mid-August, Chaohu Lake's temperature in 2019 reached its annual maximum and the average daily sunshine hours were all over 8 h. However, due to the hysteresis effect [71] of temperature on the response of HAB in Chaohu Lake, the precipitation mainly occurred from June to mid-July. Much rain in June transports the nutrients from the catchment area as the non-point source. The algae in July with the highest maximum area is due to the inflow during June. The effect of nutrient supply appears with a time lag because the controlling factor is temperature. Even with a high concentration of nutrients, insufficient temperature regulates blooming.

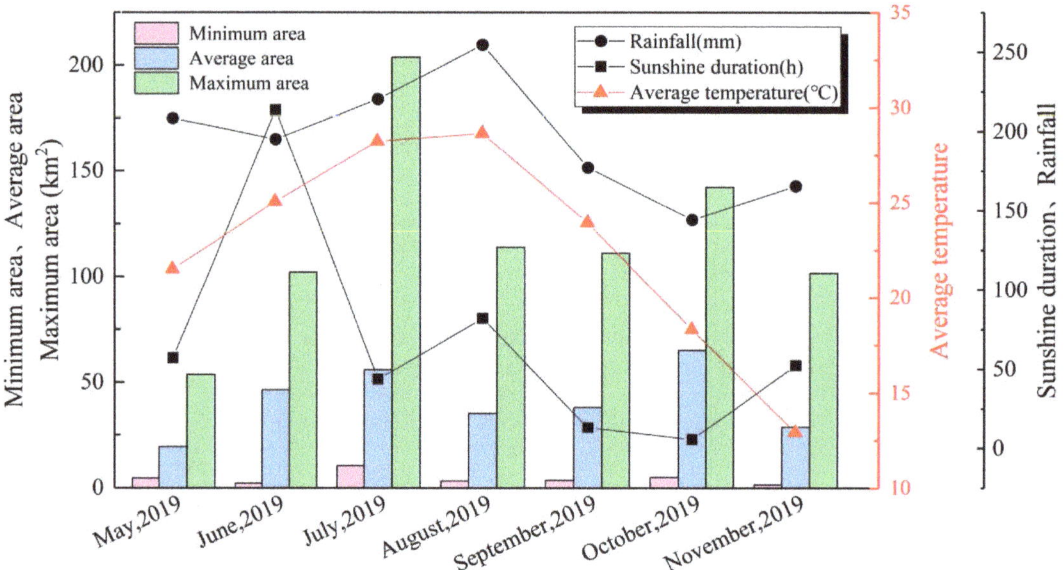

Figure 13. Chart of the minimum area, average monthly area, maximum area, average monthly temperature, sunshine hours, and precipitation of Chaohu Lake HABs.

The impact of wind speed on HAB showed a highly significant, positive correlation ($R^2 = 0.383$, $p < 0.01$). The wind direction map of Chaohu Lake in 2019 can be seen in Figure 14. A previous study revealed that when the average wind speed was larger than 3.8 m/s, the wind waves stirred the algal particles, causing them to sink, and reduced HAB concentration [72,73]. During the study period, only two days of HAB occurred with average wind speed greater than or equal to 3.8 m/s. The HAB area on August 12 was 4.8 km^2 (average wind speed of 4 m/s, average temperature of 28 °C), and the next day it was 113.94 km^2 (average wind speed of 1.5 m/s, average temperature of 29 °C). The solar radiation was similar, with sufficient sunshine hours (>9 h), but the HAB area was quite different. This indicated that the wind stirred up the algae particles so that the algae could not accumulate and sink, leading to a decrease in the HAB area. Moreover, appropriate wind speed and wind direction caused the HAB on the surface of Chaohu Lake to move toward the direction of the wind and accumulate. The results show that wind speed is an essential factor for the HAB outbreak and spread in Chaohu Lake. Prevailing winds in summer cause the shore water to converge on the northwest corner. The movement of water is not conducive to the material exchange on the surface of the flow field, which makes significant differences in the eutrophication pollution of algae of the whole lake [28]. Therefore, the frequency of HAB is the highest in the northwest of Chaohu Lake. There is counter-clockwise circulation in the vicinity of Zhefu River in the eastern Chaohu Lake and clockwise circulation in the vicinity of Zhao River [28], which brings N, P, and other nutrients to the northeast of Chaohu Lake and near the middle of the lake, and the nutrients concentrate, resulting in many of HABs. Chaohu sluice, connecting the southeastern part of Chaohu Lake with Yuxi River, has a certain influence on the flow field near the eastern part of Chaohu Lake and plays a favorable role in the exchange of HAB with the outside.

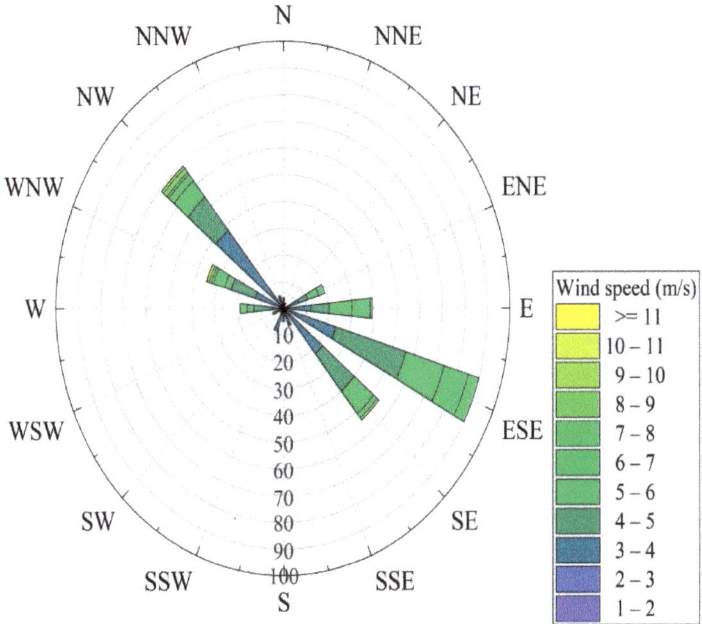

Figure 14. The wind direction map in Chaohu Lake 2019 (10, 20, 30, etc., indicate the number of days).

The average wind speed on 24 October 2019 was 1.5 m/s, which was less than the critical value (3.8 m/s) for algae aggregation and movement [74]. Additionally, the maximum wind speed was 3.8 m/s. As can be seen in the HAB distribution in Chaohu Lake detected by Terra and Aqua on October 24 (Figure 15a,c), the HAB in the central part of Chaohu Lake is gradually moving in the east–southeast direction, in line with the maximum wind speed direction 14 (that is, the west-northwest direction). On 8 November 2019, the maximum wind speed was 2.9 m/s, and the maximum wind speed direction was 3 (that is, a northeasterly). HAB areas in Chaohu Lake were 31.75 and 43.6 km^2, respectively, detected by Terra and Aqua. There was a low average wind speed (2 m/s) on Chaohu Lake on that day, which caused the algal particles to turn up and accumulate on the surface. The changes of HAB were also affected by wind waves, leading to the distribution location moving to the southwest (Figure 15b,d). Therefore, the multi-source remote sensing data can effectively monitor and reveal the diurnal change and development process of HAB.

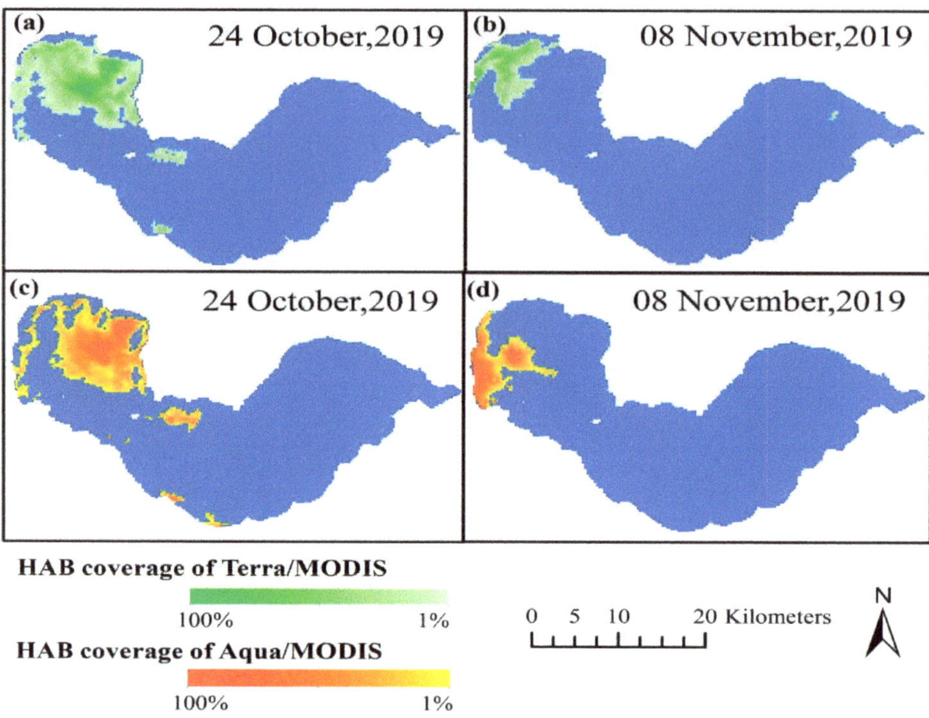

Figure 15. The intraday variations of HAB distribution: (**a**) Terra/MODIS image of the HAB distribution on 24 October, (**b**) Terra/MODIS image of the HAB distribution on 8 Novermber, (**c**) Aqua/MODIS image of the HAB distribution on 24 October, (**d**) Aqua/MODIS image of the HAB distribution on 8 Novermber.

5.2. Advantages of Multi-Source Satellite Remote Sensing

MODIS satellites with moderate spatial resolution have been widely used in monitoring HABs in large water bodies. However, the identification of HABs by moderate spatial resolution is limited in small inland water bodies or reservoirs and even has a large accuracy error. Due to the moderate spatial resolution, the boundary of a HAB identified by MODIS data is fuzzy, and the recognition ability of low-concentration HAB is low, leading to large uncertainties for monitoring HABs of a small inland lake. Sentinel-2 images, with a spatial resolution of 20 m, could significantly improve the identification accuracy and spatial details of HAB. For a concentrated outbreak area (Figure 16h), MODIS satellite has a relatively good performance in extracting HABs, but its ability to define the boundary of a HAB's area is weak. The error extraction rate is 40%, which is relatively high. Therefore, in the same timeframe, the extraction of HABs by combining multi-source data can verify and correct the extraction results of moderate-resolution images.

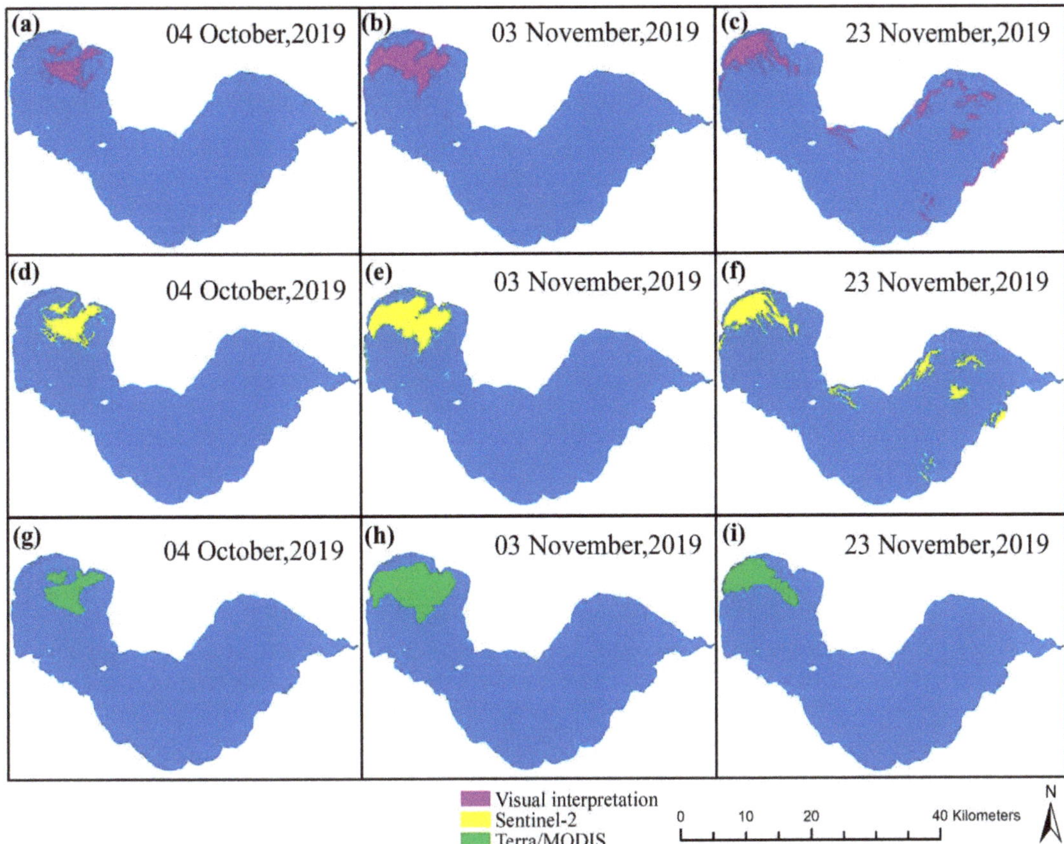

Figure 16. Comparison of the extraction results among Sentinel-2, Terra/MODIS, and visual interpretation: (**a**,**d**,**g**) Results of HAB extraction from Visual interpretation, Sentinel-2, Terra/MODIS on October 4, (**b**,**e**,**h**) Results of HAB extraction from Visual interpretation, Sentinel-2, Terra/MODIS on 3 November, (**c**,**f**,**i**) Results of HAB extraction from Visual interpretation, Sentinel-2, Terra/MODIS on 23 November.

In addition, remote sensing technology still makes it difficult to meet the requirements of high spatial-temporal resolution using a single satellite, especially for HABs with dramatic variations both spatially and temporally. To achieve both high spatial and high temporal resolution, multi-source satellite integration is an effective method to monitor the HABs in Chaohu Lake. Combined use of Terra/Aqua MODIS, Sentinel 2 MSI, and Landsat 8 OLI could provide more than three times per day monitoring of HAB, which is more efficient and accurate. For instance, parts of HAB information would be missed if only one satellite dataset was used; e.g., on 23 November 2019, some areas of HAB on the eastern part of Chaohu Lake would have been ignored by Terra image. By making full use of the advantages of multi-source images and monitoring the diurnal or long time scale changes of HAB in Chaohu Lake, they can learn from each other and make up for their shortcomings. Compared with single remote sensing data, more objective and accurate results were obtained.

6. Conclusions

Satellite remote sensing provides great potential to contribute significantly to the need for monitoring the HABs at a large scale; however, a multi-source remote sensing-based approach is preferred to achieve high temporal and spatial resolution observations of the

HABs, such as the integration of Terra/Aqua MODIS, Landsat 8 OLI, and Sentinel-2A/B MSI. With the advantage of the high temporal resolution, MODIS data are efficient in tracking the inter-monthly variations and distributions of HABs. In contrast, the integrated multi-satellite data provide the possibility to grasp the breakout and spread, especially the diurnal change of a given HAB, which is more objective and accurate than results from one single satellite's monitoring, as shown in the case of the Chaohu Lake. To obtain reliable HAB monitoring results, it is crucial to determine an appropriate HAB detection method considering the spectral characteristics of HABs and the band settings of different satellite sensors, and our study proved that NDVI is suitable for MODIS; NDVI and *FAI* combined for Landsat 8 OLI; and the NDVI and ρ_{chl} combined for Sentinel-2 MSI data. Besides, analysis of driving forces of HAB, including environmental and meteorological factors of temperature, rainfall, sunshine hours, and wind, indicated that higher temperatures and light rain favored HAB. The wind is the main factor in boosting a HAB's growth. Multi-source remote sensing provides higher measurement frequency and more detailed spatial information on the HAB, particularly the HAB's long-short term variations. The results can be used as baseline data to evaluate the lake's HAB and water quality management in the future.

Author Contributions: Conceptualization, J.M. and S.J.; methodology, J.M., J.L., Y.H. and S.J.; software, J.M.; validation, J.L. and S.J.; formal analysis, J.M.; investigation, J.M.; resources, J.M. and S.J.; data curation, J.M.; writing—original draft preparation, J.M. and S.J.; writing—review and editing, J.L., W.S. and S.J.; visualization, J.M.; supervision, S.J.; project administration, S.J.; funding acquisition, S.J. All authors have read and agreed to the published version of the manuscript.

Funding: This work was funded by the Strategic Priority Research Program Project of the Chinese Academy of Sciences (grant number XDA23040100), the Startup Foundation for Introducing Talent of NUIST (grant number 2018R037), and the Jiangsu Province Distinguished Professor Project (grant number R2018T20).

Institutional Review Board Statement: Not applicable.

Informed Consent Statement: Informed consent was obtained from all subjects involved in the study.

Data Availability Statement: The data presented in this study are available on request from the corresponding website.

Acknowledgments: The authors would like to acknowledge the data provided by the National Aeronautics and Space Administration (NASA), European Space Agency (ESA), and Planet Labs.

Conflicts of Interest: The authors declare no conflict of interest.

References

1. Wang, K. Current status of lake ecological environmental quality in China and countermeasures and suggestions. *World Environ.* **2018**, *2*, 16–18.
2. Jeremy, T.W.; Kevin, H.W.; Jason, C.D.; Rubenstein, E.M.; Rober, A.R. Hot and toxic: Tempera-ture regulates microcystin release from cyanobacteria. *Sci. Total Environ.* **2018**, *610–611*, 786.
3. Xiao, C.; Xi, P.; Tang, T.; Chen, L. Influence of the mid-route of south-to-north water transfer project on the water quality of the mid-lower reach of Hanjiang river. *J. Saf. Environ.* **2009**, *9*, 82–84. (In Chinese)
4. Li, S.M.; Liu, J.P.; Song, K.S.; Liang, C.; Gao, J. Analysis of temporal and spatial variation characteristics and driving factors of blue alga blooms in Chaohu Lake based on Landsat images. *Resour. Environ. Yangtze Basin* **2019**, *28*, 205–213.
5. Allison, A.O.; Randy, A.D.; Michae, L.D. The upside-down river: Reservoirs, algal blooms, and tributaries affect temporal and spatial patterns in nitrogen and phosphorus in the Klamath river, USA. *J. Hydrol.* **2014**, *519*, 164–176.
6. Guo, N.N.; Qi, Y.K.; Meng, S.L.; Chen, J.C. Research progress of eutrophic lake restoration technology. *Chin. Agric. Sci. Bull.* **2019**, *35*, 72–79.
7. Lu, W.K.; Yu, L.X.; Ou, X.K.; Li, F.R. Relationship between the occurrence frequency of cyanobacteria bloom and meteorological factors in Dianchi Lake. *J. Lake Sci.* **2017**, *29*, 534–545.
8. Huang, C.C.; Zhang, L.L.; Li, Y.M.; Lin, C.; Huang, T.; Zhang, M.; Zhu, A.; Yang, H.; Wang, X. Carbon and nitrogen burial in a plateau lake during eutrophication and phytoplankton blooms. *Sci. Total Environ.* **2018**, *616–617*, 296–304. [CrossRef]

9. Howarth, R.W.; Billen, G.; Swaney, D.; Townsend, A.; Jaworski, N.; Lajtha, K.; Downing, J.A.; Elmgren, R.; Caraco, N.; Jordan, T.; et al. Regional nitrogen budgets and riverine N&P fluxes for the drainages to the North Atlantic ocean: Natural and human influences. *Biogeochemistry* **1996**, *35*, 75–139.
10. Paerl, H.W.; Xu, H.; McCarthy, M.J.; Zhu, G.; Qin, B.; Li, Y.; Gardner, W.S. Controlling harmful cyanobacterial blooms in a hyper-eutrophic lake (lake Taihu, China): The need for a dual nutrient (N&P) management strategy. *Water Res.* **2011**, *45*, 1973–1983.
11. Xavier, S.P.; Vicente, E.; Urrego, P.; Pereira-Sandoval, M.; Ruíz-Verdú, A.; Delegido, J.; Soria, J.M.; Moreno, J. Remote sensing of cyanobacterial blooms in a hypertrophic lagoon (Albufera of València, Eastern Iberian Peninsula) using multitemporal Sentinel-2 images. *Sci. Total Environ.* **2019**, *698*, 134305.
12. Qian, K.; Liu, X.; Chen, Y. Effects of water level fluctuation on phytoplankton succession in Poyang lake, China–A five-year study. *Ecohydrol. Hydrobiol.* **2016**, *16*, 175–184. [CrossRef]
13. Dai, G.F.; Peng, N.Y.; Zhong, J.Y.; Yang, P.; Zou, B.; Chen, H.; Lou, Q.; Fang, Y.; Zhang, W. Effect of metals on microcystin abundance and environmental fate. *Environ. Pollut.* **2017**, *226*, 154–162. [CrossRef] [PubMed]
14. Sinha, E.; Michalak, A.M.; Balaji, V. Eutrophication will increase during the 21st century as a result of precipitation changes. *Science* **2017**, *357*, 405–408. [CrossRef] [PubMed]
15. Sha, H.M.; Li, X.S.; Yang, W.B.; Li, J.L. Preliminary study on MODIS satellite remote sensing monitoring of cyanobacteria in Taihu lake. *Trans. Oceanol. Limnol.* **2009**, *3*, 9–16.
16. Zhang, D.Y.; Yin, X.; She, B.; Ding, Y.W.; Liang, D.; Huang, L.S.; Zhao, J.L.; Gao, Y.B. Study on the monitoring of blue algae to bloom in Chaohu Lake with multi-source satellite remote sensing data. *Infrared Laser Eng.* **2019**, *48*, 303–314.
17. Xia, X.R. Information Extraction of Cyanobacteria Blooms in Taihu Lake from Domestic Satellite Remote Sensing Images. Master's Thesis, Nanjing Normal University, Nanjing, China, 2014.
18. Reinart, A.; Kutser, T. Comparison of different satellite sensors in detecting cyanobacterial bloom events in the Baltic sea. *Remote Sens. Environ.* **2006**, *102*, 74–85. [CrossRef]
19. Hu, C.; Lee, Z.; Ma, R.; Yu, K.; Li, D.Q.; Shang, S.L. Moderate resolution imaging spectroradiometer (MODIS) observations of cyanobacteria blooms in Taihu lake, China. *J. Geophys. Res. Oceans.* **2010**, *115*, C04002. [CrossRef]
20. Lin, Y.; Pan, C.; Chen, Y.Y.; Ren, W.W. Blue algae bloom recognition method based on remote sensing image spectrum analysis. *J. Tongji Univ. Nat. Sci.* **2011**, *39*, 1247–1252.
21. Randolph, K.; Wilson, J.; Tedesco, L.; Li, L.; Pascual, D.L.; Soyeux, E. Hyperspectral remote sensing of cyanobacteria in turbid productive water using optically active pigments, chlorophyll a, and phycocyanin. *Remote Sens. Environ.* **2008**, *112*, 4009–4019. [CrossRef]
22. Kahru, M.; Mitchell, B.G.; Diaz, A. Using MODIS medium-resolution bands to monitor harmful algal blooms. *Proc. SPIE-Int. Soc. Opt. Eng.* **2005**, *5885*. [CrossRef]
23. Vincent, R.K.; Qin, X.; Mckay, R.M.L.; Miner, J.; Czajkowski, K.; Savino, J. Phycocyanin detection from LANDSAT TM data for mapping cyanobacterial blooms in Lake Erie. *Remote Sens Environ.* **2004**, *89*, 381–392. [CrossRef]
24. Jin, Y.; Zhang, Y.; Niu, Z.C.; Jiang, S. The application of CCD data from the HJ-1 satellite in remote sensing monitoring of cyanobacteria blooms in Taihu Lake. *Admin. Tech. Environ. Monit.* **2010**, *22*, 53–56, 66.
25. Chen, Y.; Dai, J.F. Recognition method of cyanobacteria blooms in Taihu lake based on remote sensing data. *J. Lake Sci.* **2008**, *2*, 179–183.
26. Duan, H.T.; Zhang, S.X.; Zhang, Y.Z. Remote sensing monitoring method for cyanobacteria blooms in Taihu lake. *J. Lake Sci.* **2008**, *2*, 145–152.
27. Lekki, J.; Deutsch, E.; Sayers, M.; Bosse, K.; Anderson, R.; Tokarsa, R.; Sawtell, R. Determining remote sensing spatial resolution requirements for the monitoring of harmful algal blooms in the Great Lakes. *J. Great Lak. Res.* **2019**, *45*, 434–443. [CrossRef]
28. Zhang, J.; Chen, L.Q.; Chen, X.L.; Tian, L.Q. HJ-1B and Landsat satellite cyanobacteria bloom monitoring capability evaluation –taking Erhai Lake as an example. *J. Water Res. Water Eng.* **2016**, *27*, 38–43.
29. Zhang, M.; Shi, X.L.; Yang, Z.; Chen, K.N. Analysis of the variation trend of water quality in Chaohu lake from 2012 to 2018 and suggestions on cyanobacteria prevention and control. *J. Lake Sci.* **2020**, *32*, 11–20.
30. Shang, G.P.; Shang, J.C. Spatial and temporal variations of eutrophication in western Chaohu lake, China. *Environ. Monit. Assess.* **2007**, *130*, 99–109. [CrossRef]
31. Zhang, M.; Kong, F.X. The eutrophication process, spatial distribution, and management strategy of Chaohu lake (1984–2013). *J. Lake Sci.* **2015**, *27*, 791–798.
32. Yin, H.B.; Deng, J.C.; Shao, S.G.; Gao, F.; Gao, J.F.; Fan, C.X. Distribution characteristics and toxicity assessment of heavy metals in the sediments of lake Chaohu, China. *Environ. Monit. Assess.* **2011**, *179*, 431–442. [CrossRef] [PubMed]
33. Jiang, Y.J.; He, W.; Liu, W.X.; Qin, N. The seasonal and spatial variations of phytoplankton community and their correlation with environmental factors in a large eutrophic Chinese lake (Lake Chaohu). *Ecol. Indic.* **2014**, *40*, 58–67. [CrossRef]
34. Zhang, M.; Zhang, Y.; Yang, Z.; Wei, L.; Yang, W.; Chen, C.; Kong, F. Spatial and seasonal shifts in bloom-forming cyanobacteria in lake Chaohu: Patterns and driving factors. *Phycol. Res.* **2016**, *64*, 44–55. [CrossRef]
35. Wang, J.; Guo, X.S.; Wang, Y.Q.; Ding, S.W. Characteristics of runoff loss of farmland nutrients under different tillage and fertilization methods in Chaohu lake Basin. *J. Soil Water Conserv.* **2012**, *26*, 6–11.

36. Wu, C. Quantitative Analysis of Influencing Factors of Eutrophication in the West Half of Chaohu Lake. Master's Thesis, Anhui Agricultural University, Anhui, China, 2009.
37. Han, X.Y.; Sun, P. *Discussion on the Causes of Eutrophication of Chaohu Lake*; Anhui Science Association Annual Meeting and Anhui Water Conservancy Forum: Anhui, China, 2009; pp. 42–46.
38. Benjamin, P.; Abhishek, K.; Deepak, R. A novel cross-satellite based assessment of the spatio-temporal development of a cyanobacterial harmful algal bloom. *Int. J. Appl. Earth Obs. Geoinf.* **2018**, *66*, 69–81.
39. Li, X.W.; Niu, Z.C.; Jiang, S.; Jin, Y. Satellite image-based remote sensing intensity index and grade division algorithm design for cyanobacterial blooms in Taihu lake. *Adm. Techn. Envir. Monit.* **2011**, *23*, 23–30.
40. Dong, D.D.; Sun, J.Y.; Huang, Z.J.; Song, Q.; Wang, C.L. Temporal and spatial distribution of cyanobacterial blooms in Chaohu lake based on MODIS data. *Yangtze River* **2019**, *50*, 49–51.
41. He, X.; Bai, Y.; Pan, D.; Tang, J.; Wang, D. Atmospheric correction of satellite ocean color imagery using the ultraviolet wavelength for highly turbid waters. *Opt. Express* **2012**, *20*, 20754. [CrossRef]
42. Wang, M.H. Remote sensing of the ocean contributions from ultraviolet to near-infrared using the shortwave infrared bands: Simulations. *Appl. Optics* **2007**, *46*, 1535–1547. [CrossRef]
43. Kaire, T.; Tiit, K.; Alo, L.; Margot, S.; Birgot, P.; Tiina, N. First experiences in mapping lake water quality parameters with sentinel-2 MSI imagery. *Remote Sens.* **2016**, *8*, 640.
44. Su, W.; Zhang, M.Z.; Jiang, K.P.; Zhu, D.H.; Huang, J.X.; Wang, P.X. Sentinel-2 satellite image atmospheric correction method. *Acta Opt. Sin.* **2018**, *38*, 322–331.
45. Gu, J.; Yan, M.; Zhang, X.J.; Zou, S.J.; Cao, H.P.; Zhang, L. Study on the method of extracting water information by Gaofen-1 image. *Agricult. Technol.* **2018**, *38*, 24–26.
46. Ma, R.H.; Kong, W.J.; Duan, H.T.; Zhang, S.X. Estimation of phycocyanin content during the outbreak of cyanobacteria in Taihu lake based on MODIS images. *Chin. J. Environ. Sci.* **2009**, *3*, 32–38.
47. Ma, M.X. High-Precision Calculation Method and Application of Cyanobacterial Bloom Area in Taihu Lake Based on MODIS image. Master's Thesis, Nanjing University, Nanjing, China, 2014.
48. Oyama, Y.; Fukushima, T.; Matsushita, B.; Matsuzaki, H.; Kamiya, K.; Kobinata, H. Monitoring levels of cyanobacterial blooms using the visual cyanobacteria index (VCI) and floating algae index (FAI). *Int. J. Appl. Earth Obs.* **2015**, *38*, 335–348. [CrossRef]
49. Prangsma, G.J.; Roozekrans, J.N. Using NOAA AVHRR imagery in assessing water quality parameters. *Int. J. Remote Sens.* **1989**, *10*, 811–818. [CrossRef]
50. Kahru, M.; Leppänen, J.M.; Rud, O. Cyanobacterial blooms cause heating of the sea surface. *Mar. Ecol. Prog. Ser.* **1993**, *101*, 1–7. [CrossRef]
51. Ho, J.C.; Stumpf, R.P.; Bridgeman, T.B.; Michalak, A.M. Using Landsat to extend the historical record of lacustrine phytoplankton blooms: A lake Erie case study. *Remote Sens. Environ.* **2017**, *191*, 273–285. [CrossRef]
52. Rouse, J.W. Monitoring the Vernal Advancement and Retrogradation (Green Wave Effect) of Natural Vegetation. Nasa/gsfct Type Final Rep; 1974; p. E74-10113. Available online: https://ntrs.nasa.gov/citations/19740004927 (accessed on 3 December 2020).
53. Wynne, T.T.; Stumpf, R.P.; Tomlinson, M.C.; Warner, R.A.; Tester, P.A.; Dyble, J.; Fahnenstiel, G.L. Relating spectral shape to cyanobacterial blooms in the Laurentian great lakes. *Int. J. Remote Sens.* **2008**, *29*, 3665–3672. [CrossRef]
54. Huete, A.; Didan, K.; Miura, T.; Rodriguez, E.P.; Gao, X.; Ferreira, L.G. Overview of the radiometric and biophysical performance of the MODIS vegetation indices. *Remote Sens. Environ.* **2002**, *83*, 195–213. [CrossRef]
55. Li, X.W. Landsat-7 SLC-OFF ETM remote sensing data download and its application in cyanobacteria bloom monitoring in Taihu Lake. *Adm. Tech. Environ. Monit.* **2009**, *3*, 54–57.
56. Binding, C.E.; Greenberg, T.A.; Bukata, R.P. The MERIS Maximum Chlorophyll Index; its merits and limitations for inland water algal bloom monitoring. *J. Great Lakes Res.* **2013**, *39* (Suppl. 1), 100–107. [CrossRef]
57. Liu, H.Q.; Huete, A. A feedback based modification of the NDVI to minimize canopy background and atmospheric noise. *IEEE Trans. Geosci. Remote Sens.* **1995**, *33*, 457–465. [CrossRef]
58. Hu, C. A novel ocean color index to detect floating algae in the global oceans. *Remote Sens. Environ.* **2009**, *113*, 2118–2129. [CrossRef]
59. Hu, C.; Cannizzaro, J.; Carder, K.L.; Muller-Karger, F.E.; Hardy, R. Remote detection of Trichodesmium blooms in optically complex coastal waters: Examples with MODIS full-spectral data. *Remote Sens. Environ.* **2010**, *114*, 2048–2058. [CrossRef]
60. Shanmugam, P.; Ahn, Y.H.; Ram, P.S. SeaWiFS sensing of hazardous algal blooms and their underlying mechanisms in shelf-slope waters of the Northwest Pacific during summer. *Remote Sens. Environ.* **2008**, *112*, 3248–3270. [CrossRef]
61. Binding, C.E.; Greenberg, T.A.; Mccullough, G.; Watson, S.B.; Page, E. An analysis of satellite-derived chlorophyll and algal bloom indices on lake Winnipeg. *J. Great Lakes Res.* **2018**, *44*, 436–446. [CrossRef]
62. Shi, K.; Zhang, Y.L.; Qin, B.Q.; Zhou, B. Remote sensing of cyanobacterial blooms in inland waters: Present knowledge and future challenges. *Sci. Bull.* **2019**, *64*, 1540–1556. [CrossRef]
63. Baillarin, S.; Aimé, M.; Cécile, D.; Martimort, P.; Petrucci, B.; Lachérade, S.; Isola, C.; Duca, R.; Spoto, F.; Henry, P.; et al. Sentinel-2 level 1 products and image processing performances. In Proceedings of the ISPRS—International Archives of the Photogrammetry Remote Sensing and Spatial Information Sciences, Munich, Germany, 22–27 July 2012; pp. 197–202.
64. Wei, G.F.; Tang, D.L.; Wang, S. Distribution of chlorophyll and harmful algal blooms (HABs): A review on space-based studies in the coastal environments of Chinese marginal seas. *Adv. Space Res.* **2008**, *41*, 12–19. [CrossRef]

65. Zhang, M.; Duan, H.T.; Shi, X.L.; Yu, Y.; Kong, F. Contributions of meteorology to the phenology of cyanobacterial blooms: Implications for future climate change. *Water Res.* **2011**, *46*, 442–452. [CrossRef]
66. Zhang, Y.L.; Shi, K.; Liu, J.J.; Deng, J.M.; Qin, B.Q.; Zhu, G.; Zhou, Y. Meteorological and hydrological conditions driving the formation and disappearance of black blooms, an ecological disaster phenomena of eutrophication and algal blooms. *Sci. Total Environ.* **2016**, *569–570*, 1517–1529. [CrossRef] [PubMed]
67. Zhang, Y.C.; Ma, R.H.; Zhang, M.; Duan, H.; Loiselle, S.; Xu, J. Fourteen-year record (2000–2013) of the spatial and temporal dynamics of floating algae blooms in lake Chaohu, observed from time series of modis images. *Remote Sens.* **2015**, *7*, 10523–10542. [CrossRef]
68. He, S.Y.; Ma, X.S.; Wu, Y.L. Spatial-temporal variation analysis of Chaohu Lake bloom time series by combined multi-source optical and radar remote sensing. *Environ. Monit. Manag. Technol.* **2019**, *31*, 10–15.
69. Zhang, H.; Huang, Y.; Yao, Y.; Ma, X.Q. Chaohu lake algae remote sensing monitoring and meteorological factor analysis. *Environ. Sci. Technol.* **2009**, *32*, 118–121.
70. Paerl, H.W.; Huisman, J. Climate. Blooms like it hot. *Science* **2008**, *320*, 57–58. [CrossRef]
71. Zhang, H.; Huang, Y.; Li, K. Remote sensing monitoring and analysis of the effect of surface temperature on Chaohu Lake bloom. *Environ. Sci.* **2012**, *33*, 3323–3328.
72. Justin, D.C.; Douglas, D.K.; Alex, J. Effectiveness of a fixed-depth sensor deployed from a buoy to estimate water-column cyanobacterial biomass depends on wind speed. *J. Environ. Sci.* **2020**, *93*, 23–29.
73. Fan, Y.X.; Jin, S.J.; Zhou, P. Analysis of distribution characteristics and meteorological conditions of cyanobacteria blooms in Chaohu lake. *J. Anhui Agric. Sci.* **2015**, *43*, 191–193, 198.
74. Xun, S.P.; He, B.F.; Wu, W.Y.; Liu, H.M.; Zhang, H.Q.; Yang, Y.J. Discussion on meteorological factors of Cyanobacteria outbreak potential forecast. In Proceedings of the Healthy Lakes and Beautiful China-The Third China Lake Forum and the Seventh Hubei Science and Technology Forum, Hubei, China, 24 October 2013; pp. 356–365.

Article

Hindcast and Near Real-Time Monitoring of Green Macroalgae Blooms in Shallow Coral Reef Lagoons Using Sentinel-2: A New-Caledonia Case Study

Maële Brisset [1,*], Simon Van Wynsberge [1], Serge Andréfouët [2], Claude Payri [2], Benoît Soulard [1], Emmanuel Bourassin [1], Romain Le Gendre [1] and Emmanuel Coutures [3]

1 Institut Français de Recherche pour l'Exploitation de la Mer, UMR 9220 ENTROPIE (Institut de Recherche pour le Développement, Université de la Réunion, IFREMER, Université de la Nouvelle-Calédonie, Centre National de la Recherche Scientifique), Nouméa 98800, New Caledonia; Simon.Van.Wynsberge@ifremer.fr (S.V.W.); benoit.soulard@ifremer.fr (B.S.); Emmanuel.Bourassin@ifremer.fr (E.B.); rlegendr@ifremer.fr (R.L.G.)

2 Institut de Recherche pour le Développement, UMR 9220 ENTROPIE (Institut de Recherche pour le Développement, Université de la Réunion, IFREMER, Université de la Nouvelle-Calédonie, Centre National de la Recherche Scientifique), BP A5, Nouméa CEDEX 98848, New Caledonia; serge.andrefouet@ird.fr (S.A.); claude.payri@ird.fr (C.P.)

3 Direction du Développement Durable des Territoires, Province Sud, Nouméa 98849, New Caledonia; emmanuel.coutures@province-sud.nc

* Correspondence: maele.brisset@ifremer.fr

Abstract: Despite the necessary trade-offs between spatial and temporal resolution, remote sensing is an effective approach to monitor macroalgae blooms, understand their origins and anticipate their developments. Monitoring of small tropical lagoons is challenging because they require high resolutions. Since 2017, the Sentinel-2 satellites has provided new perspectives, and the feasibility of monitoring green algae blooms was investigated in this study. In the Poé-Gouaro-Déva lagoon, New Caledonia, recent Ulva blooms are the cause of significant nuisances when beaching. Spectral indices using the blue and green spectral bands were confronted with field observations of algal abundances using images concurrent with fieldwork. Depending on seabed compositions and types of correction applied to reflectance data, the spectral indices explained between 1 and 64.9% of variance. The models providing the best statistical fit were used to revisit the algal dynamics using Sentinel-2 data from January 2017 to December 2019, through two image segmentation approaches: unsupervised and supervised. The latter accurately reproduced the two algal blooms that occurred in the area in 2018. This paper demonstrates that Sentinel-2 data can be an effective source to hindcast and monitor the dynamics of green algae in shallow lagoons.

Keywords: Ulva; Sentinel-2; satellite; remote sensing; algal bloom; coral reefs; Pacific lagoons

Citation: Brisset, M.; Van Wynsberge, S.; Andréfouët, S.; Payri, C.; Soulard, B.; Bourassin, E.; Gendre, R.L.; Coutures, E. Hindcast and Near Real-Time Monitoring of Green Macroalgae Blooms in Shallow Coral Reef Lagoons Using Sentinel-2: A New-Caledonia Case Study. *Remote Sens.* **2021**, *13*, 211. https://doi.org/10.3390/rs13020211

Received: 26 November 2020
Accepted: 23 December 2020
Published: 9 January 2021

Publisher's Note: MDPI stays neutral with regard to jurisdictional claims in published maps and institutional affiliations.

1. Introduction

Green tides, or massive blooms of green macroalgae that wash up on the shores, have spread around the world over the past decades, sometimes leading to significant economic, health and ecologic impacts [1,2]. Among the most significant events recorded are green tides in European locations such as Brittany and Venice Lagoon and in several Iberian estuaries [3–6]. They are also reported in the Yellow Sea on Chinese coasts [7] and in the Southeast Gulf of California in North America [8].

Green tides of macroalgae result from an excessive growth and proliferation (blooms) of competitive algae, usually from *Ulva* and *Cladophora* genera, when inorganic nutrients exceed the assimilative capacity of the surrounding ecosystem [9]. Algae biomass accumulations along coastlines, and their mats washed ashore, impact human activities and tourism. Beach cleaning and protection represent a significant cost and are not always possible [10]. The proliferation of macroalgae also disturbs the equilibrium of coastal ecosystems, both temperate and tropical, via light reduction and via macroalgae decomposition that may generate anoxic conditions and hypoxia [11–13]. Beyond ecological damage, the accumulation of macroalgae biomass along coastlines can release hydrogen sulfide (H_2S) when decomposing under anoxic conditions, which may represent a serious risk for human and animal health.

To prevent or limit the impacts of green tides on society and ecosystems, some of the regions highly impacted by these events have developed expertise in risk assessment. Specifically, methods for monitoring and mapping macroalgae blooms were developed to understand their origin and attempt to anticipate mass beaching. Among the methods that have already been implemented are remote sensing approaches (see Scanlan et al., 2007 [14], for a review of existing strategies). Remote sensing studies carried out to monitor macroalgae focused on emerged or washed-ashore algae patterns (e.g., green tides in Brittany [15]), and on floating, rafting mats (e.g., [16–20]). Specifically, satellite imagery acquired at high temporal resolution (1 to 15 days) has allowed the detection of episodic blooms and provides pictures of these events which are spatially more representative than those provided by an often limited in situ number of stations and transects [14]. Furthermore, satellite archive data can be used to track the dynamic of algal blooms and identify their origins [17–21]. Ultimately, at high temporal resolution, images can be used to detect blooms in near real-time and predict their trajectory to assist the management of beach stranding events [22,23]. The main sensors used so far for these approaches were the Geostationary Ocean Color Imager (GOCI) at 500-m spatial resolution and 1 h revisit time [17,20,22,23]; the Moderate Resolution Imaging Spectroradiometer (MODIS) at 1200-m spatial resolution and 1 day revisit time [17,19–21]; the Ocean and Land Color Instrument (OLCI) at 300-m spatial resolution and 2 days revisit time [20]; the Wide Field of view (WFV) at 16-m spatial resolution and 4 days revisit time [20]; and the Ocean Land Imager (OLI) at 30-m spatial resolution and 16 days revisit time [17,20].

Green tides can occur in small tropical lagoons where they often originate from algal biomass growing on the lagoon floor. When becoming unattached, they form blooms of loose or drift macroalgae. Proliferations are limited in sizes according to the lagoon scale, but they still create nuisances. Green algal blooms that develop in oligotrophic lagoons are much smaller in size compared to those recorded in eutrophic waters and, therefore, require higher spatial resolution satellite imagery to be detected. High revisit frequency satellites (i.e., ~1 day) have the temporal resolution suitable to monitor blooms in these ecosystems, but their spatial resolution (i.e., $\geq$100 m) is too coarse for the lagoon and algal mats dimensions to provide accurate estimates. Bloom monitoring in small lagoons ideally requires both high temporal and spatial resolutions and, ideally, at low cost. Satellites with Very High Spatial Resolution (VHR, i.e., <4 m) are suitable to detect fine spatial features, but their on-request tasking procedures have a significant cost and can prevent real-time monitoring. An overview of existing sensors suggests that few sensors provide adequate capacities. Several studies have shown the potential of VHR satellite imagery to map and estimate long term biomass of macroalgae in shallow lagoons [24–26], but they did not focus on green algae. These studies also show that these recent satellite missions provide limited temporal coverage to hindcast the blooms of macroalgae, which can be very episodic.

In 2014, the European Space Agency (ESA) started the Copernicus program and launched the two Sentinel-2 satellites, which supply since 2017 free data at 10-m resolution with a 5-day revisiting time. The Multi Spectral Instrument (MSI) used for this program offers an interesting trade-off between cost and spatial and temporal resolutions that should be more suitable than previous remote sensing commercial products for characterizing the dynamics of green algae blooms in tropical lagoons. Thus, Sentinel-2 brings new perspectives to monitor these events. Here, we tested the ability of Sentinel-2 imagery to characterize green macroalgae dynamics within a shallow coral reef lagoon in New Caledonia recently impacted by green tides. First, using images quasi-concurrent with field data, we fitted a relationship between in situ algal abundance and various radiometric indices calculated from Sentinel-2 MSI reflectances. Then, the best model was used to extrapolate algal dynamics over a three-year period, from January 2017 to December 2019. The method has been evaluated regarding its ability in capturing two documented *Ulva* spp. blooms that occurred in 2018 and 2019 in the studied area [27,28].

2. Materials and Methods

2.1. Study Site

New Caledonia is a French overseas territory, located in the south-western Pacific Ocean, ~1500 km east of Australia (Figure 1a). It comprises one main large island, known as Grande Terre, bordered by a 1600-km long barrier reef, which delimits a 23,400 km^2 lagoon [29]. Due to their good health conditions and high biodiversity, six different areas, or clusters, of New Caledonia reefs and lagoons were listed as UNESCO World Heritage Areas in 2008 [30]. New Caledonia is not spared from green tides, with occasional events of short duration recorded since 1996 in the bays near the town of Nouméa, and induced by the Japanese algae *Ulva ohnoi*.

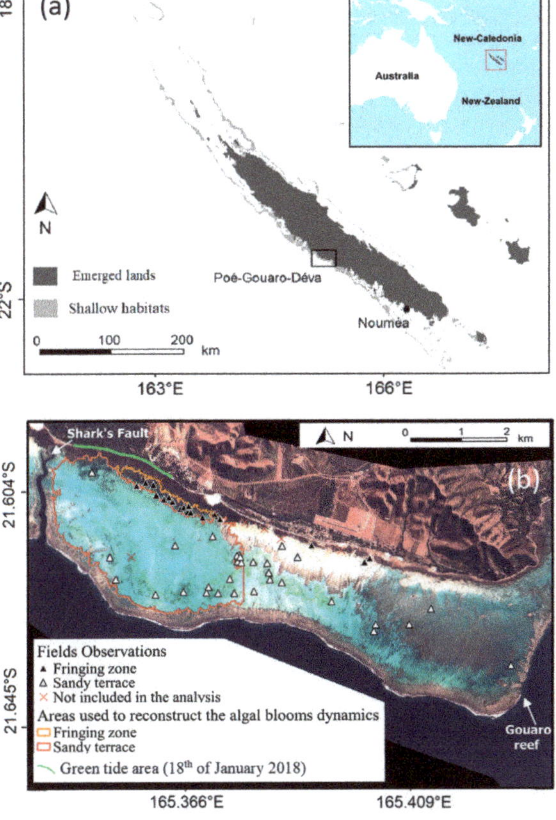

Figure 1. Study site description. (**a**) Geographical location of the Poé-Gouaro-Deva (PGD) area on New Caledonia's main island, Grande Terre. The shallow habitat layer is from the Millennium Coral Reef Mapping Project [31]. (**b**) Location of field observations and boundaries of the two areas considered in this study to reconstruct the green algal dynamics, namely the fringing zone and the sandy terrace. The image shown here is a very high-resolution Pléiades remote sensing image from 17 July 2017, used to select field sampling stations (see Section 2.2).

The present study focuses on the Poé-Gouaro-Deva (PGD) zone (21°37′10″S, 165°23′16″E) located on the west coast of Grande Terre. The studied area is located on the coastal area in the Bourail region, in the South Province of New Caledonia. It is bounded by the Gouaro reef in the southeast, and by the so-called Shark's Fault (Faille aux requins) in the northwest. The area contains a Marine Protected Area (MPA) and is included in the UNESCO areas. The coral reef complex borders the coastline (Figure 1b), forming a narrow (<3 km) and shallow lagoon (maximum depth ~4 m). Tides in this area can be classified as micro-tidal, mixed and mainly semi-diurnal, with a tidal range that can reach a maximum of 1.8 m during spring tide. The northwest part of the lagoon (~10 km^2) is mainly made of sandy soft-bottom habitat, while the southeast part (~9 km^2) is dominated by hard ground and coral communities and patch reefs. A several tens of meters-wide fringing seagrass bed (~3 km^2) borders the shoreline [31]. This seagrass belt is not homogeneous and can be intertwined with patch of coral communities (branching *Montipora* and *Acropora*) and sand.

The PGD area is a hotspot of tourism activities, with increasing human pressures on the reef and lagoon ecosystems. Tourist frequentation has increased over recent years, with about 4000 Deva visitors in 2017 and more than 6500 in 2019 (data from Maison de Deva). Developing lands have also tripled in surface area from 1998 (~100 ha) to 2014 (~350 ha, [32]). In early 2018, a green tide occurred in the PGD lagoon, with green macroalgae washed onshore in significant amounts on 18 January (~500 m^3 collected in three clean-up operations, Coutures personal observations), mostly along a 2.5-km long coastline located in the northwest part of the study zone (Figure 1b). The algae decaying on the beach prevented tourism activities and could induce health threats. Another stranding event, although of a smaller intensity than the first event, impacted the beach on 1 June 2019.

2.2. In Situ Measurements of Algal Abundance

Field investigations were first carried out in February 2019 and May 2019, at 24 and 42 stations, respectively. Among the 24 stations visited in February 2019, most of them were revisited in May 2019. All stations were preliminarily selected using a remote sensing-driven site selection method, similar to that in Andréfouët and Wantiez (2010) [33], to best capture the diversity of habitat in the area (seagrass bed, sand, coral reef and mixed habitats). For station selection, we used a multispectral Pléiades image ([34], Airbus DS Geo distribution, resolution ~70 cm) dated from 17 July 2017.

Second, three field trips took place in July 2019, September 2019 and November 2019, with a sampling dedicated to the monitoring of one single station (B21) where the biomass observed in February and May was particularly high (Figure 2a, Figure 3b).

Figure 2. Variation in green algae cover observed in the Poé-Gouaro-Déva lagoon during the study. (**a**) Algal cover observed in May 2019, in the fringing area (abundance index I_a = 5); (**b**) and (**c**) algal cover observed in May 2019, in the sandy terrace (abundance index I_a = 3 and I_a = 1, respectively).

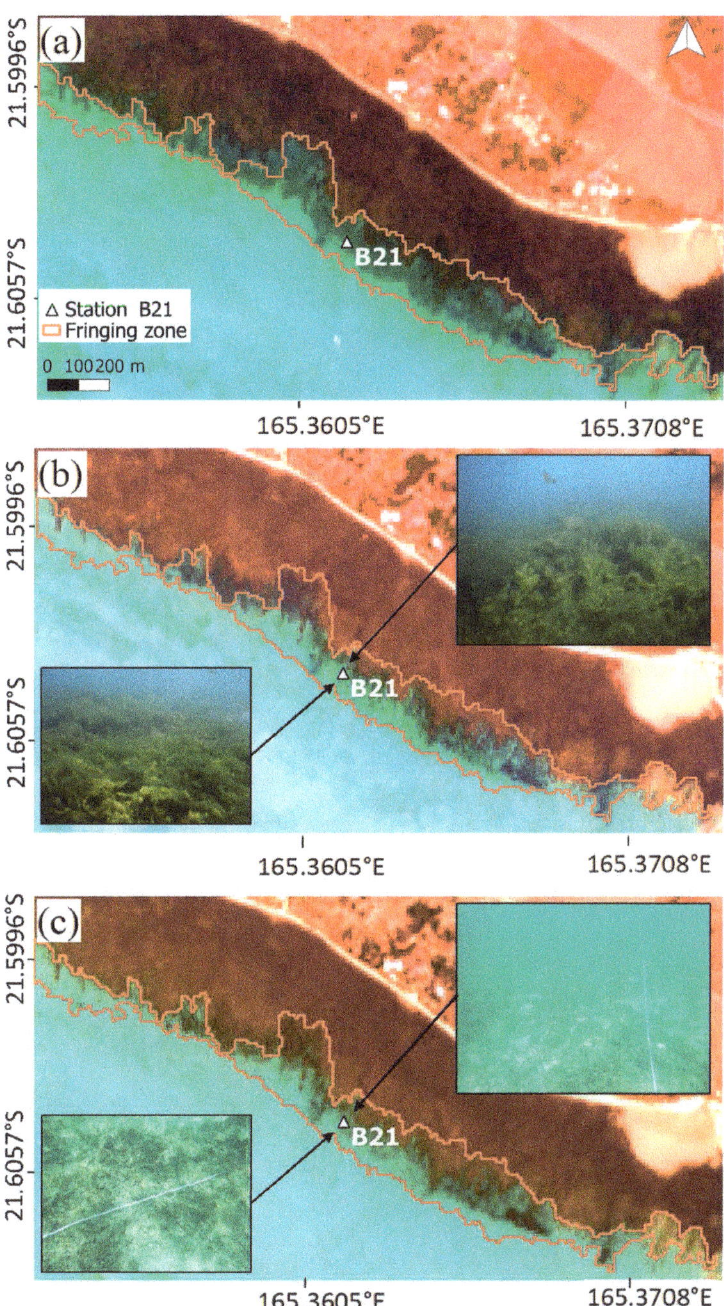

Figure 3. Correspondence between the Sentinel-2 images and the abundance of green algae in the fringing area (**a**) just before the January 2018 green tide (Sentinel-2 image dated from 4 January 2018), (**b**) during high biomass period (Sentinel-2 image dated from 13 February 2019, photographs at station B21 from 20 February 2019) and (**c**) during low abundance period (Sentinel-2 reference image from 25 November 2019 and photographs at station B21 from 18 November 2019).

Finally, 11 new additional stations were opportunistically sampled during 4 surveys dedicated to seawater samplings in December 2018, March 2019, July 2019 and October 2019. These stations did not specifically target areas with known past high abundances of green algae.

For each survey and each station, the cover of green algae was estimated semi-quantitatively using the Braun-Blanquet cover-abundance scale [35], ranging from 0 (no green algae) to 5 (cover exceeding 75%) from wide-angle photos and videos taken during the field investigations (Table 1; Figure 2). The method is similar to the Medium-Scale Approach [36] used to describe fish habitats using a semi-qualitative scale for each benthic descriptor. The results are estimates of green algae abundance (index $\overline{I_a}$) at a scale compatible with the resolution of Sentinel-2 data (100 m^2 pixels, see next section). In total, 9 samplings campaigns carried out from December 2018 to November 2019 provided 97 measurements of green algae abundances, with 57 along the fringing and seagrass reef and 40 spread over the sandy terrace (Figure 1b). This method was favored compared to a quantitative approach (using transects or quadrats) because it allows for covering a large area with more stations. Although this approach provided less precise estimates of algae abundance for each pixel individually, it could survey a large panel of pixels in different configurations.

Table 1. Semi-quantitative index (I_a) used in this study to quantify green algae abundance, based on the Braun-Blanquet index of abundance dominance [35].

I_a	Green Algae Cover (%)
0	<1
1	1–5
2	5–25
3	25–50
4	50–75
5	>75

2.3. Sentinel-2 Data Sets

Cloud-free Sentinel-2 images (10 m resolution) available for the study area were retrieved from the Data Hub website (© Data Hub System 2.4.1, [37]; Figure 4). Data downloads provided a time-series of images from 2017 to 2019. This consisted of Level 2A images (i.e., images corrected for the atmospheric signal) when available. Otherwise, Level 1C images (i.e., images without atmospheric correction) were used and were atmospherically corrected using the Sen2Cor processor from SNAP (© SNAP, 7.0) to create Level 2A products. This open-source program developed by the ESA provides surface, bottom of atmosphere (BOA) reflectance similar to the Level 2A products provided by the ESA/Copernicus program. Since we aim to apply statistical segmentation, classification and empirical correction to estimate algal abundance (see below), BOA reflectance was deemed adequate and no further correction was applied, for instance, to attempt to correct the air–water interface in the lagoon.

The Sentinel-2 MSI instrument measures reflected radiance in 13 spectral bands. Four of them have a 10-m spatial resolution, namely the blue (490 nm); green (560 nm); red (670 nm) and near-infrared (850 nm) bands. Only spectral bands in the blue (*B*2) and in the green (*B*3) zones were considered for this study, because red and especially near-infrared signals are attenuated quickly in the water column [38].

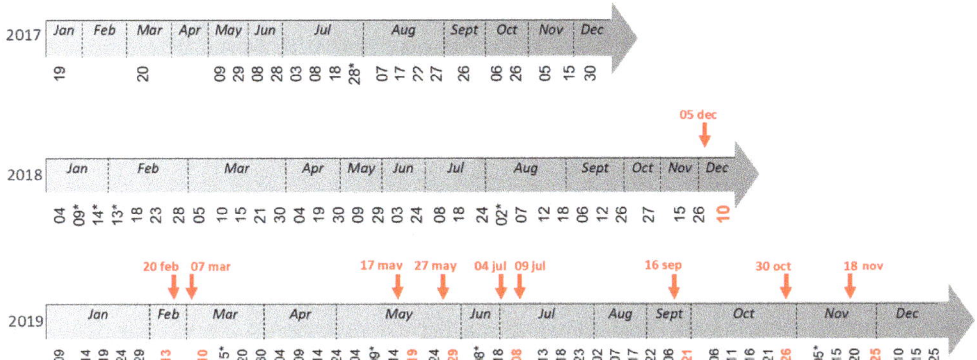

Figure 4. Historical timeline of the Sentinel-2 images downloaded from the Data Hub Copernicus website and used in this study. The dates annotated with '*' are used for the fringing area only due to a significant cloud cover above the sandy terrace. The timeline associates the fieldwork dates (red arrows) to the dates of the Sentinel-2 images used to calibrate the model (red numbers).

Prior to performing a change detection analysis, two methods were implemented to empirically inter-calibrate BOA reflectance between images. These corrections used a Sentinel-2 image dated from 25 November 2019 as a radiometric reference because the concurrent field trip performed in this period (18 November 2019) highlighted a very low algal cover in the studied area (Figure 3c). At the time of the analysis, the 25 November 2019 image was the closest in time available to this field trip. The first method (hereafter named "type 1 correction") is an empirical band-to-band and image-to-image correction which required identifying, across the time-series, 105 pixels considered as pseudo-invariant features on the ground. The method relies on the hypothesis that the targeted objects (e.g., roads and roofs) are radiometrically invariant during the studied period. Any instability in the reflectance of these objects, therefore, can create biases due to environmental conditions (e.g., wetness on ground) or noise at the time of processing after image acquisition (resampling). To inter-calibrate the images, a band-specific linear relationship was built between the BOA reflectances of the pseudo-invariant features for each pair of images (reference image and image to be processed) using Equations (1) and (2) [39].

$$R(B2)_{cor1} = a \times R(B2) + b \quad (1)$$

$$R(B3)_{cor1} = c \times R(B3) + d \quad (2)$$

where a, b, c and d are the coefficients of the linear relationships built from the pseudo-invariant pixels and $R(B2)$ and $R(B3)$, the BOA reflectances of the spectral bands in the blue and the green.

The second correction aimed at minimizing the differences caused by the seabed composition (other than green macroalgae) within an image. $B2$ and $B3$ reflectances associated to each pixel were standardized following Equations (3) and (4). This correction was first applied alone (hereafter named "type 2 correction"), and we also applied this correction followed by the first one ("type 3 correction").

$$R(B2)_{cor2} = \left(\frac{R(B2)_{ref} - R(B2)}{R(B2)_{ref}} \right) + \overline{R(B2)_{ref}} \quad (3)$$

$$R(B3)_{cor2} = \left(\frac{R(B3)_{ref} - R(B3)}{R(B3)_{ref}} \right) + \overline{R(B3)_{ref}} \quad (4)$$

where $R(B2)_{ref}$ and $R(B3)_{ref}$ are the BOA reflectances in the blue and the green bands, respectively, associated to the reference image (25 November 2019); and $\overline{R(B2)_{ref}}$ and $\overline{R(B3)_{ref}}$ are the arithmetic means of reflectance across all pixels associated to the field stations of this image. These constants were added to avoid generating negative radiometric values, which are not compatible with the calculation of some of the radiometric indices (cf. next section).

Note that correcting for co-registration errors was not performed in our study since the L1C processing chain already implements a geometric step which aims at improving the repetitiveness of the image geolocation in order to reach a multi-temporal geolocation target (<0.5 pixel at 95% confidence). When performing the type 1 correction, we could further confirm that the set of Sentinel-2 images we used was affected by negligible co-registration errors, since locations of the pseudo-invariant pixels were stable, and no offset of these pixels over time could be noticed.

2.4. Relationship Between Reflectance and Green Algae Abundances

To calibrate a biomass vs. reflectance (or spectral index, see below) relationship, for each sampling campaign the analysis was performed on the closest Sentinel-2 image in time. Offset between field work and image acquisition was always inferior to 7 days (Figure 4). GPS measurements of the field stations were deemed precise at pixel resolution after comparison with remarkable features visible on ground and in images.

The blue and green BOA values were then used to compute six radiometric indices.

- Two mono-band indices (designated ρ_b and ρ_g) used either blue or green BOA values, with or without the correction types.
- Four multispectral additional indices (designated θ1 to θ4) were generated from different combinations of ρ_b and ρ_g, with or without the corrections (Equations (5)–(8)). These indices were inspired from spectral vegetation indices used in the literature, such as the Simple Ratio (SR) index for θ1 [40,41], the Normalized Difference Vegetation Index (NDVI) for θ2 [42], the Modified Simple Ratio (MSR) for θ3 [43] and an attempt to combine the merits of NDVI and SR (NDVI × SR) for θ4 [44].

$$\theta 1 = \frac{\rho g}{\rho b} \tag{5}$$

$$\theta 2 = \frac{\rho g - \rho b}{\rho g + \rho b} \tag{6}$$

$$\theta 3 = \frac{\left(\frac{\rho g}{\rho b}\right) - 1}{\sqrt{\frac{\rho g}{\rho b}} + 1} \tag{7}$$

$$\theta 4 = \frac{\rho g * (\rho g - \rho b)}{\rho b * (\rho g + \rho b)} \tag{8}$$

For the two types of substratum encountered in the study area, namely sandy bottoms vs. mix substrates along the fringing area (made of unconsolidated calcareous and terrigenous sand, coral and seagrass), the semi-quantitative abundance of green algae from field observations was expressed as a function of each radiometric index, using a non-parametric smooth function calibrated via the R studio mgcv package (©2009–2019 R studio v1.2.1578). The performance of the models was compared on the basis of the percentage of variance explained and using the generalized cross-validation (GCV) criteria.

2.5. Mapping of Green Algal Dynamics

We hindcasted the dynamics of green algae over the period covered by the Sentinel-2 imagery (from 19 January 2017 to 12 December 2019) for two shallow areas of interest in the PGD area (Figure 1b). These areas were selected after field work considering they were prone to macroalgae proliferation or accumulation. The first area (hereafter named 'the fringing area'; ≈48 ha) included very shallow (<2-m depth at low tide) stations located along the littoral in the west part of the lagoon. It is characterized by a mosaic of benthic components (seagrass, sand and corals). The second area (hereafter named 'the sandy terrace'; ≈780 ha) was also located in the west part of the lagoon but included stations located on the sandy terrace away from the littoral and in slightly deeper water (2–4 m deep at low tide).

For each area of interest, two different methods were tested to extrapolate the abundance of green algae. The first approach (hereafter 'the unsupervised approach') used the best fitted model selected in the previous section to predict, for each pixel located in the area of interest, the algal abundance (I_a) from the $R(B2)$ and $R(B3)$ radiometric values. This was realized using the predict.gam function with the R studio mgcv package. The reconstruction of green algae dynamics was performed for the two areas of interest, computing, for each area, the arithmetic mean value of algae abundance per pixel ($\overline{I_a}$) (Equation (9)).

$$\overline{I_a} = \frac{\sum_{i=1}^{n} Ia_i}{n} \tag{9}$$

where n refers to the number of pixels in the area of interest.

The second approach (hereafter 'the supervised approach') involves a photo interpretation step. First, the two areas of interest were segmented in order to define homogeneous polygons in terms of blue and green radiometric values. The segmentation was performed using the automatized meanshift algorithm of the Orfeo ToolBox (OTB) plugin from QGIS software (© Qgis version 3.10.4). Then, algal-free polygons were visually identified by one of the authors (M.B.) and dismissed from the analysis. For each remaining polygon, an abundance-dominance green algae index was predicted, following the same procedure as for the unsupervised approach. The reconstruction of green algae dynamics was performed by computing, for each area of interest, the arithmetic weighted mean value of algae abundance associated to each polygon (I_a), weighted by its surface area (Equation (10)).

$$\overline{I_a} = \frac{\sum_{i=1}^{n} (Ia_i \times A_i)}{\sum_{i=1}^{n} A_i} \tag{10}$$

where n refers to the number of polygons in the area of interest and A_i the surface area of polygon i.

The various time-series of green algae abundances ($\overline{I_a}$) obtained from the combination of different image inter-calibration procedures, spectral indices and mapping procedures were compared regarding their ability to reproduce the two green tides recorded in the study area in 2018 and 2019.

3. Results

3.1. Algal Diversity

Several species of green algae (Chlorophyta) have been observed in the different study areas in the lagoon, including *Boodlea*, *Chaetomorpha*, *Cladophora* and *Ulva*. *Ulva* was dominant, with several species having a filamentous and tubular morphology. A molecular taxonomical study currently being completed has shown the coexistence of four genetically different species. Three of them are attached to the substratum mainly in sandy areas, while the blooms consist of free-moving masses of strongly entangled filaments which accumulate along the coastline in the seagrass areas. The bloom-forming alga is a highly branched species of *Ulva* (*Ulva* sp3) which morphologically resembles *Ulva prolifera* but is genetically distinct from it. In some samples from the blooms, there could be several taxa (*Boodlea* and *Cladophora*), but most of the time, the dominant species was *Ulva* sp3.

3.2. *In Situ Estimation of Algal Abundance*

In February 2019, green algae were abundant ($2 < I_a < 5$) at two stations located in the fringing zone in the western part of the lagoon. Relatively high abundance-dominance indices ($2 < I_a < 3$) were also obtained for three stations on the sandy terrace. The abundance of algae was poor or null ($I_a \leq 1$) at all other stations. In May 2019, the spatial distribution of green algae was similar to February 2019, with abundance-dominance indices from 0 to 4 at stations located in the sandy terraces and from 0 to 5 at stations located in the fringing zone. Along this area, a significant variability in abundance ($0 < I_a < 5$) was also found between stations during all of the 2019 surveys, except in November 2019, when abundance was low ($I_a = 0$) all across the fringing zone.

The wide range of I_a values across the different field surveys and stations allowed for calibrating a representative and robust relationship between I_a and the radiometric indices derived from remote sensing images (see next section).

3.3. *Relationship between Radiometric Indices and Algal Abundance*

Without any correction applied (neither type 1, 2 nor 3), the multispectral indices θ1, θ2, θ3 and θ4 better explained variance than the ρ_b and ρ_g indices (Figure 5). This trend was obtained both for stations located in the fringing zone (i.e., between 48.6% and 50.3% of variance explained by θ1, θ2, θ3 and θ4 vs. 27.2% and 31.5% of variance explained by ρ_b and ρ_g) and for stations located in the sandy zone stations (i.e., 32.3–32.5% vs. 6% and 1%).

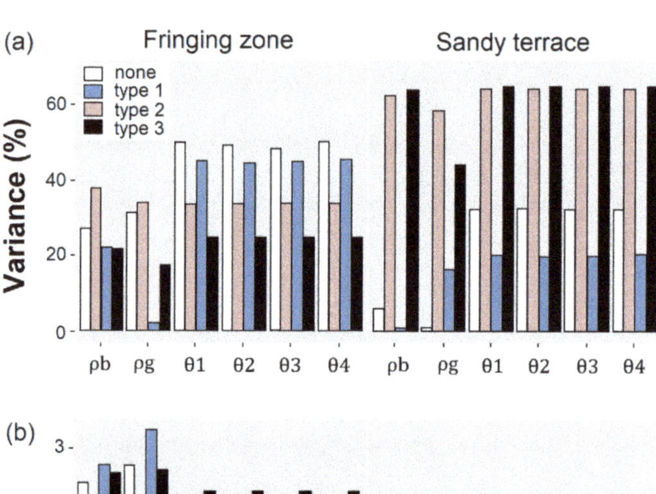

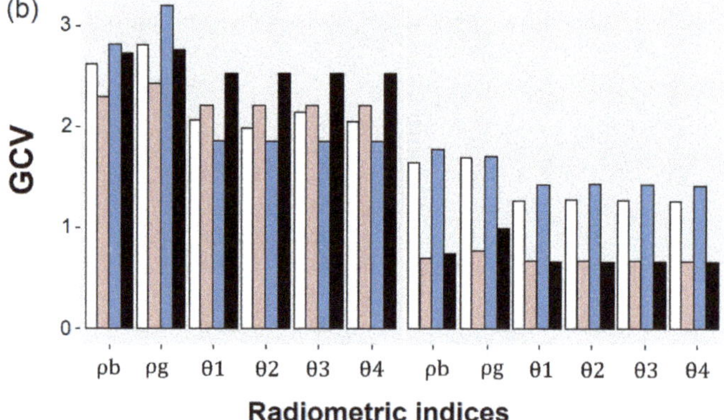

Figure 5. Performance of the fitted relationships between the green algae abundance and the radiometric indices, depending on the type of correction (type 1, type 2 or type 3). (**a**) Comparison of the relationships regarding the variance in algae abundance they explained. (**b**) Comparison of the relationships regarding the generalized cross-validation (GCV) criteria.

For the stations located on the sandy terrace, the type 2 and 3 corrections improved the goodness-of-fit test results (Figure 5). Specially for the θ4 index, the explained variance increased from 32.3% without correction to 64.3% with type 2 correction and to 64.9% with type 3 correction. However, applying the type 2 or the type 3 correction degraded the fit for the fringing zone (Figure 5).

Eventually, the θ4 index based on $R(B2)$ and $R(B3)$ without any corrections provided the best model for the fringing zone (i.e., 50.3% of variance, GCV = 2.06, Figure 6b). For the sandy terrace, θ4 also provided the best model (i.e., 64.9% of variance, GCV = 0.66, Figure 6a) but only when the type 3 correction was applied (Figure 5). These indices and correction types were therefore selected to extrapolate the algal abundance across the Sentinel-2 time-series (next section).

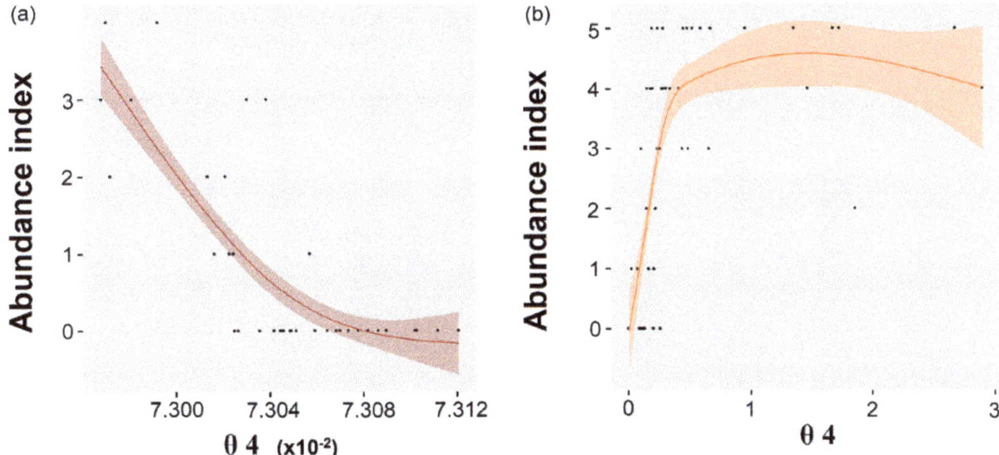

Figure 6. Calibration of a smooth function (non-parametric relationship) between the green algal abundance and the radiometric index θ4. (**a**) Fitted relationship for the stations located in the sandy terrace (n = 40; variance explained by the model = 64.9%; GCV = 0.66); radiance was corrected with type 3 correction. (**b**) Fitted relationship for the stations located in the fringing zone (n = 57; variance explained by the model = 50.3%; GCV = 2.06), without any radiance correction.

3.4. Mapping and Dynamics of Green Algae Abundances Using Sentinel-2

For the fringing area, using the supervised approach always yielded lower $\overline{I_a}$ values than with the unsupervised approach ($\overline{I_a}$ = 0.63 on average over the period of 2017–2019 against 2.00; Figure 7a). Despite this difference, both approaches could identify the January 2018 and June 2019 algal blooms, both producing significant macroalgae pile-ups on the beach. The abundance variations hindcasted on the sandy terrace area were not always synchronous with those hindcasted in the fringing area. Unlike the fringing zone, the supervised and unsupervised approaches were more in agreement to each other in this habitat (Figure 7b). The dynamics of green algae hindcasted for both areas by the supervised and unsupervised approaches suggested high algal abundance throughout the year, except during the dry season (October and November; Figure 7). In particular, periods of high algal abundance were predicted around July 2017, July 2018 and July 2019 for both fringing and sandy terrace areas, although no green tide was reported, to our knowledge, during these periods.

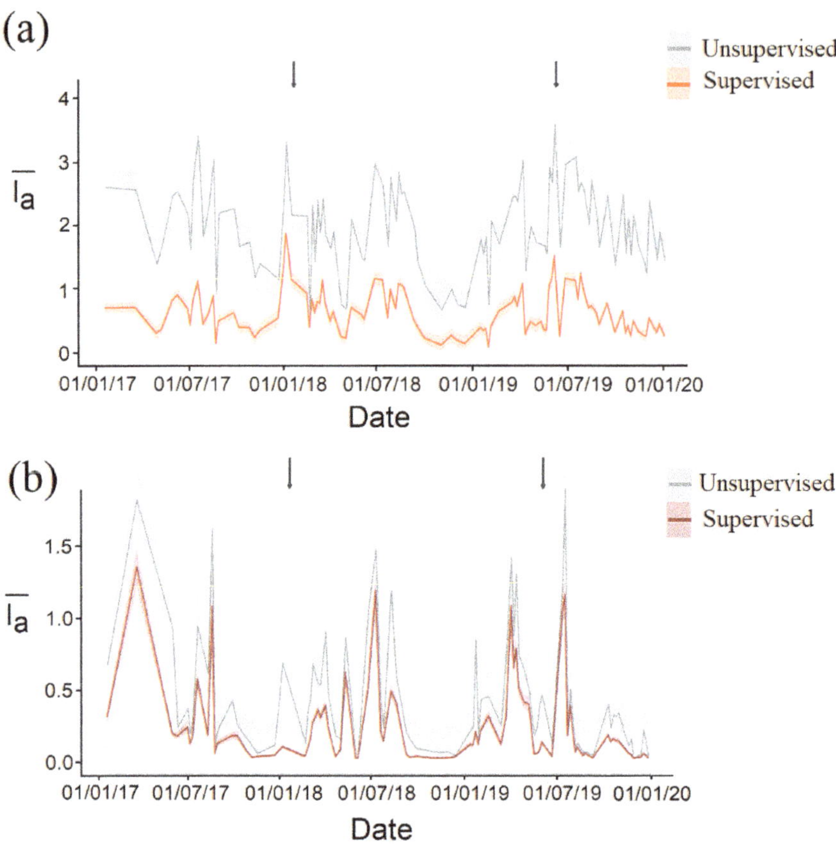

Figure 7. Variation in mean algal abundances ($\overline{I_a}$) over time, from 2017 to 2019, based on Sentinel-2 images, depending on the two different applied methodologies (unsupervised vs. supervised) (**a**) for the fringing area and (**b**) for the sandy terrace area. Arrows marks the dates of reported stranding events on the PGD beach.

4. Discussion

4.1. Innovative Nature of the Approach and Proposed Method

This paper demonstrates that the Sentinel-2 imagery is a valuable tool to hindcast the dynamics of benthic green algae in shallow coral reef lagoons, which is required to understand the causes of the algal tides when confronted with other data sets (e.g., lagoon temperature; solar irradiance; rainfall or streamflow from nearby rivers). Here, we focused on the benthic biomass of green macroalgae during blooming conditions in a shallow lagoon. This is in contrast with most recent remote sensing studies which looked at green and brown algae once they had beached and algal rafts while drifting on the ocean surface [16–21].

To date, most of the multispectral indices proposed in the literature for detecting macroalgae used the red visible and/or near-infrared bands [19,20,23,45–47]. These spectral bands have been commonly used because they are suitable to capture chlorophyll absorption peaks and the overall vegetation spectral signature when it is above water or just at the ocean surface. The NDVI, which uses a ratio of the red visible and near-infrared bands, has been widely used as an indicator of floating algae mats' presence/absence [20,48–50]. The NDVI has also been used successfully for mapping and monitoring phanerogams [51]. Despite the potential of the red and near-infrared bands to distinguish vegetation, these radiations are quickly attenuated in water, and spectral indices based on these bands are very sensitive to depth variations, either due to bathymetry or tide [52,53]. For instance, Andréfouët et al. [26] showed that even in a low-depth environment (<2 m), red and near-infrared bands were not ideal to map the biomass of benthic *Sargassum polycystum* brown algae. Although the index of floating algae based on the red and near-infrared bands may be useful to identify the presence of green algae in their final floating cycle, this index was of low interest in our case study, because the floating phase of algae's life cycle is very short and was never observed during field work. This made NDVI a poor proxy of algal abundance in our case study (Figure S1). In this paper, a modified version of the NDVI, switching the red and infrared bands with blue and green bands, provided more conclusive results to quantify the abundance of benthic green algae. This agreed with previous studies that used NDVI-like indices with a variety of band combinations [53,54].

In our study, multispectral radiometric indices (i.e., θ1, θ2, θ3 and θ4) provided more convincing results than the green band alone (i.e., ρ_g). These findings are in contrast with the work of Andréfouët et al. [26] for *S. polycystum* in an atoll lagoon. For this brown alga, the green band provided a fit with biomass nearly as good as our multispectral indices and was used to hindcast its dynamics from 2002 to 2015 using a range of high-resolution remote sensing images (QuickBird 2, IKONOS, WorldView-2, Pléiades 1B, 1A).

4.2. Influence of the Three Types of Corrections

Using the type 1 empirical approach alone to correct the radiometric data was found to be of limited interest. Two explanations can be provided. First, our study relied on images that were corrected beforehand with an ESA model in order to remove the atmospheric signal. It implies that the additional type 1 correction provided no added value to monitor the green algal proliferation, compared to using L2A images. This might be the result of using imperfect pseudo-invariant features for this correction. However, models and algorithms automatically used to correct the entire L1C images are unlikely to be enough to perfectly correct the images locally in our studied areas of interest [55,56]. Specifically, for coral reef areas, local processes can affect the atmospheric signal quality, such as ocean spray induced by waves breaking on the reef ridge.

Secondly, previous studies already highlighted the limits of the type 1 correction for areas of high algal biomass. High biomass is correlated with low radiometric value (in the green band, or for a given radiometric index [26]). Indeed, in such a configuration, the relationship applied to estimate biomass is very sensitive because a wide range of biomass corresponds to a narrow radiometric range. However, the type 1 correction for intercalibrating images is easily applicable to coastal areas where remarkable pixels can be precisely identified in all the images acquired on different dates [39].

To our knowledge, the type 2 processing proposed in this paper has never been documented before. It provided relevant results for the sandy terrace, which is made of relatively homogeneous pixels. In contrast, the type 2 correction turned out to be inefficient in the fringing zone. This result may partly be explained by tide, which may affect Equations (3) and (4) with stronger intensity in the fringing area, shallower than the sandy terrace area. Moreover, the heterogeneity of the fringing area, which is a mix of sand, corals and seagrasses, may also contribute to the poor performance of type 2 correction. Indeed, pixel resampling and misregistration may affect the signal to a greater extent in this heterogeneous area than in the sandy area. Implementing a co-registration method to correct errors and align pixels between images could be beneficial for future case-studies when the habitat is highly heterogeneous.

4.3. Supervised vs. Unsupervised Approach

The supervised and unsupervised approaches tested in this study to hindcast the algal dynamics offer different levels of cost effectiveness. Concerning 'costs' (in terms of time and expertise), the unsupervised method only involves a quick QGIS manipulation of a few minutes per image and does not require any specific expertise. The supervised approach, by contrast, requires the capacity to identify the presence/absence of algal mats on Sentinel-2 images, an expertise acquired by comparing field observations concurrent with images. The method also requires a longer analysis time (i.e., ~20 min per image for the PGD area). The cost depends on the surface area to process and can be significant for large lagoons. Although the unsupervised approach does not require any expertise (beside understanding the software), the processing time (using R software) associated to this approach increases exponentially with the number of pixels. For the entire PGD area (fringing zone and sandy terrace zone confounded, assembling 94,202 pixels), the algorithm takes about 1 h to complete (computer with 8192 Mo usable RAM and ~1.8 GHz processor). This may impair, for larger areas, the feasibility of real-time processing to anticipate beach stranding events. By contrast, following the interactive part, the processing time using R software associated to the supervised approach only takes a few minutes to complete, regardless of the size of the area, as it manipulates vectors (i.e., polygons) of several hectares instead of 100 m^2 pixels raster data.

Regarding the accuracy of estimations, the supervised approach allowed to discriminate the two bloom periods which led to beach stranding events in the PGD area. The unsupervised approach, by contrast, could not identify these specific events. The unsupervised approach also overestimated the average algal abundance ($\overline{I_a}$) over the study period (i.e., from 2017 to 2019). Indeed, the field survey carried out in November 2019 reported a fringing zone free of algae, while the algal abundance hindcasted on the reference image by the unsupervised approach was high (mean $\overline{I_a}$ value equal to 2.39). The supervised approach, on the other hand, provided more plausible estimates ($\overline{I_a}$ = 0.53 on 25 November 2019). We thus recommend the supervised approach over the unsupervised one when the time and level of expertise are not constrained.

4.4. Tracking the Origin of Green Tides

The abundance of green algae estimated using Sentinel-2 data enabled identifying a bloom of algae located in the fringing area before the beach stranding event reported at PGD in January 2018. Therefore, the hindcast of algal dynamics performed in this study provides clues to identify the main pathways of nutrients that induced this event. Indeed, to implement management actions, local stakeholders lacked evidence on whether the stranding event recorded at PGD in front of the fringing area was induced by an input of nutrients very locally or if it was the result of the drift of algae biomass that originated elsewhere. Here, we provide evidence of the existence of a peak of algal abundance on the fringing area starting several weeks before the beaching event. By contrast, the abundance hindcasted in the sandy terrace before the beach stranding event of 18 January 2018 is quite low ($\overline{I_a}$ < 1 for both approaches, see Figure S2). These results suggest that this stranding event was mainly due to algae biomass from the fringing zone. This hypothesis is consistent with the identification of two distinct species of *Ulva* genus in the PGD lagoon, which respectively dominate the sandy terrace and the fringing area [27,28]. The morphology of the species dominating the fringing area (i.e., forming dense unattached mats of long filamentous thallus) is more likely to generate green tides compared to the species that dominates the sandy terrace (species forming a smaller thallus attached to the substratum) [27,28].

It is worth noting that to track the origin of green tides in the PGD lagoon, we here developed an approach slightly different than those implemented in previous remote sensing studies for monitoring green macroalgae proliferations. Indeed, as previous studies focused on floating algae while drifting on the ocean surface, they aimed at developing methods that can detect these mats across wide areas of oceanic waters [16–20]. By contrast, the framework developed here aimed at hindcasting algae abundance in small areas suspected to experience green tides. Although this new framework requires sufficient knowledge on suitable habitats for algae and potential sources of nutrient to identify a priori the area that will be monitored, it is more relevant for monitoring blooms in small lagoons, where the rafting phase before beaching can be very brief and very spatially restricted within the lagoon.

4.5. Predicting the Risk of Beach Stranding Events

This study reconstructed the dynamics of green algae almost continuously over a three-year period. Similar objectives are not possible from in situ data. Although the fringing zone in the PGD lagoon is quite small and easily accessible from the beach, monitoring green algae cover in situ on a regular basis and in the long run turned out to be difficult to implement in practice, and using remote sensing is an attractive option. Eventually, we provided an indicator of risk for beach stranding events based on the estimates of green algal abundance ($\overline{I_a}$), via the supervised approach, with a 5-day periodicity (assuming favorable cloud cover). The dynamics reconstructed in this study over the period 2017–2019, with two significant bloom periods, can be used to set threshold values for $\overline{I_a}$, beyond which appropriate management actions may be needed (Table 2). Notably, the beaching risk can be considered as null when $\overline{I_a}$ in the fringing area is lower than 0.53, which is, as on 25 November 2019, the predicted value at the time of very low algal abundance confirmed in the field (Figure 3c). This baseline value, which is not equal to zero, corresponds to a background noise due to the various processing steps. Conversely, the beach stranding risk can be considered high when $1.13 < \overline{I_a} < 1.86$, thus, as in conditions observed in January 2018 and June 2019. In such cases, management actions can be preconceived to prevent and minimize the pile-up of algae on the beach (Table 2).

Table 2. Framework proposed in the context of Poé-Gouaro-Deva for linking the abundance of algae in the fringing area hindcasted with the supervised method ($\overline{I_a}$) with management actions. The $\overline{I_a}$ values are estimated from the θ4 index, without any radiometric corrections.

$\overline{I_a}$ Values	Beach Stranding Risk	Management Action Suggested
$I_a \leq 0.53$	Null	None
$0.53 \leq I_a \leq 0.95$	Low	None
$0.95 \leq I_a \leq 1.13$	Moderate	• Implementation of daily monitoring by an agent in situ
$1.13 \leq I_a \leq 1.86$	High	• Implementation of daily monitoring by an agent in situ • Launching/processing of clean-up operations on the beach
$1.86 \leq I_a$	Extreme	• Implementation of daily monitoring by an agent in situ • Check possible light obstruction for corals and seagrass and eventually consider launching field collection of macroalgae to reduce the threat for the ecosystem • Launching/processing of clean-up operations on the beach

Among the management actions that are recommended in case of moderate, high or extreme risk forecasted by the near real-time remote sensing protocol, we suggest (1) the implementation of, ideally, daily monitoring of algal cover along the fringing zone. This is recommended because the periodicity of the Sentinel-2 imagery (i.e., 5 days, assuming cloud-free weather) may be insufficient to accurately characterize the short-term evolution of each bloom. A second action recommended is (2) to plan clean-up measures on the beach. The removal of algae biomass accumulated along the shore is essential in order to avoid degradation of algae under anaerobic conditions and release of hydrogen sulfide. Finally, a third action that could be considered here in case of extreme risk of stranding events is (3) to remove a significant amount of macroalgae from the areas propitious to their growth (e.g., through field collection) in order to slow down the growth of the population. Although this action has already proven its efficiency to get rid of an already-formed bloom of green macroalgae [9], it seems difficult to implement in the PGD fringing zone without significantly damaging the coral reef and seagrass habitats.

Despite that the Sentinel-2 missions are expected to end in 2024, the development of such an indicator of risk for beach stranding events using this imagery is a promising path for environmental management. Indeed, the ESA agency has mentioned a possible 5-year extension of their products with new satellites to be launched, thus ensuring an access to Sentinel-2 imagery until 2029. Beyond the Sentinel-2 imagery, one can also expect the launch of new missions providing products at higher temporal and/or spatial resolution, which should also provide relevant data for monitoring green macroalgae blooms in shallow lagoons.

4.6. Generalization to Other Lagoons in New Caledonia

In this study we demonstrated the potential of using Sentinel-2 images to monitor green macroalgae proliferation in the PGD shallow lagoon. Statistical relationships established between the θ4 radiometric index and the green algae abundances, as well as radiometric corrections recommended for this study, were zone-sensitive (Figures 5 and 6). Even though the shallow habitats, either fringing or part of the barrier reef complex, represent wide areas at the New Caledonia scale (Figure 1), the PGD lagoon is quite an uncommon feature in New Caledonia due to its shallow depth and geomorphology [29,31]. The optical signature of the bands may vary from one site to another, depending on various parameters, including bathymetry, substrate nature, benthic components other than macroalgae and water quality. The diversity of these features is high in New Caledonia, and more work is needed at other sites to test the robustness of the methods established here.

Specifically for water quality, performing water column correction was not justified in our study since phytoplankton blooms were of lower concern in the PGD lagoon. Indeed, considering the shallow depths (between ~1.5 and 4 m) and low chl-a concentration in our study area (from 0.17 mg.m^{-3} after a dry period to 0.75 mg.m^{-3} after a heavy rain event, unpublished data), the decay of light along the water column remains fairly acceptable for the blue and green bands [57]. Moreover, the spatial patterns of algae blooms that can be observed on the imagery followed the benthic features of the area (Figure 3), indicating that the measured optical changes were driven by benthic algae, not by phytoplankton blooms. Finally, among all the methods developed to date for water column correction, none are capable of completely eliminating the water column effect [57]. All methods require some inputs that were not available with sufficient accuracy in our study area, and some others make assumptions not supported in our study site. As a result, performing a water column correction with available data was likely to induce more bias in the analysis. Applying the framework developed for PGD to other sites may, however, require an additional step of water column correction, especially for lagoons characterized by higher chl-a concentration or deeper habitats than studied here. Note that such an additional step is not trivial, as it may require the acquisition of specific data for model input, including accurate depth or bathymetric maps, or high-resolution maps of substrate and knowledge on the spectral behavior of each substrate type [57].

5. Conclusions

The capacity of Sentinel-2 imagery to hindcast the dynamics of green algae in shallow lagoons has been demonstrated for the PGD lagoon. It is a promising tool to track down the origin of green tides and identify potential sources. In particular, this paper highlights that an updated version of the NDVI, based on the blue and the green spectral bands instead of the red and near-infrared ones, can quantify the abundance of benthic green algae. However, the accuracy of the estimates is sensitive to the type of corrections applied to the radiometric values and the type of substrate. This prevents the transferability of the method to other sites without field validation. For the reconstruction of algae dynamics over specific and short periods, a supervised approach is recommended since it provides more accurate estimations of algal abundance. Hindcasting the algal dynamics over the 2017–2019 period evidenced that the massive beach stranding event that occurred in January 2018 in the northwest part of the PGD lagoon was the result of a macroalgae bloom that took place locally in this area. Based on our results, an indicator has been proposed to monitor the risk of beach stranding in near real-time and assist coastal management.

Supplementary Materials: The following are available online at https://www.mdpi.com/2072-4292/13/2/211/s1, Figure S1: Calibration of a smooth function (non-parametric relationship) between the green algal abundance and the radiometric index NDVI using the red and near-infrared bands. Figure S2: Algal abundances located in the fringing zone vs. in the sandy terraces before the beach stranding event of 18 January 2018.

Author Contributions: Conceptualization, M.B., S.V.W. and S.A.; methodology, M.B. and S.V.W.; software, M.B., S.V.W. and E.B.; validation, S.V.W., S.A., C.P., R.L.G., B.S. and E.C.; formal analysis, M.B.; investigation, M.B. and S.V.W.; resources, E.C.; data curation, M.B.; writing—original draft preparation, M.B. and S.V.W.; writing—review and editing, S.A., C.P., B.S., E.B., R.L.G. and E.C.; visualization, M.B.; supervision, S.V.W.; project administration, M.B., S.V.W., B.S. and E.C.; funding acquisition, E.C., S.V.W. and B.S. All authors have read and agreed to the published version of the manuscript.

Funding: This research was funded by: Direction du développement durable des territoires, Province Sud, New Caledonia. Grant number C.458-19.

Data Availability Statement: The data presented in this study are openly available at [https://sextant.ifremer.fr].

Acknowledgments: We are grateful to Miguel Clarke, Alan Choupeau and Romain Laigle for their logistical support during field trips; to Siloë Gobin and Thierry Jauffrais for their help in collecting macroalgae during fieldwork and macroalgae identification. We also thank Laurent Millet, Clarisse Majorel and Claire Bonneville from the "Plateforme du vivant", where genetic analyses were performed to support algae identification. We also thank the other members of the project that contributed to reviewing this work during workshops, including Hugues Lemonnier and Emmanuel Tessier.

Conflicts of Interest: The authors declare no conflict of interest.

References

1. Ye, N.H.; Zhang, X.W.; Mao, Y.Z.; Liang, C.W.; Xu, D.; Zou, J.; Zhuang, A.M.; Wang, Q.Y. 'Green tides' are overwhelming the coastline of our blue planet: Taking the world's largest example 1994. *Ecol. Res.* **2011**, *26*, 477–485. [CrossRef]
2. Gladyshev, M.I.; Gubelit, Y.I. Green Tides: New Consequences of the Eutrophication of Natural Waters (Invited Review). *Contemp. Probl. Ecol.* **2019**, *12*, 109–125. [CrossRef]
3. Piriou, J.Y.; Ménesguen, A. Les marées vertes à Ulves: Conditions nécessaires, évolution et comparaison de sites. IFREMER: Centre de Brest, BP 70, F-29263 Plouzané, France. Dans. *Estuaries Coasts Spat. Temporal. Intercomp.* **1991**, *19*, 117.
4. Runca, E.; Bernstein, A.; Postma, L.; Di Silvio, G. Control of macroalgae blooms in the Lagoon of Venice. *Ocean Coast. Manag.* **1996**, *30*, 235–257.
5. Flindt, M.R.; Kamp-Nielsen, L.; Marques, J.C.; Pardal, M.A.; Bocci, M.; Bendoricchio, G.; Salomonsen, J.; Nielsen, S.N.; Jorgensen, S.E. Description of the three shallow estuaries: Montego River (Portugal), Roskilde Fjord (Denmark) and the lagoon of Venice (Italy). *Ecol. Model.* **1997**, *102*, 17–31. [CrossRef]
6. Hernandez, I.; Peralta, G.; Pérez-Llorés, J.L.; Vergara, J.J. Biomass and dynamics of growth of *Ulva* species in Palmones river estuary. *J. Phycol.* **1997**, *33*, 764–772.
7. Liu, D.; Keesing, J.K.; Xing, Q.; Shi, P. World's largest macroalgal bloom caused by expansion of seaweed aquaculture in China. *Mar. Pollut. Bull.* **2009**, *58*, 888–895. [CrossRef]
8. Chávez-Sánchez, T.; Piñón-Gimate, A.; Serviere-Zaragoza, E.; López-Bautista, J.M.; Casas-Valdez, M. Ulva blooms in the southwestern Gulf of California: Reproduction and biomass. *Estuarine Coast. Shelf Sci.* **2018**, *200*, 202–211.
9. Ménesguen, A. *Les Marées Vertes. 40 Clés Pour Comprendre*; Edition Quae: Versailles, France, 2018; p. 128.
10. Wang, X.H.; Li, L.; Bao, X.; Zhao, L.D. Economic cost of an algae bloom cleanup in China's 2008 Olympic sailing venue. *EosTrans. Am. Geophys. Union* **2009**, *90*, 238–239.
11. Han, Q.; Liu, D. Macroalgae blooms and their effects on seagrass ecosystems. *J. Ocean Univ. China* **2014**, *13*, 791–798.
12. Zhang, Y.; He, P.; Li, H.; Li, G.; Liu, J.; Jiao, F.; Zhang, J.; Huo, Y.; Shi, X.; Su, R.; et al. *Ulva prolifera* green-tide outbreaks and their environmental impact in the Yellow Sea, China. *Nat. Sci. Rev.* **2019**, *6*, 825–838.
13. Santos, R.O.; Varona, G.; Avila, C.L.; Lirman, D.; Collado-Vides, L. Implications of macroalgae blooms to the spatial structure of seagrass seascapes: The case of the *Anadyomene* spp. (Chlorophyta) bloom in Biscayne Bay, Florida. *Mar. Pollut. Bull.* **2020**, *150*, 110742. [CrossRef] [PubMed]
14. Scanlan, C.M.; Foden, J.; Wells, E.; Best, M.A. The monitoring of opportunistic macroalgal blooms for the water framework directive. *Mar. Pollut. Bull.* **2007**, *55*, 162–171. [PubMed]
15. Perrot, T.; Rossi, N.; Ménesguen, A.; Dumas, F. Modelling green macroalgal blooms on the coasts of Brittany, France to enhance water quality management. *J. Mar. Syst.* **2014**, *132*, 38–53.

16. Qi, L.; Hu, C.; Xing, Q.; Shang, S. Long-term trend of *Ulva prolifera* blooms in the western Yellow Sea. *Harmful Algae* **2016**, *58*, 35–44. [CrossRef] [PubMed]
17. Qi, L.; Hu, C.; Wang, M.; Shang, S.; Wilson, C. Floating algae blooms in the East China Sea. *Geophys. Res. Lett.* **2017**, *44*, 501–509. [CrossRef]
18. Xing, Q.; Guo, R.; Wu, L.; An, D.; Cong, M.; Qin, S.; Li, X. High-resolution satellite observations of a new hazard of golden tides caused by floating *Sargassum* in winter in the Yellow Sea. *IEEE Geosci. Remote Sens. Lett.* **2017**, *14*, 1815–1819.
19. Gower, J.; Young, E.; King, S. Satellite images suggest a new *Sargassum* source region in 2011. *Remote Sens. Lett.* **2013**, *4*, 764–773. [CrossRef]
20. Zheng, H.; Liu, Z.; Chen, B.; Xu, H. Quantitative *Ulva Prolifera* bloom monitoring based on multi-source satellite ocean color remote sensing data. *Appl. Ecol. Environ. Res.* **2020**, *18*, 4897–4913.
21. Xing, B.; Li, J.; Zhu, H.; Wei, P.; Zhao, Y. Construction of green tide monitoring system and research on its key techniques. *Int. Arch. Photogramm. Remote Sens. Spat. Inform. Sci.* **2018**, *XLII-3*, 1963–1970.
22. Qiu, Z.; Li, Z.; Bilal, M.; Wang, S.; Sun, D.; Chen, Y. Automatic method to monitor floating macroalgae blooms based on multilayer perception: Case study of Yellow Sea using GOCI images. *Opt. Express* **2018**, *26*, 26810–26829. [CrossRef]
23. Jiang, B.; Fan, D.; Ji, Q.; Obodoefuna, D.C. Dynamic Diurnal Changes in Green Algae Biomass in the Southern Yellow Sea Based on GOCI Images. *J. Ocean Univ. China* **2020**, *19*, 811–817. [CrossRef]
24. Andréfouët, S.; Zubia, M.; Payri, C. Mapping and biomass estimation of the invasive brown algae *Turbinaria ornata* (Turner) J. Agardhand *Sargassum mangarevense* (Grunow) Setchell on heterogeneous Tahitian coral reefs using 4-meter resolution IKONOS satellite data. *Coral Reefs* **2004**, *23*, 26–38.
25. Hoang, T.C.; Cole, A.J.; Fotedar, R.K.; O'Leary, M.J.; Lomas, M.W.; Roy, S. Seasonal changes in water quality and *Sargassum* biomass in southwest Australia. *Mar. Ecol. Prog. Ser.* **2016**, *551*, 63–79. [CrossRef]
26. Andréfouët, S.; Payri, C.; Van Wynsberge, S.; Lauret, O.; Alefaio, S.; Preston, G.; Yamano, H.; Baudel, S. The timing and the scale of the proliferation of *Sargassum polycystum* in Funafuti Atoll, Tuvalu. *J. Appl. Phycol.* **2017**, *29*, 3097–3108. [CrossRef]
27. Gobin, S. Diversité Génétique et Morphologique des Ulves (Ulvophyceae, Chlorophyta) en Nouvelle-Calédonie. Master's Thesis, Université de Nantes, Nantes, France, 2019; p. 39.
28. Lagourgue, L. Genetic and Morphological Diversity of Ulva (Ulvophyceae, Chlorophyta) in New Caledonia. In prep.
29. Andréfouët, S.; Cabioch, G.; Flamand, B.; Pelletier, B. A reappraisal of the diversity of geomorphological and genetic processes of New Caledonian coral reefs: A synthesis from optical remote sensing, coring and acoustic multibeam observations. *Coral Reefs* **2009**, *28*, 691–707. [CrossRef]
30. Gairin, E.; Andréfouët, S. Role of habitat definition on Aichi Target 11: Examples from New Caledonian coral reefs. *Mar. Policy* **2020**, *116*, 103951. [CrossRef]
31. Andréfouët, S.; Chagnaud, N.; Chauvin, C.; Kranenburg, C. *Atlas of the French Overseas Coral Reefs*; Centre IRD de Nouméa: Nouméa, New Caledonia, 2008; p. 154.
32. Observatoire de l'Environnement Nouvelle Calédonie. Available online: www.oeil.nc (accessed on 5 January 2021).
33. Andréfouët, S.; Wantiez, L. Characterizing the diversity of coral reef habitats and fish communities found in a UNESCO World Heritage Site: The strategy developed for Lagoons of New Caledonia. *Mar. Pollut. Bull.* **2010**, *61*, 612–620. [CrossRef]
34. Pleiades | Le Site du Centre National d'Etudes Spatiales. Available online: https://pleiades.cnes.fr/fr (accessed on 5 January 2021).
35. Braun-Blanquet, J. *Plant. Sociology (Translation Fuller. GD and Conrad HS)*; McGrax-Hill: New York, NY, USA, 1932; p. 539.
36. Copernicus Open Access Hub. Available online: www.scihub.copernicus.eu (accessed on 5 January 2021).
37. Clua, E.; Legendre, P.; Vigliola, L.; Magron, F.; Kulbicki, M.; Sarramegna, S.; Labrosse, P.; Galzin, R. Medium scale approach (MSA) for improved assessment of coral reef fish habitat. *J. Exp. Mar. Biol. Ecol.* **2006**, *333*, 219–230. [CrossRef]
38. Hochberg, E.J.; Peltier, S.A.; Maritorena, S. Trends and variability in spectral diffuse attenuation of coral reef water. *Coral Reefs* **2020**, *39*, 1377–1389.
39. Andréfouët, S.; Muller-Karger, F.; Hochberg, E.; Hu, C.; Carder, K. Change detection in shallow coral reef environments using Landsat 7 ETM+ data. *Remote Sens. Environ.* **2001**, *79*, 150–162. [CrossRef]
40. Baret, F.; Guyot, G. Potentials and limits of vegetation indices for LAI and APAR assessment. *Remote Sens. Environ.* **1991**, *35*, 161–173.
41. Tucker, C.J. Red and photographic infrared linear combinations for monitoring vegetation. *Remote Sens. Environ.* **1979**, *8*, 127–150. [CrossRef]
42. Fassnacht, K.S.; Gower, S.T.; MacKenzie, M.D.; Nordheim, M.D.; Lillesand, T.M. Estimating the leaf area index of north central Wisconsin forests using the Landsat thematic mapper. *Remote Sens. Environ.* **1997**, *61*, 229–245. [CrossRef]
43. Chen, J.M. Evaluation of vegetation indices and a modified simple ratio for boreal applications. *Can. J. Remote Sens.* **1996**, *22*, 229–242. [CrossRef]
44. Gong, P.; Pu, R.; Biging, G.S.; Larrieu, M.R. Estimation of forest leaf area index using vegetation indices derived from Hyperion hyperspectral data. *IEEE Trans. Geosci. Remote Sens.* **2003**, *41*, 1355–1362.
45. Kiage, L.M.; Walker, N.D. Using NDVI from MODIS to monitor duckweed bloom in Lake Maracaibo, Venezuela. *Water Resour. Manag.* **2009**, *23*, 1125–1135.
46. Villa, P.; Mousivand, A.; Bresciani, M. Aquatic vegetation indices assessment through radiative transfer modeling and linear mixture simulation. *Int. J. Appl. Earth Obs. Geoinf.* **2014**, *30*, 113–127. [CrossRef]

47. Pu, R.; Bell, S.; English, D. Developing hyperspectral vegetation indices for identifying seagrass species and cover classes. *J. Coast. Res.* **2015**, *31*, 595–615. [CrossRef]
48. Qing, S.; Runa, A.; Shun, B.; Zhao, W.; Bao, Y.; Hao, Y. Distinguishing and mapping of aquatic vegetations and yellow algae bloom with Landsat satellite data in a complex shallow Lake, China during 1986–2018. *Ecol. Indic.* **2020**, *112*, 106073. [CrossRef]
49. Visitacion, M.R.; Alnin, C.A.; Ferrer, M.R.; Suñiga, L. Detection of algal bloom in the coastal waters of Boracay, Philippines, using normalized difference vegetation index (NDVI) and floating algae index (FAI). *Int. Arch. Photogramm. Remote Sens. Spat. Inform. Sci.* **2019**, *XLII-4*, 8. [CrossRef]
50. Hu, C.; He, M.X. Origin and offshore extent of floating algae in Olympic sailing area. *EosTrans. Am. Geophys. Union* **2008**, *89*, 302–303. [CrossRef]
51. Valle, M.; Palà, V.; Lafon, V.; Dehouck, A.; Garmendia, J.M.; Borja, Á.; Chust, G. Mapping estuarine habitats using airborne hyperspectral imagery, with special focus on seagrass meadows. *Estuar. Coast. Shelf Sci.* **2015**, *164*, 433–442. [CrossRef]
52. Armstrong, R.A. Remote sensing of submerged vegetation canopies for biomass estimation. *Int. J. Remote Sens.* **1993**, *14*, 621–627. [CrossRef]
53. Dierssen, H.M.; Chlus, A.; Russell, B. Hyperspectral discrimination of floating mats of seagrass wrack and the macroalgae Sargassum in coastal waters of Greater Florida Bay using airborne remote sensing. *Remote Sens. Environ.* **2015**, *167*, 247–258. [CrossRef]
54. Yang, D.; Yang, C. Detection of seagrass distribution changes from 1991 to 2006 in Xincun Bay, Hainan, with satellite remote sensing. *Sensors* **2009**, *9*, 830–844. [CrossRef]
55. Lyons, M.; Phinn, S.; Roelfsema, C. Integrating Quickbird multi-spectral satellite and field data: Mapping bathymetry, seagrass cover, seagrass species and change in Moreton Bay, Australia in 2004 and 2007. *Remote Sens.* **2011**, *3*, 42–64.
56. Ilori, C.O.; Pahlevan, N.; Knudby, A. Analyzing performances of different atmospheric correction techniques for Landsat 8: Application for coastal remote sensing. *Remote Sens.* **2019**, *11*, 469. [CrossRef]
57. Zoffoli, M.L.; Frouin, R.; Kampel, M. Water column correction for coral reef studies by remote sensing. *Sensors* **2014**, *14*, 16881–16931. [CrossRef]

Article

Improving Water Leaving Reflectance Retrievals from ABI and AHI Data Acquired Over Case 2 Waters from Present Geostationary Weather Satellite Platforms

Bo-Cai Gao * and Rong-Rong Li

Remote Sensing Division, Code 7232, Naval Research Laboratory, Washington, DC 20375, USA; rong-rong.li@nrl.navy.mil

* Correspondence: gao@nrl.navy.mil

Received: 2 September 2020; Accepted: 4 October 2020; Published: 7 October 2020

Abstract: The current generation of geostationary weather satellite instruments, such as the Advanced Baseline Imagers (ABIs) on board the US NOAA GOES 16 and 17 satellites and the Advanced Himawari Imagers (AHIs) on board the Japanese Himawari-8/9 satellites, have six channels located in the visible to shortwave IR (SWIR) spectral range. These instruments can acquire images over both land and water surfaces at spatial resolutions between 0.5 and 2 km and with a repeating cycle between 5 and 30 min depending on the mode of operation. The imaging data from these instruments have clearly demonstrated the capability in detecting sediment movements over coastal waters and major chlorophyll blooms over deeper oceans. At present, no operational ocean color data products have been produced from ABI data. Ocean color data products have been operationally generated from AHI data at the Japan Space Agency, but the spatial coverage of the products over very turbid coastal waters are sometimes lacking. In this article, we describe atmospheric correction algorithms for retrieving water leaving reflectances from ABI and AHI data using spectrum-matching techniques. In order to estimate aerosol models and optical depths, we match simultaneously the satellite-measured top of atmosphere (TOA) reflectances on the pixel by pixel basis for three channels centered near 0.86, 1.61, and 2.25 μm (or any combinations of two channels among the three channels) with theoretically simulated TOA reflectances. We demonstrate that water leaving reflectance retrievals can be made from ABI and AHI data with our algorithms over turbid case two waters. Our spectrum-matching algorithms, if implemented onto operational computing facilities, can be complimentary to present operational ocean versions of atmospheric correction algorithms that are mostly developed based on the SeaWiFS type of two-band ratio algorithm.

Keywords: remote sensing; sensors; ocean color; sediment; turbid water; chlorophyll; geostationary satellite

1. Introduction

The present geostationary weather satellite instruments, such as the Advanced Baseline Imagers (ABIs) on board the US GOES 16 and 17 satellites and the Advanced Himawari Imagers (AHIs) on board the Japanese Himawari-8/9 satellites, have six channels located in the visible to shortwave IR (SWIR) spectral range. These instruments can take images over land and water areas at spatial resolutions between 0.5 and 2 km and with a repeating cycle between 5 and 30 min depending on the mode of operation. The imaging data from these instruments and from other geostationary weather satellite instruments have clearly demonstrated the capability in detecting sediment movements over coastal waters and major chlorophyll blooms over deeper oceans [1–3].

During the early planning stage of GOES-R (now renamed as GOES-16), NASA, and NOAA envisioned the Hyperspectral Environment Suite (HES) instrument [4] to fulfill the future needs and requirements of the NESDIS (National Environmental Satellite, Data, and Information Service) Office [4]. NOAA was once very interested in the coastal waters (CW) task. HES was supposed to have an instrument like an imaging spectrometer to have hourly coverages over the US east coast, west coast, Gulf of Mexico, and Hawaii coastal waters. Later on, HES was canceled by the GOES-R Project because of cost constraints, while ABI remained on GOES-R. Due to the lack of a green band on ABI, NOAA has not generated operational ocean color data products, including suspended sediments, from ABI data. On the other hand, through visual inspection of GOES-R visible band images and movies provided on a web site at the Colorado State University (CIRA), it is easy to find interesting cases of ocean color observations with ABI. Figure 1 shows such an example. Figure 1A–D are false color images (Red: 0.64 μm; Green: 0.865 μm, Blue: 0.47 μm) acquired at UTC 1332, 1532, 1732, and 1932, respectively, on September 14, 2017 over the coastal areas of Florida. The silt stirred up by Hurricane Irma was captured in the Gulf Stream and advected northward while spinning off eddies. The original movie images at a time interval of five minutes can be found from http://rammb.cira.colostate.edu/ramsdis/online/loop_of_the_day/ for the day of 2017/09/14.

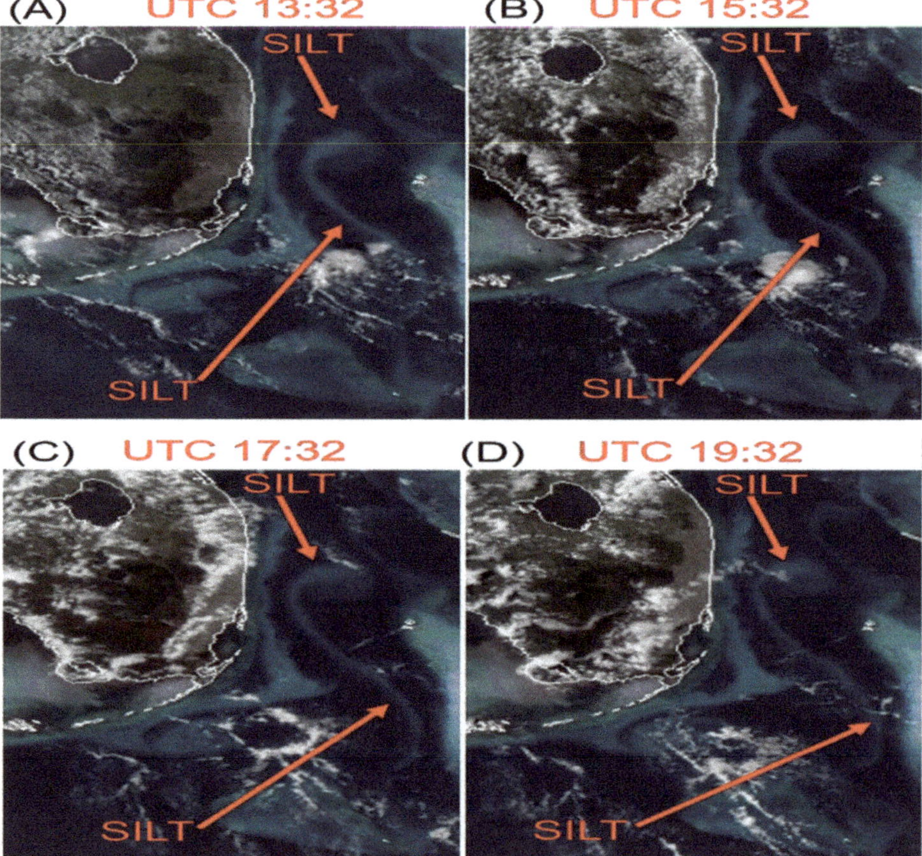

Figure 1. (**A–D**): time series of Advanced Baseline Imagers (ABI) images acquired at UTC 1332, 1532, 1732, and 1932 on September 14, 2017. The silt stirred up by Hurricane Irma was captured in the Gulf Stream and advected northward while spinning off eddies.

It should be pointed out that the silt features in Figure 1 are absent in images acquired under normal atmospheric wind conditions. With polar orbiting ocean color instruments, such as the aqua moderate resolution imaging spectroradiometer (MODIS) [5] and visible infrared imaging radiometer suite (VIIRS) [6], having a two-day global coverage, it was impossible to capture the rapid silt movement after the major hurricane event. Figure 1 also demonstrates partial coastal watch capabilities with ABI originally intended by NOAA using the HES instrument. At present, due to the lacking of a green channel on ABI, no operational ocean color data products have been produced from ABI data at NOAA. If operational ocean color data products were produced, our ability in monitoring water discharging from major US river systems to oceans and tracking turbid water movements over coastal waters could have been greatly improved because of rapid repeating cycles of ABI measurements.

The Japanese AHI instruments on board the Himawari-8/9 geostationary satellite platforms are equipped with a green band. Operational ocean color data products have been generated at the Japan Aerospace Exploration Agency (JAXA). The atmospheric correction algorithm for processing AHI data was described by Murakami [7]. This algorithm was adapted from an earlier version of the atmospheric correction algorithm [8] developed for processing the Japanese Global Imager (GLI) data, while the GLI algorithm shared the same basis with the SeaWiFS algorithm [9]. During the atmospheric correction processes using SeaWiFS-types of algorithms, the pure Rayleigh scattering effects for two bands are first subtracted out. The ratio of the two bands is then calculated. From the ratio value, an aerosol model for a given pixel is estimated. One problem with the ratio calculation for optical bands with additive noises is that it magnified the noise by a factor of 2 [10]. The noise is then propagated down to the derived water leaving reflectances of visible bands. In order to improve signal to noise ratios of measured AHI data, temporal averaging on the hourly basis is required for operational atmospheric corrections to AHI data [7]. Figure 2 shows sample ocean color results downloaded from a public Himawari web site (eroc.jaxa.jp/ptree/index.html). Figure 2a is an RGB image covering eastern part of China–Korea Peninsula, and part of Japan. The image was acquired on May 28, 2017 at UTC 0200. Land/water boundaries are drawing in yellow colored lines. Turbid coastal waters in eastern part of China and around the Korean Peninsula are seen obviously. Figure 2b is the corresponding false color chlorophyll concentration image overlaid on the RGB image. Over portions of turbid coastal water areas, such as those areas pointed by arrows, no retrievals are made. Figure 2c is another case of false color chlorophyll concentration image overlaid on the RGB image. The AHI data was acquired on April 26, 2020 at UTC 0300. In this specific case, no retrievals were made over most water surfaces in eastern coastal areas of China.

Historically, the first geostationary ocean color sensor, the geostationary ocean color imager (GOCI) [11], was launched into space in 2010. This sensor has eight SeaWiFS-like bands. Several atmospheric correction algorithms [12–16] sharing the same base as the SeaWiFS 2-band ratio algorithm but with additional assumptions about absorption and scattering properties of coastal waters in the 0.66–0.865 μm spectral range have been developed. The performance of these algorithms has been objectively evaluated [17]. The majority of multi-spectral and hyperspectral ocean versions of atmospheric correction algorithms have recently been reviewed in a lengthy paper by Frouin et al. [18]. Because GOCI lacks SWIR bands above 1 μm for atmospheric correction purposes, the spatial coverage of atmosphere-corrected GOCI data products over very turbid coastal waters is also frequently lacking.

In view of the absence of ABI ocean versions of atmospheric correction algorithms and the needs for improving spatial coverage over turbid coastal water areas with AHI ocean color retrievals, we have recently developed spectrum-matching versions of atmospheric correction algorithms using SWIR bands above 1 μm, where the turbid coastal water surfaces are often very dark, for retrieving water leaving reflectances from ABI and AHI data. In Section 2 of this article, we describe the spectral characteristics of ABI and AHI instruments and the spectrum-matching technique for water leaving reflectance retrievals. We present sample retrievals from ABI and AHI data in Section 3. We give a short discussion in Section 4. Finally, we provide a brief summary in Section 5.

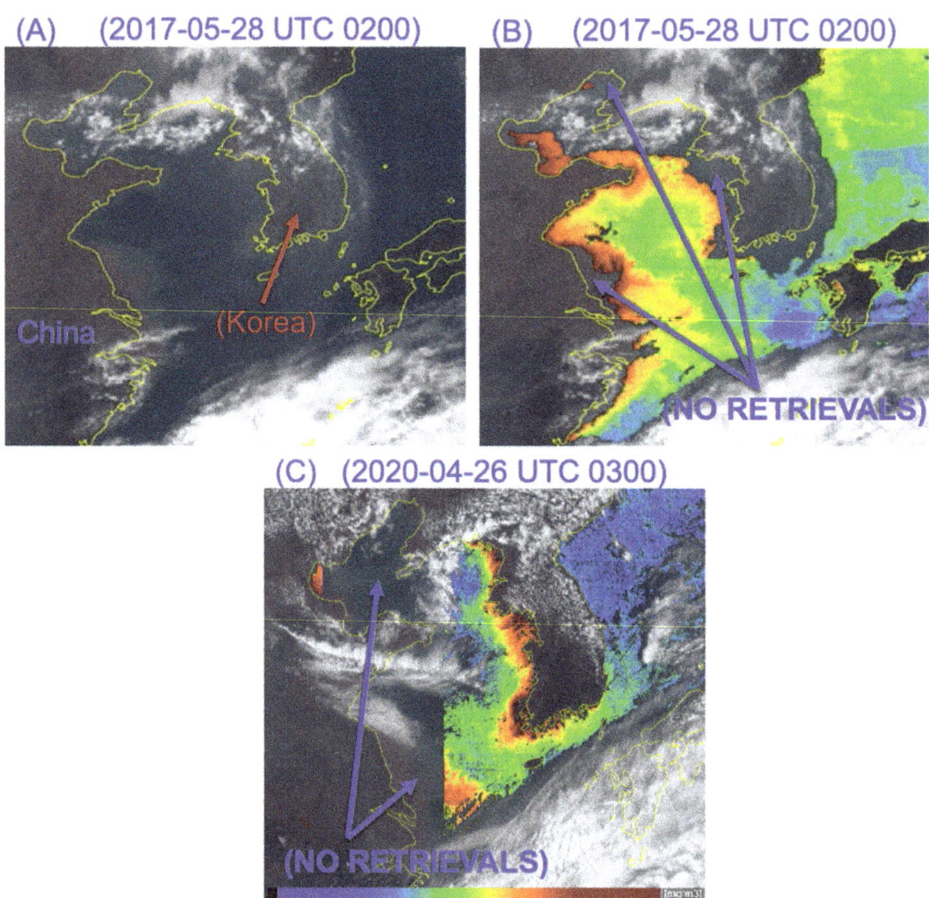

Figure 2. (**A**): A true color RGB image acquired with the Japanese Advanced Himawari Imagers (AHI) instrument over part of Japan, Korean Peninsula, and eastern part of China at UTC 0200 on May 28, 2017; (**B**): Similar to (**A**) but with a false chlorophyll concentration image overlaid; and (**C**): Another case of an AHI false color chlorophyll concentration image overlaid on an RGB image that was acquired on April 26, 2020 at UTC 0300.

2. Data and Methods

2.1. The ABI and AHI Instrument Characteristics

The ABI instruments, currently on board the NOAA GOES 16 and 17 geostationary weather satellite platforms, have six bands in the solar spectra range. These bands are centered near 0.47, 0.64, 0.865, 1.378, 1.61, and 2.25 μm. The positions and widths of these bands are illustrated in Figure 3. The bands are plotted above a green algae reflectance spectrum, which was retrieved from an AVIRIS (airborne visible infrared imaging spectrometer) [19] data set acquired over a salt pond in the San Francisco Bay in August of 2009, with a version of our hyperspectral atmospheric correction algorithm (ATREM) [20]. It is obvious to see that ABI does not have a green band to center near the 0.55 μm. To improve the ocean color capability, the Japanese AHI instruments on board the Himawari-8/9 geostationary satellite platforms are equipped with a green band centered at 0.51 μm, which is illustrated in Figure 3. The basic designs of AHI are the same as those of ABI, except filter changes. AHI replaced ABI's 1.378-μm cirrus band with the 0.51-μm green band. The addition of the green band enhances the water color information that can be obtained from the AHI data in comparison with that from the ABI data.

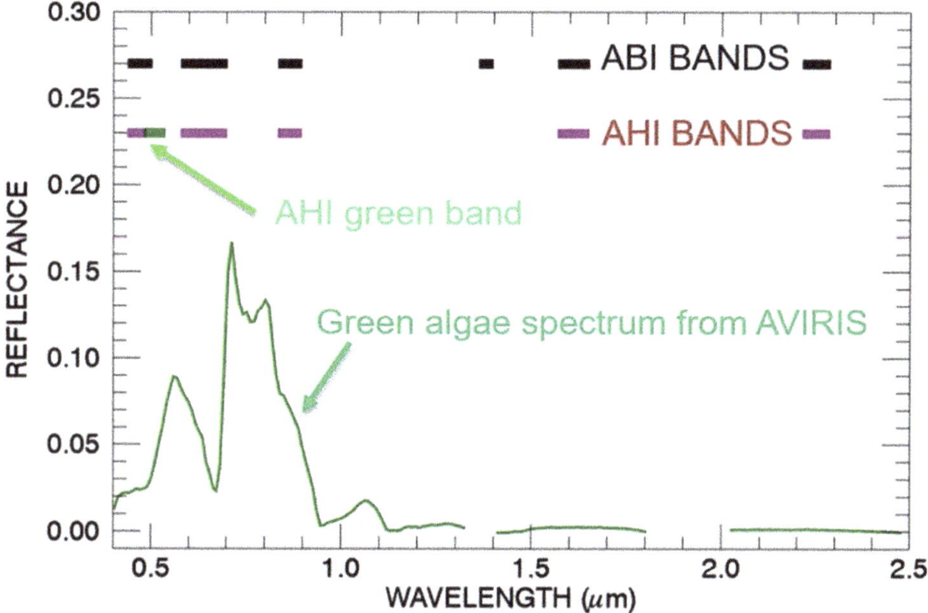

Figure 3. Bandpasses of ABI and AHI instruments. An airborne visible infrared imaging spectrometer (AVIRIS) green algae reflectance spectrum is also shown.

2.2. *Atmospheric Corrections for ABI and AHI Data*

At present, most ocean color data products have been generated from polar orbiting satellite data at various data centers. Atmospheric correction algorithms used for data product generations are mostly based on the Gordon and Wang [9] (hereafter referred as GW94) SeaWiFS types of 2-band algorithms that were designed mainly for clear deep ocean waters (case 1 waters). In the GW94 algorithm, the SeaWiFS 0.75-μm and 0.86-μm bands are used as atmospheric correction bands assuming water leaving reflectances from the two bands are close to zero. Over coastal waters, suspended sediments can contribute non-negligible water leaving reflectances to the 0.75- and 0.86-μm bands. Several investigators have tried to remove the NIR water leaving reflectance contributions from the measured top-of-atmosphere NIR apparent reflectances with different techniques [21,22], so that a "black pixel" could be provided to the SeaWiFS 2-band ratio type of atmospheric correction algorithms.

Previously, we developed an ocean version of hyperspectral atmospheric correction algorithm [20] mainly for supporting the Navy COIS (coastal ocean imaging spectrometer) project [23]. For the turbid coastal environment, the water leaving radiances in the 0.75–0.865 μm spectral range are typically not close to zero mainly because of scattering by suspended materials, particularly for the 0.75-μm band. Under these conditions, the channels in this spectral region have limited use for retrieving information on atmospheric aerosols. Because the SeaWiFS types of algorithms derive aerosol information from channels in the 0.75–0.865 μm spectral range, these algorithms cannot be easily adapted for operational retrieval of water leaving radiances over coastal waters. In view of this limitation, we designed a spectrum-matching algorithm [20] that allowed the use of channels in shortwave IR (SWIR) spectral regions between 1 and 2.5 μm, where the turbid waters are much darker, for the estimates of aerosol models and optical depths. Our algorithm was based on Robert Fraser's radiative transfer formulation and algorithm [24,25]. The adoption of the Fraser formulation permits the simultaneous matching between measured radiances of several bands centered at different wavelengths with those from theoretical simulations and results in more stable estimates of aerosol models and optical depths.

In our algorithm, we express radiances in reflectance units. We adopt the standard definition of apparent reflectance ρ^*_{obs} at a satellite level for a given wavelength as [20,26]

$$\rho^*_{obs} = \pi\, L_{obs}/(\mu_o\, E_o), \quad (1)$$

where L_{obs} is the radiance of the ocean-atmosphere system measured by a satellite instrument, μ_o the cosine of solar zenith angle, and E_o the downward solar irradiance at the top of the atmosphere when the solar zenith angle is equal to zero. Neglecting the interactions between atmospheric gaseous absorption and molecular and aerosol scattering, ρ^*_{obs} can be expressed as [20,26]

$$\rho^*_{obs} = T_g\, [\rho^*_{atm+sfc} + \rho_w\, t_d\, t_u/(1 - s\, \rho_w) \quad (2)$$

where T_g is the total atmospheric gaseous transmittance on the Sun-surface-sensor path, $\rho^*_{atm+sfc}$ the reflectance resulted from scattering by the atmosphere and specular reflection by ocean surface facets, t_d the downward transmittance (direct + diffuse), and t_u the upward transmittance, s the spherical albedo that takes account of reflectance of the atmosphere for isotropic radiance incident at its base, and ρ_w the water leaving reflectance.

Solving Equation (2) for ρ_w yields

$$\rho_w = (\rho^*_{obs}/T_g - \rho^*_{atm+sfc})/[t_d\, t_u + s\, (\rho^*_{obs}/T_g - \rho^*_{atm+sfc})] \quad (3)$$

Given a satellite measured radiance, the water leaving reflectance can be derived according to Equations (1) and (3) provided that the other quantities in the righthand side of Equation (3) can be modeled theoretically. Because of the availability of the vector radiative transfer code, the proper atmospheric layering structure in this code (up to 80 layers with proper mixing of aerosol particles and atmospheric molecules in each layer), and the treatment of wind-roughened water surfaces, we have used a modified version of Ahmad and Fraser code [27] to generate lookup tables used in our retrieving algorithm. Specifically, we use the code to generate the quantities $\rho^*_{atm+sfc}$, t_d, t_u, and s in Equation (2). Lookup tables for 14 wavelengths between 0.39 and 2.5 µm in atmospheric "window" regions, sets of aerosol models, optical depths, solar and view angles, and surface wind speeds have been generated. Aerosol models, similar to those used in the SeaWiFS algorithm, are used during our table generation. A sophisticated line-by-line based atmospheric transmittance code is used in our algorithm to calculate contiguous atmospheric gaseous transmittance spectra (T_g).

We match the satellite measured "apparent reflectances" of selected NIR and SWIR bands with those pre-computed apparent reflectances for different aerosol models and optical depths. The simulated lookup table apparent reflectances (after interpolation) have step sizes of 0.001 or smaller. The fine step sizes are smaller than the noise levels of satellite measured data. During the retrieving process for a given pixel, we search through all possible combinations of aerosol models and optical depths in lookup tables, and find the best match of the aerosol model and optical depth that minimizes the sum of the squared differences between the measured apparent reflectances (two or more NIR and SWIR bands) and those from theoretical simulations. Because of large computer memory sizes for storing lookup tables and because of the array processing capability of our codes written in Fortran 90, our table searching and minimization process is very fast. We did not need to implement an additional numerical optimization method, such as the commonly used gradient descent method, to speed up the table searching and minimization process. For example, in one test case, we processed one hyperspectral data set having 614 pixels in the cross-track direction, 1024 lines in the along-track direction, and 224 spectral bands, using a Macbook computer having a 2.9 GHz Dual-Core Intel Core i5 processor. It took about 500 megabytes of memory and 140 seconds of computer time to finish the data processing. Our hyperspectral atmospheric correction algorithm has been tested with images acquired with a number of instruments, including the airborne visible infrared imaging spectrometer (AVIRIS) [19] from an ER-2 aircraft at an altitude of 20 km.

We also adapted the hyperspectral ocean atmospheric correction algorithm for processing multi-channel imagery, such as those acquired with the Terra and Aqua MODIS instruments [28]. In the MODIS algorithm, the lookup tables corresponding to MODIS wavelengths are obtained through linear interpolation of the tables generated for the hyperspectral atmospheric correction algorithm. The lookup table quantities, $\rho^*_{atm+sfc}$, t_d, t_u, and s, are functions of wavelength, solar zenith angle, view zenith angle, relative azimuth angle, aerosol model, optical depth, relative humidity, and surface wind speed, respectively. For MODIS data, each pixel's solar zenith angle, view zenith angle, and relative azimuth angle are variable because of the large swath width (~2800 km for one scan line). In order to speed up the retrieving process, the storage order for different MODIS lookup table quantities (such as wavelengths, particle size distributions, solar and view angles, etc.) is optimized. Because MODIS has water vapor bands centered near 0.94 μm, we derive water vapor values from MODIS data on the pixel by pixel basis, and then use the derived water vapor values to calculate water vapor transmittances for other bands used for retrievals. Climatological atmospheric ozone, carbon dioxide, and methane amounts are used for modeling relevant gaseous transmittances. Similarly, we developed a VIIRS ocean version of algorithm. The VIIRS channels centered at 1.24, 1.61, and 2.25 μm with proper modeling of atmospheric CO_2 and CH_4 absorption effects and a spectrum-matching technique were used for atmospheric corrections.

Based on our experience in developing MODIS and VIIRS versions of atmospheric correction algorithms, we have recently developed spectrum-matching versions of ABI and AHI multi-channel atmospheric correction algorithms. The main tasks in the ABI and AHI algorithm development were to generate proper transmittance tables of atmospheric gases, including those of ozone, water vapor, carbon dioxide, and methane, for ABI and AHI bands in the 0.4–2.3 μm, as illustrated in Figure 2. The spectrum-matching technique is also used for atmospheric corrections and for the subsequent retrieval of water leaving reflectances from ABI and AHI data. Figure 4 illustrates the spectrum-matching technique.

Figure 4a is a false color RGB image (Red: 0.637; Green: 0.864; Blue: 0.471 μm) acquired at UTC 1100 on December 17, 2018 with ABI on board the GOES-16 satellite platform. The image covered algae blooming areas off the coast of Argentina in the Atlantic Ocean. Figure 4b illustrates the spectrum-matching technique for water leaving reflectance retrievals. The red line (marked with the plus symbol) in Figure 4b shows the apparent reflectances for 5 ABI bands for a water pixel located at the center of the red square in the right side of the Figure 4a image. The 0.86-, 1.61-, and 2.24-μm bands were used for the estimation of aerosol model and optical depth by minimizing the sum of the squared differences between the measured data and calculated data during the spectrum-matching process. The estimated aerosol model and optical depth were extracted back to the ABI blue and red bands. The green line (marked with "*") shows the estimated atmospheric path radiances in reflectance unit, $\rho^*_{atm+sfc}$ (as defined in Equation (2)), for all the 5 ABI bands. The differences between the red line and the green line are proportional to water leaving reflectances (see Equation (3)). The blue line (marked with the triangle symbol) shows the water leaving reflectances of ABI bands for the algae pixel. The water leaving reflectance value for the blue band is close to 0.1, indicating that the pixel is highly reflecting in the blue spectral region.

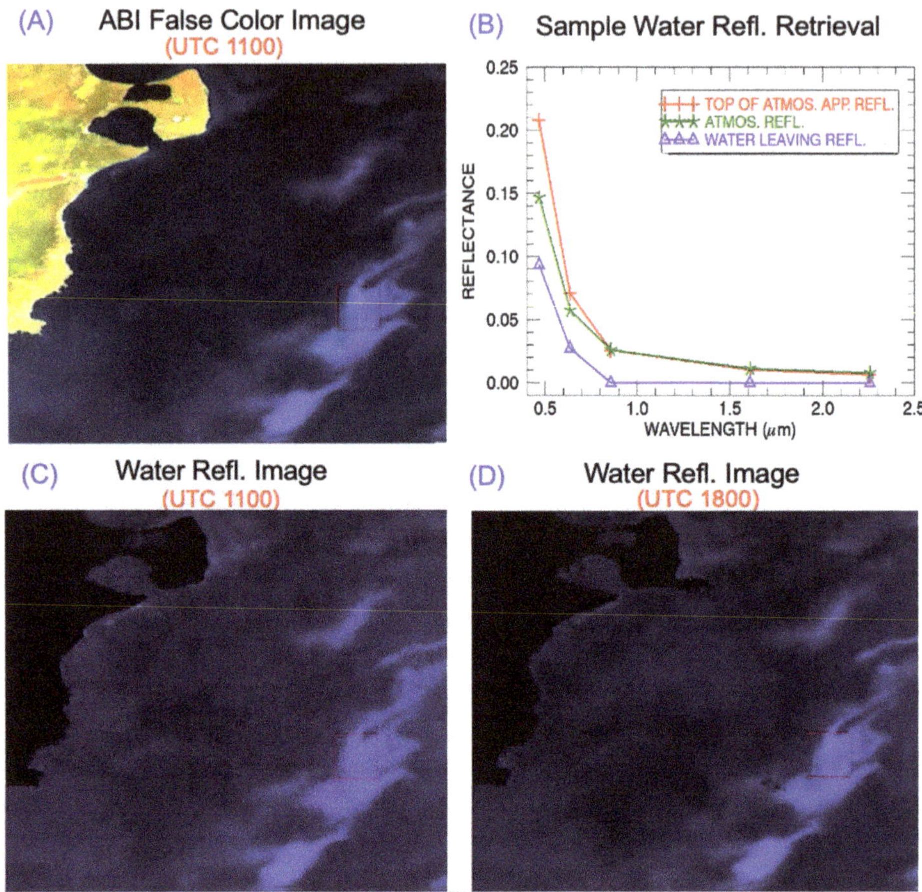

Figure 4. (**A**)—a false color ABI image acquired at UTC 1100 on December 17, 2018 off the coast of Argentina covering algae blooming regions; (**B**)—spectral plot for a bright pixel for top of atmosphere (TOA) apparent reflectances (red line), atmospheric reflectances (green line), and water leaving reflectances (blue line); (**C**)—the retrieved false color water leaving reflectance image for the UTC 1100 scene; and (**D**)—the derived false color water leaving reflectance image for the UTC 1800 scene.

3. Results

We have selected three examples to demonstrate the capabilities in deriving water leaving reflectances from ABI and AHI data. The selection of the ABI and AHI cases was based on ABI and AHI time sequences of movie images available from a public web site associated with the Cooperative Research Program at Colorado State University. The first example is for the ABI scene covering a major blooming event off the coast of Argentina (see Figure 4a). The second example is for an ABI scene over the Gulf of Mexico where large amounts of sediments were discharged from rivers into the Gulf of Mexico. The third case is for a Japanese AHI scene for the coastal areas of Japan, Korea Peninsula, and eastern part of China, where very turbid coastal waters were present.

3.1. ABI Scene Over Coastal Area of Argentina, December 17, 2018

The ABI instrument on GOES-16 captured a major chlorophyll blooming event occurred in the middle of December of 2018. As described in Section 2, Figure 4a is a false color RGB image acquired at UTC 1100 on December 17 over algae blooming areas off the coast of Argentina in the Atlantic Ocean.

Figure 4b shows the results of matching technique for water leaving reflectance retrievals from ABI data. Because ABI didn't have a true green band, it is not possible to observe the green reflectance peak centered near 0.55 μm in the Figure 4b water leaving reflectance curve. Figure 4c is the false color RGB image processed from the retrieved water leaving reflectances of the Figure 4a UTC 1100 scene. In principle, water leaving reflectance images, similar to the one illustrated in Figure 4c, at a time interval of approximately 10 minutes can be obtained from ABI data because of the rapid repeating cycle of ABI measurements. Figure 4d shows another false color ABI water leaving reflectance image for UTC 1800. Although the NASA MODIS instrument onboard the Aqua Spacecraft acquired image at UTC 1755 over the same area and covered the same chlorophyll patches, the NASA operational MODIS algorithm didn't make ocean color retrievals over areas containing the chlorophyll features in the MODIS scene. As a result, it is not possible to make quantitative comparisons between our retrieval results with those from the NASA operational algorithm. Nevertheless, the images and line plots in Figure 4 demonstrate that water leaving reflectance images can be retrieved from ABI data with our ABI version of multi-channel atmospheric correction algorithm.

3.2. ABI Scene Over Coastal Area of Gulf of Mexico, March 20, 2019

Figure 5 shows another example of water leaving reflectance retrieval from ABI data. The data sets were acquired on March 20, 2019 off the coastal area of Louisiana and Texas. At the time, large amounts of sediments were discharged from rivers into the Gulf of Mexico. Figure 5a,b is a false color RGB image (R: 0.637; G: 0.864; B: 0.471 μm) acquired at UTC 1730 and 1930, respectively. The brownish-colored areas contained large amount of sediments. Figure 5c,d is the corresponding false color water leaving reflectance images. The sediment-dominated areas were picked up very nicely in the Figure 5c,d. The ABI 1.61-, and 2.24-μm bands were used during atmospheric corrections. Figure 5e illustrates the atmospheric correction for one sediment-dominated pixel located near the center of the rectangle outlined in the Figure 5b,d images. The six points marked with "*" are the top of atmosphere (TOA) apparent reflectances for the six ABI bands. The five points marked with small green-colored triangles are the retrieved water leaving reflectances for the pixel. The atmosphere-corrected 0.637-μm red band reflectance is about 0.12, which indicates that the sediment-dominated pixel is highly reflecting in the red band. Although the Aqua MODIS instrument had a good overpass over the same area at UTC 1925, the sediment-dominated areas were masked out by the NASA MODIS atmospheric correction algorithm and no water leaving reflectance retrievals were made. Figure 5 demonstrates that ABI data can be used for water leaving reflectance retrievals over sediment-dominated areas. The data products from ABI can be complimentary to the existing NASA polar-orbiting dominated ocean color data products, particularly over turbid coastal waters.

For comparison and validation purposes, we have also retrieved water leaving reflectances from a MODIS data set acquired at UTC 1925 on the same day over similar areas. Figure 6a is a true color MODIS RGB image. Figure 6b is a false color image (R: 0.645 μm; G: 0.845 μm; B: 0.466 μm) of the same scene. Figure 6b covers nearly the same area as that of Figure 5b. Major land and water surface features in both images are approximately the same. However, because the Aqua spacecraft has a declination angle of 98.5 degree and because no spatial registration is made between the MODIS image and the ABI image, the small turbid water areas in the upper right portion of Figure 6b are not covered in Figure 5b. Figure 6c is the false color water leaving reflectance image (R: 0.645 μm; G: 0.845 μm; B: 0.466 μm). Based on visual inspection of images in Figures 6c and 5c, we selected nearly same spatial areas and extracted spectral reflectances. Figure 6d shows water leaving reflectances derived from ABI (dashed green curve) and from MODIS data (solid line) using narrow bands near 0.86, 1.61, and 2.25 μm. By comparing the two curves in Figure 6d, it is seen that the ABI band reflectance values agreed quite well with those of the corresponding MODIS band values. The good agreement partially validated our ABI version of the atmospheric correction algorithm. It should be pointed out that, because MODIS has more narrow bands, the MODIS curve provides more spectral information than the ABI curve. For example, MODIS has a band centered near 0.52 μm and this band has much larger reflectance value than the two nearby bands, while ABI does not have a band centered near 0.52 μm and has less capability in charactering the spectral reflectance properties of water surfaces.

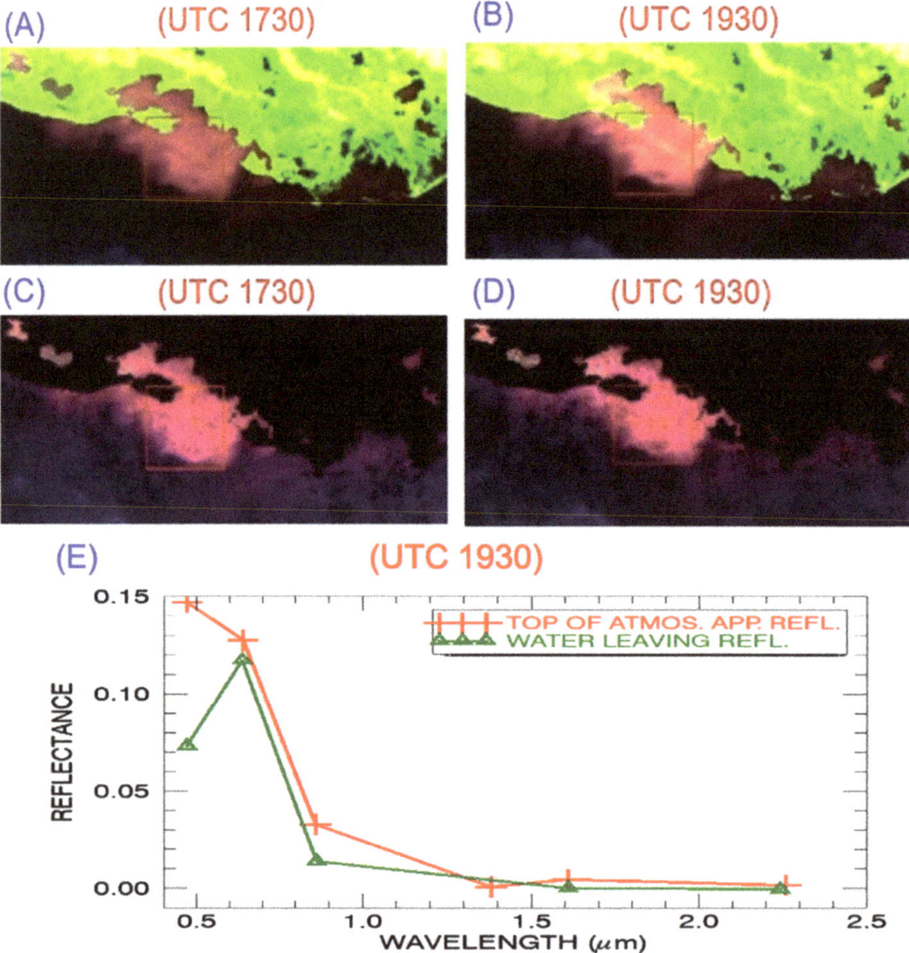

Figure 5. (**A**,**B**): false color ABI images acquired at UTC 1730 and 1930, respectively on March 20, 2019 near the Gulf of Mexico; (**C**,**D**): the corresponding derived water surface reflectance images; (**E**): spectral plot for a pixel before and after atmospheric corrections.

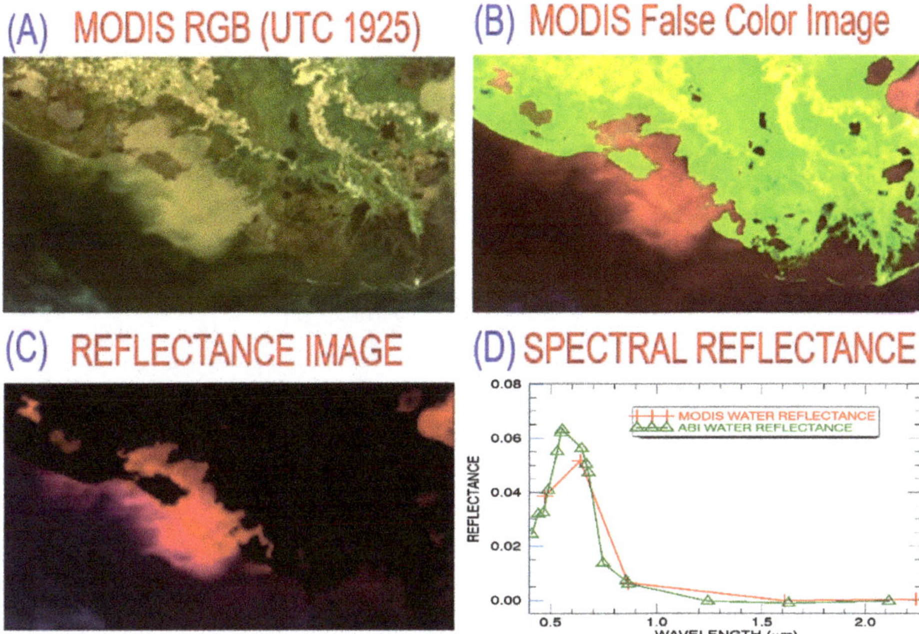

Figure 6. (**A**,**B**)—true color and false color moderate resolution imaging spectroradiometer (MODIS) images acquired near the Gulf of Mexico at UTC 1925 on March 20, 2019; (**C**)—the corresponding derived water surface reflectance images; and (**D**)—water leaving reflectances derived from ABI (dashed green curve) and from MODIS data (solid line).

3.3. AHI Scene Over Eastern Asia, May 28, 2017

Figure 7 shows an example of water leaving reflectance retrievals from the Japanese AHI data measured on May 28, 2017 over part of Japan, Korea, and eastern China. Figure 7a is the RGB image (R: 0.64-; G: 0.51-; B: 0.47-μm) acquired at UTC 0200. It can be difficult to distinguish between bright coastal waters in these visible band images, particularly in the left portion of the scene. Figure 7b shows the AHI image for the 1.61-μm band. A large inland lake (Lake Taihu located near the lower left corner), rivers, and coastal waters are quite dark in this image because of strong liquid water absorption at 1.61 μm. Through several trials and tests, we decided to select a 1.61-μm band threshold. Pixels with the 1.61-μm band TOA apparent reflectance values greater than 0.03 were classified as land or cloudy pixels. These pixels were masked out in our water leaving reflectance data product. Figure 7c shows the retrieved water leaving reflectance image for the scene. With the SWIR spectrum-matching algorithm, we are able to do retrievals over the deep ocean waters, bright coastal waters, and inland lakes and rivers. By comparing Figure 7c with Figure 2b, it seen that we have greatly expanded the retrievals over very turbid water areas in eastern part of China and around Korea Peninsula. In particular, we are able to do retrievals over Lake Taihu, which is seen in the lower left part of Figure 7c. Figure 7d illustrates the atmospheric correction for one turbid water pixel in west coastal area of the Korea Peninsula using two narrow channels centered near 1.61 and 2.25 μm for aerosol retrievals. The six points marked with the "+" symbol are the TOA apparent reflectances for the six AHI bands. The six points marked with small diamonds are the retrieved water leaving reflectances for the pixel. The near zero surface reflectance values for the AHI bands centered near 0.865, 1.61, and 2.25 μm demonstrate that proper minimization is made during the spectrum-matching process.

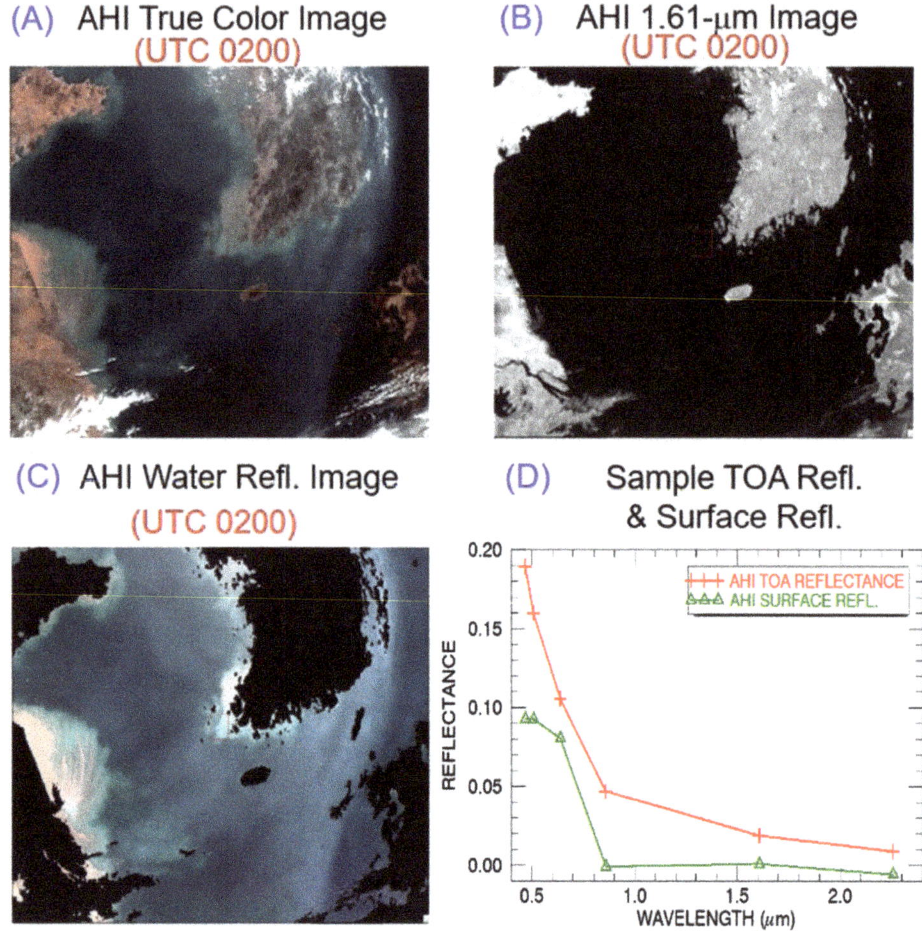

Figure 7. (**A**)—a true color AHI image acquired at UTC 0200 on May 28, 2017 over part of Japan, Korea, and eastern China; (**B**)—the 1.61-μm AHI B/W image; (**C**)—the retrieved true color water leaving reflectance image; and (**D**)—spectral plot for a bright pixel for TOA apparent reflectances marked with the "+"symbols and for the corresponding surface reflectances marked with small diamonds.

4. Discussions

It should be pointed out that water leaving reflectance retrievals, such as those showing in Figures 4, 5 and 7, are made on the pixel by pixel basis with our spectrum-matching versions of ABI and AHI atmospheric correction algorithms. No spatial and temporal averaging of ABI and AHI data is required during the retrievals. Similar retrievals with the SeaWiFS-type of algorithm adopted for AHI data processing by Murakami [7] required temporal averaging on the hourly basis. In the original operational SeaWiFS algorithm, spatial averaging of the 0.75- and 0.86-μm bands was made for improving signal to noise ratios of the two atmospheric correction bands. The calculation of two band ratios magnifies the noise by a factor of two [10]. The spectrum-matching algorithms described here do not need the calculation of two-band ratios. In fact, if two bands are used in the spectrum-matching process (e.g., Figures 5 and 7), the signal to noise ratio is increased by a factor of square root of two. If three bands are used in the matching process (e.g., Figure 4), the signal to noise ratio is increased by square root of three [10]. Therefore, the use of spectrum-matching algorithms for atmospheric corrections and for derivation of water leaving reflectances in the visible requires less signal to noise ratios of the original satellite-measured data than the use of SeaWiFS-type of two-band ratio algorithms. This fine point is very important to atmospheric corrections for the present generation of geostationary weather satellite data, because the relevant instruments were not geared for ocean color measurements with exceedingly high signal to noise ratios. In addition, performing spatial and temporal averaging to improve signal to noise ratios can sometimes be difficult. For example, during a time interval of one hour for the day May 28, 2017 (see Figure 6), the aerosol properties over a given area in the Figure 6 scene changed quite a bit because of the movement of a dust storm through the area. Temporal averaging is not quite suited for improving signal to noise ratios for atmospheric correction purposes for this particular day.

The details on our radiative transfer simulations [27], generation of lookup tables, and spectrum-matching processes were previously described in our hyperspectral ocean version of atmospheric correction algorithm [20] and multi-channel MODIS algorithm [28]. The water leaving reflectance retrieving results from MODIS data were verified with measurements with a portable field spectrometer over surface stations in Bahamas Banks and Florida Bay area [28]. The results were also compared with those from the NASA operational MODIS algorithm. Because the lookup tables used in our ABI and AHI algorithms were obtained through interpolation of the same hyperspectral lookup tables simulated with the Ahmed and Fraser radiative transfer code [27] and because our hyperspectral and MODIS multi-spectral retrieving results were previously validated, we feel that it is not absolutely necessary to have additional validation of our several sample retrieving cases from ABI and AHI data. In addition, we are now having practical difficulties in obtaining additional ABI and AHI data sets for more case studies, because the public ftp access is no longer permitted at our computing facility.

In the commonly used SeaWiFS types of ocean version of atmospheric correction algorithms, the atmospheric path radiances in reflectance unit after subtraction of Rayleigh scattering effects are stored in lookup tables, such as the tables used in the NASA operational SeaWiFS, MODIS, and VIIRS algorithms. Under very clear atmospheric conditions, the Rayleigh-subtracted atmospheric path reflectances for bands centered near 1.24, 1.61, and 2.25 μm can be very small and close to zero. The extension of the SeaWiFS types of 2-band ratio algorithms to SWIR band pairs, such as the 1.24- and 1.61-μm band pair and the 1.24- and 2.25-μm band pair, can encounter the problems of zero divided by zero or zero divided by negative numbers due to the presence of noises in measured SWIR band data. As a result, the use of 2-SWIR band ratio technique for the estimates of aerosol models and optical depths during atmospheric correction processes can be problematic. The multi-channel ocean color research community does not seem to well understand the intrinsic problems [29].

5. Summary

We have developed atmospheric correction algorithms for retrieving water leaving reflectances from ABI and AHI data using spectrum-matching techniques. Our spectrum-matching algorithms reduced significantly the signal to noise ratio requirements for atmospheric correction channels in comparison with those required with the widely used SeaWiFS types of two-channel ratio algorithms by the current multi-channel ocean color research community. We have demonstrated using three case studies that water leaving reflectance retrievals can be made from satellite-measured ABI and AHI data with our spectrum-matching versions of atmospheric correction algorithms. Our algorithms, if implemented onto operational computing facilities, can be complimentary to current operational ocean versions of atmospheric correction algorithms in the area of improving spatial coverage over turbid coastal waters.

Author Contributions: B.-C.G. made initial observations of chlorophyll blooming features and suspended sediment features from ABI and AHI data sets, and developed spectrum-matching versions of algorithms for retrieving water leaving reflectances from ABI and AHI data. R.-R.L. carried out retrievals and detailed data analysis. All authors have read and agreed to the published version of the manuscript.

Funding: This research is partially supported by the US Office of Naval Research and by a research grant managed by Jared K. Entin at the Science Mission Directorate of National Aeronautics and Space Administration.

Acknowledgments: The authors are grateful to Weile Wang affiliated with California State University in Monterey and NASA Ames Research Center for providing the ABI and AHI data in HDF4 format used in this study. The authors are also grateful to Pengwang Zhai of University of Maryland at Baltimore County for useful discussions.

Conflicts of Interest: The authors declare no conflict of interest.

References

1. Chen, X.; Shang, S.; Lee, Z.; Qi, L.; Yan, J.; Li, Y. High-frequency observation of floating algae from AHI on Himawari-8. *Remote. Sens. Environ.* **2019**, *227*, 151–161. [CrossRef]
2. Neukermans, G.; Ruddick, K.; Bernard, E.; Ramon, D.; Nechad, B.; Deschamps, P.-Y. Mapping total suspended matter from geostationary satellites: A feasibility study with SEVIRI in the Southern North Sea. *Opt. Express* **2009**, *17*, 14029–14047. [CrossRef] [PubMed]
3. Neukermans, G.; Ruddick, K.G.; Greenwood, N. Diurnal variability of turbidity and light attenuation in the southern North Sea from the SEVIRI geostationary sensor. *Remote. Sens. Environ.* **2012**, *124*, 564–580. [CrossRef]
4. Martin, G.; Criscione, J.C.; Cauffman, S.A.; Davis, M.A. Geostationary Operational Environmental Satellites (GOES): R series hyperspectral environmental suite (HES) overview. *Remote Sens.* **2004**, *5570*, 173–183. [CrossRef]
5. King, M.D.; Menzel, W.P.; Kaufman, Y.J.; Tanré, D.; Gao, B.C.; Platnick, S.; Hubanks, P.A. Cloud and aerosol properties, precipitable water, and profiles of temperature and humidity from MODIS. *IEEE Trans. Geosci. Remote Sens.* **2003**, *41*, 442–458. [CrossRef]
6. Lee, T.F.; Nelson, C.S.; Dills, P.; Riishojgaard, L.P.; Jones, A.; Li, L.; Hoffman, C. NPOESS: Next-generation operational global earth observations. *Bull. Amer. Meteorol. Soc.* **2010**, *91*, 727–740. [CrossRef]
7. Murakami, H. Ocean color estimation by Himawari-8/AHI. In Proceedings of the of SPIE Asia-Pacific Remote Sensing, New Delhi, India, 4–7 April 2016; Volume 9878, p. 987610.
8. Fukushima, H.; Toratani, M.; Tanaka, A.; Chen, W.-Z.; Murakami, H.; Frouin, R.J.; Mitchell, B.G.; Kahru, M. ADEOS-II/GLI ocean-color atmospheric correction: Early phase result. In Proceedings of theOptical Science and Technology, SPIE's 48th Annual Meeting, San Diego, CA, USA, 3–8 August 2003; Volume 5155, pp. 91–99. [CrossRef]
9. Gordon, H.R.; Wang, M. Retrieval of water-leaving radiance and aerosol optical thickness over the oceans with SeaWiFS: A preliminary algorithm. *Appl. Opt.* **1994**, *33*, 443. [CrossRef]
10. Bevington, P.R.; Robinson, D.K. *Data Reduction and Error Analysis for the Physical Sciences*, 3rd ed.; McGraw-Hill Inc.: New York, NY, USA, 2003; pp. 36–46.
11. Ryu, J.H.; Han, H.-J.; Cho, S.; Park, Y.-J.; Ahn, Y.-H. Overview of geostationary ocean color imager (GOCI) and GOCI fata processing system (GDPS). *Ocean Sci. J.* **2012**, *47*, 223–233. [CrossRef]
12. Ahn, J.-H.; Park, Y.-J.; Ryu, J.-H.; Lee, B.; Oh, I.S. Development of atmospheric correction algorithm for Geostationary Ocean Color Imager (GOCI). *Ocean Sci. J.* **2012**, *47*, 247–259. [CrossRef]
13. Bailey, S.W.; Franz, B.A.; Werdell, P.J. Estimation of near-infraredwater-leaving reflectance for satellite ocean color data processing. *Opt. Express* **2010**, *18*, 7521–7527. [CrossRef]
14. Shi, W.; Wang, M. Satellite views of the Bohai Sea, Yellow Sea, and East China Sea. *Prog. Oceanogr.* **2012**, *104*, 30–45. [CrossRef]
15. Ruddick, K.G.; Ovidio, F.; Rijkeboer, M. Atmospheric correction of SeaWiFS imagery for turbid coastal and inland waters. *Appl. Opt.* **2000**, *39*, 897–912. [CrossRef] [PubMed]
16. Doxaran, D.; Lamquin, N.; Park, Y.-J.; Mazeran, C.; Ryu, J.-H.; Wang, M.; Poteau, A. Retrieval of the seawater reflectance for suspended solids monitoring in the East China Sea using MODIS, MERIS and GOCI satellite data. *Remote. Sens. Environ.* **2014**, *146*, 36–48. [CrossRef]
17. Huang, X.; Zhu, J.; Han, B.; Jamet, C.; Tian, Z.; Zhao, Y.; Li, J.; Li, T. Evaluation of Four Atmospheric Correction Algorithms for GOCI Images over the Yellow Sea. *Remote. Sens.* **2019**, *11*, 1631. [CrossRef]
18. Frouin, R.J.; Franz, B.A.; Ibrahim, A.; Knobelspiesse, K.; Ahmad, Z.; Cairns, B.; Chowdhary, J.; Dierssen, H.M.; Tan, J.; Dubovik, O.; et al. Atmospheric Correction of Satellite Ocean-Color Imagery During the PACE Era. *Front. Earth Sci.* **2019**, *7*. [CrossRef]

19. Green, R.O.; Eastwood, M.L.; Sarture, C.M.; Chrien, T.G.; Aronsson, M.; Chippendale, B.J.; Faust, J.A.; Parvi, B.E.; Chovit, C.J.; Solis, M.; et al. Imaging spectrometry and the Airborne Visible/Infrared Imaging Spectrometer (AVIRIS). *Remote Sens. Environ.* **1998**, *65*, 227–248. [CrossRef]
20. Gao, B.; Montes, M.J.; Ahmad, Z.; Davis, C.O. Atmospheric correction algorithm for hyperspectral remote sensing of ocean color from space. *Appl. Opt.* **2000**, *39*, 887–896. [CrossRef]
21. Siegel, D.A.; Wang, M.; Maritorena, S.; Robinson, W. Atmospheric correction of satellite ocean color imagery: The black pixel assumption. *Appl. Opt.* **2000**, *39*, 3582–3591. [CrossRef]
22. Arnone, R.A.; Martinolich, P.; Gould, R.W.; Syder, M.; Stumps, R.P. Coastal optical properties using SeaWiFS. In Proceedings of the Ocean Optics XIV Conference, Kailua-Kona, Hawaii, 10–13 November 1998.
23. Davis, C.O.; Bowles, J.; Leathers, R.A.; Korwan, D.; Downes, T.V.; Snyder, W.A.; Rhea, W.J.; Chen, W.; Fisher, J.; Bissett, P.; et al. Ocean PHILLS hyperspectral imager: Design, characterization, and calibration. *Opt. Express* **2002**, *10*, 210–221. [CrossRef]
24. Fraser, R.S.; Mattoo, S.; Yeh, E.-N.; McClain, C.R. Algorithm for atmospheric and glint corrections of satellite measurements of ocean pigment. *J. Geophys. Res. Space Phys.* **1997**, *102*, 17107–17118. [CrossRef]
25. Fraser, R.S.; Ferrare, R.A.; Kaufman, Y.J.; Markham, B.; Mattoo, S. Algorithm for atmospheric corrections of aircraft and satellite imagery. *Int. J. Remote. Sens.* **1992**, *13*, 541–557. [CrossRef]
26. Tanre, D.; DeRoo, C.; Duhaut, P.; Herman, M.; Morcrette, J.J.; Perbos, J.; Deschamps, P.Y. Technical note Description of a computer code to simulate the satellite signal in the solar spectrum: The 5S code. *Int. J. Remote Sens.* **1990**, *11*, 659–668. [CrossRef]
27. Ahmad, Z.; Fraser, R.S. An Iterative Radiative Transfer Code for Ocean-Atmosphere Systems. *J. Atmospheric Sci.* **1982**, *39*, 656–665. [CrossRef]
28. Gao, B.-C.; Montes, M.J.; Li, R.-R.; Dierssen, H.M.; Davis, C.O. An atmospheric correction algorithm for remote sensing of bright coastal waters using MODIS land and ocean channels in the solar spectral region. *IEEE Trans. Geosc. Remote Sens.* **2007**, *45*, 1835–1843. [CrossRef]
29. Wang, M. Remote sensing of the ocean contributions from ultraviolet to near-infrared using the shortwave infrared bands: Simulations. *Appl. Opt.* **2007**, *46*, 1535–1547. [CrossRef]

Article

Synergy between Satellite Altimetry and Optical Water Quality Data towards Improved Estimation of Lakes Ecological Status

Ave Ansper-Toomsalu [1,*], Krista Alikas [1], Karina Nielsen [2], Lea Tuvikene [3] and Kersti Kangro [1,3]

[1] Tartu Observatory, University of Tartu, Observatooriumi 1, Tõravere, 61602 Tartu, Estonia; krista.alikas@ut.ee (K.A.); kersti.kangro@ut.ee (K.K.)
[2] Division of Geodesy and Earth Observation, DTU Space, National Space Institute, 2800 Kgs. Lyngby, Denmark; karni@space.dtu.dk
[3] Chair of Hydrobiology and Fishery, Institute of Agricultural and Environmental Sciences, Estonian University of Life Sciences, Kreutzwaldi 5, 51006 Tartu, Estonia; lea.tuvikene@emu.ee
* Correspondence: ave.ansper-toomsalu@ut.ee

Abstract: European countries are obligated to monitor and estimate ecological status of lakes under European Union Water Framework Directive (2000/60/EC) for sustainable lakes' ecosystems in the future. In large and shallow lakes, physical, chemical, and biological water quality parameters are influenced by the high natural variability of water level, exceeding anthropogenic variability, and causing large uncertainty to the assessment of ecological status. Correction of metric values used for the assessment of ecological status for the effect of natural water level fluctuation reduces the signal-to-noise ratio in data and decreases the uncertainty of the status estimate. Here we have explored the potential to create synergy between optical and altimetry data for more accurate estimation of ecological status class of lakes. We have combined data from Sentinel-3 Synthetic Aperture Radar Altimeter and Cryosat-2 SAR Interferometric Radar Altimeter to derive water level estimations in order to apply corrections for chlorophyll a, phytoplankton biomass, and Secchi disc depth estimations from Sentinel-3 Ocean and Land Color Instrument data. Long-term in situ data was used to develop the methodology for the correction of water quality data for the effects of water level applicable on the satellite data. The study shows suitability and potential to combine optical and altimetry data to support in situ measurements and thereby support lake monitoring and management. Combination of two different types of satellite data from the continuous Copernicus program will advance the monitoring of lakes and improves the estimation of ecological status under European Union Water Framework Directive.

Keywords: ecological status class of lakes; European Union Water Framework Directive (2000/60/EC); water quality parameters; water level; Sentinel-3; Cryosat-2; shallow lakes; synergy; altimetry data; optical data

Citation: Ansper-Toomsalu, A.; Alikas, K.; Nielsen, K.; Tuvikene, L.; Kangro, K. Synergy between Satellite Altimetry and Optical Water Quality Data towards Improved Estimation of Lakes Ecological Status. *Remote Sens.* **2021**, *13*, 770. https://doi.org/10.3390/rs13040770

Academic Editor: Giacomo De Carolis

Received: 30 December 2020
Accepted: 17 February 2021
Published: 19 February 2021

1. Introduction

Consistent monitoring and estimation of the ecological status of lakes is essential for having an overview of current conditions and prediction of water quality (WQ) [1,2]. As lakes regulate the climate and carbon cycle, provide habitat for biota, drinking water, and other ecosystem services, it is important to monitor their ecological status and indications of their decline. To implement water management for sustainable environment, it is essential to evaluate the influence of human activity to lake ecosystems and also take into account variability of natural factors [3,4]. The European Union (EU) has established Water Framework Directive (2000/60/EC) [5] (WFD) for all European countries to achieve at least "good" ecological status class in lakes through consistent monitoring and water policy by the latest 2027 [6,7]. In total, five ecological quality classes from "high" to "bad" have been determined for every defined natural water body type, according to different biological (e.g., phytoplankton), physico-chemical (e.g., transparency) and hydro-morphological (e.g., depth variations) quality elements [5]. The main emphasis is given to the biological quality elements while the physico-chemical and hydro-morphological quality elements only affect the decision indirectly through their effect on the state of biological elements [5].

Traditional in situ measurements (e.g., collection of water samples, laboratory analyses, gauging stations, etc.) tend to be money and time consuming, and tend to take a lot of resources [8]. It is difficult to collect data from remote areas, where infrastructure is missing and access to the water body is complicated [9]. In situ gauging stations provide more frequent data than traditional measurements, but transmission of the data to the further workflow is rather slow and time consuming [10–12]. These kinds of systems are usually expensive to buy and difficult to maintain, therefore the number of gauging stations has decreased in recent years [11–14]. As lakes are mostly categorized as heterogeneous bodies, the biggest disadvantage of traditional in situ measurements is providing only point measurements, which do not represent spatial and temporal variations of the entire water body [1,11,15]. The combination of Earth Observation (EO) and in situ measurements complement each other by fulfilling their shortages and also helps to optimize and improve the selection of sampling locations and surveying times [16–18]. Earth Observation gives opportunity to have large scale data for spatial analysis and synchronized view of group of lakes [8,19]. In addition, digital data forms accelerates data reading and processing time and helps to produce historical and long time series [15]. The European Space Agency's (ESA) Copernicus program [20] provides full, free, open access, high quality near-real-time EO data on a global level for all users. Since the launch of the first Sentinel mission satellite in 2014, the purpose of the program is to launch nearly 20 satellites by the year of 2030 to provide long-term time series for various applications, e.g., environmental protection, climate change monitoring, and civil security [21]. The continuity of data allows to develop and advance methods to better integrate EO data into environmental monitoring, management, and policy.

As the satellite era has lasted almost 40 years by now, many algorithms and methods have been developed for different domains in lake research [10,22–31]. As the WFD requirements are obligatory to every EU country, it has been indicated that the in situ methods are not feasible to cover the monitoring expectations, therefore many EO-based applications have been developed for implementation of requirements of WFD [32]. According to the WFD, phytoplankton is the key parameter of estimating ecological status of water [33–36], which also indicates eutrophication status of the lake [37]. Estimation of the concentration of the photosynthetic pigment chlorophyll a (chl a) has been one of the main goals in all ocean color missions starting from the first National Aeronautics and Space Administration (NASA) Coastal Zone Color Scanner on board Nimbus-7 in 1978. Additionally, water transparency is one of the key indicators for lake's ecological status class according to the WFD [38–40], where also EO-based assessment methods have shown the suitability to obtain additional information [7,25,32,41]. Covering various ocean color missions, the approaches and algorithms have been developed to provide trophic status based estimations [31,42–48] to obtain chl a and transparency estimates. However, estimation is based only on certain parameters (e.g., chl a and transparency), which do not take into account any external factors [34,35,38,49]. Ranges of different ecological status classes of lakes may be very narrow, which is the main reason why accurate WQ parameter estimation is essential. Water level (WL) estimation, using satellite altimetry data over inland water bodies, has become more advanced during recent years [10,16,17,50–52], which allows the monitoring of its influence on the physical environment, which in turn changes the biota and whole lake ecosystem [53], especially in the case of shallow lakes.

In general, the ecological status of lakes is affected by anthropogenic factors, such as agricultural pollution by nutrients and industrial waste and various chemical elements [54], nevertheless natural factors, such as diurnal and seasonal changes, and prolonged changes in atmospheric circulations should be taken into account as well [55–57]. Water level is one of the natural factors, which affects WQ directly, worsening WQ parameters in low WL periods [58,59]. Especially in large, shallow lakes, the influence of lake depth variations on biota is remarkable, due to large seasonal and inter-annual WL fluctuations. The periods of extremely low WL are the most critical for the ecological status in such lakes, strengthening sediment resuspension, increasing concentrations of ions, phytoplankton biovolume and chl a concentration [60]. In contrast, the phytoplankton biomass (FBM) is significantly lower due to light limitation in years of high WL [61]. Such high variation in biological quality indicators, caused by natural factors, may over-mask human induced effects on a water body [62]. Based on the long-term in situ data, Tuvikene et. al. [62] showed the significant influence of WL while estimating the ecological status class according to the chl a, FBM and transparency in case of large shallow lakes. It was shown that, while phytoplankton species composition and traditional WQ parameters indicated different ecological status class, correction of metrics for WL enabled more consistent assessment of the ecological status of the lake.

As assessment of ecological status of lakes is currently parameter-based, thus accounting for the external natural factors is needed for more precise estimation. In the increasing constellation of ocean color satellites, carrying passive optical sensor usable for mapping various optical WQ parameters, and altimeters for estimating WL, the synergy of data could potentially improve the ecological status assessment over large shallow lakes. Therefore, the aim of this paper is to (1) create synergy between the optical and altimetry data in order to, (2) derive chl a, total FBM and transparency based ecological status classes taking into account the dependence of biological status indicators on the changing WL over large shallow lakes.

2. Materials and Methods

2.1. Study Sites Description

Data from two large and shallow lakes are used for this study. Võrtsjärv is situated in the central part of Estonia (Figure 1) with surface area 270 km^2 (length 35 km, width ~14 km). With the average depth of 2.8 m (the maximum depth is 6 m), the lake has strong WL fluctuation as the annual mean amplitude is 1.4 m and absolute range is 3.1 m, causing up to 3.2-fold difference in the lake volume [61]. Water level is generally characterized by a high period in spring, due to snow melting until end of April or beginning of May, a low period in winter and summer, and increasing WL from middle of the October due to autumn rainfalls. Võrtsjärv is a eutrophic lake with low water transparency (<1 m) [63]. According to the Estonian regulation of Water Act [64], Võrtsjärv belongs to independent ecological status class Type 6 (Table 1) (non-stratified, light water color, low amount of chlorides, moderate hardness, surface area 100–300 km^2). Annually, national monitoring is performed at one station once in a month and from additional 2 or 9 stations in August [65].

Peipsi is a transboundary lake situated in the border of Estonia and Russia (Figure 1) with surface area 3555 km^2 (length 152 km, width 48 km). The average depth is 7.1 m (the maximum depth is 15.3 m) and the annual mean amplitude of WL change is 1.5 m. Water level of Peipsi is characterized the highest in spring during the ice melting period and is the lowest in October [66]. It is a eutrophic lake, which consists of three parts: Peipsi *sensu stricto* (*s.s.*, 2611 km^2 average depth 8.3 m), Lämmijärv (236 km^2 average depth 2.5 m) and Pihkva (708 km^2, average depth 3.8 m). According to the Estonian regulation of Water Act [8], Peipsi belongs also to the independent ecological status class Type 7 (Table 1) (non-stratified, light water color, low amount of chloride, moderate hardness, surface area >1000 km^2) and is monitored 7 times per year (in March, when the lake is ice-covered, and from May to October) from 7 stations [65].

The both lakes are eutrophic and shallow, with similar WL fluctuation in terms of the annual mean (Võrtsjärv 1.4 m and Peipsi 1.5 m), but with different average depth (Võrtsjärv 2.8 m and Peipsi 7.1 m) leading to differences in the effect of WL. In addition, both lakes were considered unique, having no undisturbed reference sites for comparison, therefore, pressure-based approach, morphoedaphic index, and historical data were used to derive site-specific reference conditions and classification criteria for them [67].

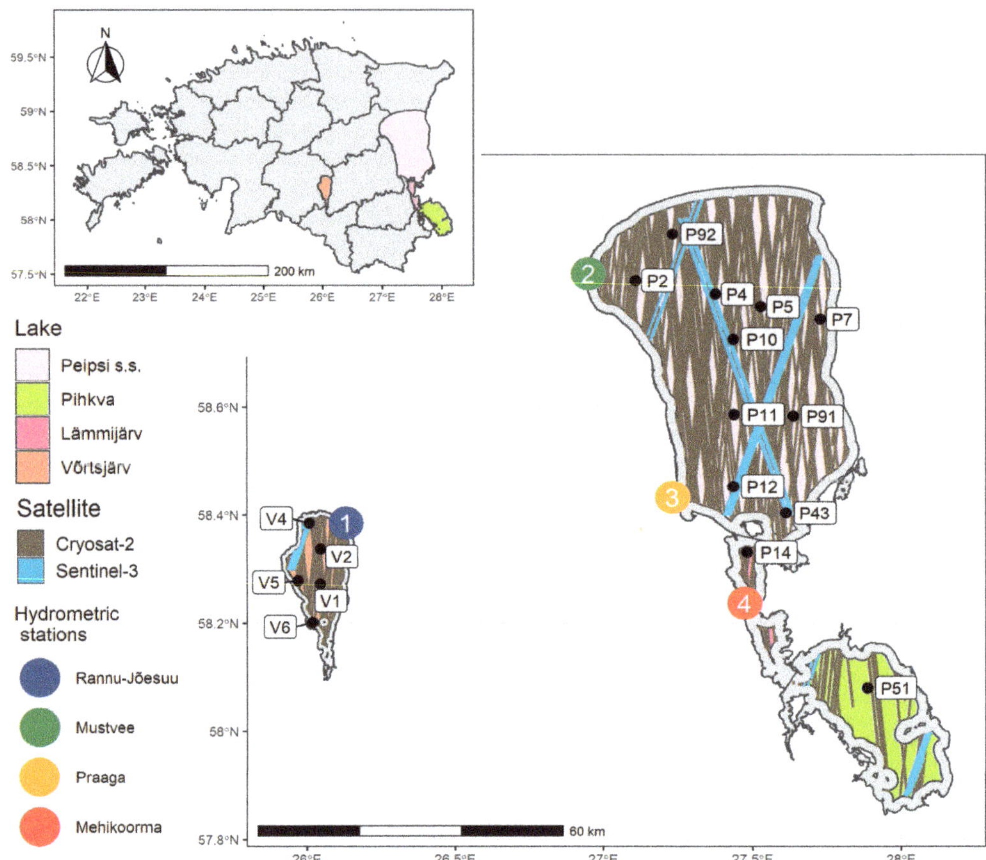

Figure 1. Locations of in situ stations in Võrtsjärv (Vx) and Peipsi (Px). Peipsi consists of three parts: Peipsi s.s. (light purple), Lämmijärv (pink), Pihkva (light green); Võrtsjärv is colored as orange. Buffers of 1 and 2 km from the shore are created for Võrtsjärv and Peipsi respectively. Sentinel-3- altimetry tracks are shown in light blue, whereas Cryosat-2 tracks are shown in brown. Locations of hydrometric stations are numbered as 1 Rannu-Jõesuu (blue), 2 Mustvee (green), 3 Praaga (yellow), 4 Mehikoorma (red).

Table 1. Thresholds for chlorophyll a (chl a) (mg/m^3), phytoplankton biomass (FBM) (g/m^3) and Secchi disk depth (SDD) (m) estimating ecological status class in Võrtsjärv and Peipsi according to Estonian regulation of Water Act [64]. Each color represents ecological status class defined by Water Framework Directive (WFD) [5] and the range of measured values for each parameter during the study period from both lakes are shown in the last column, whereas the number in the brackets indicates average of the measured values.

	Very Good	Good	Moderate	Bad	Very Bad	Study Site 2016–2019
			Võrtsjärv			
Chl a (mg/m^3)	≤24	>24–38	>38–45	>45–51	>51	5–81 (33)
FBM (g/m^3)	–	–	–	–	–	–
SDD (m)	≥0.9	<0.9–0.7	<0.7–0.6	<0.6–0.5	<0.5	0.4–2.15 (0.9)
			Peipsi			
Chl a (mg/m^3)	≤3 (Peipsi s.s.) ≤6 (Lämmijärv, Pihkva)	>3–8 (Peipsi s.s.) >6–13 (Lämmijärv, Pihkva)	>8–20 (Peipsi s.s.) >13–37 (Lämmijärv, Pihkva)	>20–38 (Peipsi s.s.) >37–75 (Lämmijärv, Pihkva)	>38 (Peipsi s.s.) >75 (Lämmijärv, Pihkva)	5–89 (19) (Peipsi s.s.) 4–90 (46) (Lämmijärv, Pihkva)

Table 1. *Cont.*

	Very Good	Good	Moderate	Bad	Very Bad	Study Site 2016–2019
FBM (g/m^3)	≤1 (Peipsi s.s.) ≤2.6 (Lämmijärv, Pihkva)	>1–2.6 (Peipsi s.s.) >2.6–6.4 (Lämmijärv, Pihkva)	>2.6–9.4 (Peipsi s.s.) >6.4–16.1 (Lämmijärv, Pihkva)	>9.4–17.3 (Peipsi s.s.) >16.1–37 (Lämmijärv, Pihkva)	>17.3 (Peipsi s.s.) >37 (Lämmijärv, Pihkva)	1.4–98.5 (7.1) (Peipsi s.s.) 2.0–74.4 (18.6) (Lämmijärv, Pihkva)
SDD (m)	≥3.5 (Peipsi s.s.) ≥2.0 (Lämmijärv, Pihkva)	<3.5–2.5 (Peipsi s.s.) <2.0–1.5 (Lämmijärv, Pihkva)	<2.5–1.5 (Peipsi s.s.) <1.5–1.0 (Lämmijärv, Pihkva)	<1.5–1.0 (Peipsi s.s.) <1.0–0.7 (Lämmijärv, Pihkva)	<1.0 (Peipsi s.s.) <0.7 (Lämmijärv, Pihkva)	0.12–3.1 (1.0) (Peipsi s.s.) 0.13–1.1 (0.5) (Lämmijärv, Pihkva)

2.2. In Situ Data

2.2.1. Water Level

Water level was measured by the Vaisala automatic weather station named as MAWS110, which sensor was located near the bottom of the lake and it measured the pressure of the water column, which is converted to the WL in centimeters [68]. Measurements were performed every hour, but WL was averaged over a day for the study. To derive WL above sea level (in Estonian Height system (EH2000), which corresponds to European Vertical Reference System), height of the sensor from sea level (national geoid named as EST-GEOID2017 (Estonian national geoid model)) [69] was added. Hydrometric stations were located near the shore or in the mouth of the river (Figure 1). Water level data was available in Rannu–Jõesuu (hereafter Rannu) hydrometric station from 1927, in Mustvee and Praaga from 1921 and in Mehikoorma from 1949 until 2019.

2.2.2. Chl a, FBM, and SDD

Chl a and FBM samples of integral water at discrete depths (0.5 m, 1 m, etc., until 0.5 m from the bottom) were collected from different stations and analyzed in a laboratory from the 1960s. Chl a samples were filtered through Whatman glass-microfiber filters (GF/F) ø 25 mm filter and pigment extraction was done with 5 mL 96% ethanol [70]. Chl a was measured spectrophotometrically with Hitachi U-3010 (430–750 nm) and the concentrations of chl a were derived according to Jeffrey and Humphrey [71]. For determination of FBM, up to 400 counting units were counted and measured with an inverted microscope based on Utermöhl technique [72]. SDD values were measured with white Secchi disc from the 1950s.

2.3. Satellite Data

Sentinel-3A/3B (S3A/S3B) and CryoSat-2 (C2) data from 2016 to 2019 covering the period from May to October were used. Buffers of 1 km (Võrtsjärv) and 2 km (Peipsi) were created to reduce the possible errors in the vicinity of land in case of optical and altimetry data (Figure 1). Depending on the cloud cover, 300 m resolution data from S3 Ocean and Land Colour Instrument (OLCI) is available daily in the study location. Surface elevations from the Synthetic Aperture Radar Altimeter (SRAL) sensor onboard S3A and S3B are available every 27 days for a given track. Lake Peipsi is measured by both S3A and S3B, hence, data are available from 2016 and onwards. Lake Võrtsjärv is only measured by S3B, hence data are available from the launch in 2018. C2, carrying Synthetic Aperture Radar/Interferometric Radar Altimeter (SIRAL) sensor, is operating in a geodetic orbit with a repeat period of 369 days. S3 and C2 tracks during the study period are shown on the Figure 1 in blue and gray, respectively.

2.3.1. Optical Data

S3 OLCI Level-1 (L1B) Full Resolution "Non Time Critical" images were used from Estonian National Satellite images database named as ESTHub [73]. Data were used from the beginning of S3 operation, respectively, S3A since 2016, and additional data from S3B since 2018. L1B data was filtered by quality flags ("bright", "sun_glint_risk", "duplicated", "invalid") and phytoplankton related parameters were derived via lake specific empirical relationship (Equations (2) and (3)) [74] applied on Maximum Chlorophyll Index (MCI) (Equation (1)) [75]:

$$MCI = L_{709} - L_{681} - \frac{709 - 681}{753 - 681} \times (L_{753} - L_{681}) \tag{1}$$

where L_x corresponds to top-of-atmosphere radiance at certain wavelength.

$$Chl\ a = 10.9 \times MCI + 15.3 \tag{2}$$

$$FBM = 5.8 \times MCI + 5.4 \tag{3}$$

For water transparency estimates, OLCI L1B data was processed with Polymer v4.10 for retrieving water-leaving reflectance (ρ_w). Processed data were filtered by quality flag Bitmask ("0", "1024", "1023"). For the SDD estimates, first the weighting function (Equation (4)) [39] was applied on ρ_w which defined the appropriate empirical algorithm to derive diffuse attenuation coefficient of downwelling irradiance ($K_d(490)$) (Equation (5)).

$$W = 5.098 - 2.2099\left(\frac{\rho(560)}{\rho(709)}\right) \tag{4}$$

$$K_d(490) = (1 - W) \times \exp(-0.9 * (\log_{10}\left(\frac{\rho_{490}}{\rho_{709}}\right) + 0.4) + 0.02) + W \times \exp(-1.2 * (\log_{10}\left(\frac{\rho_{560}}{\rho_{709}}\right) + 1.4) + 0.02) \tag{5}$$

Calculated $W \leq 0$ corresponds to more transparent waters and algorithm (ρ_{490}/ρ_{709}) was used by assigning $W = 0$. Calculated $W \geq 1$ corresponds to more turbid waters and the algorithm ρ_{560}/ρ_{709} was used by assigning $W = 1$. When the W is between 0 and 1, then the algorithm merges the retrievals from both (ρ_{490}/ρ_{709} and ρ_{490}/ρ_{709}) algorithms [39].

Secondly, the SDD was derived based on the empirical algorithm from $K_d(490)$:

$$SDD_{Vortsjarv} = -0.038 \times K_d(490) + 0.9448 \tag{6}$$

SDD for Peipsi *s.s.* (Equation (7)) was calculated using band ratio $K_d(490)$ algorithm (Equation (5)) [38]:

$$SDD_{Peipsi\ s.s.} = 3.42 \times K_d(490)^{-0.82} \tag{7}$$

SDD for Lämmijärv and Pihkva was calculated using empirical algorithm (Equation (8)) [38]:

$$SDD_{Lammijarv,Pihkva} = \left(\frac{\rho(490)}{\rho(709)}\right)^{0.7} \times 2.14 \tag{8}$$

2.3.2. Altimetry Data

To construct water surface elevations from S3A and S3B the ESA "Non Time Critical" Level-2 product "Enhanced measurements" was used. This product consists of 20Hz along-track measurements of, e.g., tracker range, satellite altitude, atmospheric and geophysical corrections, and waveforms. To identify the nadir water surface, the waveforms were retracked with a narrow primary peak threshold retracker [76]. For C2 we applied the ESA L1B product from baseline D. This product also contains 20Hz measurements of, e.g., waveforms, tracker range, satellite altitude and corrections. The same retracker was applied to the C2 waveforms.

The water surface elevations are derived from the following relation (Equation (9))

$$H = h - R - N, \tag{9}$$

where h is the altitude of the satellite, and N is the geoid height, which here is the EST-GEOID2017 national geoid model. R is the range, i.e., the distance from the satellite to the surface which can be expressed as (Equation (10))

$$R = R_{tracker} + R_{retrack} + R_{atm} + R_{geo}. \tag{10}$$

Here $R_{tracker}$ is the distance to the presumed surface measured by the onboard tracker and $R_{retrack}$ is the retracking correction. R_{atm} represents the atmospheric corrections consisting of the wet and dry troposphere and ionosphere. R_{geo} represents the geophysical corrections including solid earth tide, pole tide, and ocean loading tide. All corrections are here taken from the altimetry products.

The altimetry-based WL time series based for Peipsi and Võrtsjärv were reconstructed using the "R" package "tsHydro" available from https://github.com/cavios/tshydro (accessed on 14 October 2020). The software package is based on an implementation of a simple state-space model where the observations are assumed to be described by a random walk plus noise. The observation noise is described by a mixture of a Normal and the heavy tailed Cauchy distributions making it more robust against erroneous observation. In the solution, bias between the different missions is also estimated. The model is described in detail in [10]. Here, one time series based on all available altimetry data is constructed for each lake.

2.4. Data Correction for the Effects of Water Level

EO-derived WQ variables (chl a, FBM, SDD) were corrected for the effects of changing WL by the methodology developed by Tuvikene et al. [62]. The methodology is based on the long-term in situ data, where statistically significant relationships (p-value < 0.05) between the specific WQ variables and WL were found (Table A1). Corrected WQ variables were calculated using regression equations according to the long-term monthly mean WL and daily measured WL. If the WL of the sampling day corresponded to the long-term average for that period, no correction was needed, but if it differed, the correction either up or down was proportional to the deviation of the WL from the mean.

3. Results

3.1. Inter-Annual Dynamics in Water Level

Comparison of long-term WL measured in three hydrometric stations (Mustvee, Mehikoorma, Praaga) in Peipsi (Figure 2) show similar dynamics and very small spatial differences, e.g., the mean difference between Praaga and Mustvee is 0.02 m, and between Mehikoorma and Mustvee is 0.04 m. In both lakes, the higher and lower WL periods vary similarly and long-term monthly mean WL is highest in May and decreases towards October. While the annual difference can exceed the long-term monthly mean WL (Figure 2) in some years more than 1 meter, its effect is more pronounced in shallower Võrtsjärv (average depth 2.8 m) compared to Peipsi (average depth 7.1 m).

Water levels based on altimetry data and WL measured at the Mustvee hydrometric station show very similar dynamics (Figure 3) with a bias of 0.4 m. Bias corrected Root Mean Standard Error (RMSE) is 0.07 m, whereas the Pearson correlation between WL measured in Mustvee hydrometric station, and derived from altimetry data, is 0.99. Comparison between altimetry data and Rannu hydrometric station in Võrtsjärv shows similar results, with a bias 0.4 m. Bias corrected RMSE is 0.08, whereas the Pearson correlation between WL measured in Rannu hydrometric station and WL, derived from altimetry data, is 0.98. The combination of altimetry data from C2 and S3 missions allows to obtain much more frequent time series, especially since 2018 due to the launch of S3B (Figure 3).

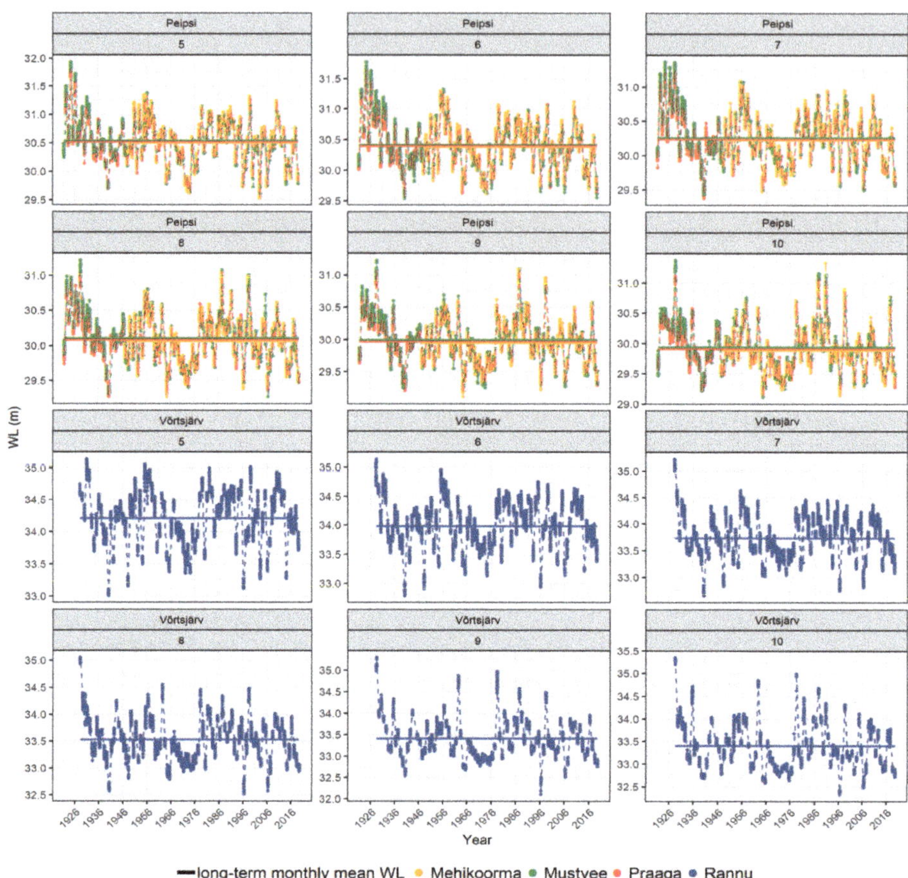

Figure 2. Long-term water level (WL) in Rannu, Mustvee, Praaga, and Mehikoorma hydrometric stations from May (5) to October (10). Long-term monthly mean WL is showed in solid line.

3.2. Inter-Annual Dynamics of Water Quality Parameters

Long-term in situ time series of chl a, FBM, and SDD show variation in seasonal dynamics and pronounced inter-annual differences in Võrtsjärv (Figure 4A). For example, chl a shows increasing trend from 1995 until 2008, then it started to decrease again (Figure 4A). FBM shows increasing trend from the early 2000's. SDD shows seasonal variation from less than 0.5 to 1.5 m. As WQ parameters are in situ measured once during the vegetation period in Võrtsjärv station named as V2 (Figure 4B), the seasonal dynamics are better obtained with more frequent satellite data resulting on average 50 points, which doubled with S3B launch. This allows to acquire more representative data to analyze the WQ parameters over specific time period and have better knowledge of the seasonal and inter-annual differences.

The dynamics of satellite data corresponded well to in situ measured WQ parameters, showing similar dynamics during vegetation period with a bias for chl a 4.9 ± 11.8 mg/m^3, for FBM 2.7 ± 14.2 g/m^3, and for SDD was 0.3 ± 0.8 m. Chl a derived from OLCI corresponds to in situ data with Pearson correlation 0.7, whereas FBM 0.6 and SDD 0.4.

Depending on the monitoring system, some areas are not covered with sufficient monitoring frequency (Figure A1 Pihkva A) or include gaps (Figure A1 Peipsi s.s A., Figure A1 Lämmijärv A) in its long-term time series. Satellite data covers these areas 2–3 times per week and with two satellites the amount of data is doubled from 2018.

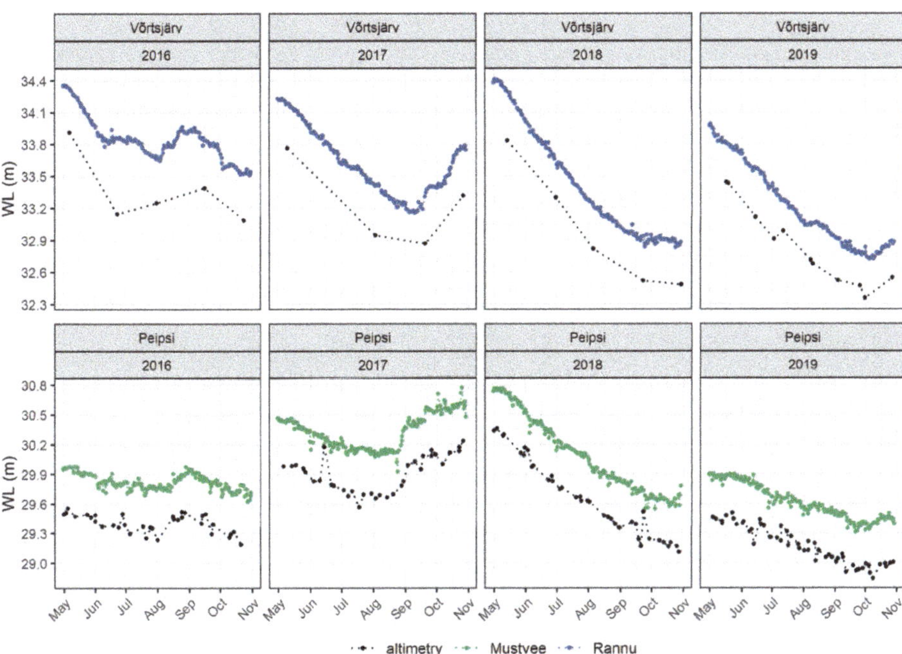

Figure 3. Water level (WL) measured in Mustvee and Rannu hydrometric stations from May to October. WL derived from altimetry data (S3A, S3B, C2).

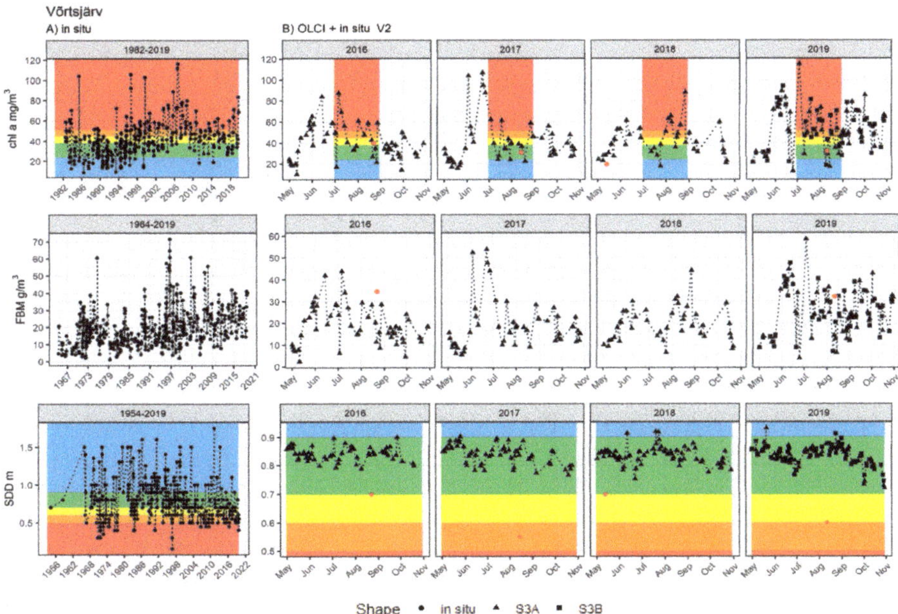

Figure 4. Long-term time series of in situ measured chl a, FBM, and SDD in Võrtsjärv in station named as V2 (**A**) and the same parameters measured from S3 Ocean and Land Color Instrument (OLCI) (**B**). Dots indicate in situ measured values. Triangles and squares indicate S3A and S3B values, respectively. Different colors represent ecological status classes according to European Union (EU) WFD, while FBM is not considered as important parameter to assess the ecological status in Võrtsjärv according to the WFD.

3.3. Correction of chl a, FBM, and SDD Based on the Water Evel

Relationships between WL and WQ parameters vary in both lakes (Table A1). In Võrtsjärv, chl a was significantly related to WL ($p < 0.05$) in July, August, and October, FBM was significantly related in July, whereas SDD was related in all months (from May to October). In Peipsi *s.s.*, WL was significantly related only with FBM in June and September and in Pihkva, FBM was related in August. In Lämmijärv, chl a was significantly related in July and August, FBM was related in June and SDD in June, July, and August. When the relationship was significant between WL and specific WQ parameter, correction was applied on a daily basis according to the WL value in respect to long-term monthly mean WL. When WL was higher than long-term monthly mean WL, WQ parameter values increased after correction, if the WL was lower, then WQ parameters were corrected to lower values. This changed the mean value of the month, as well as minimum and maximum values.

In Võrtsjärv, WL was higher than long-term monthly mean WL in 2016 (up to 0.3 m) and lower in 2017 until 2019, especially low in 2018 and 2019 towards autumn, when WL was lower by more than 0.5 m (Figure 5). As WL was high in 2016, the mean value of chl a and SDD increased after applying the WL correction, whereas, in 2017 until 2019, mean values decreased. Due to the decrease in mean values, ecological status class of WQ parameters also changed. In 2017 and 2018, the ecological status class of chl a changed from "moderate" to "good", and in 2019 from "bad" to "moderate", whereas the ecological status class remained the same in 2016, although mean value of chl a increased. The ecological status class of SDD, changed from "good" to "very good" in 2018 and 2019, when the WL was the lowest compared to long-term monthly mean WL (Figure 5).

In Peipsi, there was less significant relationships between WL and WQ parameters than in Võrtsjärv (Table A1). In Peipsi *s.s.*, there was no statistically significant relationships between WL and WQ parameters, except for FBM in June and September, which only leads to a small change of mean values of FBM (Figure A2 Peipsi *s.s.* FBM). As WL was lower than long-term monthly mean WL in 2016, 2018, and 2019, mean FBM value slightly decreased without changing the ecological status class. More significant relationships between WL and WQ parameters appeared in Lämmijärv (Figure A2 Lämmijärv), which also changed the ecological status class of SDD. As WL was more than 0.4 m lower than long-term monthly mean WL in 2019, then ecological status class of SDD changed from "moderate" to "good", and the mean value of SDD in 2016 also decreased. In Pihkva, the amount of in situ data was mostly insufficient for statistical analysis (significant relationship was found only in August for FBM) and therefore, there were no changes in ecological status class estimations of WQ parameters (Figure A2 Pihkva).

As Võrtsjärv and Lämmijärv are shallower than Peipsi *s.s.*, statistically significant relationships between WL and specific WQ parameters tend to occur more frequently there on a monthly basis. More pronounced changes in WL (difference approximately from 0.4 m) from long-term monthly mean WL were influencing not only the mean values of WQ parameters, but even changed the ecological status class from one to another. Smaller changes in WL in respective to long-term monthly mean influenced only the mean values of WQ, not the ecological status class. However, it is still important, considering long-term monitoring of the lake.

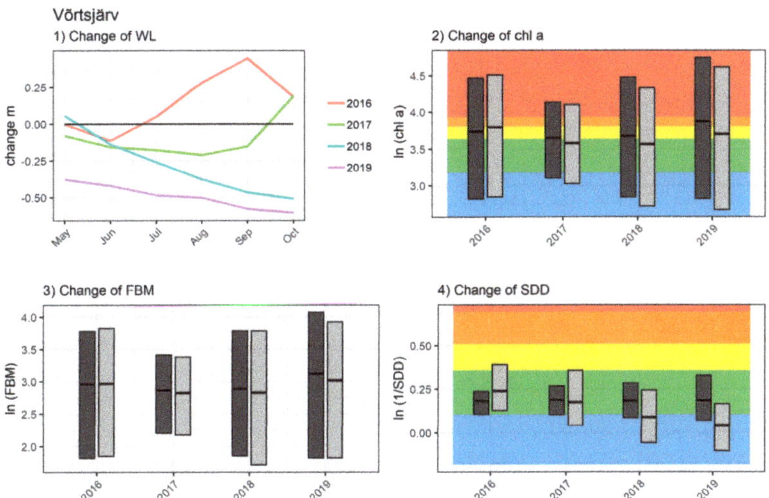

Figure 5. Water level change from long-term monthly mean WL during study period in Võrtsjärv (**1**). Value 0 indicates long-term monthly mean WL. Mean, maximum and minimum values of EO-derived WQ parameters are represented on the graphs (**2–4**), where dark gray indicates values before and light gray indicates values after applying WL correction. Different colors represent ecological status classes according to EU WFD.

4. Discussion

Long-term time series of WQ parameters in lakes are important to obtain information about the seasonal dynamics, their past and present situation and form the basis for future predictions [77,78]. The variety of measured parameters allow the partition of the influences by anthropogenic, as well as natural factors. Moreover, long-term time series help to understand the necessity for current management improvements for the future [79]. Water level as a natural factor, has an impact on lakes ecosystem and on WQ parameters, which in turn have a high impact on the ecological status of lakes [80,81]. Changes in the lake ecosystem influence biota existence and survival complexity in the lake, which affect ecosystem structure and future perspective [82–84]. Water level plays an important role in the lake ecosystem, especially in large shallow lakes [85], where monthly WL differences can be highly different from long-term observations, therefore removing WL influence from WQ parameters is beneficial for fair status assessment [62]. This study shows the possibility to create synergy between optical and altimetry data for lakes research, especially in terms of WFD reporting needs. The removal of one natural aspect will improve the estimation of the ecological status class of a lake, which will allow to keep the focus on the anthropogenic changes in the lake ecosystem.

The study shows high variations between seasonal and long-term WL in large and shallow lakes. Low WL periods vary with higher WL periods in both Võrtsjärv and Peipsi, which are caused by variation in precipitation, hydrological inflows, and evaporation in different years [86]. Long-term monthly mean WL decrease, from May towards October, characterizing higher WL after snow and ice melting, and lower WL in summer due to less inflow and higher rate of evaporation during the warm period [86,87]. Not only seasonal temperature and precipitation variability influences WL, but climate change has also direct effect to lake ecosystem [88–91], which has been even pointed out by The Intergovernmental Panel on Climate Change [92] referring to a global problem. The decrease in WL in some areas is often not compensated by the precipitation, which is caused by the higher annual mean air temperature, which, in turn, causes intensive evaporation and higher water temperature, causing a decrease in ice cover thickness and duration [93,94]. On the other hand, extreme events (flooding, rainfalls, storms, etc.) due to climate change are causing an abrupt increase in the WL in some areas, causing also more apparent WL fluctuations [81,86].

As WL is sensitive to external factors, it is important to monitor it continuously [86,90,95–97]. As the number of in situ gauging stations is decreasing due to their cost and maintenance needs, the use of altimetry data becomes an important supplement to measure lake level variations. The use of altimetry has already provided promising results over lakes [10,11,16,50,98–105]. This is confirmed in this study, where a good agreement between the altimetry-based WL and WL measured at hydrometric stations with a correlation up to 0.99 in both lakes, with RMSE for Võrtsjärv 0.07 and for Peipsi 0.08, respectively. By applying measurements from different altimetry satellites the amount of data is increased, hence, providing a better understanding of WL dynamics and its changes [106–108]. However, combining data located at different tracks may introduce errors due to unresolved geoid signals, or other effects that causes the lake surface so be spatially different such as wind or land signals near the shoreline, but could be combined when the spatial differences in WL are small inside the water body. For example, comparing WL in Peipsi hydrometric stations, seasonal dynamics of the WL were similar and differences remained between 0.02 and 0.04 m, which shows possible WL homogeneity in the water body. Here, S3 SRAL and C2 SIRAL data were combined to increase the data frequency, but other missions could also have been used. Additionally, S3 and C2 provide SAR mode data in the study area, which give higher along-track resolution than conventional altimetry, with dense C2 spatial tracks suitable for lakes. The continuation of S3 missions, and the recently launched S6 mission, ensure data for future monitoring, which allows to continue developing EO-based applications to monitor WL and its implication for ecosystems and their services.

The possibility to derive certain WQ parameters from satellites helps to collect data more frequently and from places, which are not easily accessible. Satellite data corresponded to in situ data showing correlation 0.7 for chl a and lower than 0.6 for FBM and SDD, whereas bias for chl a was 4.9 ± 11.8 mg/m^3, for FBM 2.7 ± 14.2 g/m^3, and for SDD 0.3 ± 0.8 m. More data from satellites gives an extra information to in situ data, which enables to detect seasonal dynamics and rapid changes in different parts of a lake, especially in terms of estimating chl a, FBM, SDD based ecological status class. In some cases, satellite data overestimated (Figure 4 Võrtsjärv B SDD, Figure A1 Lämmijärv B SDD) or underestimated (Figure A1 Peipsi *s.s.* B chl a) in comparison of in situ data. This can be due to spatial differences (point measurement vs. 300 m $\times$ 300 m pixel), measurement technique (integral sample vs upper water column captured by satellite), but still, both in situ and EO data show similar trends during the vegetation period and complement each other. Different WQ parameters have been used for estimating ecological status of lakes from satellite data by many authors already for a while [7,25,33,36,41,109], which confirms necessity of this work for WFD.

Based on the long-term in situ data, the study shows significant relationships ($p < 0.05$) between WL and WQ more frequently in Võrtsjärv than in Peipsi. It is caused by the fact, that Võrtsjärv is shallower than Peipsi and high seasonal WL amplitude affects its biota in greater extent. In addition, the relationships between WL and WQ parameters were more often significant in the shallower parts of Peipsi (e.g., in Lämmjärv). Correcting WQ parameters with WL results in the change of mean annual WQ parameter values and even in changes of their ecological status class. The impact was more pronounced in Võrtsjärv, where WL was much lower than long-term monthly mean WL (approximately from 0.5 m), which did not only decrease the mean value of WQ parameters, but even changed the ecological status estimation to a higher class. As WL has a high impact to WQ parameters and the whole ecosystem, it is encouraging to see that EO data can be used to remove WL effect from data of WQ estimates in large shallow lakes, thus it could be one of the research topics deserving more attention due to its applicability to other large and shallow lakes. Moreover, the variety of past, present, and future satellite missions allows to apply the demonstrated methodology to extend the temporal and also spatial extent of the results. This could be done by combining satellite sensors for WQ, e.g., Environmental Satellite

(ENVISAT) Medium Resolution Imaging Spectrometer (MERIS), Moderate Resolution Imaging Spectroradiometer (MODIS) to estimate WQ parameters [35,40,110–114], with Topex Poseidon and Jason mission satellites to estimate WL data [51,104,105,115,116]. Additionally, several databases like Hydroweb [105,117], Database for Hydrological Time Series of Inland Waters (DAHITI) [98,118], Global Reservoirs and Lakes Monitor database (G-REALM) of United States Department of Agriculture (USDA) [119] provide water level data of lakes, whereas Globolakes [120,121] and Climate Change Initiative (CCI) [122] provide data of WQ parameters from different satellites. However, Sentinel missions will provide data at least for 20 years, which gives a possibility to gather continuous data from now.

As sustainable lake ecosystems are essential in the present and in the future, it is important to ensure frequent monitoring system and to develop EO-based approaches to improve monitoring and assessment of the ecological status class of lakes, leading, finally, to the better management of lakes. In situ monitoring is still essential to gather data to monitor the ecosystem, but also it is basis for developing and improving new tools and approaches, while satellite data would support it with better spatial and temporal resolution for certain parameters. Synergy between altimetry and optical data, gives possibility to estimate WL, as well as WQ parameters for more precise analysis. Results of this study show that, WL can have a strong impact on WQ parameters in large shallow lakes, where amplitude in WL change is high. This effect can be corrected with the proposed method to improve the estimation of ecological status class of such lakes.

5. Conclusions

EU WFD obligates European countries to monitor WQ of lakes and estimate their ecological status according to reference condition. The ecological status of lakes is influenced by anthropogenic, as well as by natural factors. To decrease the effect of natural factors, WQ parameters should be corrected for WL in case of shallow lakes with changing WL for more accurate identification of human impact and estimation of ecological status. Traditional monitoring methods give valuable long-term time series, however often with constrained spatial and temporal coverage, while satellites provide more frequent data. Many water quality parameters (e.g., chl a, FBM, and SDD) and WL can be derived using EO based data. The development and validation of EO algorithms allow to create synergy between different types of satellite data which enables to analyze and assess the lake ecosystems and provide complementary information to improve the monitoring and management decisions.

Long-term in situ data show inter-annual and seasonal dynamics of indicators characterizing lake ecosystems. In large and shallow lakes such as Võrtsjärv and Peipsi, amplitude of seasonal WL changes is high (more than 1 m), which influences WQ parameters, especially in shallower Võrtsjärv. The results support previous studies [10,99,101,104] in the fact that WL can be estimated with high accuracy (bias 0.4 m, R 0.99) over lakes with EO data. Correcting WQ parameters with the WL data derived from satellite, the mean values of the parameters and resulting ecological status classes changed in some cases. The highest changes occurred in Võrtsjärv during the years with lower WL compared to the long-term monthly mean. In this case the correction of the values of WQ parameters changed the resulting ecological status to better quality class. Therefore, taking into account the variability of WL and correcting its effect in case of shallow lakes is important to decrease the influence of the natural factors on the assessment of the lakes ecological status class in terms of EU WFD.

Author Contributions: Conceptualization, A.A.-T. and K.A.; methodology, A.A.-T. and K.A.; validation, A.A.-T. and K.A.; formal analysis, A.A.-T. and K.A.; resources, A.A.-T., K.A., K.N., L.T., and K.K.; writing—original draft preparation, A.A.-T.; writing—review and editing, A.A.-T., K.A., K.N., L.T., and K.K.; visualization, A.A.-T.; supervision, K.A., K.N., and L.T. All authors have read and agreed to the published version of the manuscript.

Funding: This research was funded by (1) Estonian Research Council research and development program grant number RITA1/02-52-01; (2) EU's Horizon 2020 research and innovation program (grant agreement no. 730066, EOMORES) and; (3) Estonian Research Council personal start-up grant named as PSG10 and team grant named as PRG709.

Institutional Review Board Statement: Not applicable.

Informed Consent Statement: Not applicable.

Data Availability Statement: Data available on request.

Acknowledgments: We are grateful to anonymous reviewers for their feedback and suggestions that helped to improve the manuscript. This study was financially supported by (1) the European Regional Development Fund within National Program for Addressing Socio-Economic Challenges through R&D (project named as RITA1 grant number RITA1/02-52-01); (2) European Union's Horizon 2020 research and innovation program under grant agreement no. 730066; (3) Estonian Research Council grants PSG10 and PRG709. Authors thank people from Centre for Limnology providing in situ water quality data and Estonian Weather Service providing in situ water level data.

Conflicts of Interest: The authors declare no conflict of interest.

Appendix A

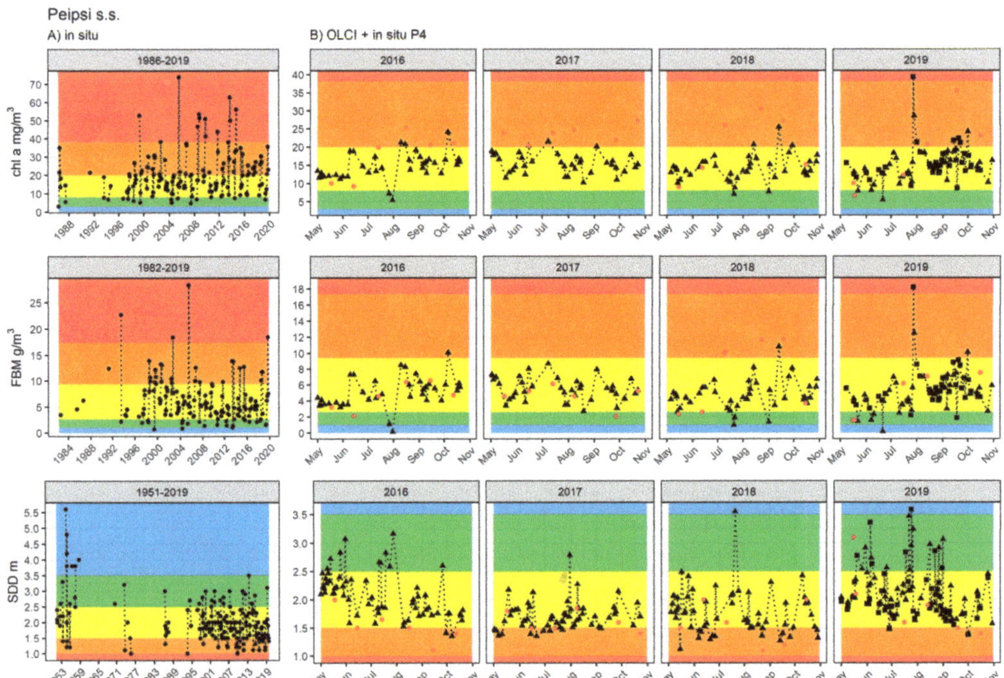

Figure A1. *Cont.*

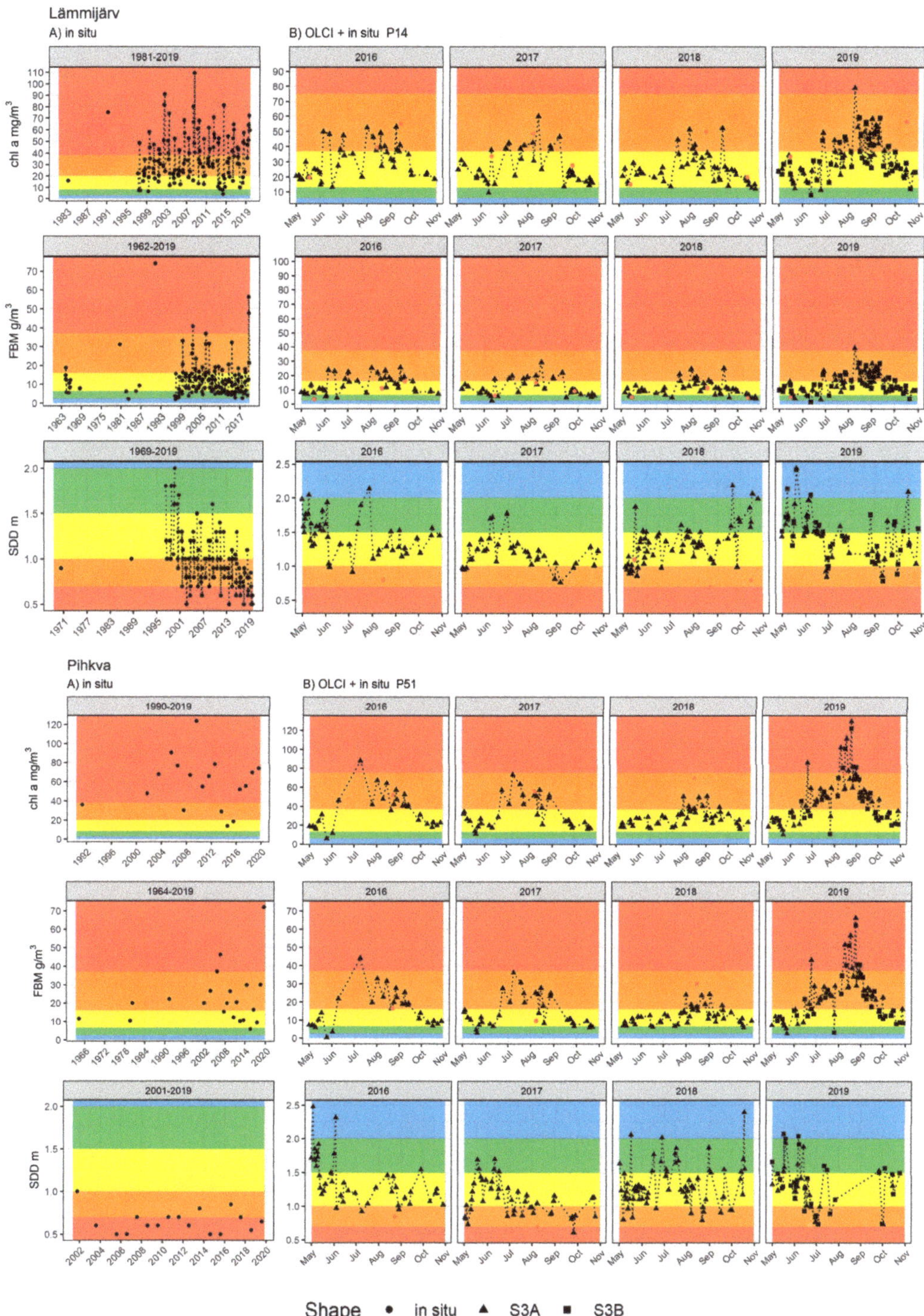

Figure A1. Long-term time series of in situ measured chl a, FBM, and SDD in Peipsi s.s., Lämmijärv, and Pihkva (**A**) and same parameters measured from S3 OLCI (**B**). Dots indicate in situ measured values. Triangles and squares indicates S3A and S3B values, respectively. Different colors represent ecological status classes according to EU WFD.

Appendix B

Table A1. Relationship between in situ measured WL (cm) and water quality parameters from May (5) to October (10), where bold values indicate statistically significant relationships ($p < 0.05$), N indicates number of observations during the study period and "-" indicates not enough observations for statistical analysis.

Parameter	Month	*p*-Value	r	Intercept	Slope	N
			Võrtsjärv			
Ln (chl a)	5	0.12	−0.20	8.76	−0.002	61
	6	0.67	−0.06	5.85	−0.001	58
	7	**0.01**	−0.35	15.12	−0.003	57
	8	**0.01**	−0.31	14.36	−0.003	65
	9	0.19	−0.17	9.37	−0.002	57
	10	**0.02**	−0.31	12.17	−0.002	57
Ln (FBM)	5	0.61	0.06	−0.88	0.001	78
	6	0.48	−0.08	6.83	−0.001	74
	7	**0.01**	−0.28	16.00	−0.004	76
	8	0.23	−0.14	9.59	−0.002	76
	9	0.17	−0.16	9.06	−0.002	78
	10	0.15	−0.17	8.99	−0.002	70
Ln (1/SDD)	5	**0.02**	−0.24	4.37	−0.001	85
	6	**0.02**	−0.27	6.56	−0.002	78
	7	**0.00**	−0.41	8.66	−0.002	72
	8	**0.00**	−0.54	11.46	−0.003	77
	9	**0.00**	−0.58	14.77	−0.004	78
	10	**0.00**	−0.48	10.82	−0.003	75
			Peipsi s.s.			
Ln (chl a)	5	0.45	−0.15	9.85	−0.002	29
	6	0.11	−0.35	25.58	−0.008	22
	7	0.88	0.03	1.34	0.000	25
	8	0.20	−0.26	16.39	−0.004	26
	9	0.13	−0.30	18.30	−0.005	27
	10	0.90	0.03	1.97	0.000	24
Ln (FBM)	5	0.59	0.11	−5.32	0.002	28
	6	**0.00**	−0.58	33.07	−0.011	22
	7	0.95	0.01	0.60	0.000	25
	8	0.52	−0.13	11.26	−0.003	26
	9	**0.01**	−0.48	33.50	−0.011	25
	10	0.76	0.06	−1.74	0.001	27
Ln (1/SDD)	5	1.00	0.00	−0.89	0.000	53
	6	0.07	−0.27	5.97	−0.002	46
	7	0.48	−0.09	2.91	−0.001	62
	8	0.48	−0.11	3.96	−0.002	47
	9	0.71	−0.06	1.00	−0.001	47
	10	0.45	−0.11	1.82	−0.001	47
			Lämmijärv			
Ln (chl a)	5	0.33	−0.20	11.25	−0.003	26
	6	1.00	0.00	2.97	0.000	22
	7	**0.00**	−0.62	26.65	−0.008	22
	8	**0.03**	−0.46	23.17	−0.006	23
	9	0.38	−0.20	13.54	−0.003	22
	10	0.45	−0.17	11.97	−0.003	21
Ln (FBM)	5	0.54	−0.11	7.21	−0.002	31
	6	**0.04**	−0.44	23.63	−0.007	22
	7	0.21	−0.26	9.39	−0.002	25
	8	0.09	−0.34	23.96	−0.007	26
	9	0.05	−0.40	21.75	−0.006	24
	10	0.20	−0.28	14.40	−0.004	23

Table A1. *Cont.*

Parameter	Month	*p*-Value	r	Intercept	Slope	N
Ln (1/SDD)	5	0.21	−0.27	4.82	−0.002	23
	6	**0.00**	−0.61	11.25	−0.004	22
	7	**0.01**	−0.56	13.38	−0.004	21
	8	**0.04**	−0.43	12.46	−0.004	23
	9	0.17	−0.31	8.22	−0.003	21
	10	0.10	−0.37	8.53	−0.003	21
			Pihkva			
Ln (chl a)	5	-	-	-	-	-
	6	-	-	-	-	-
	7	-	-	-	-	-
	8	0.05	−0.47	27.75	−0.008	18
	9	-	-	-	-	-
	10	-	-	-	-	-
Ln (FBM)	5	0.96	−0.02	3.14	0.000	7
	6	-	-	-	-	-
	7	-	-	-	-	-
	8	**0.00**	−0.67	46.28	−0.014	17
	9	-	-	-	-	-
	10	-	-	-	-	-
Ln (1/SDD)	5	0.17	−0.64	19.19	−0.006	6
	6	0.63	0.19	−10.91	0.003	9
	7	0.31	0.50	−7.50	0.002	6
	8	0.14	−0.34	14.57	−0.005	20
	9	0.36	−0.46	27.64	−0.009	6
	10	0.26	−0.62	15.38	−0.005	5

Appendix C

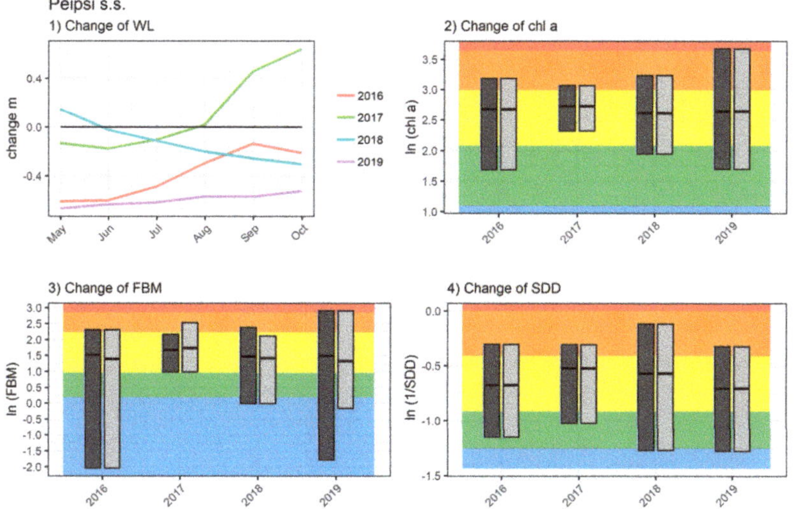

Figure A2. *Cont.*

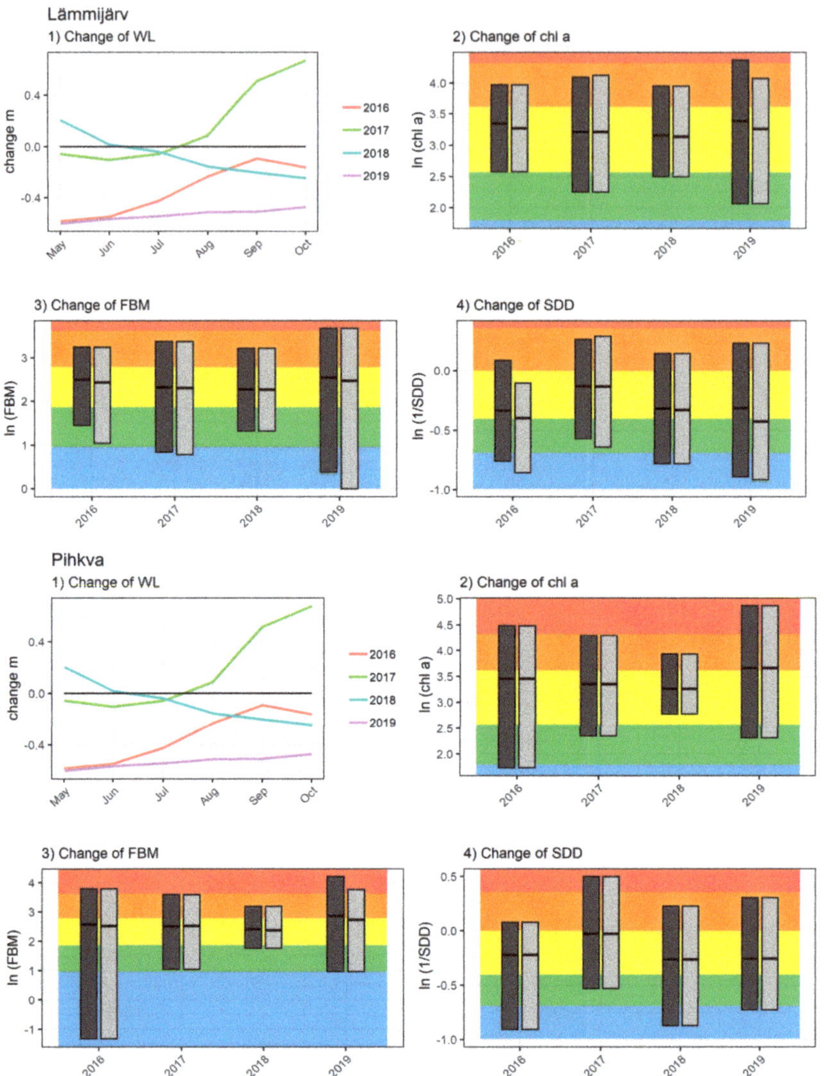

Figure A2. Water level change from long-term monthly mean WL during the study period in Peipsi s.s., Lämmijärv, and Pihkva (**1**). Value 0 indicates long-term monthly mean WL. Mean, maximum, and minimum values of water quality parameters are represented on the graphs (**2–4**), where dark gray indicates values before and light gray indicates values after applying WL correction. Different colors represent ecological classes according to EU WFD.

References

1. Ritchie, J.C.; Zimba, P.V.; Everitt, J.H. Remote Sensing Techniques to Assess Water Quality. *Photogramm. Eng. Remote Sens.* **2003**, *69*, 695–704. [CrossRef]
2. Gray, N.F. *Drinking Water Quality. Problems and Solutions*; Cambridge University Press: Cambridge, UK, 2008.
3. Munné, A.; Prat, N. Ecological aspects of the Water Framework Directive. In *The Water Framework Directive in Catalonia*; Generalitat de Catalunya: Barcelona, Spain, 2006; pp. 53–75.
4. Tranvik, L.J.; Downing, J.A.; Cotner, J.B.; Loiselle, S.A.; Striegl, R.G.; Ballatore, T.J.; Dillon, P.; Finlay, K.; Fortino, K.; Knoll, L.B.; et al. Lakes and reservoirs as regulators of carbon cycling and climate. *Limnol. Oceanogr.* **2009**, *54*, 2298–2314. [CrossRef]

5. European Comission. WFD 2000/60/EC of the European Parliament and of the Council of 23 October 2000 establishing a framework for Community action in the field of water policy. *Off. J. Eur. Parliam.* **2000**, *L327*, 1–82. [CrossRef]
6. Ferreira, J.G.; Vale, C.; Soares, C.V.; Salas, F.; Stacey, P.E.; Bricker, S.B.; Silva, M.C.; Marques, J.C. Monitoring of coastal and transitional waters under the E.U. water framework directive. *Environ. Monit. Assess.* **2007**, *135*, 195–216. [CrossRef]
7. Arle, J.; Mohaupt, V.; Kirst, I. Monitoring of Surface Waters in Germany under the Water Framework Directive—A Review of Approaches, Methods and Results. *Water* **2016**, *8*, 217. [CrossRef]
8. Giardino, C.; Bresciani, M.; Stroppiana, D.; Oggioni, A.; Morabito, G. Optical remote sensing of lakes: An overview on Lake Maggiore. *J. Limnol.* **2014**, *73*, 201–214. [CrossRef]
9. Gao, Q.; Makhoul, E.; Escorihuela, M.; Zribi, M.; Quintana Seguí, P.; García, P.; Roca, M. Analysis of Retrackers' Performances and Water Level Retrieval over the Ebro River Basin Using Sentinel-3. *Remote Sens.* **2019**, *11*, 718. [CrossRef]
10. Nielsen, K.; Stenseng, L.; Andersen, O.B.; Villadsen, H.; Knudsen, P. Validation of CryoSat-2 SAR mode based lake levels. *Remote Sens. Environ.* **2015**, *171*, 162–170. [CrossRef]
11. Jiang, L.; Schneider, R.; Andersen, O.B.; Bauer-Gottwein, P. CryoSat-2 altimetry applications over rivers and lakes. *Water* **2017**, *9*, 211. [CrossRef]
12. Ayana, E.K. Validation of Radar Altimetry Lake Level Data And It's Application in Water Resource Management. Master's Thesis, International Institute for Geo-information Science and Earth Observation, Enschede, The Netherlands, 2007.
13. Sheffield, J.; Wood, E.F.; Pan, M.; Beck, H.; Coccia, G.; Serrat-Capdevila, A.; Verbist, K. Satellite Remote Sensing for Water Resources Management: Potential for Supporting Sustainable Development in Data-Poor Regions. *Water Resour. Res.* **2018**, *54*, 9724–9758. [CrossRef]
14. Avisse, N.; Tilmant, A.; François Müller, M.; Zhang, H. Monitoring small reservoirs' storage with satellite remote sensing in inaccessible areas. *Hydrol. Earth Syst. Sci.* **2017**, *21*, 6445–6459. [CrossRef]
15. Gholizadeh, M.H.; Melesse, A.M.; Reddi, L. A Comprehensive Review on Water Quality Parameters Estimation Using Remote Sensing Techniques. *Sensors* **2016**, *16*, 1298. [CrossRef]
16. Göttl, F.; Dettmering, D.; Müller, F.L.; Schwatke, C. Lake Level Estimation Based on CryoSat-2 SAR Altimetry and Multi-Looked Waveform Classification. *Remote Sens.* **2016**, *8*, 885. [CrossRef]
17. Dettmering, D.; Schwatke, C.; Boergens, E.; Seitz, F. Potential of ENVISAT Radar Altimetry for Water Level Monitoring in the Pantanal Wetland. *Remote Sens.* **2016**, *8*, 596. [CrossRef]
18. Ismail, K.; Boudhar, A.; Abdelkrim, A.; Mohammed, H.; Mouatassime, S.; Kamal, A.; Driss, E.; Idrissi, E.; Nouaim, W. Evaluating the potential of Sentinel-2 satellite images for water quality characterization of artificial reservoirs: The Bin El Ouidane Reservoir case study (Morocco). *Meteorol. Hydrol. Water Manag.* **2019**, *7*, 31–39. [CrossRef]
19. Salmaso, N.; Mosello, R. Limnological research in the deep southern subalpine lakes: Synthesis, directions and perspectives. *Adv. Oceanogr. Limnol.* **2010**, *1*, 29–66. [CrossRef]
20. European Commission. *Copernicus: New Name for European Earth Observation Programme*; European Commission: Brussels, Belgium, 2012.
21. European Commission. Copernicus Programme. Available online: https://www.copernicus.eu/en/about-copernicus/copernicus-brief (accessed on 12 June 2020).
22. van der Woerd, H.J.; Wernand, M.R. True Colour Classification of Natural Waters with Medium-Spectral Resolution Satellites: SeaWiFS, MODIS, MERIS and OLCI. *Sensors* **2015**, *15*, 25663–25680. [CrossRef]
23. Merchant, C.J.; Embury, O.; Bulgin, C.E.; Block, T.; Corlett, G.K.; Fiedler, E.; Good, S.A.; Mittaz, J.; Rayner, N.A.; Berry, D.; et al. Satellite-based time-series of sea-surface temperature since 1981 for climate applications. *Sci. Data* **2019**, *6*, 223. [CrossRef]
24. Matthews, M.W. A current review of empirical procedures of remote sensing in Inland and near-coastal transitional waters. *Int. J. Remote Sens.* **2011**, *32*, 6855–6899. [CrossRef]
25. Alikas, K.; Kangro, K.; Randoja, R.; Philipson, P.; Asuküll, E.; Pisek, J.; Reinart, A. Satellite-based products for monitoring optically complex inland waters in support of EU Water Framework Directive. *Int. J. Remote Sens.* **2015**, *36*, 4446–4468. [CrossRef]
26. Riffler, M.; Lieberherr, G.; Wunderle, S. Lake surface water temperatures of European Alpine lakes (1989–2013) based on the Advanced Very High Resolution Radiometer (AVHRR) 1 km data set. *Earth Syst. Sci. Data* **2015**, *7*, 1–17. [CrossRef]
27. Clark, J.M.; Schaeffer, B.A.; Darling, J.A.; Urquhart, E.A.; Johnston, J.M.; Ignatius, A.R.; Myer, M.H.; Loftin, K.A.; Werdell, P.J.; Stumpf, R.P. Satellite monitoring of cyanobacterial harmful algal bloom frequency in recreational waters and drinking water sources. *Ecol. Indic.* **2017**, *80*, 84–95. [CrossRef]
28. Leshkevich, G.A.; Nghiem, S.V. Satellite SAR Remote Sensing of Great Lakes Ice Cover, Part 2. Ice Classification and Mapping. *J. Great Lakes Res.* **2007**, *33*, 736–750. [CrossRef]
29. Iestyn Woolway, R.; Merchant, C.J. Amplified surface temperature response of cold, deep lakes to inter-annual air temperature variability. *Sci. Rep.* **2017**, *7*, 4130. [CrossRef]
30. Duguay, C.R.; Bernier, M.; Gauthier, Y.; Kouraev, A. Remote sensing of lake and river ice. In *Remote Sensing of the Cryosphere*; Tedesco, M., Ed.; John Wiley & Sons, Ltd.: Hoboken, NJ, USA, 2015; pp. 273–306. ISBN 9781118368909.
31. Simis, S.G.H.; Peters, S.W.M.; Gons, H.J. Remote sensing of the cyanobacterial pigment phycocyanin in turbid inland water. *Limnol. Oceanogr.* **2005**, *50*, 237–245. [CrossRef]

32. Papathanasopoulou, E.; Simis, S.; Alikas, K.; Ansper, A.; Anttila, S.; Jenni, A.; Barillé, A.-L.; Barillé, L.; Brando, V.; Bresciani, M.; et al. Satellite-assisted monitoring of water quality to support the implementation of the Water Framework Directive. *EOMORES White Pap.* **2019**, 1–28. [CrossRef]
33. Ansper, A.; Alikas, K. Retrieval of Chlorophyll a from Sentinel-2 MSI Data for the European Union Water Framework Directive Reporting Purposes. *Remote Sens.* **2019**, *11*, 64. [CrossRef]
34. Bresciani, M.; Stroppiana, D.; Odermatt, D.; Morabito, G.; Giardino, C. Assessing remotely sensed chlorophyll-a for the implementation of the Water Framework Directive in European perialpine lakes. *Sci. Total Environ.* **2011**, *409*, 3083–3091. [CrossRef]
35. Attila, J.; Kauppila, P.; Kallio, K.Y.; Alasalmi, H.; Keto, V.; Bruun, E.; Koponen, S. Applicability of Earth Observation chlorophyll-a data in assessment of water status via MERIS—With implications for the use of OLCI sensors. *Remote Sens. Environ.* **2018**, *212*, 273–287. [CrossRef]
36. Gohin, F.; Saulquin, B.; Oger-Jeanneret, H.; Lozac'h, L.; Lampert, L.; Lefebvre, A.; Riou, P.; Bruchon, F. Towards a better assessment of the ecological status of coastal waters using satellite-derived chlorophyll-a concentrations. *Remote Sens. Environ.* **2008**, *112*, 3329–3340. [CrossRef]
37. Matthews, M.W.; Bernard, S.; Winter, K. Remote sensing of cyanobacteria-dominant algal blooms and water quality parameters in Zeekoevlei, a small hypertrophic lake, using MERIS. *Remote Sens. Environ.* **2010**, *114*, 2070–2087. [CrossRef]
38. Alikas, K.; Kratzer, S. Improved retrieval of Secchi depth for optically-complex waters using remote sensing data. *Ecol. Indic.* **2017**, *77*, 218–227. [CrossRef]
39. Alikas, K.; Kratzer, S.; Reinart, A.; Kauer, T.; Paavel, B. Robust remote sensing algorithms to derive the diffuse attenuation coefficient for lakes and coastal waters. *Limnol. Oceanogr. Methods* **2015**, *13*, 402–415. [CrossRef]
40. Liu, X.; Lee, Z.; Zhang, Y.; Lin, J.; Shi, K.; Zhou, Y.; Qin, B.; Sun, Z. Remote Sensing of Secchi Depth in Highly Turbid Lake Waters and Its Application with MERIS Data. *Remote Sens.* **2019**, *11*, 2226. [CrossRef]
41. Free, G.; Bresciani, M.; Trodd, W.; Tierney, D.; O'Boyle, S.; Plant, C.; Deakin, J. Estimation of lake ecological quality from Sentinel-2 remote sensing imagery. *Hydrobiologia* **2020**, *847*, 1423–1438. [CrossRef]
42. Uudeberg, K.; Ansko, I.; Põru, G.; Ansper, A.; Reinart, A. Using Optical Water Types to Monitor Changes in Optically Complex Inland and Coastal Waters. *Remote Sens.* **2019**, *11*, 2297. [CrossRef]
43. Soomets, T.; Uudeberg, K.; Jakovels, D.; Zagars, M.; Reinart, A.; Brauns, A.; Kutser, T. Comparison of lake opticalwater types derived from sentinel-2 and sentinel-3. *Remote Sens.* **2019**, *11*, 2883. [CrossRef]
44. Xue, K.; Ma, R.; Wang, D.; Shen, M. Optical classification of the remote sensing reflectance and its application in deriving the specific phytoplankton absorption in optically complex lakes. *Remote Sens.* **2019**, *11*, 184. [CrossRef]
45. González Vilas, L.; Spyrakos, E.; Torres Palenzuela, J.M. Neural network estimation of chlorophyll a from MERIS full resolution data for the coastal waters of Galician rias (NW Spain). *Remote Sens. Environ.* **2011**, *115*, 524–535. [CrossRef]
46. Odermatt, D.; Pomati, F.; Pitarch, J.; Carpenter, J.; Kawka, M.; Schaepman, M.; Wüest, A. MERIS observations of phytoplankton blooms in a stratified eutrophic lake. *Remote Sens. Environ.* **2012**, *126*, 232–239. [CrossRef]
47. Ioannou, I.; Gilerson, A.; Gross, B.; Moshary, F.; Ahmed, S. Deriving ocean color products using neural networks. *Remote Sens. Environ.* **2013**, *134*, 78–91. [CrossRef]
48. Fink, G.; Burke, S.; Simis, S.G.H.; Kangur, K.; Kutser, T.; Mulligan, M. Management Options to Improve Water Quality in Lake Peipsi: Insights from Large Scale Models and Remote Sensing. *Water Resour. Manag.* **2020**, *34*, 2241–2254. [CrossRef]
49. Gitelson, A.A.; Dall'Olmo, G.; Moses, W.; Rundquist, D.C.; Barrow, T.; Fisher, T.R.; Gurlin, D.; Holz, J. A simple semi-analytical model for remote estimation of chlorophyll-a in turbid waters: Validation. *Remote Sens. Environ.* **2008**, *112*, 3582–3593. [CrossRef]
50. Nielsen, K.; Stenseng, L.; Andersen, O.B.; Knudsen, P. The Performance and Potentials of the CryoSat-2 SAR and SARIn Modes for Lake Level Estimation. *Water* **2017**, *9*, 374. [CrossRef]
51. Troitskaya, Y.I.; Rybushkina, G.V.; Soustova, I.A.; Balandina, G.N.; Lebedev, S.A.; Kostyanoi, A.G.; Panyutin, A.A.; Filina, L.V. Satellite Altimetry of Inland Water Bodies. *Water Resour.* **2012**, *39*, 169–185. [CrossRef]
52. Fernandes, M.J.; Lázaro, C.; Nunes, A.L.; Scharroo, R. Atmospheric corrections for altimetry studies over inland water. *Remote Sens.* **2014**, *6*, 4952–4997. [CrossRef]
53. Leira, M.; Cantonati, M. Effects of water-level fluctuations on lakes: An annotated bibliography. *Hydrobiologia* **2008**, *613*, 171–184. [CrossRef]
54. Wang, X.L.; Lu, Y.L.; Han, J.Y.; He, G.Z.; Wang, T.Y. Identification of anthropogenic influences on water quality of rivers in Taihu watershed. *J. Environ. Sci.* **2007**, *19*, 475–481. [CrossRef]
55. Nõges, P.; Van De Bund, W.; Cardoso, A.C.; Heiskanen, A.S. Impact of climatic variability on parameters used in typology and ecological quality assessment of surface waters—Implications on the Water Framework Directive. *Hydrobiologia* **2007**, *584*, 373–379. [CrossRef]
56. Koff, T.; Vandel, E.; Marzecová, A.; Avi, E.; Mikomägi, A. Assessment of the effect of anthropogenic pollution on the ecology of small shallow lakes using the palaeolimnological approach. *Est. J. Earth Sci.* **2016**, *65*, 221–233. [CrossRef]
57. Khatri, N.; Tyagi, S. Influences of natural and anthropogenic factors on surface and groundwater quality in rural and urban areas. *Front. Life Sci.* **2015**, *8*, 23–39. [CrossRef]
58. Wang, M.; Cheng, W.; Huang, J.; Shi, L.; Yu, B. Effects of water-level on water quality of reservoir in numerical simulated experiments. *Chem. Eng. Trans.* **2016**, *51*, 733–738. [CrossRef]

59. Chen, M.Q.; Li, Y.; Li, K.F.; Li, R.; Li, J. Effects of water-level decline on water quality of reservoir. *Sichuan Daxue Xuebao (Gongcheng Kexue Ban)/J. Sichuan Univ. Eng. Sci. Ed.* **2012**, *44*, 32–36.
60. Nõges, P.; Nõges, T.; Laas, A. Climate-related changes of phytoplankton seasonality in large shallow Lake Võrtsjärv, Estonia. *Aquat. Ecosyst. Health Manag.* **2010**, *13*, 154–163. [CrossRef]
61. Nõges, T.; Nõges, P.; Laugaste, R. Water level as the mediator between climate change and phytoplankton composition in a large shallow temperate lake. In *Hydrobiologia*; Springer: Dordrecht, The Netherlands, 2003; Volume 506–509, pp. 257–263.
62. Tuvikene, L.; Nõges, T.; Nõges, P. Why do phytoplankton species composition and "traditional" water quality parameters indicate different ecological status of a large shallow lake? *Hydrobiologia* **2011**, *660*, 3–15. [CrossRef]
63. Nõges, T.; Nõges, P. The effect of extreme water level decrease on hydrochemistry and phytoplankton in shallow eutrophic lake. *Hydrobiologia* **1999**, *408–409*, 277–283. [CrossRef]
64. Ministry of Environment. Pinnaveekogumite Moodustamise Kord ja Nende Pinnaveekogumite Nimestik, Mille Seisundiklass Tuleb Määrata, Pinnaveekogumite Seisundiklassid ja Seisundiklassidele Vastavad Kvaliteedinäitajate Väärtused Ning Seisundiklasside Määramise kord-RT I. 25 November 2010. Available online: https://www.riigiteataja.ee/akt/125112010015 (accessed on 3 September 2020).
65. Keskkonnaministeerium. Keskkonnaseire Infrosüsteem. Available online: https://kese.envir.ee/kese/listProgramAndPublicReport.action (accessed on 27 February 2020).
66. Mäemets, H.; Laugaste, R.; Palmik, K.; Haldna, M. Response of primary producers to water level fluctuations of Lake Peipsi. *Proc. Est. Acad. Sci.* **2018**, *67*, 231–245. [CrossRef]
67. Nõges, P.; Nõges, T. Indicators and criteria to assess ecological status of the large shallow temperate polymictic lakes Peipsi (Estonia/Russia) and Võrtsjärv (Estonia). *Boreal Environ. Res.* **2006**, *11*, 67–80.
68. Eesti Ilmateenistus. Hydrological Measurements. Available online: http://www.ilmateenistus.ee/ilmatarkus/mootetehnika/hudroloogiliste-vaatluste-mootetehnika/ (accessed on 9 March 2020).
69. Ellmann, A.; Märdla, S.; Oja, T. The 5 mm geoid model for Estonia computed by the least squares modified Stokes's formula. *Surv. Rev.* **2020**, *52*, 352–372. [CrossRef]
70. Republic of Estonia Environment Agency. *Annex 4. National Environmental Monitoring Program Surface Water Monitoring SUB-Program*; Republic of Estonia Environment Agency: Tallinn, Estonia, 2019.
71. Jeffrey, S.W.; Humphrey, G.F. New spectrophotometric equations for determining chlorophylls a, b, c1 and c2 in higher plants, algae and natural phytoplankton. *Biochem. Physiol. Pflanz.* **1975**, *167*, 191–194. [CrossRef]
72. Utermöhl, H. Zur Vervollkommnung der quantitativen Phytoplankton-Methodik. *Int. Vereinigung für Theor. Angew. Limnol. Mitt.* **1958**, *9*, 1–38. [CrossRef]
73. Respublic of Estonia Land Board. ESTHub Processing Platform. Available online: https://ehcalvalus.maaamet.ee/calest/calvalus.jsp (accessed on 10 December 2019).
74. Alikas, K.; Kangro, K.; Reinart, A. Detecting cyanobacterial blooms in large North European lakes using the Maximum Chlorophyll Index. *Oceanologia* **2010**, *52*, 237–257. [CrossRef]
75. Gower, J.; King, S.; Borstad, G.; Brown, L. Use of the 709 nm band of meris to detect intense plankton blooms and other conditions in coastal waters. *Int. J. Remote Sens.* **2005**, *26*, 2005–2012. [CrossRef]
76. Villadsen, H.; Andersen, O.B.; Stenseng, L.; Nielsen, K.; Knudsen, P. CryoSat-2 altimetry for river level monitoring—Evaluation in the Ganges-Brahmaputra River basin. *Remote Sens. Environ.* **2015**, *168*, 80–89. [CrossRef]
77. Taheri Tizro, A.; Ghashghaie, M.; Georgiou, P.; Voudouris, K. Time series analysis of water quality parameters. *J. Appl. Res. Water Wastewater* **2014**, *1*, 43–52.
78. Monteiro, M.; Costa, M. A time series model comparison for monitoring and forecasting water quality variables. *Hydrology* **2018**, *5*, 37. [CrossRef]
79. Wang, F.; Wang, X.; Zhao, Y.; Yang, Z.F. Long-term water quality variations and chlorophyll a simulation with an emphasis on different hydrological periods in Lake Baiyangdian, Northern China. *J. Environ. Inform.* **2012**, *20*, 90–102. [CrossRef]
80. Wantzen, K.M.; Rothhaupt, K.-O.; Mörtl, M.; Cantonati, M.; László, G.; Fischer, P. *Ecological Effects of Water-Level Fluctuations in Lakes*; Springer: Dordrecht, The Netherlands, 2008.
81. Gownaris, N.J.; Rountos, K.J.; Kaufman, L.; Kolding, J.; Lwiza, K.M.M.; Pikitch, E.K. Water level fluctuations and the ecosystem functioning of lakes. *J. Great Lakes Res.* **2018**, *44*, 1154–1163. [CrossRef]
82. Maihemuti, B.; Aishan, T.; Simayi, Z.; Alifujiang, Y.; Yang, S. Temporal Scaling of Water Level Fluctuations in Shallow Lakes and Its Impacts on the Lake Eco-Environments. *Sustainability* **2020**, *12*, 3541. [CrossRef]
83. Liu, X.; Yang, Z.; Yuan, S.; Wang, H. A novel methodology for the assessment of water level requirements in shallow lakes. *Ecol. Eng.* **2017**, *102*, 31–38. [CrossRef]
84. Stefanidis, K.; Papastergiadou, E. Effects of a long term water level reduction on the ecology and water quality in an eastern Mediterranean lake. *Knowl. Manag. Aquat. Ecosyst.* **2013**, *411*. [CrossRef]
85. Coops, H.; Beklioglu, M.; Crisman, T.L. The role of water-level fluctuations in shallow lake ecosystems—Workshop conclusions. *Hydrobiologia* **2003**, *506–509*, 23–27. [CrossRef]
86. Zohary, T.; Ostrovsky, I. Ecological impacts of excessive water level fluctuations in stratified freshwater lakes. *Inl. Waters* **2011**, *1*, 47–59. [CrossRef]

87. Górniak, A.; Piekarski, K. Seasonal and Multiannual Changes of Water Levels in Lakes of Northeastern Poland. *Pol. J. Environ. Stud.* **2002**, *11*, 349–354.
88. Jeppesen, E.; Meerhoff, M.; Davidson, T.A.; Trolle, D.; Søndergaard, M.; Lauridsen, T.L.; Beklioğlu, M.; Brucet, S.; Volta, P.; González-Bergonzoni, I.; et al. Climate change impacts on lakes: An integrated ecological perspective based on a multi-faceted approach, with special focus on shallow lakes. *J. Limnol.* **2014**, *73*, 84–107. [CrossRef]
89. Mooij, W.M.; Hülsmann, S.; De Senerpont Domis, L.N.; Nolet, B.A.; Bodelier, P.L.E.; Boers, P.C.M.; Dionisio Pires, L.M.; Gons, H.J.; Ibelings, B.W.; Noordhuis, R.; et al. The impact of climate change on lakes in the Netherlands: A review. *Aquat. Ecol.* **2005**, *39*, 381–400. [CrossRef]
90. Torabi Haghighi, A.; Kløve, B. A sensitivity analysis of lake water level response to changes in climate and river regimes. *Limnologica* **2015**, *51*, 118–130. [CrossRef]
91. Whitehead, P.G.; Wilby, R.L.; Battarbee, R.W.; Kernan, M.; Wade, A.J. A review of the potential impacts of climate change on surface water quality. *Hydrol. Sci. J.* **2009**, *54*, 101–123. [CrossRef]
92. Oppenheimer, M.; Glavovic, B.C.; Hinkel, J.; van de Wal, R.; Magnan, A.K.; Abd-Elgawad, A.; Cai, R.; Cifuentes-Jara, M.; DeConto, R.M.; Ghosh, T.; et al. Sea Level Rise and Implications for Low-Lying Islands, Coasts and Communities. In *IPCC Special Report on the Ocean and Cryosphere in a Changing Climate*; Pörtner, H.-O., Roberts, D.C., Masson-Delmotte, V., Zhai, P., Tignor, M., Poloczanska, E., Mintenbeck, K., Alegría, A., Nicolai, M., Okem, A., et al., Eds.; IPCC: Geneva, Switzerland, 2019.
93. Vincent, W.F. Effects of Climate Change on Lakes. In *Encyclopedia of Inland Waters*; Elsvier: Amsterdam, The Netherlands, 2009; pp. 55–60. ISBN 9780123706263.
94. Kiani, T.; Ramesht, M.H.; Maleki, A.; Safakish, F. Analyzing the Impacts of Climate Change on Water Level Fluctuations of Tashk and Bakhtegan Lakes and Its Role in Environmental Sustainability. *Open J. Ecol.* **2017**, *7*, 158–178. [CrossRef]
95. Fan, Z.; Wang, Z.; Li, Y.; Wang, W.; Tang, C.; Zeng, F. Water Level Fluctuation under the Impact of Lake Regulation and Ecological Implication in Huayang Lakes, China. *Water* **2020**, *12*, 702. [CrossRef]
96. Nhu, V.-H.; Mohammadi, A.; Shahabi, H.; Shirzadi, A.; Al-Ansari, N.; Ahmad, B.B.; Chen, W.; Khodadadi, M.; Ahmadi, M.; Khosravi, K.; et al. Monitoring and Assessment of Water Level Fluctuations of the Lake Urmia and Its Environmental Consequences Using Multitemporal Landsat 7 ETM + Images. *Int. J. Environ. Res. Public Health* **2020**, *17*, 4210. [CrossRef] [PubMed]
97. Jalili, S.; Hamidi, S.A.; Namdar Ghanbari, R. Climate variability and anthropogenic effects on Lake Urmia water level fluctuations, northwestern Iran. *Hydrol. Sci. J.* **2016**, *61*, 1759–1769. [CrossRef]
98. Schwatke, C.; Dettmering, D.; Bosch, W.; Seitz, F. DAHITI—An innovative approach for estimating water level time series over inland waters using multi-mission satellite altimetry. *Hydrol. Earth Syst. Sci.* **2015**, *19*, 4345–4364. [CrossRef]
99. Berry, P.A.M.; Garlick, J.D.; Freeman, J.A.; Mathers, E.L. Global inland water monitoring from multi-mission altimetry. *Geophys. Res. Lett.* **2005**, *32*, 16. [CrossRef]
100. Liibusk, A.; Kall, T.; Rikka, S.; Uiboupin, R.; Suursaar, Ü.; Tseng, K.H. Validation of copernicus sea level altimetry products in the baltic sea and estonian lakes. *Remote Sens.* **2020**, *12*, 4062. [CrossRef]
101. Nielsen, K.; Andersen, O.B.; Ranndal, H. Validation of sentinel-3a based lake level over US and Canada. *Remote Sens.* **2020**, *12*, 2835. [CrossRef]
102. Crétaux, J.F.; Bergé-Nguyen, M.; Calmant, S.; Jamangulova, N.; Satylkanov, R.; Lyard, F.; Perosanz, F.; Verron, J.; Montazem, A.S.; Guilcher, G.L.; et al. Absolute calibration or validation of the altimeters on the Sentinel-3A and the Jason-3 over Lake Issykkul (Kyrgyzstan). *Remote Sens.* **2018**, *10*, 1679. [CrossRef]
103. Birkett, C.M. Radar altimetry: A new concept in monitoring lake level changes. *Eos Trans. Am. Geophys. Union* **1994**, *75*, 273–275. [CrossRef]
104. Crétaux, J.F.; Birkett, C. Lake studies from satellite radar altimetry. *Comptes Rendus-Geosci.* **2006**, *338*, 1098–1112. [CrossRef]
105. Crétaux, J.F.; Jelinski, W.; Calmant, S.; Kouraev, A.; Vuglinski, V.; Bergé-Nguyen, M.; Gennero, M.C.; Nino, F.; Abarca Del Rio, R.; Cazenave, A.; et al. SOLS: A lake database to monitor in the Near Real Time water level and storage variations from remote sensing data. *Adv. Space Res.* **2011**, *47*, 1497–1507. [CrossRef]
106. Song, C.; Ye, Q.; Sheng, Y.; Gong, T. Combined ICESat and CryoSat-2 Altimetry for Accessing Water Level Dynamics of Tibetan Lakes over 2003–2014. *Water* **2015**, *7*, 4685–4700. [CrossRef]
107. Wu, Y.J.; Qiao, G.; Li, H.W. Water Level Changes Of Nam-Co Lake Based On Satellite Altimetry Data Series. *Int. Arch. Photogramm. Remote Sens. Spatial Inf. Sci.* **2017**, *XLII-2/W7*, 1555–1560. [CrossRef]
108. Li, P.; Li, H.; Chen, F.; Cai, X. Monitoring long-term lake level variations in middle and lower yangtze basin over 2002-2017 through integration of multiple satellite altimetry datasets. *Remote Sens.* **2020**, *12*, 1448. [CrossRef]
109. Chen, Q.; Zhang, Y.; Ekroos, A.; Hallikainen, M. The role of remote sensing technology in the EU water framework directive (WFD). *Environ. Sci. Policy* **2004**, *7*, 267–276. [CrossRef]
110. Moses, W.J.; Gitelson, A.A.; Berdnikov, S.; Povazhnyy, V. Estimation of chlorophyll-a concentration in case II waters using MODIS and MERIS data—Successes and challenges. *Environ. Res. Lett.* **2009**, *4*, 045005. [CrossRef]
111. Salem, S.I.; Strand, M.H.; Higa, H.; Kim, H.; Kazuhiro, K.; Oki, K.; Oki, T. Evaluation of MERIS Chlorophyll-a Retrieval Processors in a Complex Turbid Lake Kasumigaura over a 10-Year Mission. *Remote Sens.* **2017**, *9*, 1022. [CrossRef]
112. Wozniak, M.; Bradtke, K.M.; Krezel, A. Comparison of satellite chlorophyll a algorithms for the Baltic Sea. *J. Appl. Remote Sens.* **2014**, *8*, 083605. [CrossRef]

113. Chavula, G.; Brezonik, P.; Thenkabail, P.; Johnson, T.; Bauer, M. Estimating chlorophyll concentration in Lake Malawi from MODIS satellite imagery. *Phys. Chem. Earth* **2009**, *34*, 755–760. [CrossRef]
114. Wu, M.; Zhang, W.; Wang, X.; Luo, D. Application of MODIS satellite data in monitoring water quality parameters of Chaohu Lake in China. *Environ. Monit. Assess.* **2009**, *148*, 255–264. [CrossRef]
115. Zhang, J.; Xu, K.; Yang, Y.; Qi, L.; Hayashi, S.; Watanabe, M. Measuring water storage fluctuations in Lake Dongting, China, by Topex/Poseidon satellite altimetry. *Environ. Monit. Assess.* **2006**, *115*, 23–37. [CrossRef]
116. Birkett, C.M. The contribution of TOPEX/POSEIDON to the global monitoring of climatically sensitive lakes. *J. Geophys. Res.* **1995**, *100*, 25179–25204. [CrossRef]
117. Hydroweb. Available online: http://hydroweb.theia-land.fr/ (accessed on 1 February 2021).
118. Deutsches Geodätisches Forschungsinstitut der Technischen Universität München Database for Hydrological Time Series of Inland Waters. Available online: https://dahiti.dgfi.tum.de/en/ (accessed on 1 February 2021).
119. Global Reservoirs and Lakes Monitor (G-REALM). Available online: https://ipad.fas.usda.gov/cropexplorer/global_reservoir/ (accessed on 1 February 2021).
120. Politi, E.; MacCallum, S.; Cutler, M.E.J.; Merchant, C.J.; Rowan, J.S.; Dawson, T.P. Selection of a network of large lakes and reservoirs suitable for global environmental change analysis using Earth Observation. *Int. J. Remote Sens.* **2016**, *37*, 3042–3060. [CrossRef]
121. Globolakes. Available online: http://www.globolakes.ac.uk/index.html (accessed on 1 February 2021).
122. Sathyendranath, S.; Brewin, R.J.W.; Brockmann, C.; Brotas, V.; Calton, B.; Chuprin, A.; Cipollini, P.; Couto, A.B.; Dingle, J.; Doerffer, R.; et al. An ocean-colour time series for use in climate studies: The experience of the ocean-colour climate change initiative (OC-CCI). *Sensors* **2019**, *19*, 4285. [CrossRef]

Article

Spatiotemporal Dynamics and Environmental Controlling Factors of the Lake Tana Water Hyacinth in Ethiopia

Abeyou W. Worqlul [1,*], Essayas K. Ayana [2], Yihun T. Dile [2], Mamaru A. Moges [3], Minychl G. Dersseh [3], Getachew Tegegne [4,5] and Solomon Kibret [6]

1 Blackland Research Center, Texas A & M AgriLife Research, Temple, TX 76502, USA
2 Spatial Sciences Laboratory, Texas A & M University, College Station, TX 77843, USA; ekk45@cornell.edu (E.K.A.); yihundile@tamu.edu (Y.T.D.)
3 Faculty of Civil and Water Resource Engineering, Bahir Dar Institute of Technology, Bahir Dar University, Bahir Dar P.O. Box 26, Ethiopia; mamarumoges@gmail.com (M.A.M.); minychl2009@gmail.com (M.G.D.)
4 Department of Civil & Environmental Engineering, Seoul National University, Gwanak-ro 1, Gwanak-gu, Seoul 151-742, Korea; hydro@snu.ac.kr
5 Department of Earth and Environment, Florida International University, Miami, FL 33199, USA; gdamtew@fiu.edu
6 Program in Public Health, University of California Irvine, Irvine, CA 92602, USA; sbirhani@uci.edu
* Correspondence: aworqlul@brc.tamus.edu; Tel.: +1-254-774-6020

Received: 9 July 2020; Accepted: 17 August 2020; Published: 21 August 2020

Abstract: The largest freshwater lake in Ethiopia, Lake Tana, has faced ecological disaster due to water hyacinth (*Eichhornia crassipes*) infestation. The water hyacinth is a threat not only to the ecology but also to the socioeconomic development of the region and cultural value of the lake, which is registered as a UNESCO reserve. This study aims to map the spatiotemporal dynamics of the water hyacinth using high-resolution PlanetScope satellite images and assesses the major environmental variables that relate to the weed spatial coverage dynamics for the period August 2017 to July 2018. The plausible environmental factors studied affecting the weed dynamics include lake level, water and air temperature, and turbidity. Water temperature and turbidity were estimated from the moderate resolution imaging spectroradiometer (MODIS) satellite image and the water level was estimated using Jason-1 altimetry data while the air temperature was obtained from the nearby meteorological station at Bahir Dar station. The results indicated that water hyacinth coverage was increasing at a rate of 14 ha/day from August to November of 2017. On the other hand, the coverage reduced at a rate of 6 ha/day from December 2017 to June 2018. However, the length of shoreline infestation increased significantly from 4.3 km in August 2017 to 23.4 km in April 2018. Lake level and night-time water temperatures were strongly correlated with water hyacinth spatial coverage ($p < 0.05$). A drop in the lake water level resulted in a considerable reduction of the infested area, which is also related to decreasing nutrient levels in the water. The water hyacinth expansion dynamics could be altered by treating the nutrient-rich runoff with best management practices along the wetland and in the lake watershed landscape.

Keywords: Lake Tana; water hyacinth; waterbody temperature; turbidity; lake level; Planetscope

1. Introduction

Aquatic invasive species threaten ecological and socioeconomic integrity by affecting the productivity and functionality of freshwater systems. Water hyacinth (*Eichhornia crassipes*) is one of the dangerous invasive aquatic weeds, which is native to the Amazon Basin [1–3]. When a conducive environment exists, water hyacinth can produce an annual biomass of 50–60 t/ha and doubles its area coverage every 6–15 days [2,4]. Its high tolerance to different environmental conditions such as pH and nutrient level [5,6] combined with its rapid growth and formation of dense impenetrable mats make it a unique weed with a severe potential negative effect to freshwater ecosystems. Water hyacinth exists as a major weed in more than 50 countries such as Zimbabwe, South Africa, and Ethiopia [7–9], where it is prevalent in tropical and subtropical regions [10]. In Ethiopia, it was observed the first time in 1974 in the Koka hydropower reservoir [11]. In Lake Tana, the water hyacinth was observed in 2011 [12,13]. The alarming expansion of the weed in Lake Tana since 2017 has attracted wider scientific and public attention.

Lake Tana has a significant ecological, religious, historic, and economic importance. The lake is home to more than 37 islands with historical churches and more than 20,000 inhabitants [14]. Due to the lake's rich biodiversity and cultural and historical significance, it is registered as the United Nations Educational, Scientific, and Cultural Organization's (UNESCO) biosphere reserve in 2014. This helps to facilitate the protection of the lake biodiversity, conduct research, and attract more visitors. However, a recent rapid expansion of water hyacinth has posed a major threat to the lake's biodiversity and socioeconomics activities. The weed has severely disrupted ecological and socioeconomic functions such as aquatic food chain, nutrient cycling, agricultural activities, tourism, transportation, cultural and religious practices, and public health issues [15,16]. The weed's impact is, in fact, substantial in Lake Tana where the ecological, socioeconomic, and cultural activities are highly intertwined. Understanding the spatial and temporal dynamics of water hyacinth, and factors affecting the spatial dynamics are critical in implementing mitigation measures to limit the expansion, and eventual elimination before the lake's biodiversity and socioeconomics become in jeopardy.

Managing and controlling the extent of water hyacinth in Lake Tana is paramount to save its unique biodiversity. The management should include effective and frequent monitoring of the water hyacinth expansion both spatially and temporally. The monitoring of water hyacinth can be conducted using conventional methods and remote sensing approaches. Conventional approaches monitor the water hyacinth situation on the ground using tracking devices such as a global positioning system (GPS). The conventional approach requires extensive manpower, funding, and infrastructural setup, which are difficult to sustain in developing countries. On the other hand, freely available remote sensing data and GIS applications can make monitoring of water hyacinth and other aquatic weeds convenient and less costly [17,18].

In the recent past, various studies have addressed the water hyacinth infestation [12,19–23]. The preliminary survey of water hyacinth infestation by Tewabe [12] between October and November of 2011 indicated the earliest infestation of the water hyacinth at the mouth of the Megech River. The effect of water hyacinth infestation on fishing in Lake Tana was evaluated by Asmare [19], which reported the impact of the weed on the fishing industry. The study indicated how the entangling of fishing nets and boat propeller had made the fishing task exhaustive and affected the livelihood quality of the fishermen [19]. Dersseh, et al. [20] mapped water hyacinth infestation hotspot using a GIS-based multi-criteria evaluation technique and estimated the water hyacinth infestation susceptible area. Dersseh, et al. [21] used a Sentinel-2 satellite image to determine the spatial coverage of water hyacinth in the lake. This study builds on previous studies using finer resolution satellite images from PlanetScope to map the spatial distribution of water hyacinth as well as to evaluate major environmental factors affecting the water hyacinth infestation in Lake Tana. The specific objectives of this study are to (1) determine the spatial and temporal distribution of water hyacinth in Lake Tana and (2) identify the major environmental factors that are related to the water hyacinth temporal and spatial dynamics. The findings of this study will provide timely knowledge that is helpful to reduce the water hyacinth infestation in the lake and thereby improving the health of Lake Tana to ensure sustainable ecosystem services.

2. Materials and Methods

2.1. Study Area

Lake Tana is the largest freshwater lake in Ethiopia, and it is the source of the Upper Blue Nile River. The Upper Blue Nile River contributes more than 50% of annual streamflow to the Nile River [24,25]. Lake Tana has a catchment area of 15,000 km^2, out of which the lake covers 20%. The lake is located in the north-west highland of Ethiopia; its exact geographic coordinate extends between 10°45′54.1′′ N, 36°10′24.9′′ E and 12°50′15.9′′ N, 38°50′54.48′′ E (Figure 1).

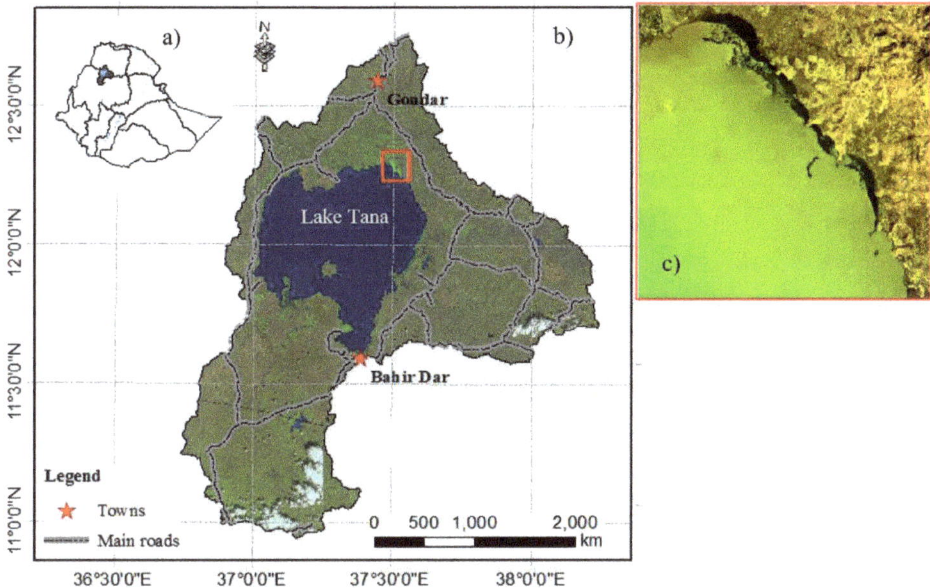

Figure 1. Study area location. (**a**) The left insert is the Ethiopian map with its major river basins; the dark color represents the location of Lake Tana catchment. (**b**) A true-color composite of Sentinel-2 satellite image taken on November 28, 2017 (**c**) Extent of water hyacinth in the northeast shore of Lake Tana (PlanetScope image, January 20, 2018).

The lake provides substantial socioeconomic services. For example, more than 2 million people live in the Lake Tana catchment depending for their livelihood directly or indirectly on the lake and its resources [26,27]. The Lake Tana catchment, due to its potential suitable land and climate, it is considered as one of the agricultural growth corridors by the government of Ethiopia [28].

2.2. Lake Tana Hydrology

In general, the lake has two main seasons. The main rainfall season is called "Kremt" and it extends from June to September and accounts for 90% of the annual rainfall (Figure 2). The dry season extends from October to May and it is characterized by high daily temperature. According to data from the Bahir Dar meteorological station, the annual average rainfall (1994–2016) in the area is 1490 mm. The average minimum and maximum temperature at Bahir Dar meteorological station (1994–2016) were 9 and 25 °C, respectively (Figure 2). In the Lake Tana watershed, four major rivers contribute about 93% of the streamflow into the lake [29]. Among the largest tributary river, Gilgel Abay contributes 50% of the streamflow followed by Gumara 32%, Ribb 12%, and Megech 6%. These rivers carry a substantial amount of sediment, nutrient, and other pollutants into the lake [30–32]. The lake annual evaporation is reported in the range of 1500–1690 mm [29,33], which is greater than the annual rainfall. The lake is considered as shallow with a mean and maximum depth of 12 and 17 m, respectively [30,34].

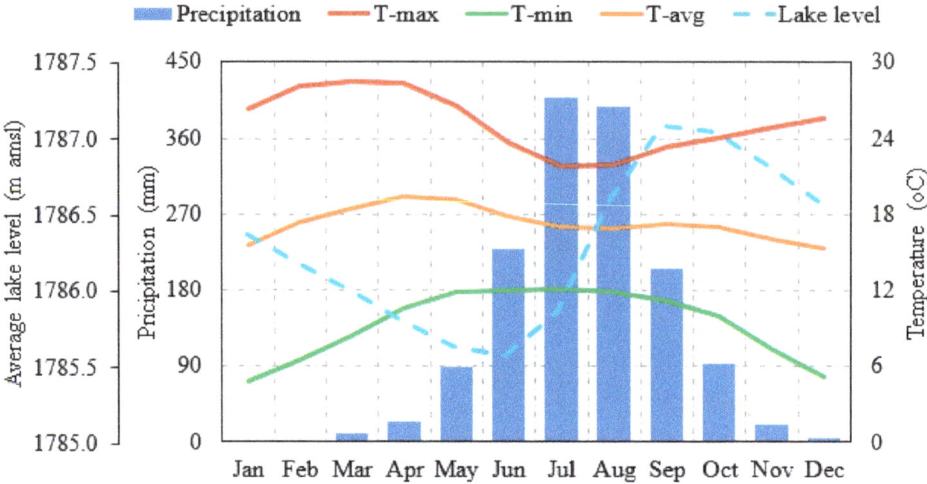

Figure 2. Average monthly rainfall (mm), minimum, and maximum temperature (°C) of Bahir Dar Station (1994–2016) and average monthly lake level (m; 1994–2010).

2.3. Methods

The spatial and temporal distribution of the water hyacinth was mapped from August 2017 to July 2018 using high-resolution satellite images from the PlanetScope (https://www.planet.com/). Details of the satellite image processing are presented in the following section. The potential environmental factors that affect the water hyacinth spatial dynamics studied were lake water level, air and water body temperature, and lake water turbidity. The water body temperature and lake water turbidity were estimated from the moderate resolution imaging spectroradiometer (MODIS) images. The air temperature was obtained from the Bahir Dar weather station, which is the closest meteorological station to the lake. The observed lake level was assimilated with altimetry lake level data from Jason-1 and Jason-2. The mapped spatial coverage of water hyacinth was compared with the environmental factors to identify the factors that played a role in the spatial dynamics of the water hyacinth coverage. The applied methodology to map water hyacinth coverage, identify major factors plausibly contributing to the weed dynamics, and determine their association is presented in Figure 3.

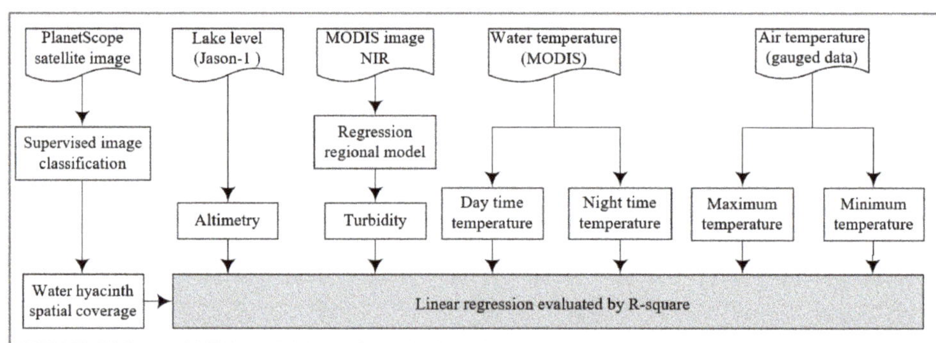

Figure 3. The framework developed to map the water hyacinth spatiotemporal distribution and determine the environmental factors attributing to the water hyacinth spatial coverage dynamics.

2.3.1. Spatiotemporal Water Hyacinth Distribution Mapping

Effective water hyacinth management requires close monitoring of the spatiotemporal coverage. The conventional way of monitoring using a tracking device requires a sustained funding necessary infrastructure and operational setups. Therefore, it is economically and practically expensive to have effective conventional monitoring for a vast lake such as Lake Tana. Rather, the use of satellite data and GIS applications have become more practical and economically viable to map the spatial distribution of aquatic weeds [35,36]. This study uses a high-resolution satellite image from Planet Constellation to capture the spatial and temporal distribution of the water hyacinth in Lake Tana. Since 2013, Planet Constellation images the entire Earth and the climate every day to make the global change detectable, accessible, and actionable [37,38]. The Planet Constellation operates PlanetScope and RapidEye Earth-imaging constellations [39]. With more than 175 satellites in orbit, it has the largest fleet of Earth observation satellites [40]. Since the PlanetScope has a relatively finer resolution than RapidEye, this study used PlanetScope to map the spatial distribution of the weed coverage. PlanetScope has a daily revisit with a spatial resolution of 3–5 m in the visible (455–515 nm, 500–590 nm, and 590–670 nm) and near-infrared (780–860 nm) spectrum [37].

The PlanetScope satellite image for the Lake Tana area was collected since 9 August 2017. Between August 2017 and July 2018, twelve images free of cloud cover were chosen as presented in the annex (Table S1). Emphasis was given to the northeast shore of the lake, where there is extensive water hyacinth infestation. PlanetScope provides geometrically corrected images using highly accurate distributed ground control points [41,42]. To maintain consistency between sensory and overpass time, the metadata provides assets to evaluate the quality of the pixel and coefficients necessary to convert the radiance to reflectance. In this study, the selected satellite images were preprocessed to convert the raw data to reflectance using the coefficients provided. The preprocessed satellite image was classified with a maximum likelihood supervised classification method [43]. The maximum likelihood classification algorithm is exceptionally suitable for our study in that the inverse matrix of the variance–covariance matrix is very stable due to the very low correlation between the bands used in the classification. The representative training samples distributed across the selected images were collected on individual images. The supervised classification on 25 June 2017 was validated using water hyacinth coverage tracked with a boat and handheld GPS and then after validation a similar supervised classification was applied on the selected images.

2.3.2. Major Factors Affecting Water Hyacinth Spatial Coverage Dynamics

Plant growth is affected by a complex interaction of environmental factors such as climatic conditions and nutrient availability [44–46]. Suitable climate and abundant nutrient availability ensure convenient conditions for vegetation growth in the aquatic environment [47,48]. The environmental factors studied to assess their impact on water hyacinth growth in the Lake Tana include air and lake water temperature, lake level, and lake water turbidity.

Lake depth: Morphometric characteristics influence the ecological processes in a lake. The lake depth, for example, affects the thermal stratification, which in turn influences dissolved oxygen levels. Depth also limits the type of weed growing in a portion of the lake. Floating weeds can survive in a deeper section of the lake and avoid nutrient competition with shallow-rooted crops. Water hyacinth thrives both as a free-floating or rooted weed as it adjusts to the depth of the water. The depth of the water hyacinth infested area was characterized based on a recent bathymetry survey conducted by the Tana Sub-basin Organization (TaSBO) in 2011. This survey is better than the surveys conducted by Pietrangeli [49] and Ayana [50] since it has more data points that enabled a robust depth interpolation. The bathymetry survey was interpolated with Inverse Distance Weighting (IDW) and overlayed with the historical water hyacinth infested area to evaluate the lake depth infested with water hyacinth.

Air and lake water temperature: Temperature affects plant development and growth by regulating biochemical processes and photosynthesis activity [51]. Air temperature regulates water body temperature. In fact, its effect is significant over shallow lakes [52]. Plants have a defined range of minimum and maximum temperatures within which growth occurs, and at optimum temperature, the plants grow at a faster rate [53,54]. Kasselmann [55] reported that the minimum (base) and optimum temperatures for water hyacinth growth are 10 °C and 25–30 °C, respectively. The daily air temperature was obtained from the closest weather station to the lake, which is located in Bahir Dar city. The day and night water temperature were obtained for the period 2010–2018 from a moderate resolution imaging spectroradiometer (MODIS) MOD11A2 product. MOD11A2, which is available at 1 km spatial and 8-days temporal resolution. MOD11A2 product has been used for the monitoring of inland water bodies [56–58]. Since the water hyacinth infested area is covered by the green vegetation, water body temperature was extracted at a 2 km buffer distance from the water hyacinth infestation. Since short vegetation over water can be expected to alter the water surface temperature relative to the adjacent non infested water surface [59–61], the nearby water temperature is assumed to be representative. The monthly average day and night-time water temperature, and minimum and maximum air temperature were compared with the water hyacinth spatial coverage to assess if they impact the spatial dynamics of the water hyacinth.

Lake level: The lake level was measured at two diametrically opposite locations since 1960 at Bahir Dar and Gorgora (Figure 4). However, both of these lake level measuring stations were far from the location where a major water hyacinth infestation was observed. The water level gauged data from these stations may be susceptible to bias due to the backwater curve effect from the Chara-Chara weir, which was built in 1996 to regulate the outflow of the lake through the Blue Nile river. The lake water level data is available until 2006 and it does not cover the study period. These drawbacks compelled us to use remotely sensed lake level data from TOPEX/POSEIDON, Jason-1, and Jason-2. These data are available at 10-day intervals since 1993 [62]. The altimeter tracks over Lake Tana were the closest water level measurement to the study site (Figure 4). The altimetry data were compared with gauged lake level data, which was measured before the construction of the weir (i.e., 1993–1996). The validated altimetry data, which is available at near-real-time was used to evaluate the association of the lake level on weed dynamics during the study period (August 2017 to July 2018).

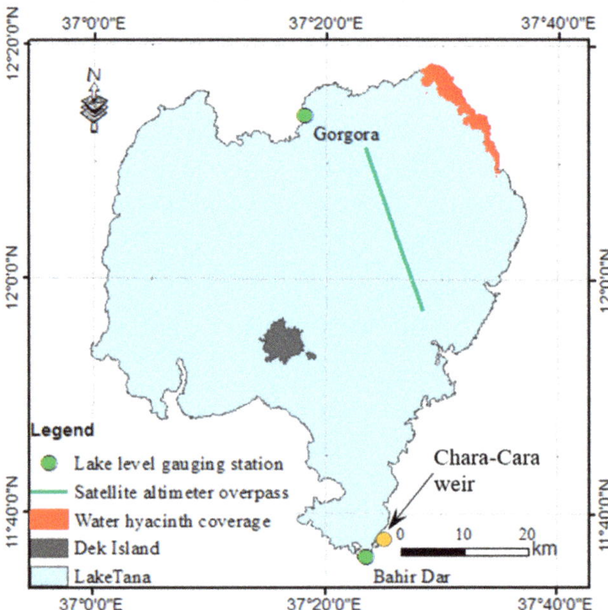

Figure 4. Ground lake level observation stations of Lake Tana, location of Chara-Chara weir, and footprint of Jason-1 and Jason-2 altimeter overpass over Lake Tana.

Turbidity: Turbidity is the most widely used parameter to evaluate physical water quality. It is used to evaluate the cloudiness of given water in comparison to a clear water standard [63]. A lake gets turbid because of the inflow of suspended sediment and dissolved materials through the major tributary rivers. Higher turbidity is considered as an indicator of higher suspended sediment concentration and consequently higher nutrient inflow. Excess availability of nutrients may result in the growth of aquatic crops [64]. Comparing the change in turbidity to the change in spatial coverage of the water hyacinth dynamics may provide insight if the nutrient inflow is the cause for the expansion of the water hyacinth in Lake Tana.

Lake water turbidity data is not available for the Lake Tana. Rather, calibration coefficients from previous studies [31,65,66] were used to estimate Lake Tana's turbidity from MODIS satellite images. Ayana, et al. [31] showed a strong linear association between in situ measurements of turbidity (NTU) and reflectance of MODIS near-infrared channel (NIR) of the electromagnetic spectrum (620–670 nm). Turbidity estimation using NIR has been widely applied for monitoring the water quality of water bodies in different regions and provides a spatial perspective of the water quality of large water bodies [31,65–70].

This study used a 500 m resolution and 8-days revisiting time MODIS satellite NIR images to generate the long-term (2010–2016) average spatial turbidity of the lake. The average monthly historical turbidity for the study period was also estimated in the proximity of the historical water hyacinth infestation area at 2 km buffer distance and used to compare with the weed spatial extent for the study duration (August 2017 to July 2018).

3. Results

3.1. Water Hyacinth Spatiotemporal Dynamics

The visual inspection of the PlanetScope images showed a large infestation of aquatic weed in the northeast shore of Lake Tana. Figure 5 presents the reflectance of the water hyacinth and waterbody at 56 selected pixels on the satellite images across the study area over the selected satellite images (Table S1). The one-way analysis of variance (ANOVA) on the reflectance of the aquatic weed and waterbody indicated a significant difference in reflectance between band 1, 2, 3, and 4 ($p < 0.05$). The difference in the near-infrared wavelength (band 4) was very high since water has relatively lower reflectance while water hyacinth (green vegetation) has higher reflectance in this range. Therefore, a supervised classification was implemented by developing an image mosaic of bands 4, 3, and 2 to map the spatial extent of the water hyacinth.

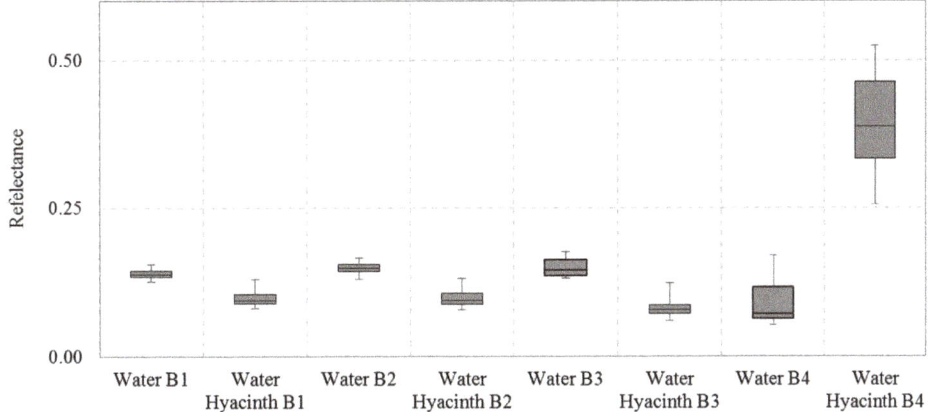

Figure 5. The reflectance of the water hyacinth and water body for the four bands (B1, B2, B3, and B4) of the twelve selected PlanetScope images.

The selected PlanetScope imagery (Table S1) was classified using a supervised classification technique to determine the spatial coverage and length of shoreline infestation (Figure 6). The water hyacinth spatial coverage estimated with a supervised classification on October 25, 2017, was compared with the water hyacinth coverage tracked at the ground with GPS attached on a boat. The comparison of tracked water hyacinth coverage collected with a handheld GPS (June 25, 2017) was compared with the corresponding water hyacinth coverage mapped with supervised classification on October 24, 2017, the result indicated more than 84% match (Supplementary Information (SI), Figure S1). The difference in the agreement was due to the one-day overpass gap between the satellite image and GPS tracking, as well as the fast growth and movability of the water hyacinth.

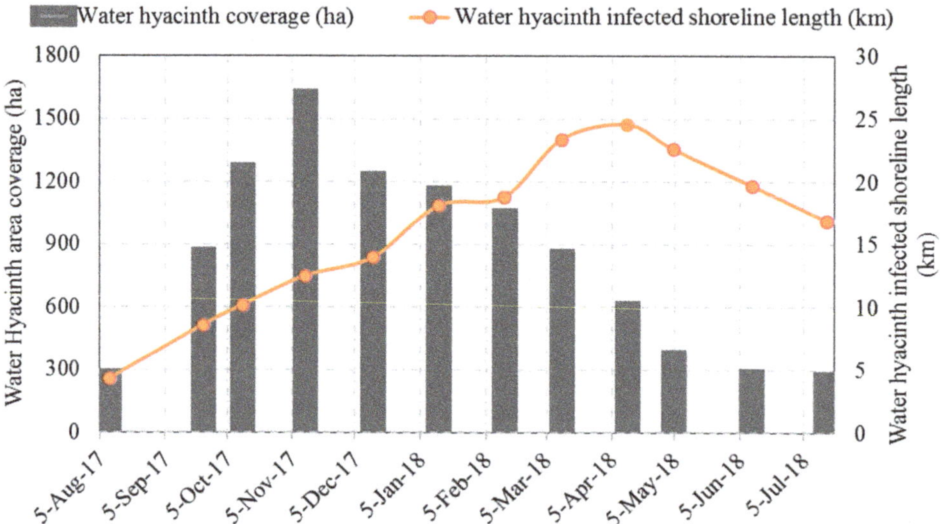

Figure 6. Water hyacinth area coverage and length of infested shoreline in the northeastern shore of Lake Tana (August 2017 to April 2018).

Figure 7 presents the spatial coverage of the water hyacinth of the selected images during the study period (Table S1). The water hyacinth was first observed in the northern shore of the lake (Figure 7a). A significant proportion of the weed moved to the northeast and expands to the eastern shore of Lake Tana (Figure 7a–e). According to the satellite images, for the period August to November 2017, the water hyacinth was expanding at a rate of 14 ha/day and covered 1600 ha of the lake. On the other hand, the coverage decreased at a rate of 6 ha/day from December 2017 to June 2018. The lowest water hyacinth coverage was observed in June 2018 at 300 ha. Despite the area reduction for the period August 2017 to July 2018, the length of water hyacinth infestation was increasing across the lake shoreline. The maximum infestation length was 23.8 km in April 2018 (Figure 6).

3.2. Environmental Factors Plausibly Contributing to the Water Hyacinth Dynamics

The water hyacinth spatial dynamics in Lake Tana is influenced by environmental and human factors. Identifying the major factor requires a detailed analysis of field data. The environmental factors that could affect the water hyacinth dynamics may include air and lake water temperature, turbidity, lake level, and the human factors (i.e., manual and mechanical harvesting of water hyacinth). In this study, the plausible environmental factors associated with the water hyacinth dynamics were studied.

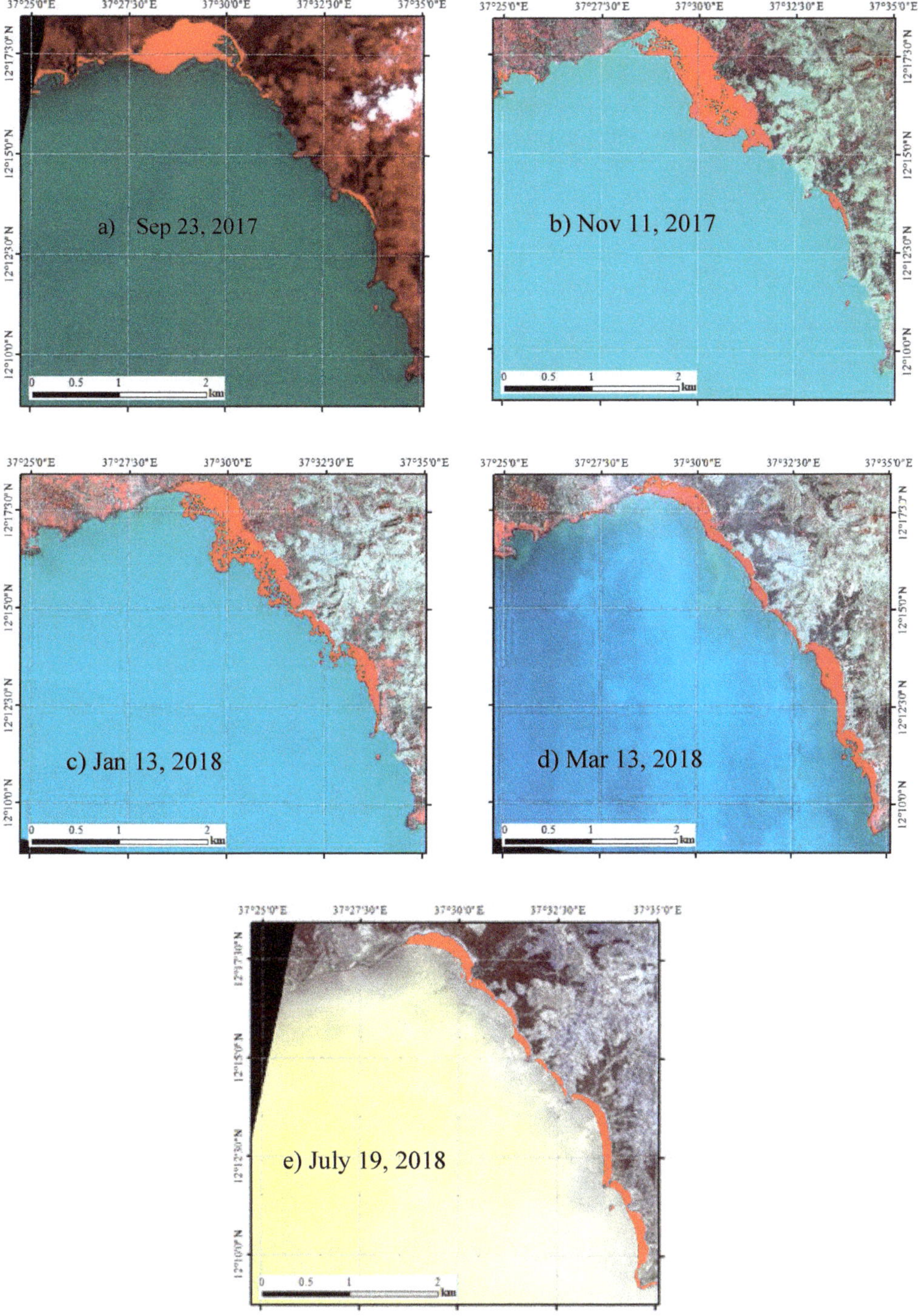

Figure 7. A false-color composite of PlanetScope image and spatial location of the water hyacinth infested area. Supervised classification of (**a**) September 23, 2017, (**b**) November 11, 2017, (**c**) January 13, 2018, (**d**) March 13, 2018, and (**e**) July 19, 2018. The red color represents the spatial area coverage of the water hyacinth infestation.

3.2.1. Lake Tana Bathymetric Survey

The bathymetric survey by TaSBO interpolated with IDW revealed a new maximum depth higher than the previously reported depth of 14 m [30,34,71,72]. The bathymetric survey indicated a maximum and average depth of the lake as 16.8 and 12 m, respectively. The depth map (Figure 8) indicated that the lake depth across the historical water hyacinth infestation area for the study period varied between 0 and 8.4 m. Approximately 28% and 37% of the water hyacinth floats on the lake depth less than 0.6 and 1.0 m, respectively (Figure 8b). This suggests that more than half of the water hyacinth could only be harvested with a mechanical harvester since manual harvesting by hand on a water depth greater than one meter could be dangerous. Therefore, integration of manual and mechanical harvesting controls is vital for the removal of water hyacinth in Lake Tana.

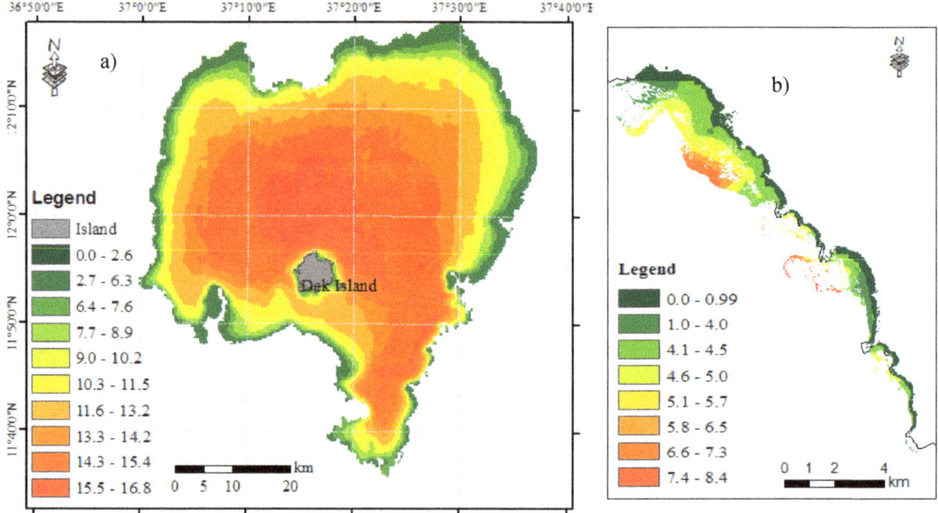

Figure 8. Depth of the lake and water hyacinth infestation area. (**a**) Lake Tana depth map. (**b**) Depth of Lake Tana where water hyacinth infestation is observed (2017 and 2018). The depth of the lake was measured at a lake level of 1786 m amsl (above mean sea level).

3.2.2. Air and Water Temperature

The average annual day and night-time water temperature for Lake Tana were estimated from the MODIS satellite image for the period 2010–2016 (Figure 9a,b). For this period, the annual average daytime water temperature ranged from 20.7 to 33.9 °C, averaging at 22 °C. The northern and central parts of the lake were cooler in the daytime (Figure 9a). The daytime temperature indicated higher spatial variability with a standard deviation of 1.9 °C compared to that of the night-time temperature, which had a standard deviation of 0.6 °C. The night-time temperature was warmer in the northeast shore of the lake where there is water hyacinth infestation (Figure 9b), and ranges between 14 and 19.9 °C, averaging at 18.4 °C. The northeastern side of the lake is also characterized by shallow depth with a gentle slope.

The monthly average day and night-time water temperatures were estimated for the 2 km buffer zone along with the water hyacinth coverage for the period August 2017 to July 2018 (Figure 10). Since water is a slow conductor of heat compared to air, the night and daytime water temperature were within the maximum and minimum air temperature (Figure 10). The water temperature was above the base temperature of the water hyacinth throughout the year; however, the minimum air temperature was below the base temperature of the water hyacinth during the period December to April. Perhaps, the lower minimum air temperature from December to April may contribute to the decrease in water hyacinth coverage although the decrease of the water hyacinth coverage continues until June-July (Figure 6).

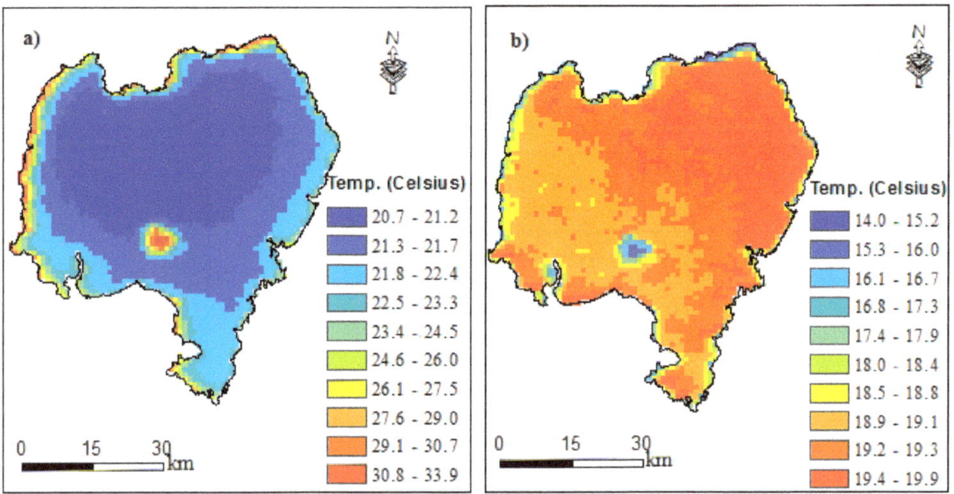

Figure 9. Average annual day and night-time water temperature for the period 2010–2016. (**a**) Daytime average annual water temperature and (**b**) night-time average annual water temperature.

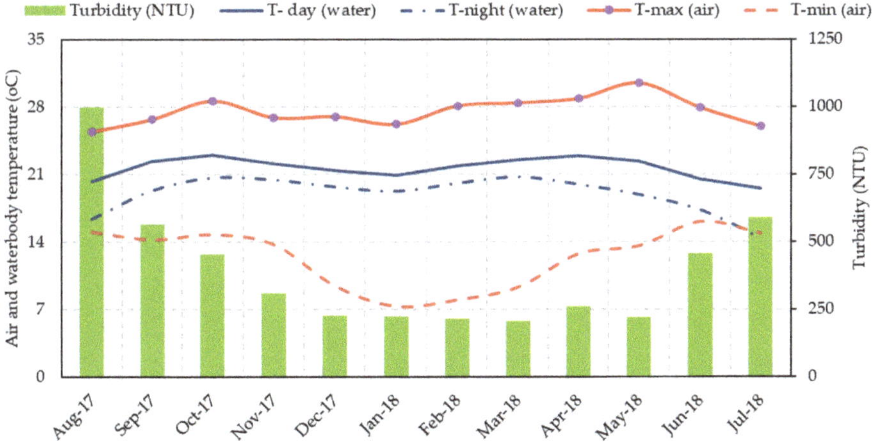

Figure 10. Minimum and maximum air temperature, day and night-time water temperature (primary y-axis), and turbidity in the proximity of water hyacinth infestation area (August 2017 to July 2018).

3.2.3. Turbidity

Turbidity in Lake Tana is highly variable in both space and time. The average annual turbidity estimated for the period 2010–2016 is presented in Figure 11. The average annual turbidity of the lake was 348 nephelometric turbidity units (NTU) with a standard deviation of 30 NTU. Turbidity less than 25 NTU is relatively considered as clear and at 2000 NTU is completely opaque [73]. The highest turbidity was observed in the southwest shore of the lake near Gilgel Abay river mouth followed by the northeast part of the lake (Figure 11). The higher turbidity is due to sediment carried with the tributary rivers. For example, a Landsat image in December 1984 and December 2017 indicated that the river mouth delta at the entrance of Gilgel Abay to the Lake Tana increased from 12.7 to 21.8 km^2

(an increase by 72%; see Supplementary Information, Figure S2). Likewise, the river entrance deltas for Gumara and Ribb rivers were also increasing, but at a lesser extent. Since Gumara and Ribb rivers drop the sediment at the Fogera floodplain, the sediment contribution of both rivers was relatively smaller [74,75]. The floodplains play a critical role by intercepting and trapping nutrient and sediment loadings that are transported with streamflow. Protection of the lake ecosystem should consider implementing best management practices along the watershed landscape and protecting the wetland.

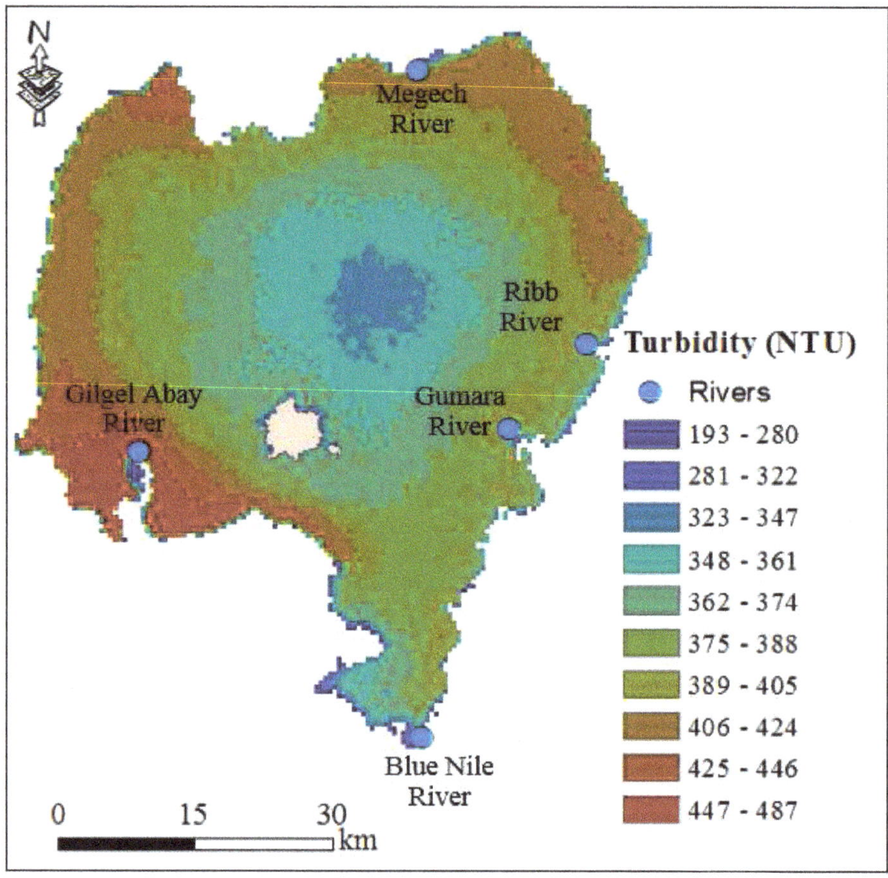

Figure 11. Long-term annual average turbidity of Lake Tana (2010–2016).

The monthly average turbidity of the lake is shown in Figure 12. The highest turbidity was observed during the major rainfall season (June–September) where the rivers drain to the lake bringing the sedimentation laden water from upland agricultural watersheds. The turbidity of the lake is very low during the dry season (November–May) with average turbidity ranging between 225 and 330 NTU. In the northeast corridor, turbidity becomes high in August after two months from the beginning of the main rainfall season. The monthly average turbidity (August 2017 to July 2018) within the proximity (2 km buffer) of the water hyacinth infestation (Figure 10) was estimated from MODIS images.

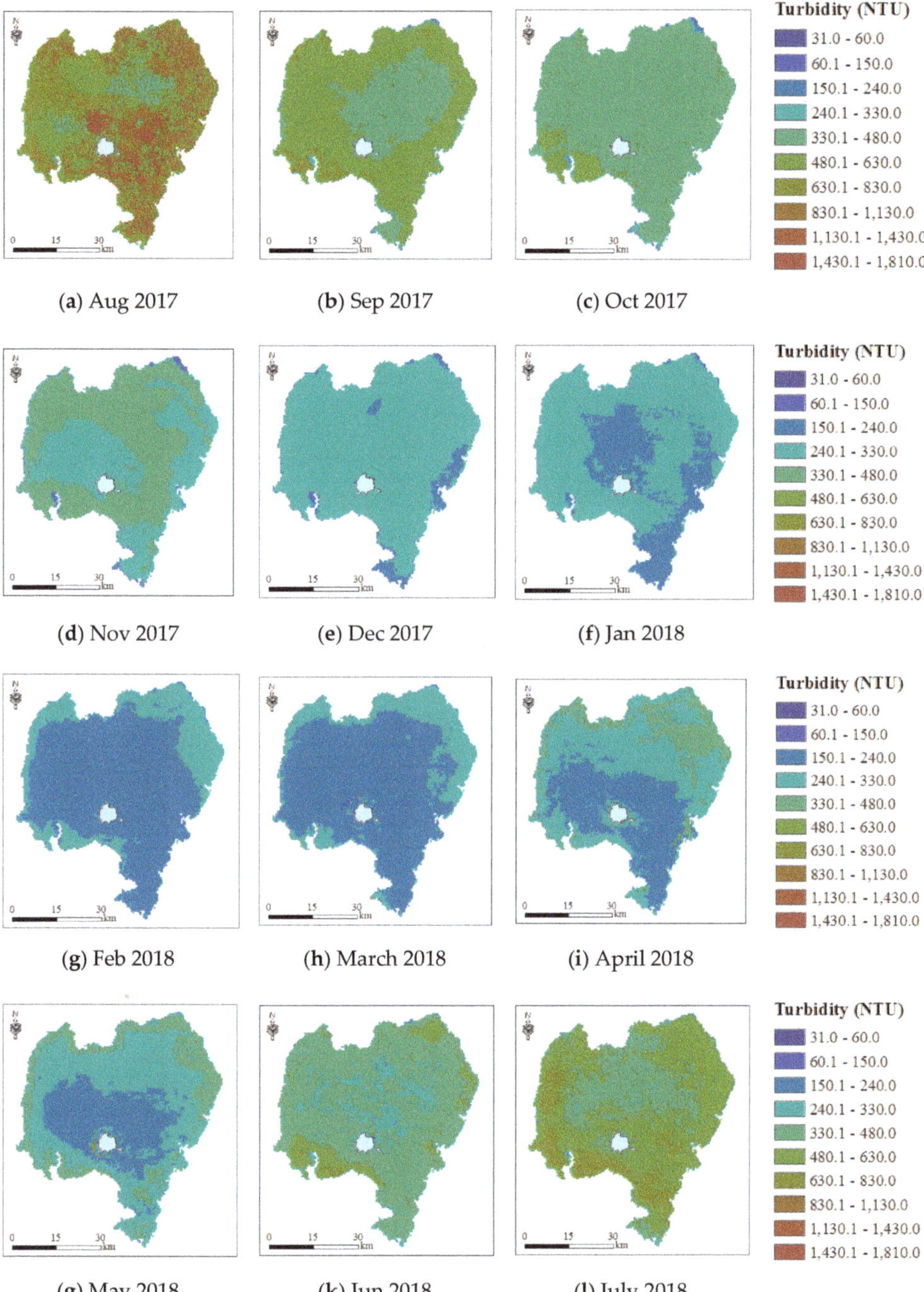

Figure 12. Spatial and temporal distribution of average monthly turbidity for the study period (August 2017 to July 2018).

3.2.4. Lake Level Variation

Comparison between ground-based and that of Jason-1 altimetry mission water level measurements before the construction of the Chara-Chara weir (i.e., 1993–1996) showed a very strong agreement (R-square of 0.96, $p < 0.05$, Figure 13) with a root mean square error of 15 cm. Jason-1 altimetry data has also captured well the observed lake level of several lakes of different size [50,76,77]. The altimetry data showed an annual average lake level fluctuation of 1.5 m for the period 1994–2010. This fluctuation was primarily driven by lake rainfall and streamflow during the major rainfall season. The lake level started rising in June, where rainfall started and reached a maximum level in September and declines linearly towards June of the following year (Figure 2). Given the Jason-1 altimetry data represented the lake level, and that it has data for the study period (August 2017 to July 2018), it was used to understand the relationship between lake level and water hyacinth spatial coverage dynamics.

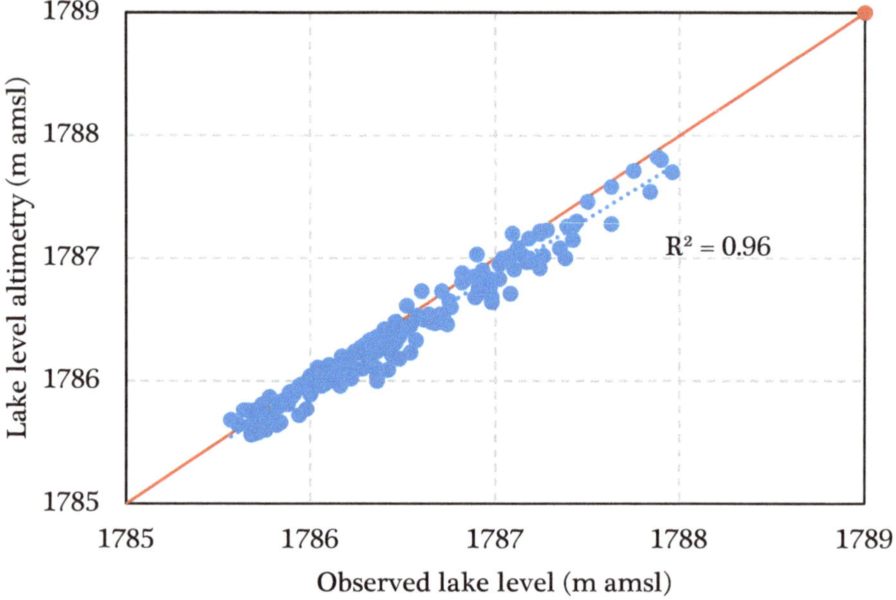

Figure 13. Comparison of observed lake level and Jason-1 lake level altimetry for Lake Tana (1993–1996).

3.3. *The Interplay between Environmental Factors and Water Hyacinth Coverage*

The environmental factors that plausibly contribute to the dynamics of the water hyacinth during the study period were compared with the spatial coverage of the water hyacinth (Figure 14). Among the factors studied, water hyacinth spatial dynamics showed a strong association with the lake level and night-time water temperature (Figure 14a,f). The lake level and night-time temperature relationship to water hyacinth expansion are explained by an R-square of 0.75 and 0.59, respectively. The remaining studied factors showed a weak linear relationship with the spatial coverage of the weed dynamics (Figure 14b–e). Turbidity, however, indicated a strong correlation with the spatial coverage of the water hyacinth with a 2-month lag (R-square of 0.74).

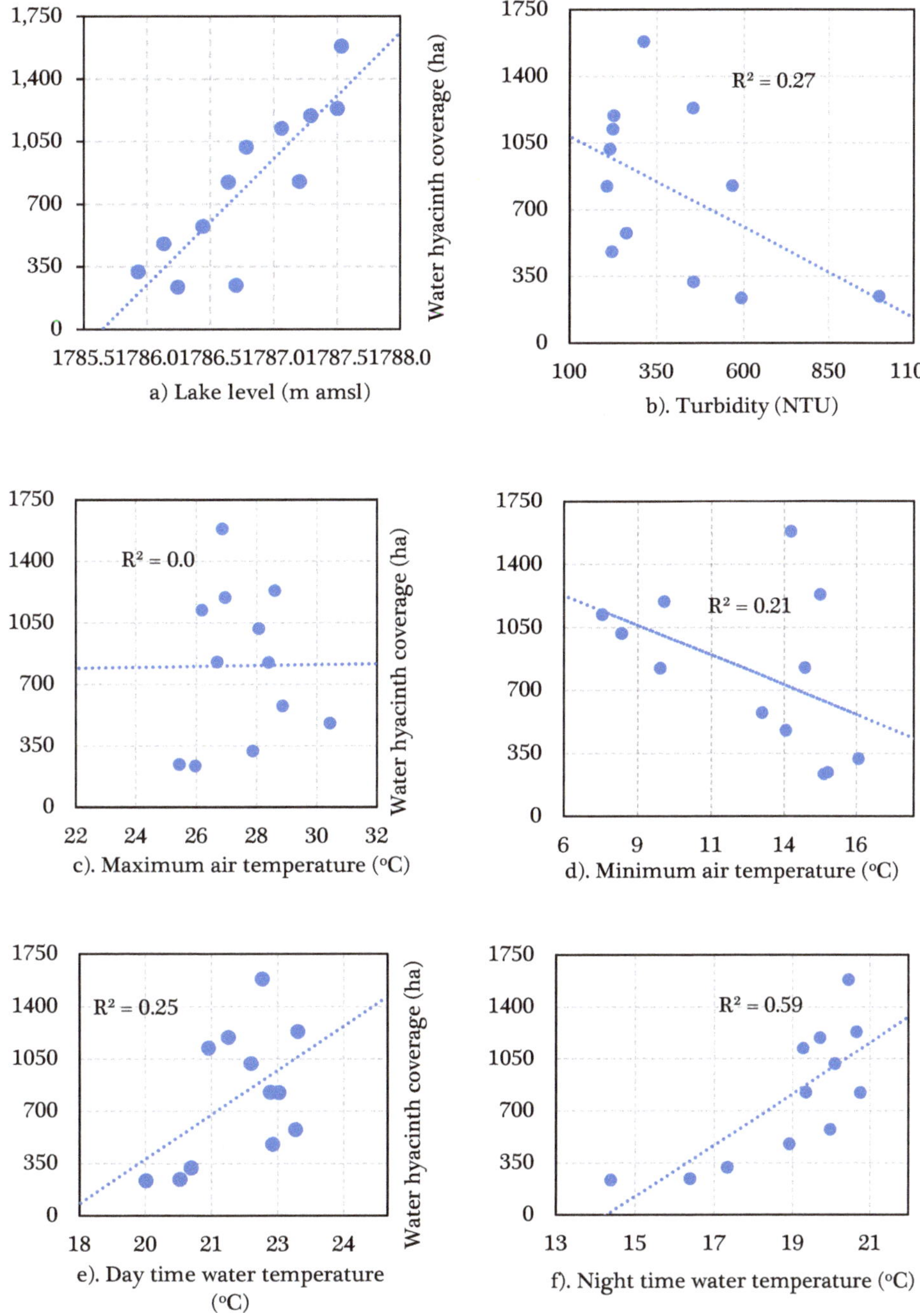

Figure 14. Correlation between water hyacinth spatial coverage and the environmental factor.

4. Discussion

The relationship between the spatiotemporal dynamics of the water hyacinth and some of the studied environmental factors revealed a strong association providing some insight to mitigate the expansion, and eventual elimination before the lake's biodiversity and socioeconomics become in jeopardy. During the main rainfall season, the surface runoff carried significant sediment and other non-point source pollutants into the lake. During this season, the lake level increased flooding the lake shoreline making it convenient for water hyacinth growth on the floodplain. As it is shown in Figure 14a, the change in the lake level captured 75% of the water hyacinth dynamics. The lake level decreased as the streamflow to the lake decrease in the dry season, which led the lake water to retreat from the shoreline and leaving the water hyacinth to remain on the dry land. Hence, the water hyacinth coverage reduced in spatial coverage due to water stress leading to a slow death of the weed. Moreover, the reduction in streamflow during the dry season is also accompanied by a reduction in the total suspended sediment inflow, which carries nutrients such as nitrogen, phosphorus, and organic matter. The night-time temperature indicated a strong association with the water hyacinth growth capturing 59% of the water hyacinth spatial coverage variability. During the water hyacinth expansion period (August to September) the night-time temperature was warming consistently. The higher night-time temperature in the root zone improves root growth due to increased metabolic activity of root cells and the development of lateral roots enabling the weed to effectively utilize the fertilizer nutrient washed from the upland agricultural fields [78]. Overall, the night-time water temperature helped the water hyacinth to convert the stored energy to root and above-ground biomass.

The minimum and maximum air temperature did not show a strong correlation with the water hyacinth spatial dynamics for the study period. The minimum and maximum temperatures were within the base/minimum and optimal temperature for 67% of the study period. The night-time water temperature showed a strong correlation with the water hyacinth spatial dynamics while the daytime water temperature showing a weaker correlation (with R-square of 0.25). Even though there was a relatively higher rate of water hyacinth areal expansion followed by a reduction during the dry season, however, the water hyacinth areal coverage increased year after year.

Due to data limitation, this study did not consider some other factors that could potentially affect the water hyacinth spatial dynamics. For example, lake water pH, dissolved oxygen level, wind, nutrient, hydrodynamics, etc. In regards to pH, normal water hyacinth growth occurs at a pH of 6.5–8.5 [79]. Growth is hampered at a pH of 4.5 and decline in growth will occur from a pH of 10.5 [80]. However, since pH and other environmental variables such as dissolved oxygen vary over the year, which thereby may contribute to the seasonal water hyacinth coverage fluctuations. Likewise, wind also affects the nutrient dynamics in the lake, it moves the floating water hyacinth to new areas and triggers sediment resuspension that increases internal nutrient release in the lake. As more data is available future studies should consider such hydro-environmental dynamics to fully understand the water hyacinth expansion, and suggest appropriate measures to control the weed infestation.

While this study does not exhaustively explore environmental variables that could determine water hyacinth expansion, the relationship established with studied variables would give a glance at the factors affecting the water hyacinth dynamics in Lake Tana.

5. Conclusions

The invasion of habitats by non-native species is a global phenomenon with serious negative ecological, economic, and social consequences. This paper studied the spatial and temporal dynamics of water hyacinth growth in Lake Tana using high-resolution satellite images. The study also assessed the relationship between the water hyacinth dynamics and environmental variables such as lake water level, turbidity, and water and air temperatures.

A strong linear correlation between the lake water level and water hyacinth expansion was found. Although lag time was observed, water turbidity also showed a strong relationship. This suggests that the water hyacinth expansion was occurring in the rainy season and following the rainy season where streamflow, sediment and nutrient flux to the lake were increasing. Therefore, the harvesting effect should focus on the water hyacinth hovering on the deeper section of the lake (deeper than one meter), the weed hovering at shallow depth will end up on the dry land during the dry season. The bathymetric survey indicated approximately 63% of the water hyacinth floats on a lake depth of more than one meter.

The night-time temperature also indicated a strong association with the water hyacinth expansion, it helped the plant to convert the day time stored energy through photosynthesis to growth. The higher night-time temperature in the root zone improves root growth due to increased metabolic activity of root cells and the development of lateral roots enabling the weed to effectively utilize the fertilizer nutrient washed from the upland agricultural fields. Restoring the nutrient-rich wetland along the shore which otherwise serves as a perfect breeding ground for the weed could essentially alter the water hyacinth expansion dynamics. Implementation of best watershed management practices in the watershed (e.g., nutrient management in upland farmlands, riparian buffer strips, stone bunds, terraces, etc.) may reduce nutrient influx into the lake, which thereby limits the expansion of the water hyacinth. Identifying the environmental variables that determine the water hyacinth expansion is helpful to plan proper water hyacinth management options.

Supplementary Materials: The following are available online at http://www.mdpi.com/2072-4292/12/17/2706/s1, Figure S1: GPS tracks of June 25, 2017, overlain on a false-color composite image of June 24, 2017, PlanerScope, Table S1: List of satellite images used to map the spatiotemporal distribution of water hyacinth and major environmental factors plausibly affecting the weed spatial dynamics.

Author Contributions: A.W.W. develop the methodology, data analysis, and write up, E.K.A. and Y.T.D. contributed to the development of the method and write up, M.A.M. and M.G.D. did the field data collection, G.T. and S.K. reviewed the work. All authors have read and agreed to the published version of the manuscript.

Funding: This research received no external funding.

Acknowledgments: The authors would like to thank Planet Scope imagery for providing the high-resolution satellite data free of charge. The editor and the three anonymous reviewers gratefully acknowledged for their valuable comments on our manuscript.

Conflicts of Interest: The authors declare no conflict of interest.

References

1. Penfound, W.T.; Earle, T.T. The Biology of the Water Hyacinth. *Ecol. Monogr.* **1948**, *18*, 447–472. [CrossRef]
2. Reddy, K. Water hyacinth (Eichhornia crassipes) biomass production in Florida. *Biomass* **1984**, *6*, 167–181. [CrossRef]
3. Malik, A. Environmental challenge vis a vis opportunity: The case of water hyacinth. *Environ. Int.* **2007**, *33*, 122–138. [CrossRef] [PubMed]
4. Chopra, R.; Verma, V.; Sharma, P.K. Mapping, monitoring and conservation of Harike wetland ecosystem, Punjab, India, through remote sensing. *Int. J. Remote Sens.* **2001**, *22*, 89–98. [CrossRef]
5. Gong, Y.; Zhou, X.; Ma, X.; Chen, J. Sustainable removal of formaldehyde using controllable water hyacinth. *J. Clean. Prod.* **2018**, *181*, 1–7. [CrossRef]
6. Shrestha, S.; Fonoll, X.; Khanal, S.K.; Raskin, L. Biological strategies for enhanced hydrolysis of lignocellulosic biomass during anaerobic digestion: Current status and future perspectives. *Bioresour. Technol.* **2017**, *245*, 1245–1257. [CrossRef]
7. Brendonck, L.; Maes, J.; Rommens, W.; Dekeza, N.; Nhiwatiwa, T.; Barson, M.; Callebaut, V.; Phiri, C.; Moreau, K.; Gratwicke, B.; et al. The impact of water hyacinth (Eichhornia crassipes) in a eutrophic subtropical impoundment (Lake Chivero, Zimbabwe). II. Species diversity. *Arch. Hydrobiol.* **2003**, *158*, 389–405. [CrossRef]
8. Cilliers, C. Biological control of water hyacinth, Eichhornia crassipes (Pontederiaceae), in South Africa. *Agric. Ecosyst. Environ.* **1991**, *37*, 207–217. [CrossRef]
9. Fessehaie, R. Water hyacinth (Eichhornia crassipes): A Review of its weed status in Ethiopia. *Arem (Ethiopia)* **2005**, *6*, 105–106.
10. Byrne, M.; Hill, M.; Robertson, M.; King, A.; Jadhav, A.; Katembo, N.; Wilson, J.; Brudvig, R.; Fisher, J. Integrated Management of Water Hyacinth in South Africa: Development of an integrated management plan for water hyacinth control, combining biological control, herbicidal control and nutrient control, tailored to the climatic regions of South Africa. *WRC Rep.* **2010**, *454*, 302.

11. Mesfin, M.; Tudorancea, C.; Baxter, R.M. Some limnological observations on two Ethiopian hydroelectric reservoirs: Koka (Shewa administrative district) and Finchaa (Welega administrative district). *Hydrobiologia* **1988**, *157*, 47–55. [CrossRef]
12. Tewabe, D. Preliminary Survey of Water Hyacinth in Lake. *Glob. J. Allergy* **2015**, *1*, 13–18. [CrossRef]
13. Anteneh, W.; Tewabe, D.; Assefa, A.; Zeleke, A.; Tenaw, B.; Wassie, Y. Water Hyacinth Coverage Survey Report on Lake Tana Biosphere Reserve; Technical Report Series 2. Available online: https://welkait.com/wp-content/uploads/2017/06/Water-hacinth_Lake-Tana_Report-Series-2.pdf (accessed on 20 August 2020).
14. Gebremedhin, S.; Getahun, A.; Anteneh, W.; Bruneel, S.; Goethals, P. A Drivers-Pressure-State-Impact-Responses Framework to Support the Sustainability of Fish and Fisheries in Lake Tana, Ethiopia. *Sustainability* **2018**, *10*, 2957. [CrossRef]
15. Wilson, J.R.; Richardson, D.M.; Rouget, M.; Procheş, Ş.; Amis, M.A.; Henderson, L.; Thuiller, W. Residence time and potential range: Crucial considerations in modelling plant invasions. *Divers. Distrib.* **2007**, *13*, 11–22. [CrossRef]
16. Priya, P.; Nikhitha, S.; Anand, C.; Nath, R.D.; Bhaskaran, K. Biomethanation of water hyacinth biomass. *Bioresour. Technol.* **2018**, *255*, 288–292. [CrossRef] [PubMed]
17. Venugopal, G. Monitoring the Effects of Biological Control of Water Hyacinths Using Remotely Sensed Data: A Case Study of Bangalore, India. *Singap. J. Trop. Geogr.* **1998**, *19*, 91–105. [CrossRef]
18. Verma, R.; Singh, S.; Raj, K.G. Assessment of changes in water-hyacinth coverage of water bodies in northern part of Bangalore city using temporal remote sensing data. *Curr. Sci.* **2003**, *84*, 795–804.
19. Asmare, E. Current Trend of Water Hyacinth Expansion and Its Consequence on the Fisheries around North Eastern Part of Lake Tana, Ethiopia. *J. Biodivers. Endanger. Species* **2017**, *5*, 5. [CrossRef]
20. Dersseh, M.G.; Kibret, A.A.; Tilahun, S.A.; Worqlul, A.W.; Moges, M.A.; Dagnew, D.; Abebe, W.B.; Melesse, A.M. Potential of Water Hyacinth Infestation on Lake Tana, Ethiopia: A Prediction Using a GIS-Based Multi-Criteria Technique. *Water* **2019**, *11*, 1921. [CrossRef]
21. Dersseh, M.G.; Tilahun, S.A.; Worqlul, A.W.; Moges, M.A.; Abebe, W.B.; Mihret, D.A.; Melesse, A.M. Spatial and Temporal Dynamics of Water Hyacinth and Its Linkage with Lake-Level Fluctuation: Lake Tana, a Sub-Humid Region of the Ethiopian Highlands. *Water* **2020**, *12*, 1435. [CrossRef]
22. Teshome, G.; Getahun, A.; Mengist, M.; Hailu, B. Some biological aspects of spawning migratory Labeobarbus species in some tributary rivers of Lake Tana, Ethiopia. *Int. J. Fish. Aquat. Stud.* **2015**, *3*, 136–141.
23. Gezie, A.; Assefa, W.W.; Getnet, B.; Anteneh, W.; Dejen, E.; Mereta, S.T. Potential impacts of water hyacinth invasion and management on water quality and human health in Lake Tana watershed, Northwest Ethiopia. *Boil. Invasions* **2018**, *20*, 2517–2534. [CrossRef]
24. Conway, D. The Climate and Hydrology of the Upper Blue Nile River. *Geogr. J.* **2000**, *166*, 49–62. [CrossRef]
25. Dile, Y.T.; Tekleab, S.; Ayana, E.K.; Gebrehiwot, S.G.; Worqlul, A.W.; Bayabil, H.K.; Yimam, Y.T.; Tilahun, S.A.; Daggupati, P.; Karlberg, L.; et al. Advances in water resources research in the Upper Blue Nile basin and the way forward: A review. *J. Hydrol.* **2018**, *560*, 407–423. [CrossRef]
26. Vijverberg, J.; Sibbing, F.A.; Dejen, E. Lake Tana: Source of the Blue Nile. In *The Nile*; Springer: Dordrecht, The Netherlands, 2009; pp. 163–192. [CrossRef]
27. Ligdi, E.E.; El Kahloun, M.; Meire, P. Ecohydrological status of Lake Tana—A shallow highland lake in the Blue Nile (Abbay) basin in Ethiopia: Review. *Ecohydrol. Hydrobiol.* **2010**, *10*, 109–122. [CrossRef]
28. Zur Heide, F. Feasibility Study for a Lake Tana Biosphere Reserve, Ethiopia. Bundesamt für Naturschutz, BfN. 2012. Available online: https://en.nabu.de/imperia/md/images/nabude/projekteaktionen/international/aethiopien/nabu-f-zur-heide-feasability_study.pdf (accessed on 20 August 2020).
29. Wale, A.; Rientjes, T.H.M.; Gieske, A.S.; Getachew, H.A. Ungauged catchment contributions to Lake Tana's water balance. *Hydrol. Process.* **2009**, *23*, 3682–3693. [CrossRef]
30. Kebede, S.; Travi, Y.; Alemayehu, T.; Marc, V. Water balance of Lake Tana and its sensitivity to fluctuations in rainfall, Blue Nile basin, Ethiopia. *J. Hydrol.* **2006**, *316*, 233–247. [CrossRef]
31. Ayana, E.K.; Worqlul, A.W.; Steenhuis, T.S. Evaluation of stream water quality data generated from MODIS images in modeling total suspended solid emission to a freshwater lake. *Sci. Total Environ.* **2015**, *523*, 170–177. [CrossRef]
32. Zimale, F.A.; Moges, M.A.; Alemu, M.L.; Ayana, E.K.; Demissie, S.S.; Tilahun, S.A.; Steenhuis, T.S. Budgeting suspended sediment fluxes in tropical monsoonal watersheds with limited data: The Lake Tana basin. *J. Hydrol. Hydromech.* **2018**, *66*, 65–78. [CrossRef]

33. Rientjes, T.H.; Perera, B.U.J.; Haile, A.T.; Reggiani, P.; Muthuwatta, L.P. Regionalisation for lake level simulation—The case of Lake Tana in the Upper Blue Nile, Ethiopia. *Hydrol. Earth Syst. Sci.* **2011**, *15*, 1167–1183. [CrossRef]
34. SMEC, I. Hydrological Study of The Tana-Beles Sub-Basins. "part 1.". *Sub-Basins Groundw. Investig. Rep.* **2007**, 5089018.
35. Yang, C.; Everitt, J.H. Mapping three invasive weeds using airborne hyperspectral imagery. *Ecol. Inform.* **2010**, *5*, 429–439. [CrossRef]
36. Underwood, E.C.; Mulitsch, M.J.; Greenberg, J.; Whiting, M.L.; Ustin, S.L.; Kefauver, S.C. Mapping Invasive Aquatic Vegetation in the Sacramento-San Joaquin Delta using Hyperspectral Imagery. *Environ. Monit. Assess.* **2006**, *121*, 47–64. [CrossRef] [PubMed]
37. Altena, B.; Mousivand, A.; Mascaro, J.; Kääb, A. Potential and limitations of photometric reconstruction through a flock of dove cubesats. *ISPRS Int. Arch. Photogramm. Remote Sens. Spat. Inf. Sci.* **2017**, 7–11. [CrossRef]
38. Planet Team. Planet Application Program Interface: In Space for Life on Earth. Available online: https://www.planet.com (accessed on 20 August 2020).
39. Planet Labs. Planet Imagery Product Specifications. Available online: https://www.planet.com (accessed on 20 August 2020).
40. McCabe, M.F.; Aragon, B.; Houborg, R.; Mascaro, J. CubeSats in Hydrology: Ultrahigh-Resolution Insights into Vegetation Dynamics and Terrestrial Evaporation. *Water Resour. Res.* **2017**, *53*, 10017–10024. [CrossRef]
41. Dobrinić, D.; Gašparović, M.; Župan, R. Horizontal Accuracy Assessment of PlanetScope, RapidEye and WorldView-2 Satellite Imagery. In Proceedings of the 18th International Multidisciplinary Scientific Geoconference SGEM 2018, Albena, Bulgaria, 30 June–9 July 2018.
42. Hang, N.T.T.; Hoa, N.T.; Son, T.P.H.; Nguyen-Ngoc, L. Vegetation Biomass of Sargassum Meadows in An Chan Coastal Waters, Phu Yen Province, Vietnam Derived from PlanetScope Image. *J. Environ. Sci. Eng. B* **2019**, *8*, 81–92. [CrossRef]
43. Richards, J.A.; Richards, J. *Remote Sensing Digital Image Analysis*; Springer: Berlin/Heidelberg, Germany, 1999; Volume 3.
44. Kaur, M.; Kumar, M.; Sachdeva, S.; Puri, S. Aquatic weeds as the next generation feedstock for sustainable bioenergy production. *Bioresour. Technol.* **2018**, *251*, 390–402. [CrossRef]
45. Bowman, W.D.; Theodose, T.A.; Schardt, J.C.; Conant, R.T. Constraints of Nutrient Availability on Primary Production in Two Alpine Tundra Communities. *Ecology* **1993**, *74*, 2085–2097. [CrossRef]
46. Santamaría, L. Why are most aquatic plants widely distributed? Dispersal, clonal growth and small-scale heterogeneity in a stressful environment. *Acta Oecologica* **2002**, *23*, 137–154. [CrossRef]
47. A Davis, M.; Grime, J.P.; Thompson, K. Fluctuating resources in plant communities: A general theory of invasibility. *J. Ecol.* **2000**, *88*, 528–534. [CrossRef]
48. Fageria, N.K.; Baligar, V.C.; Jones, C.A. *Growth and Mineral Nutrition of Field Crops*; CRC Press: Boca Raton, FL, USA, 2010. [CrossRef]
49. Pietrangeli, S. *Bathymetry of Lake Tana*; Unpublished Report; Studio Pietrangeli: Rome, Italy, 1998.
50. Ayana, E.K. Validation of Radar Altimetry Lake Level Data and It's Application in Water Resource Management. ITC. 2007. Available online: https://webapps.itc.utwente.nl/librarywww/papers_2007/msc/wrem/kaba.pdf (accessed on 20 September 2019).
51. Carr, G.M.; Duthie, H.C.; Taylor, W.D. Models of aquatic plant productivity: A review of the factors that influence growth. *Aquat. Bot.* **1997**, *59*, 195–215. [CrossRef]
52. Belding, D.L. Water Temperature and Fish Life. *Trans. Am. Fish. Soc.* **1928**, *58*, 98–105. [CrossRef]
53. Hatfield, J.L.; Boote, K.J.; Kimball, B.A.; Ziska, L.H.; Izaurralde, R.C.; Ort, D.; Thomson, A.M.; Wolfe, D. Climate Impacts on Agriculture: Implications for Crop Production. *Agron. J.* **2011**, *103*, 351–370. [CrossRef]
54. Imaoka, T.; Teranishi, S. Rates of nutrient uptake and growth of the water hyacinth [Eichhornia crassipes (mart.) Solms]. *Water Res.* **1988**, *22*, 943–951. [CrossRef]
55. Kasselmann, C. *Aquarienpflanzen*; Ulmer, E., Ed.; Eugen Ulmer: Stuttgart, Germany, 1995; ISBN 978-3-8186-0699-2. Available online: https://www.ulmer.de/usd-1557211/aquarienpflanzen-.html (accessed on 10 October 2019).
56. Zhang, G.; Yao, T.; Xie, H.; Qin, J.; Ye, Q.; Dai, Y.; Guo, R. Estimating surface temperature changes of lakes in the Tibetan Plateau using MODIS LST data. *J. Geophys. Res. Atmos.* **2014**, *119*, 8552–8567. [CrossRef]

57. Luo, Y.; Zhang, Y.; Yang, K.; Yu, Z.; Zhu, Y. Spatiotemporal Variations in Dianchi Lake's Surface Water Temperature From 2001 to 2017 Under the Influence of Climate Warming. *IEEE Access* **2019**, *7*, 115378–115387. [CrossRef]
58. Yang, K.; Yu, Z.; Luo, Y.; Zhou, X.; Shang, C. Spatial-Temporal Variation of Lake Surface Water Temperature and its Driving Factors in Yunnan-Guizhou Plateau. *Water Resour. Res.* **2019**, *55*, 4688–4703. [CrossRef]
59. Mildrexler, D.J.; Zhao, M.; Running, S.W. A global comparison between station air temperatures and MODIS land surface temperatures reveals the cooling role of forests. *J. Geophys. Res. Space Phys.* **2011**, *116*, 116. [CrossRef]
60. Ayana, E.K.; Zimale, F.A.; Collick, A.S.; Tilahun, S.A.; Elkamil, M.; Philpot, W.D.; Steenhuis, T.S. Monitoring State of Biomass Recovery in the Blue Nile Basin Using Image-Based Disturbance Index. In *Nile River Basin*; Springer: Berlin/Heidelberg, Germany, 2014; pp. 237–252. [CrossRef]
61. Nemani, R.R.; Running, S.W. Estimation of Regional Surface Resistance to Evapotranspiration from NDVI and Thermal-IR AVHRR Data. *J. Appl. Meteorol.* **1989**, *28*, 276–284. [CrossRef]
62. Birkett, C.; Reynolds, C.; Beckley, B.; Doorn, B. From research to operations: The USDA global reservoir and lake monitor. In *Coastal Altimetry*; Springer: Berlin/Heidelberg, Germany, 2011; pp. 19–50. ISBN 978-3-642-12796-0.
63. Lloyd, D.S. Turbidity as a Water Quality Standard for Salmonid Habitats in Alaska. *N. Am. J. Fish. Manag.* **1987**, *7*, 34–45. [CrossRef]
64. Funge-Smith, S.; Briggs, M.R. Nutrient budgets in intensive shrimp ponds: Implications for sustainability. *Aquaculture* **1998**, *164*, 117–133. [CrossRef]
65. Fraser, R.N. Hyperspectral remote sensing of turbidity and chlorophyll a among Nebraska Sand Hills lakes. *Int. J. Remote Sens.* **1998**, *19*, 1579–1589. [CrossRef]
66. Chen, Z.; Hu, C.; Muller-Karger, F. Monitoring turbidity in Tampa Bay using MODIS/Aqua 250-m imagery. *Remote Sens. Environ.* **2007**, *109*, 207–220. [CrossRef]
67. Dall'Olmo, G.; Gitelson, A.A.; Rundquist, D.C.; Leavitt, B.; Barrow, T.; Holz, J.C. Assessing the potential of SeaWiFS and MODIS for estimating chlorophyll concentration in turbid productive waters using red and near-infrared bands. *Remote Sens. Environ.* **2005**, *96*, 176–187. [CrossRef]
68. Quang, N.H.; Sasaki, J.; Higa, H.; Huan, N.H. Spatiotemporal Variation of Turbidity Based on Landsat 8 OLI in Cam Ranh Bay and Thuy Trieu Lagoon, Vietnam. *Water* **2017**, *9*, 570. [CrossRef]
69. Dogliotti, A.; Ruddick, K.; Nechad, B.; Doxaran, D.; Knaeps, E. A single algorithm to retrieve turbidity from remotely-sensed data in all coastal and estuarine waters. *Remote Sens. Environ.* **2015**, *156*, 157–168. [CrossRef]
70. Kaba, E.; Philpot, W.D.; Steenhuis, T. Evaluating suitability of MODIS-Terra images for reproducing historic sediment concentrations in water bodies: Lake Tana, Ethiopia. *Int. J. Appl. Earth Obs. Geoinf.* **2014**, *26*, 286–297. [CrossRef]
71. Wale, A.; Rientjes, T.; Dost, R.; Gieske, A. Hydrological Balance of Lake Tana Upper Blue Nile Basin, Ethiopia. *Neth. ITC* **2008**. Available online: https://webapps.itc.utwente.nl/librarywww/papers_2008/msc/wrem/wale.pdf (accessed on 10 October 2019).
72. Marshall, M.H.; Lamb, H.F.; Huws, D.; Davies, S.J.; Bates, C.R.; Bloemendal, J.; Boyle, J.; Leng, M.J.; Umer, M.; Bryant, C. Late Pleistocene and Holocene drought events at Lake Tana, the source of the Blue Nile. *Glob. Planet. Chang.* **2011**, *78*, 147–161. [CrossRef]
73. Myre, E.; Shaw, R. The turbidity tube: Simple and accurate measurement of turbidity in the field. *Mich. Technol. Univ.* **2006**. Available online: http://serresconseil.com/WASH/Watsanmissionassistant/mainSpace/files/Turbidity-Myre_Shaw.pdf (accessed on 12 August 2019).
74. Anteneh, W.; Dejen, E.; Getahun, A. Shesher and Welala Floodplain Wetlands (Lake Tana, Ethiopia): Are They Important Breeding Habitats forClarias gariepinusand the MigratoryLabeobarbusFish Species? *Sci. World J.* **2012**, *2012*, 1–10. [CrossRef]
75. Abate, M.; Nyssen, J.; Moges, M.M.; Enku, T.; Zimale, F.A.; Tilahun, S.A.; Adgo, E.; Steenhuis, T.S. Long-Term Landscape Changes in the Lake Tana Basin as Evidenced by Delta Development and Floodplain Aggradation in Ethiopia. *Land Degrad. Dev.* **2017**, *28*, 1820–1830. [CrossRef]
76. Crétaux, J.-F.; Jelinski, W.; Calmant, S.; Kouraev, A.V.; Vuglinski, V.; Bergé-Nguyen, M.; Gennero, M.-C.; Nino, F.; Del Rio, R.A.; Cazenave, A.; et al. SOLS: A lake database to monitor in the Near Real Time water level and storage variations from remote sensing data. *Adv. Space Res.* **2011**, *47*, 1497–1507. [CrossRef]
77. Birkett, C.M.; Beckley, B. Investigating the performance of the Jason-2/OSTM radar altimeter over lakes and reservoirs. *Mar. Geod.* **2010**, *33*, 204–238. [CrossRef]

78. Puhe, J. Growth and development of the root system of Norway spruce (Picea abies) in forest stands—A review. *For. Ecol. Manag.* **2003**, *175*, 253–273. [CrossRef]
79. Santiago, C.M., Jr. Some environmental factors affecting growth of water hyacinth (Eichhornia crassipes (Mart.) Solms). *Philipp. J. Sci. (Philippines)* **1984**, *113*, 67–82.
80. El-Gendy, A.; Biswas, N.; Bewtra, J. Growth of Water Hyacinth in Municipal Landfill Leachate with Different pH. *Environ. Technol.* **2004**, *25*, 833–840. [CrossRef]

Article

Monitoring Lakes Surface Water Velocity with SAR: A Feasibility Study on Lake Garda, Italy

Marina Amadori [1,2,†], Virginia Zamparelli [3,†], Giacomo De Carolis [1], Gianfranco Fornaro [3], Marco Toffolon [2], Mariano Bresciani [1], Claudia Giardino [1] and Francesca De Santi [1,*]

1 IREA-CNR, Institute for Electromagnetic Sensing of the Environment-National Research Council of Italy, 20133 Milan, Italy; amadori.m@irea.cnr.it (M.A.); giacomo.decarolis@cnr.it (G.D.C.); bresciani.m@irea.cnr.it (M.B.); giardino.c@irea.cnr.it (C.G.)

2 Department of Civil, Environmental and Mechanical Engineering, University of Trento, 38123 Trento, Italy; marco.toffolon@unitn.it

3 IREA-CNR, Institute for Electromagnetic Sensing of the Environment-National Research Council of Italy, 80128 Naples, Italy; zamparelli.v@irea.cnr.it (V.Z.); fornaro.g@irea.cnr.it (G.F.)

* Correspondence: desanti.f@irea.cnr.it

† These authors equally contributed to this work.

Abstract: The SAR Doppler frequencies are directly related to the motion of the scatterers in the illuminated area and have already been used in marine applications to monitor moving water surfaces. Here we investigate the possibility of retrieving surface water velocity from SAR Doppler analysis in medium-size lakes. ENVISAT images of the test site (Lake Garda) are processed and the Doppler Centroid Anomaly technique is adopted. The resulting surface velocity maps are compared with the outputs of a hydrodynamic model specifically validated for the case study. Thermal images from MODIS Terra are used in support of the modeling results. The surface velocity retrieved from SAR is found to overestimate the numerical results and the existence of a bias is investigated. In marine applications, such bias is traditionally removed through Geophysical Model Functions (GMFs) by ascribing it to a fully developed wind waves spectrum. We found that such an assumption is not supported in our case study, due to the small-scale variations of topography and wind. The role of wind intensity and duration on the results from SAR is evaluated, and the inclusion of lake bathymetry and the SAR backscatter gradient is recommended for the future development of GMFs suitable for lake environments.

Keywords: SAR; Doppler Centroid Anomaly; inland waters; physical limnology; hydrodynamics

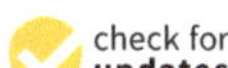

Citation: Amadori, M.; Zamparelli, V.; De Carolis, G.; Fornaro, G.; Toffolon, M.; Bresciani, M.; Giardino, C.; De Santi, F. Monitoring Lakes Surface Water Velocity with SAR: A Feasibility Study on Lake Garda, Italy. *Remote Sens.* **2021**, *13*, 2293. https://doi.org/10.3390/rs13122293

Academic Editor: Joong-Sun Won

Received: 30 March 2021
Accepted: 7 June 2021
Published: 11 June 2021

Publisher's Note: MDPI stays neutral with regard to jurisdictional claims in published maps and institutional affiliations.

1. Introduction

Microwave methodologies based on the use of Synthetic Aperture Radar (SAR) sensors have been widely developed for many Earth surface monitoring applications. In the last decade, images acquired by SAR have been increasingly exploited by the scientific community for deformation surveys [1] even at single facilities scale, that is, buildings and infrastructures [2]. More recently, increasing attention has been paid to the observation of sea surface, complementing the traditional use of optical or multispectral images.

Differently from optical sensors, SAR allows the obtaining of all weather and day–night 2D images of the illuminated area of the Earth's surface [3]. SAR remote sensing for ocean, seas and coastal applications mostly exploits the amplitude of the backscattered signal for, for example, monitoring oil-spills [4] and sea-ice [5], ship detection [6] and high-resolution wind fields retrieval [7]. However, SAR uses coherent radiation and the complementary information carried in the phase of the received complex signal can be also exploited. An example from land applications is the use of the complex information from a single SAR image to reconstruct infrastructures micro-motion [8]. By analysing the complex backscattered (received) signal, it is possible to measure the Doppler properties of the scatterers. The latter are directly related to the motion of the scatterers along the radar line of sight (LOS) in both land [9] and marine [10] contexts. In ocean applications, these properties are currently used to retrieve near surface wind speed [11] and surface current [12].

Several factors contribute to the measured Doppler frequency, such as the bulk surface current, the wind waves and the wave–current interactions at different scales [13]. The drift current induced by the wind on the sea surface (wind drift) and the sea state (more precisely the high-frequency waves) depend on the near-surface wind field. In this regard, [14] performed a global ocean analysis and and found that an increase of the wind speed (projected along the LOS) corresponds to a first-order increase of the Doppler anomaly. Such a wind-wave bias is empirically estimated and removed to retrieve the radial component of the bulk surface current. To this end, empirical Geophysical Model Functions (GMF) have been determined [12].

This analysis is going toward a consolidated state for oceanic applications, such that surface velocity maps in open waters are distributed by the European Space Agency (ESA) among the second level products of Sentinel 1 [15]. On the contrary, the extension of a similar approach to coastal zones is more delicate. At least two main assumptions of the analysis of the Doppler Centroid Anomaly (DCA) in the open ocean do not hold for coastal areas: (i) uniform surface flow field and (ii) fully developed wind-waves. In the near-shore areas, small- scale variations of topography and wind can cause higher variability in the currents and the waves to develop in fetch-limited condition.Thus, the application to other contexts of the above mentioned methods for the removal of the wind waves bias may not be straightforward as in the open ocean case, although examples along this line are given in the recent literature [10,13,14,16–19]. At the smaller spatial scale of enclosed basins (e.g., lakes), no studies have yet explored the possibility of inferring surface currents from SAR Doppler frequencies.

More generally, SAR images of lakes have been seldom exploited for several reasons, among which the minor intensity of hydrodynamic processes (e.g., smaller wave heights and currents than in open seas and oceans) and the effect of surrounding topography (e.g., foreshortening and layover effect [20]).

Currently, SAR applications for lakes are limited to the analysis of the backscattered signal amplitude. SAR amplitude images have been used in support of optical or thermal imagery for identifying and mapping inland waters surface features, for example, ice [21] and cyanobacteria blooms [22,23]. A few tests also showed the potentialities of using SAR images for retrieving the size of surface eddies in large lakes [24] and spatially distributed information on wind intensity [25].

The SAR Doppler analysis could represent a useful way to retrieve surface velocity in lakes and would compensate for the lack of synoptic measurements of the key physical quantities determining lake hydrodynamics. Indeed, while the use of remote sensing products is well consolidated to infer lake surface water temperature (LSWT) [26], water velocity, as well as all wind velocity, is traditionally monitored with in-situ point measurements. In most cases, traditional instrumentation requires direct access to the lake, with strong operational limitations, and provide data at single locations. Remote sensing techniques provide instead synoptic coverage, fine spatial detail and repeated regular sampling.

Periodic snapshots of lake surface currents, at the present day, cannot be achieved alternatively than with remote sensing instruments, for example, airborne or space, the latter being traditionally exploited in marine applications [27–29]. Few attempts towards the reconstruction of the surface transport from thermal infrared imagery [30] showed encouraging results in this direction, but such approach requires high temporal resolution, which is not always guaranteed. Moreover, the use of thermal infrared remote sensing is hindered by cloud cover. In this regard, the use of SAR images combined with the DCA technique could overcome the technical issues typically encountered for reconstructing the surface flow field in lakes, asides from being rather innovative.

This study aims to investigate the feasibility of extracting the surface velocity from DCA in lakes by considering as test case Lake Garda, a large and deep lake in northern Italy. The test case is a clear oligotrophic lake, thus allowing for neglecting turbidity and algal blooms which might affect the SAR signal.

We analyze and process images acquired by the ENVISAT C-Band sensor. Results are compared with the outputs of a three-dimensional (3D) numerical model validated for Lake Garda [31]. Water surface temperature products from the MODIS (Moderate Resolution Imaging Spectroradiometer) Terra sensor are also considered to further validate the simulation results. The numerical results are used to interpret the SAR retrieved signal and to evaluate the different factors influencing the Doppler shift analysis in the case study. Based on this, we draw some guidelines on the variables to be considered for a future definition of GMFs for lakes.

2. Materials and Methods

In this section we describe the test site (Section 2.1), the sensors and datasets for SAR and thermal infrared images (Section 2.2) , the procedure for the analysis of the DCA (Section 2.3) and the adopted numerical models (Section 2.4).

2.1. Case Study

Lake Garda (Figure 1a) is one of the large European perialpine lakes and is the largest lake in Italy by surface extension and volume. It covers an area of 370 km^2, with a water volume of 50 km^3 and a maximum depth of 346 m. The lake represents an essential water supply for many sectors of the local economy (e.g., agriculture, industry, fishing and drinking [32]). Thanks to its attractive landscape, mild climate and water quality, Lake Garda is also an important resource for recreation and tourism.

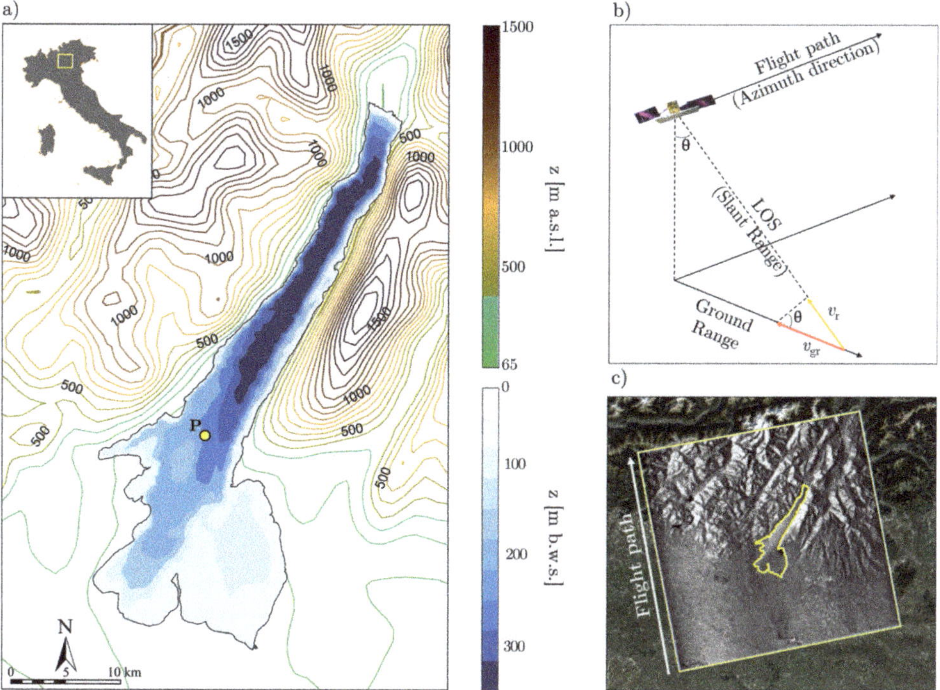

Figure 1. (**a**) Location of Lake Garda in northern Italy; orography of the surrounding area in m above sea level (m a.s.l.); lake bathymetry in m below water surface (m b.w.s.). (**b**) Scheme of the SAR acquisition geometry. (**c**) Example of an amplitude focused SAR image of an ENVISAT acquisition centered on lake. The SAR image is geocoded and superimposed on the corresponding Google Earth image.

The lake has a heterogeneous bathymetry (Figure 1a). A narrow (average width 4 km) and deep (maximum depth 346 m) northern trunk, enclosed between steep mountains, is connected to a southern large (maximum width 18 km) and shallower (average depth 65 m) basin, which is characterized by more gentle slopes and lies in a flat plain.

The typical winds in Lake Garda are thermal breezes [33]. In the northern region, the most frequent provenance direction of high speed winds is along the longitudinal axis of the lake, where winds are channeled by the steep lateral mountains (see the orography in Figure 1a). During summer, northerly and southerly breezes alternate in the night/morning and in the afternoon/evening, respectively [34]. Such breezes are characterized by moderate to high velocities ($\approx$5–10 m/s), alternating directions and not necessarily uniform spatial distribution. During winter and mid-seasons, winds are usually weaker. However, it is not uncommon for this lake to be subject to long-lasting storm winds at a synoptic scale, for example, Föhn winds. These winds frequently come from North-East, have almost uniform spatial distribution on the lake surface, reach speeds of more than 10 m/s and often last for more than one day [35].

2.2. Sensor and Dataset

In the last years, there has been a consistent proliferation of SAR systems with different technical characteristics, that is, resolution, operational frequencies and revisiting time [36]. Among them we selected the ENVISAT, following the experience gained in marine environment with C-Band sensors for the estimation of the surface current velocity. In this work, we adopt a methodology validated in the coastal area of the Gulf of Naples [16,27,37] and the Gulf of Trieste [38]. ASAR ENVISAT operated in C Band (wavelength λ equal to 5.6 cm), in multiple polarization mode (HH/VV,HH/HV,VV/VH). In stripmap mode it covered up to 100 km with a single swath. The spatial resolution of the sensor's Single Look Complex (SLC) product is 5 m in azimuth and 20 m ground range (see Figure 1b for SAR acquisition geometry). A considerable archive of data exists from the ENVISAT operative period, that is, from 2002 to 2012, on ascending and descending orbits. Within the broad landscape of SAR sensors currently available, data acquired by Sentinel 1 (S1) could also be exploited. S1 operates in C-Band and regularly provides very wide swaths of about 250 km by using the TOPSAR (Terrain Observation with Progressive Scans SAR) operation mode. However, due to the steering of the antenna along the azimuth, this operation mode complicates the estimation of the Doppler Centroid. The presence of discontinuities in the estimated Doppler Anomaly, which in turn affects the retrieved radial velocity, has been already observed [39]. Based on these considerations and on the experience we gained in marine environments, ENVISAT data were chosen for this exploratory study on Lake Garda. The ENVISAT archive was screened to eliminate images characterized by very low backscattering coefficients. We selected six images showing amplitude patterns, that is, sufficiently intense and spatially varying backscattering, potentially associated with interesting hydrodynamic features and fulfilling the wind criteria specified in detail in Section 3.2. The selected images are acquired over ascending orbit, corresponding to evening acquisitions (around 21:00 UTC), and span from 2008 to 2010 in summer, autumn and winter seasons. The choice of ascending acquisitions is due to the favourable acquisition geometry related to the orientation of the lake. In fact, the ground projection of the ascending sensor LOS is more aligned with the longitudinal axis of the lake than the descending one. We note that the longitudinal axis of the lake is expected to be the main direction of lake currents and waves propagation, as it also corresponds with the typical direction of winds (see Section 2.1). The images have been classified based on the wind velocity registered at the acquisition time. In Table 1 the selected dates are listed and their simulated geophysical characteristics are summarized. For a more detailed description of the investigated dates, we refer to Section 3.2.

For the selected dates, thermal images from MODIS were also analyzed to validate the simulations. In particular, we preferred nighttime Level 2 MODIS Terra Sea Surface Temperature (SST) products, because they are closer in time to SAR acquisitions. MODIS images are cloud free for four out of six investigated dates. The four dates are indicated with an asterisk in Table 1.

Table 1. Mean, standard deviation and maximum of the simulated key geophysical quantities for each date considered. Module of wind speed from WRF (W^{SIM}); module of the surface water velocity from Delft3D + SWAN (v^{SIM}); significant wave height (H_s) and peak wavelength (λ_{peak}) from SWAN.

Class	Date	W^{SIM} [m/s]		v^{SIM}[cm/s]		H_s [cm]		λ_{peak} [m]	
		Mean ± Std	Max	Mean ± Std	Max	Mean ± Std	Max	Mean ± Std	Max
low	29 October 2009 *	2.3 ± 1.5	5.5	8.3 ± 6.8	30.3	8.6 ± 5.6	19.6	2.04 ± 1.23	4.40
low	5 August 2010	3.1 ± 0.9	5.1	10.6 ± 5.4	33.5	11.4 ± 4.2	17.4	2.34 ± 0.94	3.74
low	14 October 2010 *	1.3 ± 0.5	2.1	3.2 ± 2.0	13.6	3.3 ± 0.9	5.3	0.81 ± 0.18	1.14
high	7 February 2008 *	8.2 ± 3.2	16.4	16.4 ± 8.1	61.0	49.7 ± 24.4	92.4	9.71 ± 5.04	19.24
high	13 November 2008	10.2 ± 5.2	22.2	33.1 ± 21.2	117.5	68.1 ± 42.7	145.1	12.79 ± 8.31	27.8
high	11 June 2009 *	7.9 ± 4.6	17.2	17.3 ± 8.4	53.2	48.9 ± 35.0	109.7	9.34 ± 7.00	29.67

* MODIS product available.

2.3. SAR Doppler Centroid Anomaly Estimation

For the retrieval of information about the ocean surface velocity from SAR data, two methodologies have been developed: the Along-Track Interferometry (ATI) [18] and the DCA method [14]. Here we exploit the technique of the DCA, which was successfully applied in open oceans [14,40,41] and ocean coastal areas [14,18,42]. The same approach was also adopted for the estimation of water currents in the Mediterranean sea, a semi-enclosed sea, where they show a smaller range of intensity, [16,27,38] . The DCA technique estimates the surface velocity of the illuminated area by measuring the variation that such a velocity produces in the spectrum of the SAR image [14]. The principle is rather well known: let us consider a SAR system acquiring data on a scene where a target is in relative motion with respect to the sensor. When the transmitted radiation hits the target, it is backscattered toward the radar along the antenna pointing direction, also called Line of Sight (LOS) or radial direction (see Figure 1b). If the velocity associated with the motion has a non-zero component along the radial direction (v_r), this component determines a shift in the azimuth spectrum of the received signal. Such a shift is called Doppler Centroid frequency f_{DC} and is proportional to the radial component of the velocity associated to the relative motion between target and sensor.

Some remarks on the formation of the Doppler shift, which is influenced by the target and sensor motion along the LOS direction, are in order. First of all, the radial direction is influenced by the platform attitude as well as by possible electronic steering introduced to point the beam in a direction different from the broadside one (squinted acquisition). Secondly, the measurement of f_{DC} from the power spectral density allows the estimation of the radial component of the (relative) target motion [43,44]. For each pixel, say with azimuth x and slant range r, $f_{DC}(x,r)$ can be considered the sum of two contributions: the Doppler Centroid $f_{DC0}(x,r)$ and Doppler Centroid Anomaly $f_{DCA}(x,r)$. The first one (f_{DC0}) corresponds to a stationary scene and is referred to as "background DC", due to a scene in which all the scatterers move at the same velocity (background motion). The second one (f_{DCA}) is the Doppler component associated with the target motion with respect to the background one.

From what is stated above, it follows that for the satellite case the background DC, $f_{DC0}(x,r)$, is contributed by both the platform movement and the Earth rotation, in a way that depends on the platform attitude and beam orientation. This term is generally a slow varying function along the space, typically well approximated by a polynomial in r. It is worth noting that, strictly speaking, ionospheric variations could also impact $f_{DC0}(x,r)$: this phenomenon is, however, more relevant for low frequency systems, for example, L- and P-Band SAR sensors [45]. The latter are usually not exploited for water surface velocity retrieval, due to the reduced sensitivity associated with the increase of the wavelength.

As for ASAR-Envisat, the term $f_{DC0}(x,r)$ can be retrieved from the ancillary data through annotated polynomial coefficients, which are computed knowing the platform attitude, the antenna pointing and the scene latitude. Previous experiences carried out with other platforms showed that this information is frequently not sufficiently accurate, especially for the range variation. This fact might be due to inaccurate attitude information, to unexpected beam bending as well as to other spurious effects. Following the strategy adopted in [16] for marine applications, we estimated $f_{DC0}(x,r)$ directly from the data via a polynomial Least Square fitting along r. More details on the estimation of the background Doppler component associated with the stationary scene are given in [16]. In the same work it is also shown how to handle unwanted azimuth oscillating components, frequently observed for ASAR-ENVISAT acquisitions and evidently not associated with any geophysical process.

The velocity at which the target proceeds toward or away from the radar, with respect to the background scene motion, can be thus obtained by removing from the measured Doppler Centroid its component associated to the stationary scene, as follows:

$$\begin{aligned} v_r(x,r) &= \frac{\lambda}{2}[f_{DC}(x,r) - f_{DC0}(x,r)] \\ &= \frac{\lambda}{2} f_{DCA}(x,r), \end{aligned} \tag{1}$$

where λ is the sensor wavelength. The scatters' surface velocity along the ground range direction v_{gr} can be finally obtained from v_r by considering that the latter is its projection along the LOS (see also Figure 1b):

$$v_{gr} = \frac{v_r}{\sin\theta}, \tag{2}$$

where θ is the incidence angle.

In order to retrieve the actual lake ground range surface velocity, any possible spurious effect, such as the wind-wave bias [12], must be precisely identified, estimated and removed from v_{gr}. Thus, v_{gr} represents a raw variable to be processed with an appropriate GMF to obtain the ground range surface velocity, hereinafter surface velocity. The identification of the spurious effects on v_{gr} is the goal of this study, where no GMFs are used, and it is performed with the help of the numerical model.

In Figure 2a ,we summarize the procedure for the generation of v_{gr} maps. After the SAR focusing operation on the raw data, the measurement of f_{DC} is performed pixel by pixel by estimating the azimuth self-correlation function along the azimuth direction [43]. The Doppler spectrum is evaluated on patches, hereafter referred to as Doppler Resolution Cell (DRC), sliding along the whole SAR image. The DRC has been chosen to guarantee a good balance between spatial and spectral resolutions. In particular, we considered patches of width 512 azimuth by 128 range pixels, corresponding to a DRC of 2.5 km × 2.5 km. In this condition the spectral resolution is 2.85 Hz, which corresponds to a velocity resolution of 7.98 cm/s for v_r.The Doppler Centroid estimation accuracy can be estimated through the calibration of the radial velocity over the land area (i.e masking out Lake Garda). For all the acquisitions here analyzed (see Table 1) v_r has zero mean and a standard deviation ranging from 27 cm/s to 30 cm/s. Note that these values account for the variability of the residual signal all over the scene with the lake only covering a small portion.

Subsequently, the compensation of the stationary component of the Centroid is performed and the final DCA is estimated (Equation (1) and then v_{gr} from Equation (2)). The DCA map is converted to surface velocity and geocoded through a SRTM DEM on a geographical grid with a spacing of about 92.7 m N and 64.9 m E [46]. At this stage, the geocoded surface velocity map is comparable with geo-referred simulations results and MODIS products.

The output product is supplemented by the basckscattering information. A standard calibration procedure is applied to the focused amplitude data to get the normalised measure of the radar cross-section per unit area on the ground, that is, the backscatter coefficient σ^0. A multi-look algorithm is then applied to reduce the noise (or "speckle") of SAR images by averaging adjacent pixels. The resulting σ^0 with resolution 100 m × 100 m is finally geo-referred (Figure 1c).

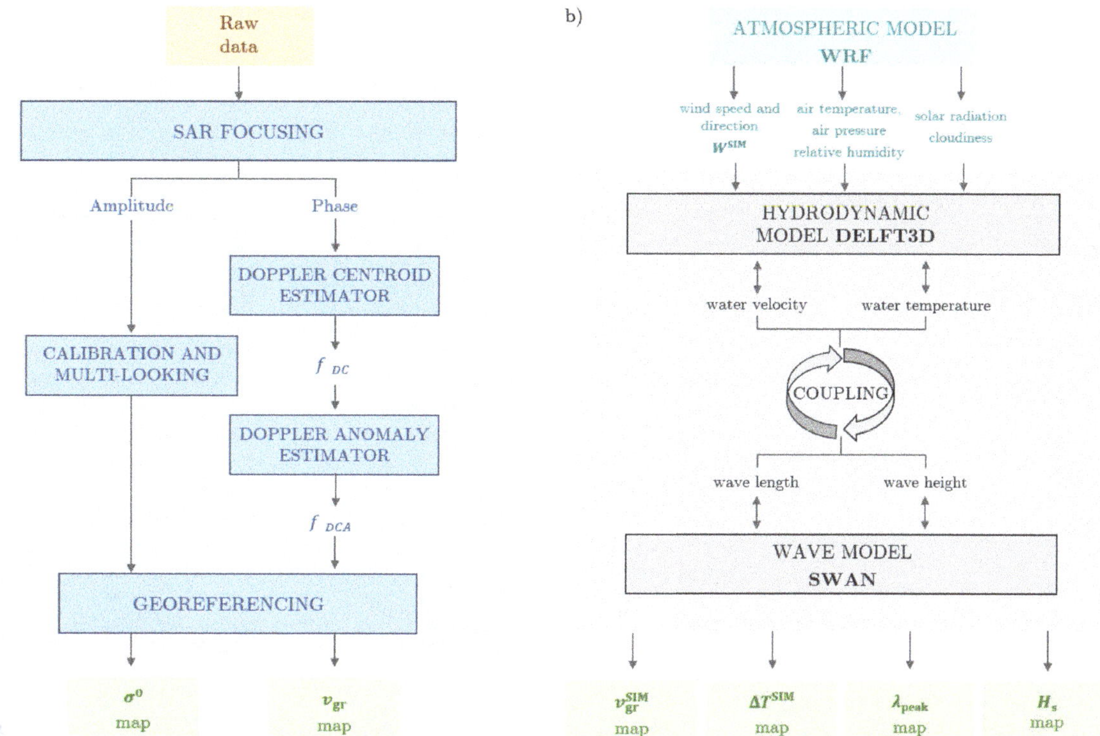

Figure 2. (**a**) Procedure for the generation of v_{gr} and σ^0 maps from SAR data; (**b**) modeling chain for the simulation of surface water velocity, temperature and waves field.

2.4. Numerical Modeling

For simulating the flow field, we use a modeling chain (Figure 2b) composed by three numerical models: an atmospheric, a hydrodynamic and a wave model. In this work, we start from the outputs of an existing model for the case study [35,47] in the configuration presented in [31], which consists of the hydrodynamic model Delft3D [48] fed by the results of the Weather Research and Forecasting (WRF) atmospheric model [48].

The WRF model provides the atmospheric variables as time and space varying fields of wind velocity, air pressure, temperature and relative humidity, incoming shortwave radiation and cloudiness at the lake surface. The resolution of the simulated weather forcing is 2 km in space and 1 h in time.

From the WRF outputs, the hydrodynamic model Delft3D computes the heat fluxes between air and water and wind stress at the lake surface, which are necessary for simulating the flow field, turbulence, heat and mass transport within the lake. The nominal horizontal resolution of the hydrodynamic model is ≈100 m, with smaller cells in the northern and larger cells in the southern part. The vertical layers are 1 m thick near the surface (over the upper 40 m) and smoothly increase their thickness up to 25 m at the deepest point. The outputs of the described model are available over the period 2004–2018 and will be henceforth referred to as "long-run data". The long-run data are saved on a daily basis at 10:00 UTC.

For the present work, the available long-run data were used to restart ad hoc simulations of the selected dates (Table 1) with the same setup as in [31]. The restart was necessary to produce outputs synchronous with the ENVISAT acquisitions (21:00 UTC) and to couple the Delft3D model with the wave model. We adopted a third-generation wind-wave model known as Simulating Waves Nearshore (SWAN) [49]. The wave spectrum employed in the simulations was composed of 101 logarithmically spaced frequencies in the range of 0.15–3 Hz. The frequency range was specifically chosen to be representative of wave field expected in Lake Garda and therefore included higher frequencies than those typically employed in coastal applications, but neglected very low frequencies that hardly occur in medium-size lakes. Since no data were available for the calibration of the model, the parameters were chosen in consistency with previous experiences in other large and deep lakes (e.g., Lake Constance [50] and Michigan [51]).

The results of the six Delft3D + SWAN simulations were saved at hourly intervals, over the whole computational domain, and covered a period of two days (the SAR image acquisition day plus one day before).

3. Results

In this work we use the model outputs with the aim of interpreting the results obtained from the SAR images. Thus, we rely here on the assumption that the model is capable of reproducing wind fields and surface transport. The reliability of the wind fields simulated by the WRF model has been extensively demonstrated by [31,52]. As for the surface transport, we base our work on [31], who showed that the Delft3D model results are coherent with in-situ and remotely sensed water temperature observations. We stress that [31] provided an indirect verification of the flow field, since in-situ measurements of water velocity are not available. In Section 3.1 we perform a similar verification by comparing the spatial patterns of surface temperature simulated by the model and reconstructed from the MODIS images. In Section 3.2, we introduce the main features of the investigated dates and in Sections 3.3 and 3.4, we evaluate the outputs of our procedure for the extraction of lake surface currents from SAR.

3.1. Model Verification Against MODIS

We provide here the comparison between the Level 2 MODIS Terra SST products (T^{MODIS}) in Celsius degrees and the simulated lake surface water temperature (T^{SIM}). Figure 3 shows that the model correctly reproduces the remotely sensed patterns of lake surface temperature and the spatial variability of this quantity in the different seasons. This confirms that the heat fluxes between atmosphere and lake are well simulated, and that the numerical flow field is coherent with the transport patterns responsible for horizontal anomalies in lake surface temperature. The difference between the spatial mean values of T^{SIM} and T^{MODIS} (reported in each panel of Figure 3) can be due to model errors as well as to the skin-bulk bias [53]. However, in all investigated dates the mean error is below 1 °C, well in line with the performances of the long-run model in [31].

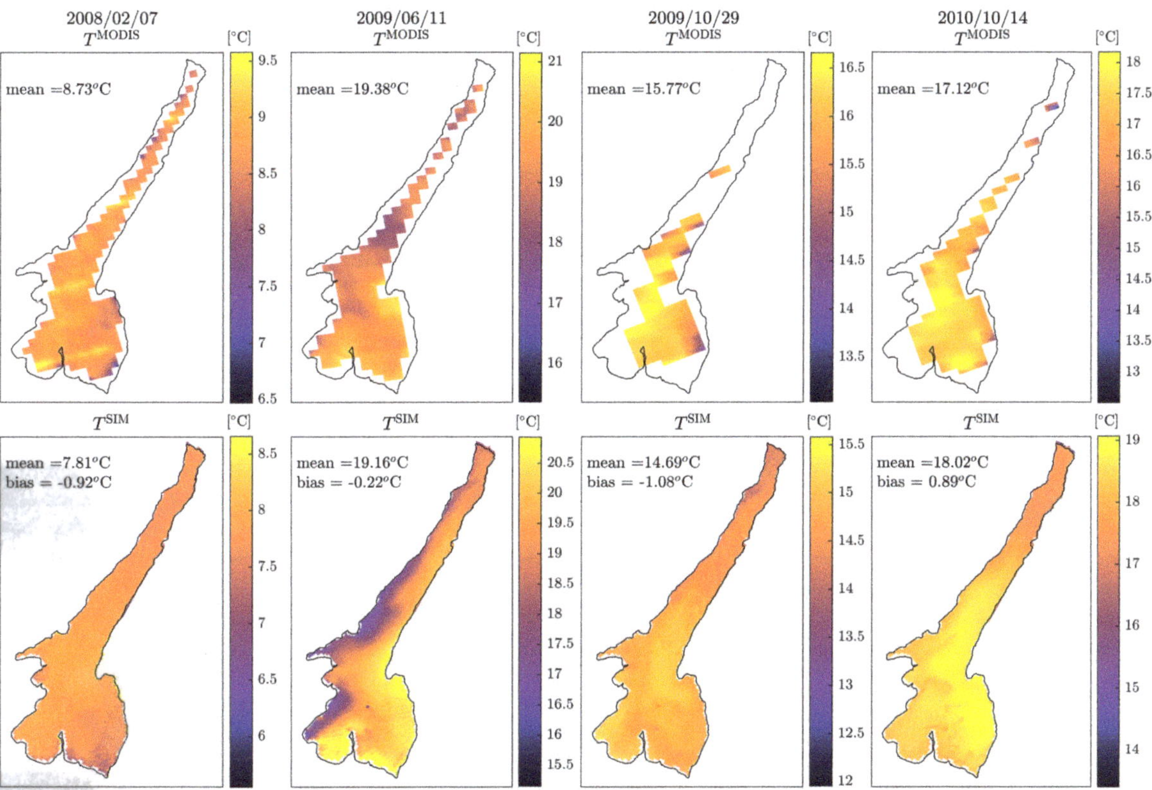

Figure 3. Top plots: maps of SST from MODIS Terra on cloud-free investigated dates. Bottom plots: maps of simulated lake surface water temperature. The spatial mean of MODIS and simulated water temperature is displayed together with the existing bias ($\overline{T^{SIM}} - \overline{T^{MODIS}}$). Note: in each panel, colorbar limits are set differently to highlight the spatial patterns. The colorbar limits of bottom plots are set as those from top plots at the corresponding date + the computed bias.

3.2. Main Features of Investigated Dates

The wind field has been recognized as playing a major role for the use of SAR in the open ocean [54]. In particular, in order generate capillary waves detectable from C-band sensors, the wind forces have to overcome the viscous ones. A wind speed threshold value for this process to occur is estimated to be about 3.25 m/s (at 10 m above the surface, [55]).

Therefore, we classify the six cases considered in this study (and listed in Table 1) based on the wind intensity. In Figure 4 the daily cycle of wind speed and direction simulated in a mid-lake point (point P in Figure 1) is displayed for the selected dates. The SAR acquisition time is indicated as a red vertical line in the figure panels.

We define "low wind" dates (Figure 4a–c) the three acquisitions in which the spatial mean of the simulated wind speed is equal or smaller than this threshold value (29 October 2009, 5 August 2010, 14 October 2010). In these dates mild evening breezes blew at 21:00 UTC over the lake surface (≈1–3 m/s). This condition is frequent in sunny and warm days, when thermal breezes develop with alternating directions during daylight hours and calm down after the sunset. Under low wind conditions, simulations predict mean surface velocities below 10 cm/s, significant wave heights H_s of the order of few cm, and peak wavelengths λ_{peak} of the order of 1 m (Table 1). Such values suggest that gravity effects do not completely dominate the dynamics as surface tension still plays a significant role [56].

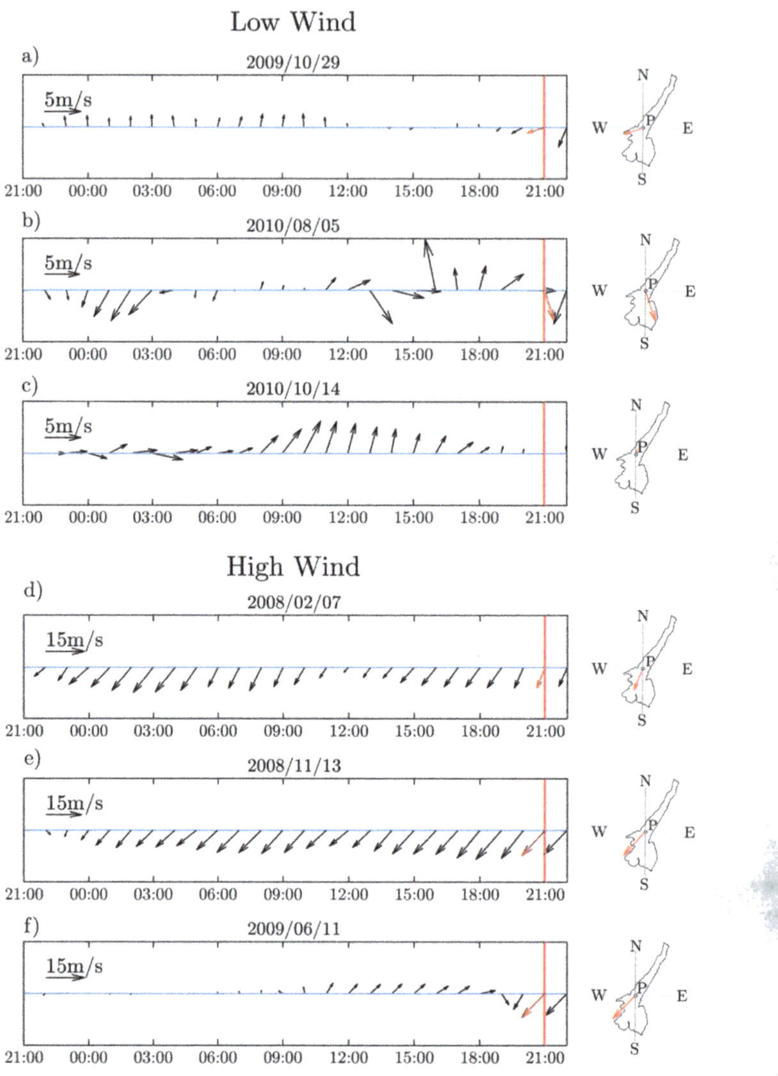

Figure 4. Simulated wind direction and speed on the investigated dates a in-lake point (point P in Figure 1a). The red arrow in each panel corresponds to the wind simulated in P at the SAR acquisition time (i.e., the red line at 21:00 UTC). The sketch at the right displays the location of P and the direction of the wind with respect to the lake geography.

The remaining three acquisitions are classified as "high wind" dates (7 February 2008, 13 November 2008, 11 June 2009, Figure 4d–f). On those days, the mean value of wind speed is above 7 m/s and its maximum value exceeds 15 m/s in large regions of the lake. Both the mean and the maximum wind speed values are significantly higher than the above-mentioned threshold. Such high speeds are often due to large scale synoptic winds, blowing almost uniformly over the lake's surface for more than 12 h. Figure 4d,e shows that the latter is indeed the case for the dates 7 February 2008 and 13 November 2008, when a northerly Föhn wind persistently blew for one day before the SAR acquisition time. On the contrary, the date 11 June 2009 (Figure 4f) coincides with the onset of a similar wind, which started one hour before the acquisition time and lasted all night long (not shown).

The simulated gravity waves on high wind speed days have H_s of the order of 50 cm on average (Table 1), overcoming 1 m in some areas of the lake, and λ_{peak} of the order of 10 m. These values are fully consistent with the gravity waves field that can be theoretically expected [57] and observed [58] in a medium-size lake.

In the next two sections, we will provide for all investigated dates the amplitude SAR backscatter image (σ_0), the ground range surface velocity retrieved from SAR DCA (Figure 2a, v_{gr}) and simulated by the numerical model (Figure 2b, v_{gr}^{SIM}), and the simulated wind field (direction and ground range component magnitude W_{gr}^{SIM}). The represented quantities are obtained as follows:

- σ_0 is processed as described at the end of Section 2.3 expressed in dB to enhance the visibility of its variation over lake;
- v_{gr} is obtained as in Equation (2).
- v_{gr}^{SIM} is obtained by computing the component of the simulated surface velocity along the ground range;
- W_{gr}^{SIM} is obtained by computing the component of the simulated wind velocity vector (at 10 m a.w.s.) along the ground range direction.

3.3. *Low Wind Dates*

The results for the "low wind" dates are reported in Figure 5. In marine/ocean observations, a low-wind regime corresponds to a low mean amplitude and a limited spatial variation of the backscatter. In our case study, Figure 5 clearly shows a different behaviour. The SAR backscatter retrieved on these dates (first column of the Figure 5, panels a, e, i) shows recurring patterns in all examined images. Higher values are detected in the northwestern basin, lower values in the south-eastern region. A high intensity feature located at the NE edge of the lake is present in all the ascending SAR images analysed. It consists in a cross-track radiometric compression, known as foreshortening effect, which occurs when the radar beam impinges on the foreslope of a mountainous area [20]. The component of surface motion retrieved from SAR DCA v_{gr} is represented in the second column of Figure 5, panels b, f, j. Red and blue colors (redshift and blueshift) indicate a motion towards and away from the radar antenna, respectively. We recall here that the surface velocity has been estimated with a DRC of 2.5 km × 2.5 km (see Section 2.3) and that any variation of the resulting v_{gr} at a smaller spatial scale has to be interpreted with care. While this is not a critical issue along the longitudinal axis of the lake and in the southeastern basin, where both the length and the width of the lake are significantly larger than the DRC size, difficulties arise in the northern trunk of the lake. Here the DRC size is comparable to the width of the lake and the estimation might be affected by the presence of portions of land inside the DRC. For such a reason we recommend caution in the interpretation of the variations of v_{gr} along the crosswise direction.

The spatial patterns of v_{gr} resemble those from σ^0. In the northern trunk of the lake, v_{gr} shows in all dates a recurring pattern: it overcomes 2 m/s in the central and western region, and is almost null along the eastern shore. This behaviour suggests a south-westward surface transport which is not confirmed by the numerical model (third column of Figure 5, panels c, g, k). In fact, the simulated field of v_{gr}^{SIM} shows significant differences from one date to another, mostly due to the differences in the spatial distribution of the wind field (fourth column of Figure 5, panels d, h, l). Moreover, the simulations show significantly lower surface velocities, below 0.1 m/s, as it is reasonably expected in a lake of the dimensions of Lake Garda under low wind conditions [59].

In the southern sub-basin, similar patterns are found in v_{gr}, retrieved on 29 October 2009 and 14 October 2010 (Figure 5b and j respectively), with high and positive velocities at the center of the sub-basin and almost null or negative velocities close to the shores. The spatial distribution of v_{gr} in this region is similar to that of v_{gr}^{SIM} on the same dates (Figure 5c,k). On 5 August 2010 v_{gr} (Figure 5f) is strongly positive over the entire southern region, while v_{gr}^{SIM} (Figure 5g) is strongly negative. The spatial distribution of v_{gr}^{SIM} on this date reveals a predominant eastward surface transport in the southern basin and an opposite tendency in the northern basin, which is the direct consequence of two divergent wind fields with respect to the ground range direction (Figure 5h).

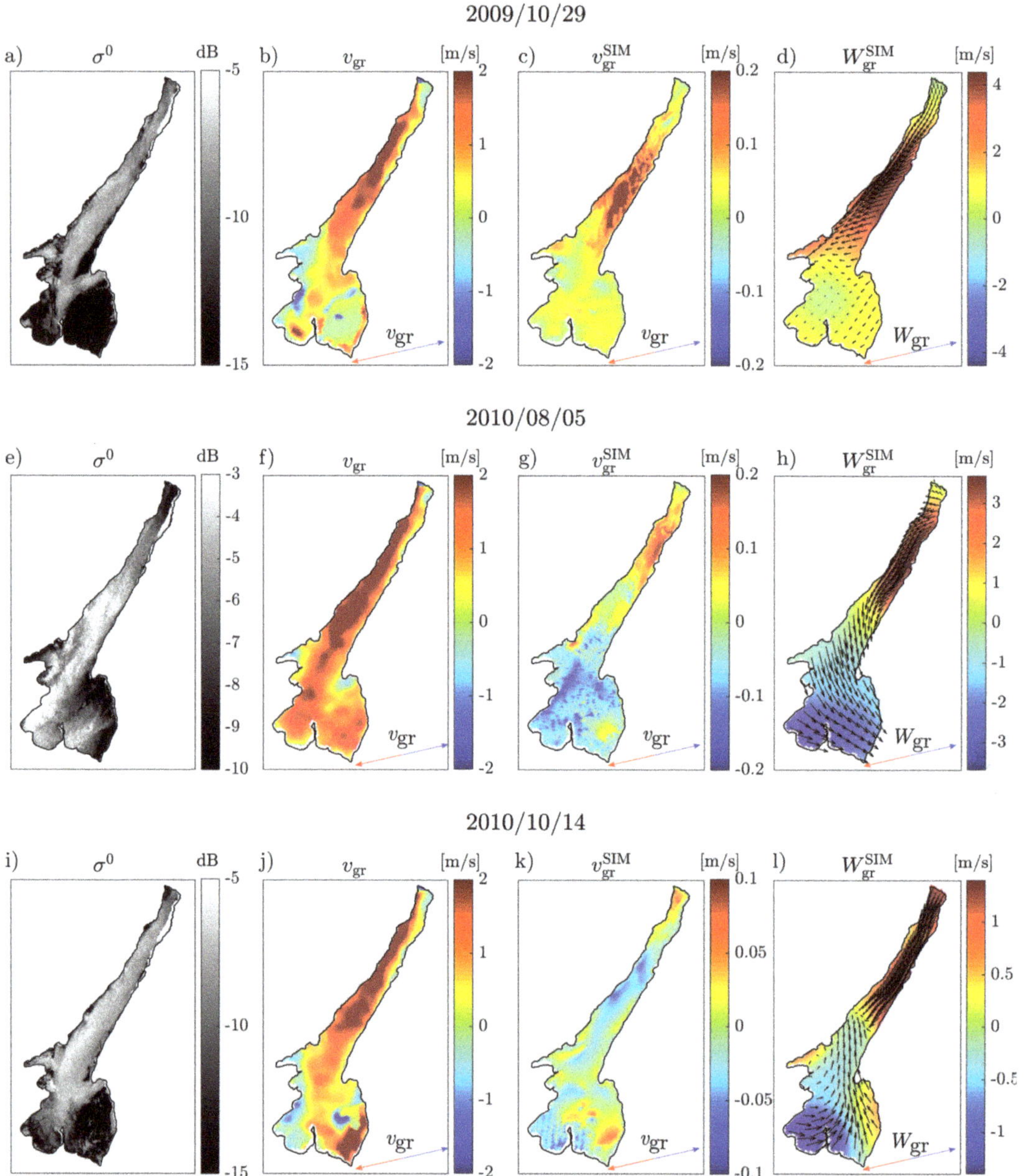

Figure 5. Maps from low wind days (**a–d**) 29 October 2009, (**e–h**) 5 August 2010, (**i–l**) 14 October 2010 of: first column (**a**,**e**,**i**) SAR bakcscatter amplitude; second column (**b**,**f**,**j**) v_{gr} estimated from SAR DCA (Equation (2)); third column (**c**,**g**,**k**) v_{gr}^{SIM} simulated by the numerical model; fourth column (**d**,**h**,**l**) ground range component of simulated wind speed W_{gr}^{SIM}. Black arrows in fourth column indicate the wind vector. The ground range direction is displayed by the blue and red arrow at the bottom-right corner of panels.

3.4. High Wind Dates

The results for the "high wind" dates are reported in Figure 6. The SAR backscatter images (Figure 6, panels a,e,i) and the surface velocity v_{gr} (Figure 6b,f,j) present similar patterns, which are in good agreement with the simulated surface velocity v_{gr}^{SIM} (c,g,k), especially in the elongated region.

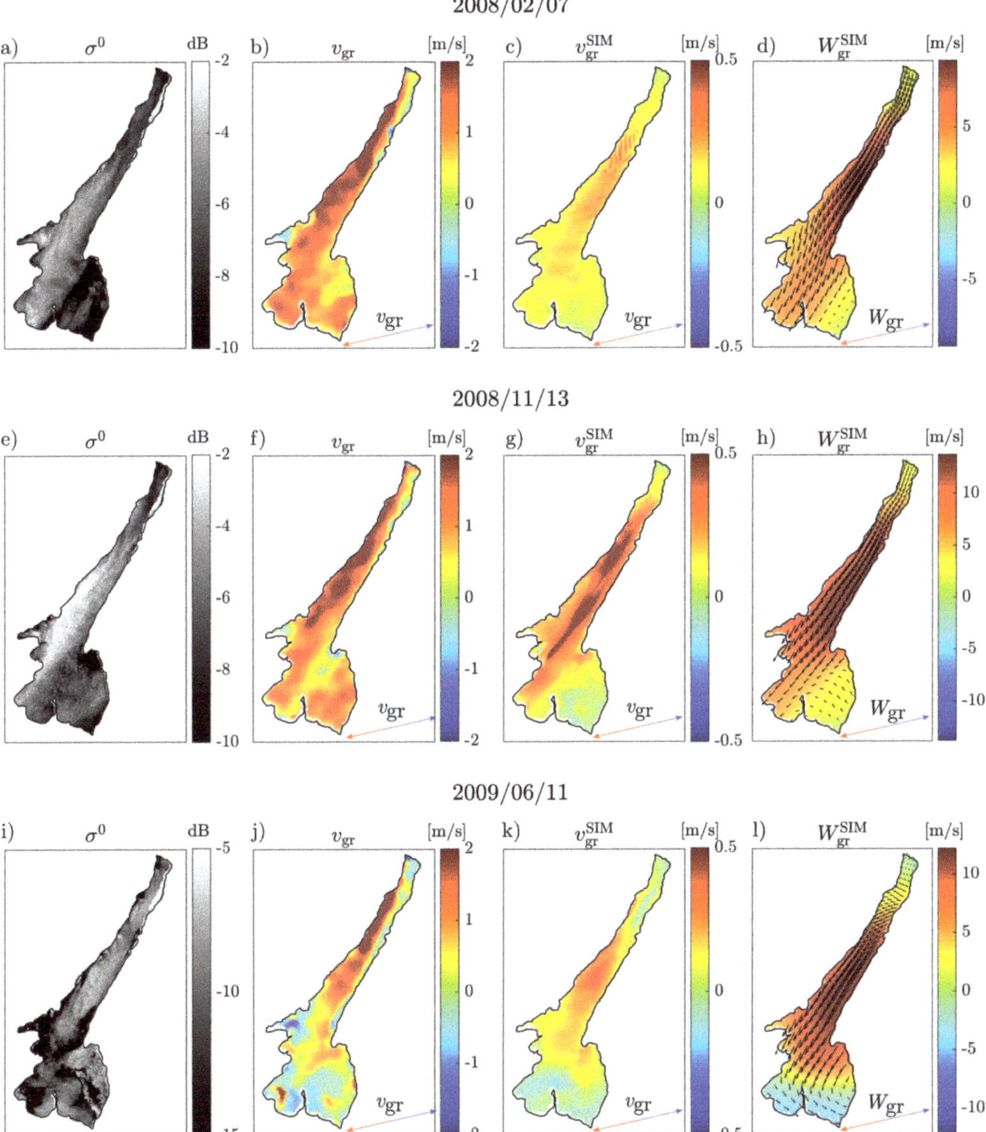

Figure 6. Maps from high wind days (**a–d**) 7 February 2008, (**e–h**) 13 November 2008, (**i–l**) 11 June 2009 of: first column (**a**,**e**,**i**) SAR backscatter amplitude; second column (**b**,**f**,**j**) v_{gr} estimated from SAR DCA (Equation (2)); third column (**c**,**g**,**k**) v_{gr}^{SIM} simulated by the numerical model; fourth column (**d**,**h**,**l**) ground range component of simulated wind speed W_{gr}^{SIM}. Black arrows in fourth column indicate the wind vector. The ground range direction is displayed by the blue and red arrow at the bottom-right corner of the panels.

On all dates σ^0 shows a pattern characterized by increasingly high values as moving south-west in the northwestern basin (Figure 6a,e,i). Such pattern can be observed also in the v_{gr} images (Figure 6b,f,j), with a redshift located mostly along the western shore indicating a south-westerly surface current moving towards the sensor. The value of v_{gr} in this area is higher than 2 m/s and slightly decreases in the southeastern part of the region. This behavior finds qualitative correspondence with the patterns of v_{gr}^{SIM} (Figure 6c,g,k), where the signature of wind waves can be detected from localized peaks of velocity up to 0.5 m/s. The black arrows displayed in Figure 6d,h,l indicate the presence of a wind blowing from north-east to south-west, which is responsible for the surface transport observed both from SAR and the model.

4. Discussion

In all cases examined, the surface ground range velocity retrieved from SAR (v_{gr}) is much higher than the simulated v_{gr}^{SIM}. Regardless of the wind intensity, v_{gr} always overcomes ± 2 m/s, which is too high for a medium-size lake environment. Despite the empirical relations linking lake surface velocity and wind speed are often site-specific, the order of magnitude of the first quantity is generally assumed to depend on the second. A wind factor can be computed as the ratio between water surface velocity and wind speed. In open waters the wind factor is typically of the order of 3%, but it can also be lower for the nearshore areas [60]. In the investigated dates, the simulated water velocity was found to be, on average, between 2% and 3.5% of the wind speed. Even assuming that the observed v_{gr} is the only component of the surface current (i.e., assuming that the water is moving precisely along the ground range direction), a value of 2 m/s would require a wind speed of more than 60 m/s, which is an absolutely unrealistic wind speed. Such wind speed would be even higher if we take into account also the other component of surface motion, which is likely to exist and contributes to the module of surface current, despite it can not be retrieved with the DCA technique. Hence, the velocity retrieved from SAR is a large overestimation of the actual lake surface velocity.

Such overestimation is also found in open ocean applications [61], where v_{gr} is affected by the surface bulk velocity and by a term of the order of the 30% the ground range component of the wind velocity W_{gr}^{SIM} [14]. The latter term represents a bias to be removed for the effective estimate of the surface ground range velocity. Such bias is typically predicted and removed using an ad hoc geophysical model function (GMF), for example, CDOP [12], which exploits the correlation between DCA and the ground range component of the wind velocity W_{gr}^{SIM}. The key geophysical process underlying this function is represented by wind waves, due to their orbital velocities.

Following this approach, the existence of a correlation between v_{gr} and W_{gr}^{SIM} is investigated. Scatter plots of v_{gr} vs W_{gr}^{SIM} are shown in the first row of Figure 7 (panels a, b from the low and high wind dates respectively). The colour scale reflects the probability density estimation from data computed via a Kernel Density Estimation (KDE) algorithm. The two variables appear not to be correlated in all cases examined here. This was reasonably expected, as in small to medium-size lakes the wind–waves field is typically underdeveloped, with complex dynamics often subject to local effects (such as topography) [57]. Thus, the validated approaches operationally used in the open ocean [10,15] cannot be applied to our case study and other quantities possibly affect v_{gr}.

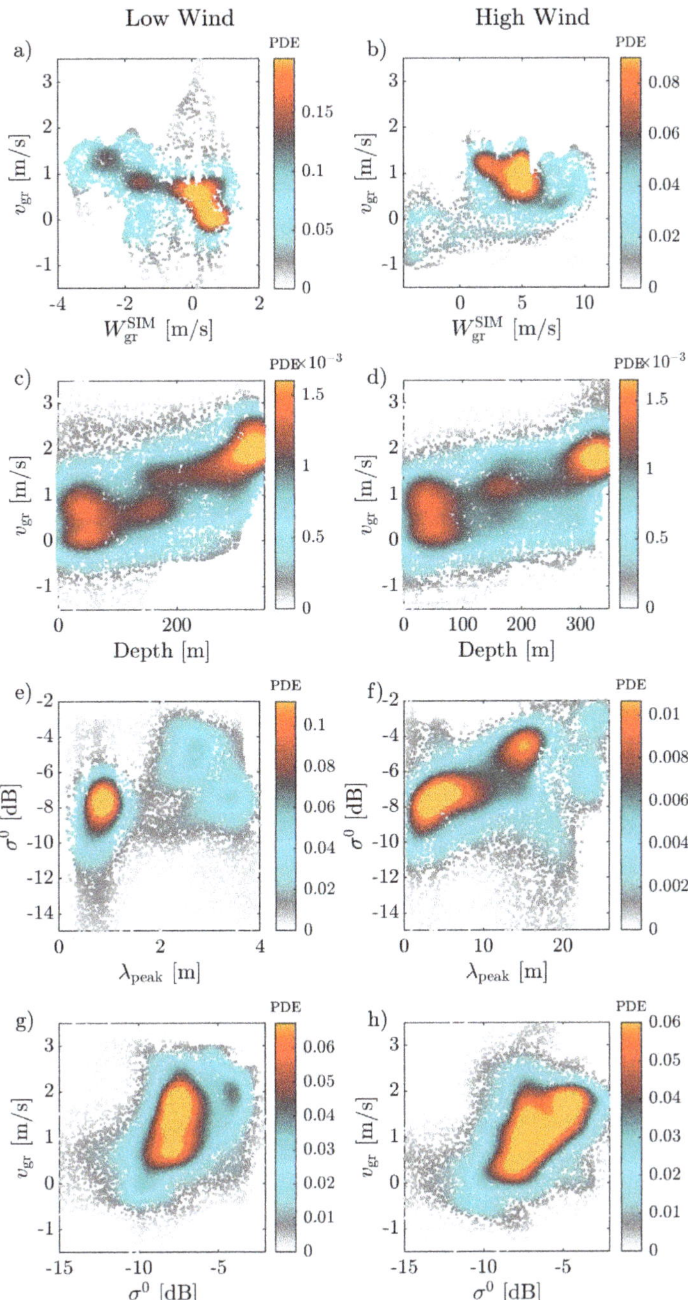

Figure 7. Scatter plots of v_{gr} vs W_{gr}^{SIM} (**a**,**b**); v_{gr} vs depth (**c**,**d**); σ^0 vs λ_{peak} (**e**,**f**); v_{gr} vs σ^0 (**g**,**h**). Panels on left column (**a**,**c**,**e**,**g**) refer to low wind cases, right column (**b**,**d**,**f**,**h**) to high wind cases. Color scale reflects the probability density estimate (PDE) computed via the Kernel smoothing function, which evaluates the spatial density of nearby points.

Patterns observed in Figures 5 and 6 for σ^0 and v_{gr} suggest a possible contribution from bathymetry to the SAR products. Figure 7c,d shows that a correlation between v_{gr} and bathymetry exists in all the investigated dates: higher correlation between v_{gr} and water depth is found the shallower areas (0 to 100 m deep), and in the deepest area (below 300 m). It is recognized that underwater bottom topography can be sensed indirectly via surface effects [62], although the electromagnetic waves emitted by the radar penetrate water only to a depth that is small in comparison to the radar wavelength. In fact, the underwater bottom topography generates short-scale surface roughness variations, which in turn modulate the radar reflectivity. As a complement to the analysis of the correlation between v_{gr} and water depth, we note that the deepest region of the lake (depth $>$ 300 m, where highly correlated values are visible in Figure 7c,d) coincide with the northern narrow area of the lake. In this area, the width of the lake ranges between 3 and 5 km, that is, between one and two times the DRC size (2.5 km). Hence, we assume that an effect of the mixed signals from water and land shall also be considered. To what extent such an effect combines with the bathymetry and the wind waves field and alters the measured v_{gr} is beyond the aims of this work and we recommend further investigation in this sense.

Given the specific geometry of Lake Garda, the observed significance of bathymetry could, in principle, also be ascribed to the presence of longer (meters scale) wind waves. As shown in Figure 7c,d there is a peak of correlated values around $\approx$170 m depth and $v_{gr} \approx 1.5$ m/s. Such a feature, in the case of high wind, is likely related to the wave field development. In this regard, in Figure 8a–c we report the peak wavelength map in the high wind dates. As it is clear from the first two days (panels a,b), λ_{peak} reaches its maximum at the end of the northern trunk, where water depth is below the two 150 m isobaths. This area also corresponds to maximum effective fetch for a wind blowing from North-East. The relative weight of the short and long waves also affects v_{gr} and is at the base of the applications of the DCA method to oceans [11,61]. Nevertheless, it cannot be easily quantified in lake applications, despite the fact that it can be inferred from σ^0 variations. According to the Bragg scattering model, the radar backscatter is proportional to the surface wave spectral density at the Bragg wavelength. As these centimeters-scale waves are tilted by longer (meters-scale) waves, σ^0 varies along with such longer wave profiles. This leads to a correlation between σ^0 and the orbital velocities of the peak waves [63]. Theoretical findings predict an average increase in the mean spectral density of the short waves due to the interaction between short "Bragg" waves and meters-scale waves [64]. Such a relationship also holds in the high wind cases analyzed here, as shown in Figure 7f, where a clear correlation exists between σ^0 and λ_{peak}. However, no correlation is found in the case of low wind dates (Figure 7e), when the wave field is not developed.

Figure 7g,h clearly shows that a tight relationship exists between v_{gr} and σ^0. A remark about the compensation of the effects of the azimuth variation of the backscattering on the DCA is now in order. The Doppler shift is estimated by exploiting the spectral tapering of the azimuth antenna beam pattern. The algorithms for the DC estimation assume a backscattering spectral flatness, that is, a flat backscattering spectral power density [43,44]. Evidently, the presence of azimuth variations of the backscattering within the Doppler Resolution Cell (DRC) can negatively impact this assumption leading to a spatially varying Doppler shift bias, that is, a disturbance in the measured DCA. In [12], this phenomenon was physically explained by referring to the unbalancing, appearing in the presence of a backscattering azimuth gradient, between the contributions to the Doppler Centroid associated with the scatterers ahead and behind the zero Doppler scatter. In [12], the authors proposed a solution to this problem, which is based on the subtraction of the contribution correlated to the azimuth backscattering gradient from the estimated DCA. A detailed analysis of the effect of azimuth backscattering variations on the estimated DCA, as well as its mitigation, is beyond the scope of the present work and is certainly worth future research, especially for the specific application to lakes. The compensation procedure proposed in [12] was not adopted here due to the possible consequences on the cancellation of the useful signal.

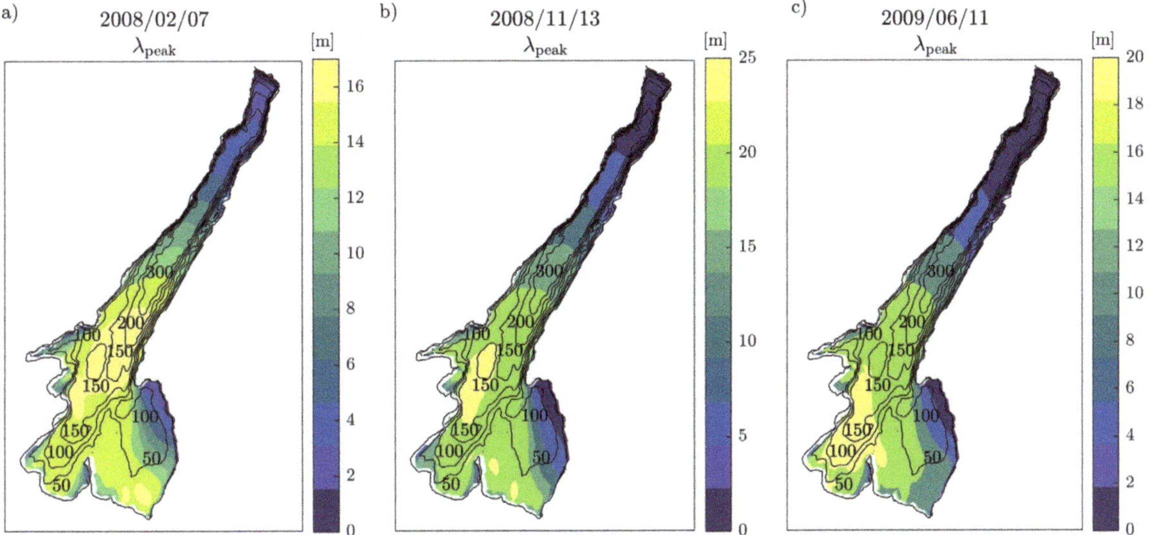

Figure 8. Maps of peak wavelength for high wind dates 7 February 2008 (**a**), 13 November 2008 (**b**), 11 June 2009 (**c**).

5. Conclusions

In recent years, the literature has reflected increasing awareness of the opportunities of satellite-based monitoring of lake surfaces. In this study, we investigate the possibility of inferring lake surface water velocity from existing methods, exploiting the SAR Doppler Centroid Anomaly. To our knowledge, this is the first application of such methodologies to inland waters. The analysis of ENVISAT SAR images and the related DCA maps for Lake Garda returns evident structures with intense and spatially varying values of backscattering and Doppler frequency. These are compared with the results of a numerical model. In the three dates when a high wind blows uniformly over the lake, the spatial patterns retrieved from SAR resemble those from the numerical model, while this is not the case for the low wind dates. In all cases, the surface velocity retrieved from DCA appears to always be significantly overestimated.

The factors potentially influencing the extraction of surface velocity on the investigated dates are identified as are the intensity of the wind forcing, the bathymetry, and the wind waves' wavelength. The simulated wind waves are also found to be well correlated with the backscatter amplitude. Some hypotheses on the potential biases to be removed are addressed, to preliminarily evaluate the applicability of existing GMFs.

Lake dynamics can be significantly different to those in open seas, especially in medium-size lakes where the response to the wind forcing varies locally. In our case study, wind space and time variability is mainly responsible for the surface flow field [59]. The cases analyzed so far suggest that the GMFs derived for inferring the surface velocity from DCA in open seas [12] cannot be applied to our case study, as they account for the wind vector solely and assume a fully developed wind waves field. Indeed, none of the cases analyzed show a clear correlation between ground range velocity measured from SAR and the correspondent wind component. On the other hand, the data processed suggest that a specific GMF could be defined for our case study, and potentially for the more general case of medium-size lakes. This function should include parameterizations handling the small scale variations associated with wind forcing, bathymetry and waves. The possibility of including the information carried by the SAR backscatter should be evaluated, and the interpretation of the environmental variables signature on the SAR backscatter should be further improved.

The application to lakes of Doppler methods commonly exploited in the open sea is not only challenging for the site related features, for example, specific environmental conditions, hydrodynamic and topographic aspects, but also from the data processing perspective. Further developments shall be carried out in the processing chain to account for possible size limitations of the lake compared to the Doppler resolution cell. Adaptive strategies based on multi-resolution Doppler estimation could be investigated, as well as approaches for the correction of the bias induced by a non-uniform backscattering within the Doppler resolution cell. Such data processing developments are considered priorities for improving the application of SAR-based current estimation methods to lakes.

The results and the interpretations provided in this work must be deeply validated and massive research activity is required, as well as dedicated field campaigns. In this sense, we hope that our contribution will pave the way for the development of a procedure for the extraction of the current signature from SAR that is adequately re-adapted for lakes.

Author Contributions: Conceptualization, M.A., V.Z., G.D.C., G.F. and F.D.S.; Data curation, M.A., V.Z. and F.D.S.; Formal analysis, M.A., V.Z. and F.D.S.; Investigation, M.A., V.Z., G.D.C., G.F. and F.D.S.; Methodology, M.A., V.Z., M.T., M.B. and C.G.; Project administration, F.D.S.; Resources, M.A. and V.Z.; Software, M.A. and V.Z.; Supervision, G.F. and F.D.S.; Visualization, M.A.; Writing—original draft, M.A., V.Z. and F.D.S.; Writing—review & editing, G.D.C., G.F., M.T., M.B. and C.G. All authors have read and agreed to the published version of the manuscript.

Funding: This study has been supported the EU Horizon 2020 project Water-ForCE (grant nr. 101004186).

Institutional Review Board Statement: Not applicable.

Informed Consent Statement: Not applicable.

Data Availability Statement: Data sharing is no applicable to this article.

Acknowledgments: We thank Lorenzo Giovanni for the WRF simulations. We are also thankful to Marco Pilotti, Giulia Valerio and Francesco Serafino for insightful discussions on the potential impacts of this preliminary work.

Conflicts of Interest: The authors declare no conflict of interest.

Abbreviations

The following abbreviations are used in this manuscript:

DCA	Doppler Centroid Anomaly
DRC	Doppler Resolution Cell
ECMWF	European Center for Medium-Range Weather Forecast
GMF	Geophysical Model Function
KDE	Kernel Density Estimation
LOS	Line Of Sight
S1	Sentinel-1
SAR	Synthetic Aperture Radar
SLC	Single Look Complex
SST	Sea Surface Temperature
SWAN	Simulating WAves Nearshore
WRF	Weather Research and Forecasting

References

1. Noviello, C.; Verde, S.; Zamparelli, V.; Fornaro, G.; Pauciullo, A.; Reale, D.; Nicodemo, G.; Ferlisi, S.; Gulla, G.; Peduto, D. Monitoring Buildings at Landslide Risk With SAR: A Methodology Based on the Use of Multipass Interferometric Data. *IEEE Geosci. Remote Sens. Mag.* **2020**, *8*, 91–119. [CrossRef]
2. Cascini, L.; Peduto, D.; Reale, D.; Arena, L.; Ferlisi, S.; Verde, S.; Fornaro, G. Detection and monitoring of facilities exposed to subsidence phenomena via past and current generation SAR sensors. *J. Geophys. Eng.* **2013**, *10*, 064001. [CrossRef]
3. Franceschetti, G.; Lanari, R. *Synthetic Aperture Radar Processing*; CRC Press: Boca Raton, FL, USA, 1999.

4. Trivero, P.; Adamo, M.; Biamino, W.; Borasi, M.; Cavagnero, M.; De Carolis, G.; Di Matteo, L.; Fontebasso, F.; Nirchio, F.; Tataranni, F. Automatic oil slick detection from SAR images: Results and improvements in the framework of the PRIMI pilot project. *Deep Sea Res. Part Top. Stud. Oceanogr.* **2016**, *133*, 146–158. [CrossRef]
5. De Carolis, G.; Olla, P.; De Santi, F. SAR image wave spectra to retrieve the thickness of grease-pancake sea ice using viscous wave propagation models. *Sci. Rep.* **2021**, *11*, 1–14. [CrossRef] [PubMed]
6. Tello, M.; Lopez-Martinez, C.; Mallorqui, J.J. A novel algorithm for ship detection in SAR imagery based on the wavelet transform. *IEEE Geosci. Remote. Sens. Lett.* **2005**, *2*, 201–205. [CrossRef]
7. Lehner, S.; Horstmann, J.; Koch, W.; Rosenthal, W. Mesoscale wind measurements using recalibrated ERS SAR images. *J. Geophys. Res. Ocean.* **1998**, *103*, 7847–7856. [CrossRef]
8. Biondi, F.; Addabbo, P.; Ullo, S.L.; Clemente, C.; Orlando, D. Perspectives on the Structural Health Monitoring of Bridges by Synthetic Aperture Radar. *Remote Sens.* **2020**, *12*, 3852. [CrossRef]
9. Cerutti-Maori, D.; Sikaneta, I.; Gierull, C.H. Optimum SAR/GMTI Processing and Its Application to the Radar Satellite RADARSAT-2 for Traffic Monitoring. *IEEE Trans. Geosci. Remote Sens.* **2012**, *50*, 3868–3881. [CrossRef]
10. Moiseev, A.; Johnsen, H.; Hansen, M.; Johannessen, J. Evaluation of Radial Ocean Surface Currents Derived from Sentinel-1 IW Doppler Shift Using Coastal Radar and Lagrangian Surface Drifter Observations. *J. Geophys. Res. Ocean.* **2020**, *125*, e2019JC015743. [CrossRef]
11. Dagestad, K.F.; Horstmann, J.; Mouche, A.; Perrie, W.; Shen, H.; Zhang, B.; Li, X.; Monaldo, F.; Pichel, W.; Lehner, S.; et al. Wind retrieval from synthetic aperture radar-an overview. In Proceedings of the 4th SAR Oceanography Workshop (SEASAR 2012), Tromsø, Norway, 18–22 June 2012; pp. 18–22.
12. Mouche, A.A.; Collard, F.; Chapron, B.; Dagestad, K.F.; Guitton, G.; Johannessen, J.A.; Kerbaol, V.; Hansen, M.W. On the use of Doppler shift for sea surface wind retrieval from SAR. *IEEE Trans. Geosci. Remote Sens.* **2012**, *50*, 2901–2909. [CrossRef]
13. Chapron, B.; Collard, F.; Kerbaol, V. Satellite synthetic aperture radar sea surface Doppler measurements. In Proceedings of the 2nd Workshop on SAR Coastal and Marine Applications, 2004, Svalbard, Norway, 8–12 September 2003; pp. 8–12.
14. Chapron, B.; Collard, F.; Ardhuin, F. Direct measurements of ocean surface velocity from space: Interpretation and validation. *J. Geophys. Res. Ocean.* **2005**, *110*.doi:10.1029/2004JC002809 [CrossRef]
15. S1 Level 2 OCN: Surface Radial Velocity (RVL) Component. Available online: https://sentinel.esa.int/web/sentinel/technical-guides/sentinel-1-sar/products-algorithms/level-2/products/surface-radial-velocity-component (accessed on 15 February 2021).
16. Zamparelli, V.; De Santi, F.; Cucco, A.; Zecchetto, S.; De Carolis, G.; Fornaro, G. Surface Currents Derived from SAR Doppler Processing: An Analysis over the Naples Coastal Region in South Italy. *J. Mar. Sci. Eng.* **2020**, *8*, 203. [CrossRef]
17. Zamparelli, V.; De Santi, F.; De Carolis, G.; Fornaro, G. On the Analysis of SAR Derived Wind and Sea Surface Currents. In Proceedings of the IGARSS 2020—2020 IEEE International Geoscience and Remote Sensing Symposium, Waikoloa, HI, USA, 26 September–2 October 2020; pp. 5721–5724. [CrossRef]
18. Romeiser, R.; Runge, H.; Suchandt, S.; Kahle, R.; Rossi, C.; Bell, P.S. Quality Assessment of Surface Current Fields from TerraSAR-X and TanDEM-X Along-Track Interferometry and Doppler Centroid Analysis. *IEEE Trans. Geosci. Remote Sens.* **2014**, *52*, 2759–2772. [CrossRef]
19. Romeiser, R.; Alpers, W. An improved composite surface model for the radar backscattering cross section of the ocean surface: 2. Model response to surface roughness variations and the radar imaging of underwater bottom topography. *J. Geophys. Res. Oceans* **1997**, *102*, 25251–25267. [CrossRef]
20. Gelautz, M.; Frick, H.; Raggam, J.; Burgstaller, J.; Leberl, F. SAR image simulation and analysis of alpine terrain. *Isprs J. Photogramm. Remote Sens.* **1998**, *53*, 17–38. [CrossRef]
21. Nghiem, S.V.; Leshkevich, G.A. Satellite SAR remote sensing of Great Lakes ice cover, Part 1. Ice backscatter signatures at C band. *J. Great Lakes Res.* **2007**, *33*, 722–735. [CrossRef]
22. Wu, L.; Wang, L.; Min, L.; Hou, W.; Guo, Z.; Zhao, J.; Li, N. Discrimination of Algal-bloom using spaceborne SAR observations of Great Lakes in China. *Remote Sens.* **2018**, *10*, 767. [CrossRef]
23. De Santi, F.; Luciani, G.; Bresciani, M.; Giardino, C.; Lovergine, F.P.; Pasquariello, G.; Vaiciute, D.; De Carolis, G. Synergistic Use of Synthetic Aperture Radar and Optical Imagery to Monitor Surface Accumulation of Cyanobacteria in the Curonian Lagoon. *J. Mar. Sci. Eng.* **2019**, *7*, 461. [CrossRef]
24. McKinney, P.; Holt, B.; Matsumoto, K. Small eddies observed in Lake Superior using SAR and sea surface temperature imagery. *J. Great Lakes Res.* **2012**, *38*, 786–797. [CrossRef]
25. Katona, T.; Bartsch, A. Estimation of wind speed over lakes in Central Europe using spaceborne C-band SAR. *Eur. J. Remote Sens.* **2018**, *51*, 921–931. [CrossRef]
26. Pareeth, S.; Bresciani, M.; Buzzi, F.; Leoni, B.; Lepori, F.; Ludovisi, A.; Morabito, G.; Adrian, R.; Neteler, M.; Salmaso, N. Warming trends of perialpine lakes from homogenised time series of historical satellite and in-situ data. *Sci. Total Environ.* **2017**, *578*, 417–426. [CrossRef] [PubMed]
27. Jackson, G.; Fornaro, G.; Berardino, P.; Esposito, C.; Lanari, R.; Pauciullo, A.; Reale, D.; Zamparelli, V.; Perna, S. Experiments of sea surface currents estimation with space and airborne SAR systems. In Proceedings of the 2015 IEEE International Geoscience and Remote Sensing Symposium (IGARSS), Milan, Italy, 26–31 July 2015; pp. 373–376.

28. Kersten, P.R.; Toporkov, J.V.; Ainsworth, T.L.; Sletten, M.A.; Jansen, R.W. Estimating Surface Water Speeds With a Single-Phase Center SAR Versus an Along-Track Interferometric SAR. *IEEE Trans. Geosci. Remote Sens.* **2010**, *48*, 3638–3646. [CrossRef]
29. Toporkov, J.V.; Perkovic, D.; Farquharson, G.; Sletten, M.A.; Frasier, S.J. Sea surface velocity vector retrieval using dual-beam interferometry: First demonstration. *IEEE Trans. Geosci. Remote Sens.* **2005**, *43*, 2494–2502. [CrossRef]
30. Steissberg, T.E.; Hook, S.J.; Schladow, S.G. Measuring surface currents in lakes with high spatial resolution thermal infrared imagery. *Geophys. Res. Lett.* **2005**, *32*. doi:10.1029/2005GL022912 [CrossRef]
31. Amadori, M.; Giovannini, L.; Toffolon, M.; Piccolroaz, S.; Zardi, D.; Bresciani, M.; Giardino, C.; Luciani, G.; Kliphuis, M.; van Haren, H.; et al. Multi-scale evaluation of a 3D lake model forced by an atmospheric model against standard monitoring data. *Environ. Model. Softw.* **2021**, *139*, 105017. [CrossRef]
32. Salmaso, N.; Mosello, R. Limnological research in the deep southern subalpine lakes: Synthesis, directions and perspectives. *Adv. Oceanogr. Limnol.* **2010**, *1*, 29–66. [CrossRef]
33. Giovannini, L.; Laiti, L.; Serafin, S.; Zardi, D. The thermally driven diurnal wind system of the Adige Valley in the Italian Alps. *Q. J. R. Meteorol. Soc.* **2017**, *143*, 2389–2402. [CrossRef]
34. Giovannini, L.; Laiti, L.; Zardi, D.; de Franceschi, M. Climatological characteristics of the Ora del Garda wind in the Alps. *Int. J. Climatol.* **2015**, *35*, 4103–4115. [CrossRef]
35. Piccolroaz, S.; Amadori, M.; Toffolon, M.; Dijkstra, H.A. Importance of planetary rotation for ventilation processes in deep elongated lakes: Evidence from Lake Garda (Italy). *Sci. Rep.* **2019**, *9*, 8290. [CrossRef]
36. Moreira, A. A golden age for spaceborne SAR systems. In Proceedings of the 2014 20th International Conference on Microwaves, Radar and Wireless Communications (MIKON), Gdansk, Poland, 16–18 June 2014; pp. 1–4. [CrossRef]
37. Zamparelli, V.; Jackson, G.; Cucco, A.; Fornaro, G.; Zecchetto, S. SAR based sea current estimation in the Naples coastal area. In Proceedings of the 2016 IEEE International Geoscience and Remote Sensing Symposium (IGARSS), Beijing, China, 10–15 July 2016; pp. 4665–4668. [CrossRef]
38. Zamparelli, V.; Fornaro, G. SAR based sea surface currents estimation: Application to the Gulf of Trieste. In Proceedings of the ISAG-2019 International Symposium on Applied Geoinformatics, Istanbul, Turkey, 7–9 November 2019.
39. Johnsen, H.; Nilsen, V.; Engen, G.; Mouche, A.A.; Collard, F. Ocean doppler anomaly and ocean surface current from Sentinel 1 tops mode. In Proceedings of the 2016 IEEE International Geoscience and Remote Sensing Symposium (IGARSS), Beijing, China , 10–15 July 2016; pp. 3993–3996. [CrossRef]
40. Hansen, M.; Kudryavtsev, V.; Chapron, B.; Johannessen, J.; Collard, F.; Dagestad, K.F.; Mouche, A. Simulation of radar backscatter and Doppler shifts of wave–current interaction in the presence of strong tidal current. *Remote Sens. Environ.* **2012**, *120*, 113–122. [CrossRef]
41. Kraemer, T.; Johnsen, H.; Brekke, C. Emulating Sentinel-1 Doppler Radial Ice Drift Measurements Using Envisat ASAR Data. *IEEE Trans. Geosci. Remote Sens.* **2015**, *53*, 6407–6418. [CrossRef]
42. Wang, L.; Zhou, Y.; Ge, J.; Johannessen, J.; Shen, F. Mapping sea surface velocities in the Changjiang coastal zone with advanced synthetic aperture radar. *Acta Oceanol. Sin.* **2014**, *33*, 141–149. [CrossRef]
43. Madsen, S. Estimating the Doppler centroid of SAR data. *IEEE Trans. Aerosp. Electron. Syst.* **1989**, *25*, 134–140. [CrossRef]
44. Bamler, R. Doppler frequency estimation and the Cramer-Rao bound. *IEEE Trans. Geosci. Remote Sens.* **1991**, *29*, 385–390. [CrossRef]
45. Meyer, F.J. Performance Requirements for Ionospheric Correction of Low-Frequency SAR Data. *IEEE Trans. Geosci. Remote Sens.* **2011**, *49*, 3694–3702. [CrossRef]
46. Sansosti, E. A simple and exact solution for the interferometric and stereo SAR geolocation problem. *IEEE Trans. Geosci. Remote Sens.* **2004**, *42*, 1625–1634. [CrossRef]
47. Ghirardi, N.; Amadori, M.; Free, G.; Giovannini, L.; Toffolon, M.; Giardino, C.; Bresciani, M. Using remote sensing and numerical modelling to quantify a turbidity discharge event in Lake Garda. *J. Limnol.* **2020**, *80*, 47–52. [CrossRef]
48. Lesser, G.R.; Roelvink, J.A.; Keste, T.M.V.; Stelling, G.S. Development and validation of a three-dimensional morphological model. *Coast. Eng.* **2004**, *51*, 883–915. [CrossRef]
49. Booij, N.; Ris, R.C.; Holthuijsen, L.H. A third-generation wave model for coastal regions: 1. Model description and validation. *J. Geophys. Res. Oceans* **1999**, *104*, 7649–7666. [CrossRef]
50. Seibt, C.; Peeters, F.; Graf, M.; Sprenger, M.; Hofmann, H. Modeling wind waves and wave exposure of nearshore zones in medium-sized lakes. *Limnol. Oceanogr.* **2013**, *58*, 23–36. [CrossRef]
51. Mao, M.; Van Der Westhuysen, A.J.; Xia, M.; Schwab, D.J.; Chawla, A. Modeling wind waves from deep to shallow waters in Lake Michigan using unstructured SWAN. *J. Geophys. Res. Oceans* **2016**, *121*, 3836–3865. [CrossRef]
52. Giovannini, L.; Antonacci, G.; Zardi, D.; Laiti, L.; Panziera, L. Sensitivity of simulated wind speed to spatial resolution over complex terrain. *Energy Procedia* **2014**, *59*, 323–329. [CrossRef]
53. Wilson, R.C.; Hook, S.J.; Schneider, P.; Schladow, S.G. Skin and bulk temperature difference at Lake Tahoe: A case study on lake skin effect. *J. Geophys. Res. Atmos.* **2013**, *118*, 10–332. [CrossRef]
54. Monaldo, F.; Kerbaol, V.; Clemente-Colon, P.; Furevik, B.; Horstmann, J.; Johannessen, J.; Li, X.; Pichel, W.; Sikora, T.; Thompson, D.; et al. The SAR Measurements of Ocean Surface Winds: A White Paper for the 2nd Workshop on Coastal and Marine Applications of SAR, Longyearbyen, Spitsbergen, Norway, 8–12 September 2003. *ESA SP* **2003**, *565*.

55. Donelan, M.A.; Pierson, W.J., Jr. Radar scattering and equilibrium ranges in wind-generated waves with application to scatterometry. *J. Geophys. Res. Oceans* **1987**, *92*, 4971–5029. [CrossRef]
56. Toffoli, A.; Bitner-Gregersen, E.M. Types of Ocean Surface Waves, Wave Classification. In *Encyclopedia of Maritime and Offshore Engineering;* American Cancer Society, Hoboken, NJ, USA : 2017; pp. 1–8.10.1002/9781118476406.emoe077. [CrossRef]
57. Wüest, A.; Lorke, A. Small scale hydrodynamics in lakes. *Annu. Rev. Fluid Mech.* **2003**, *35*, 373–412. [CrossRef]
58. Hofmann, H.; Lorke, A.; Peeters, F. The relative importance of wind and ship waves in the littoral zone of a large lake. *Limnol. Oceanogr.* **2008**, *53*, 368–380. [CrossRef]
59. Amadori, M.; Piccolroaz, S.; Giovannini, L.; Zardi, D.; Toffolon, M. Wind variability and Earth's rotation as drivers of transport in a deep, elongated subalpine lake: The case of Lake Garda. *J. Limnol.* **2018**, *77*, 505–521. [CrossRef]
60. Henderson-Sellers, B. The dependence of surface velocity in water bodies on wind velocity and latitude. *Appl. Math. Model.* **1988**, *12*, 202–203. [CrossRef]
61. Johannessen, J.A.; Chapron, B.; Collard, F.; Kudryavtsev, V.; Mouche, A.; Akimov, D.; Dagestad, K.F. Direct ocean surface velocity measurements from space: Improved quantitative interpretation of Envisat ASAR observations. *Geophys. Res. Lett.* **2008**, *35*. doi:10.1029/2008GL035709 [CrossRef]
62. Alpers, W.; Hennings, I. A theory of the imaging mechanism of underwater bottom topography by real and synthetic aperture radar. *J. Geophys. Res. Oceans* **1984**, *89*, 10529–10546. [CrossRef]
63. Wright, J. A new model for sea clutter. *IEEE Trans. Antennas Propag.* **1968**, *16*, 217–223. [CrossRef]
64. Lyzenga, D.R. Effects of intermediate-scale waves on radar signatures of ocean fronts and internal waves. *J. Geophys. Res. Oceans* **1998**, *103*, 18759–18768. [CrossRef]

Article

CDOM Optical Properties and DOC Content in the Largest Mixing Zones of the Siberian Shelf Seas

Anastasia N. Drozdova [1,*], Andrey A. Nedospasov [1], Nikolay V. Lobus [2], Svetlana V. Patsaeva [3] and Sergey A. Shchuka [1]

1 Shirshov Institute of Oceanology, Russian Academy of Sciences, 117997 Moscow, Russia; nedospasov.aa@ocean.ru (A.A.N.); shchuka@ocean.ru (S.A.S.)
2 K.A. Timiryazev Institute of Plant Physiology, Russian Academy of Sciences, 127276 Moscow, Russia; lobus.nv@ocean.ru
3 Department of Physics, Lomonosov Moscow State University, 119234 Moscow, Russia; patsaeva@physics.msu.ru
* Correspondence: adrozdova@ocean.ru

Abstract: Notable changes in the Arctic ecosystem driven by increased atmospheric temperature and ice cover reduction were observed in the last decades. Ongoing environmental shifts affect freshwater discharge to the Arctic Ocean, and alter Arctic land-ocean fluxes. The monitoring of DOC distribution and CDOM optical properties is of great interest both from the point of view of validation of remote sensing models, and for studying organic carbon transformation and dynamics. In this study we report the DOC concentrations and CDOM optical characteristics in the mixing zones of the Ob, Yenisei, Khatanga, Lena, Kolyma, and Indigirka rivers. Water sampling was performed in August–October 2015 and 2017. The DOC was determined by high-temperature combustion, and absorption coefficients and spectroscopic indices were calculated using the seawater absorbance obtained with spectrophotometric measurements. Kara and Laptev mixing zones were characterized by conservative DOC behavior, while the East Siberian sea waters showed nonconservative DOC distribution. Dominant DOM sources are discussed. The absorption coefficient a_{CDOM} (350) in the East Siberian Sea was two-fold lower compared to Kara and Laptev seawaters. For the first time we report the DOC content in the Khatanga River of 802.6 μM based on the DOC in the Khatanga estuary.

Keywords: CDOM absorbance; spectroscopic indices; DOC; Arctic; shelf seas; estuarial and coastal areas

Citation: Drozdova, A.N.; Nedospasov, A.A.; Lobus, N.V.; Patsaeva, S.V.; Shchuka, S.A. CDOM Optical Properties and DOC Content in the Largest Mixing Zones of the Siberian Shelf Seas. *Remote Sens.* **2021**, *13*, 1145. https://doi.org/10.3390/rs13061145

Academic Editor: Giacomo De Carolis

Received: 26 January 2021
Accepted: 14 March 2021
Published: 17 March 2021

1. Introduction

The oceanic dissolved organic matter (DOM) pool is one of Earth's large organic carbon reservoirs [1], and, therefore, represents an important component of marine ecosystems and the carbon cycle [2]. The main sources of DOM in the world ocean are primary production of phytoplankton and ice algae, as well as the secondary production of zooplankton. Additional DOM sources include more refractory, compared to the autochthonous material, terrestrial-derived DOM supplied by river runoff (more than 80% of terrigenous DOM), aeolian dust, and coastal abrasion [3].

The Arctic Ocean represents a unique ecosystem, which, on the one hand, is highly sensitive to climate changes occurring during previous decades, and on another, is an important feedback component of the global climate system [4]. An essential feature of the Arctic region is its exposure to large river discharge. Representing only 1% of the global ocean volume, the Arctic Ocean receives more than 10% of the global freshwater and, therefore, the vast amounts of riverine DOM [5,6]. This is a key factor in regulating biogeochemical cycles in the area. In estuarine and coastal zones, fresh waters and terrestrial material control the distribution of flora and fauna, their productivity, and consumption [7]. Here, the limitation of photosynthetic activity due to humic substances, absorbing sunlight in the blue spectral range, where chlorophyll and photosynthetic carotenoids have absorption maxima [8], is the most significant. Conversely, humic substances absorb in the UVB (280–320 nm) and/or UVA (320–400 nm) ranges [9], diminishing the negative effects of ultraviolet radiation on plankton populations [10]. High concentrations of terrigenous colored DOM (CDOM) rich in humic substances have a significant impact on penetration of light into the water column [11], water color, and its spectral features (see [12] and references therein). Thus, humic substances providing an essential component of the remotely sensed optical signal, affect estimates of chlorophyll concentration using satellite imagery [13–15] or shipborne lidar measurements [16,17].

Numerous studies have contributed to understanding biogeochemical cycling of organic carbon in the Arctic Ocean, see, for example, the PANGAEA database (https://www.pangaea.de/, accessed on 15 March 2021), the monographs and reviews [18–21], etc. At the same time, seasonal variability of DOC distribution is not quite clear: significant changes, often comparable with multiyear variations, may occur on time scales of a few weeks [19], while most sampling expeditions are restricted to a few months during summer. Thus, for example, Dai et al. [22] estimated the uncertainty of global river DOC discharge to the coastal seas as 30%. Climate change in the Arctic [23–26] has already resulted in reduced ice cover and increased flows of terrigenous DOM due to permafrost thawing and coastal erosion. These changes require monitoring of the concentration and quality of DOM [27] to better understand the Arctic ecosystem and its response to changing conditions and anthropogenic stress. The development of forecasting models for predicting river export of DOM also involves the acquisition of new field data. Thus, a great need for additional DOC data, which can only be obtained in the field, was emphasized by Harrison et al. [28]. Taking into account the recent progress in development of quantitative CDOM and DOC determination with the use of satellite remote sensing [29,30], new field studies of DOC and optical properties of natural waters might be useful for validation of remote sensing models, as well as regional algorithms for estimation of absorption coefficient of colored organic matter [31].

In this paper, we present new data on the content of DOC in the Kara, Laptev, and East Siberian seas obtained during expeditions in summer and fall of 2015 and 2017. We analyze the data on the optical characteristics of the colored DOM fraction [32] to reveal its quality and the sources of DOM input. For the first time, we present DOM concentration at the section from the mixing zone of the Khatanga River to the continental slope obtained with the high temperature combustion technique, which allowed us to estimate the content of DOC in the Khatanga River. In previous studies (see, for example, the paper of Wheeler at al. [33]), the content of DOC in the Khatanga River was evaluated on the basis of total organic carbon concentration (TOC) and the ratio DOC/TOC = 0.9, typical for the Russian rivers [34].

2. Materials and Methods

2.1. Sampling

Samples were collected during the 63rd (28 August–6 October 2015) and 69th (22 August–26 September 2017) cruises aboard the R/V *Akademik Mstislav Keldysh* (Figure 1). Exact sampling dates for each sample are given in Supplementary Table S1. In 2015, the sampling was performed along a cross-slope transect in the Laptev Sea, starting from the Lena River Delta region (station 5216) and moving along the 130° E towards the continental slope. Absorbance was measured spectrophotometrically for five surface water samples, and DOC concentration was determined in 36 water samples from different depths. In 2017, 141 samples for DOC and 135 ones for absorbance were taken in the Kara, Laptev and East Siberian seas. We consider four shelf-crossing transects from the mixing zones of the Lena, Khatanga, Indigirka, and Kolyma rivers, as well as individual sampling sites in the Kara and Laptev seas. In the Kara Sea, the samples were collected in the estuarine zone of the Ob and Yenisei rivers (station 5588), in the central part of the Kara Sea to the north of Novaya Zemlya Trough (station 5587), and in the Blagopoluchiya Bay (the entrance—5641_2 and inner part of the bay water area—5642). One more station (5586) was located close to Novaya Zemlya Trough. In previous years, this region was characterized as less affected by Ob and Yenisei riverine waters [35].

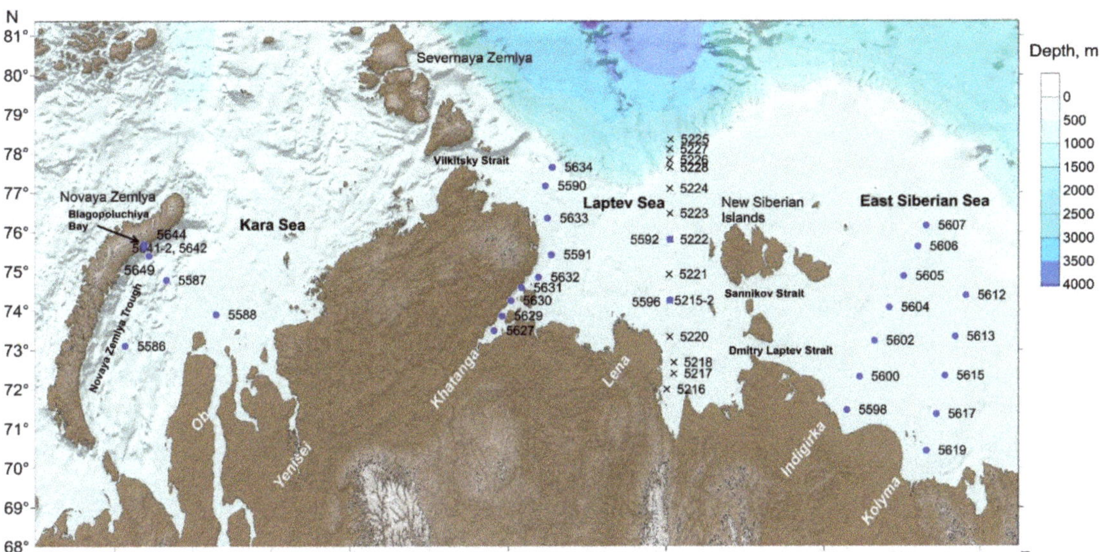

Figure 1. Sample site locations during the 63rd (black crosses) and 69th (blue dots) cruises of the R/V *Akademik Mstislav Keldysh*. The data from the study of Amante et al. [36] were used for the map plotting.

Water samples were taken using Niskin bottles of 5 L volume mounted on the CTD/rosette system at surface and discrete depths, associated with boundaries of large gradients of hydro-physical parameters. All the samples were filtered through precombustion at 450 °C Whatman GF/F filters with a nominal pore size of 0.7 µm. The filtrate was collected into the acid-cleaned 10 mL glass vials and stored under dark conditions at 4 °C until further analysis. For the DOC measurements, the filtrate was acidified up to pH = 2 before storage. Water temperature and practical salinity were derived from CTD measurements.

2.2. Flow Measurements

During the 69th cruise, we performed continuous observations of conductivity and temperature of subsurface water layer at 2.7 m depth. The measurements were carried out with a temporal resolution of 3 s (at a speed of 10 knots—15 m) using a SBE 21 SeaCAT Thermosalinograph (Sea-Bird Scientific) and a pump to supply outboard water.

2.3. DOC

DOC concentration was measured onshore by high-temperature combustion with a Shimadzu TOC-VCPH/CPN analyzer. Precision and accuracy of our measurements were determined relative to external laboratory standards, namely solutions of potassium hydrogen phthalate and sodium hydrogen carbonate diluted to different concentrations according to estimated DOC content, and amounted to ±5% and 1%, respectively.

2.4. Optical Measurements and Spectroscopic Indices

Absorbance A(λ) of water samples have been registered in the laboratory conditions at room temperature 22 ± 2 °C. Measurements were performed within the spectral range from 200–700 nm at 1 nm increments using double-beam scanning spectrophotometer Solar PB2201 with 3 or 5 cm quartz cuvettes depending on the DOC concentration and

Milli-Q water as a blank. CDOM absorbance spectra are available online as supplementary information to the study of Drozdova et al. [32]. The blank-corrected absorbance spectra were converted into the Napierian absorption coefficients $a_{CDOM}(\lambda)$ by multiplying CDOM absorbance by 2.303 and dividing by the cuvette path length taken in meters. In accordance with the study of Helms et al. [37], the spectral slope for the 275–295 nm range ($S_{275-295}$) and the ratio of $S_{275-295}$ and $S_{350-400}$ (S_R) were determined using linear regression of the log-transformed functions of absorption coefficients $a_{CDOM}(\lambda)$ defined as:

$$a_{CDOM}(\lambda) = a_{CDOM}(\lambda_0)e^{-S(\lambda-\lambda_0)}, \quad (1)$$

where λ_0 is a reference wavelength [38]. S_R index was reported by Helms et al. [37] to correlate with molecular weight and the degree of photochemical degradation of CDOM. Thus, S_R <1 is typical for terrestrial CDOM, while S_R > 1.5 indicate the presence of oceanic and photodegraded terrestrial CDOM, see also [12,39]. The value of $S_{275-295}$ is also used for the CDOM source discrimination so that $S_{275-295}$ > 20 is typical for marine waters, while $S_{275-295}$ < 16 is a characteristic of riverine waters. A specific UV absorbance (SUVA) was calculated by normalizing the decadic absorption at 254 nm to the DOC concentration in milligrams per liter (mg/L). It was shown to be a useful parameter for estimating the dissolved aromatic carbon content in aquatic systems [40]. The value of SUVA < 1.8 is an evidence of algae and bacteria CDOM predominance, and SUVA >3 suggests the terrestrial CDOM origin. Spectroscopic indices are given in our recent studies [32,41].

3. Results

3.1. Subsurface Water Temperature and Salinity

At the end of August 2017, an influx of warm and salty waters that originated from the Barents Sea, was recorded in the Kara Sea (Figure 2). The eastern type of freshened water distribution [42] was observed in the Kara Sea at the beginning of the cruise. It is characterized by the transfer of low-salinity waters, formed under the influence of Ob and Yenisei river runoff, along the coast towards the Wilkitsky Strait. To the east of 94° E and up to the Laptev Sea, the drifting ice was constantly met, therefore, the thermosalinograph was switched off. On the way back, ice did not occur, but in the northern part (north of 76° N), the subsurface temperature was still below zero. Freshened waters were shifted westward a month later, which is typical for the central type of freshened water distribution.

In the northwestern part of the Laptev Sea, to the east of the Vilkitsky Strait, surface water temperature was below zero (Figure 2). The highest values of about 5 °C were observed for the freshened waters formed under the influence of the Lena River discharge. On the way back, the temperature and salinity distribution in the central and eastern parts of the Laptev Sea did not change significantly. In the northeast, the temperature dropped by 1–1.5 °C and negative water temperature values were observed between the station 5634 and the Vilkitsky Strait.

According to the thermosalinograph data, the water temperature of the subsurface layer in the East Siberian Sea decreased from south to north (Figure 2). Negative values were observed near the ice massif, the edge of which was located above the 70 m isobath. Water freshening was observed in the shelf region. At the transect from the Kolyma River estuary to the continental slope, a pronounced frontal zone to the south of station 5615 was observed. Salinity did not change significantly to the north, while to the south it dropped sharply from 28 to 17 within 125 km distance. On the contrary, no large horizontal gradients were observed in the Indigirka section. Similar to the Laptev Sea, the distribution of river waters reaches an isobath of 25 m.

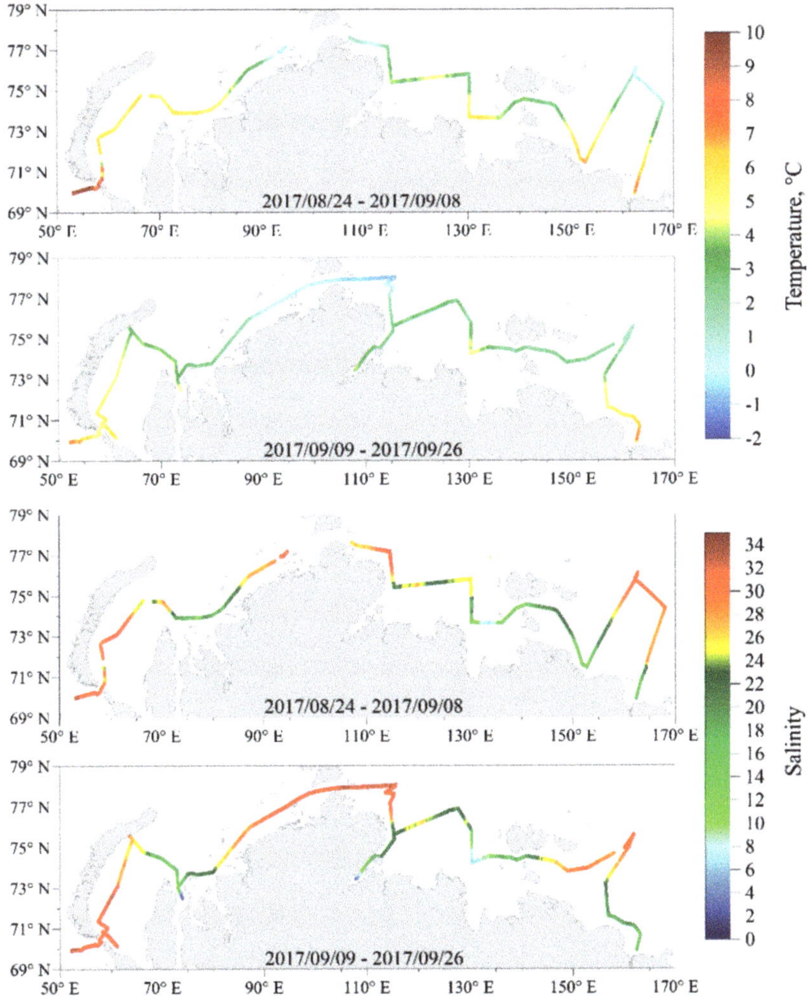

Figure 2. Water temperature and salinity at the depth of 2.7 m (thermosalinograph data) along the route of the 69th cruise of the R/V *Akademik Mstislav Keldysh* in the Kara, Laptev and East-Siberian seas.

3.2. Vertical Sensing of Hydrophysical Parameters

3.2.1. Kara Sea

In the first part of the cruise, the surface waters of the Kara Sea warmed up to 2.5–5.5 °C (Figure 3). With the exception of the stations located in the Blagopoluchiya Bay (5642, 5641, and 5644), a pronounced cold intermediate layer (CIL), formed during winter convection, was observed.

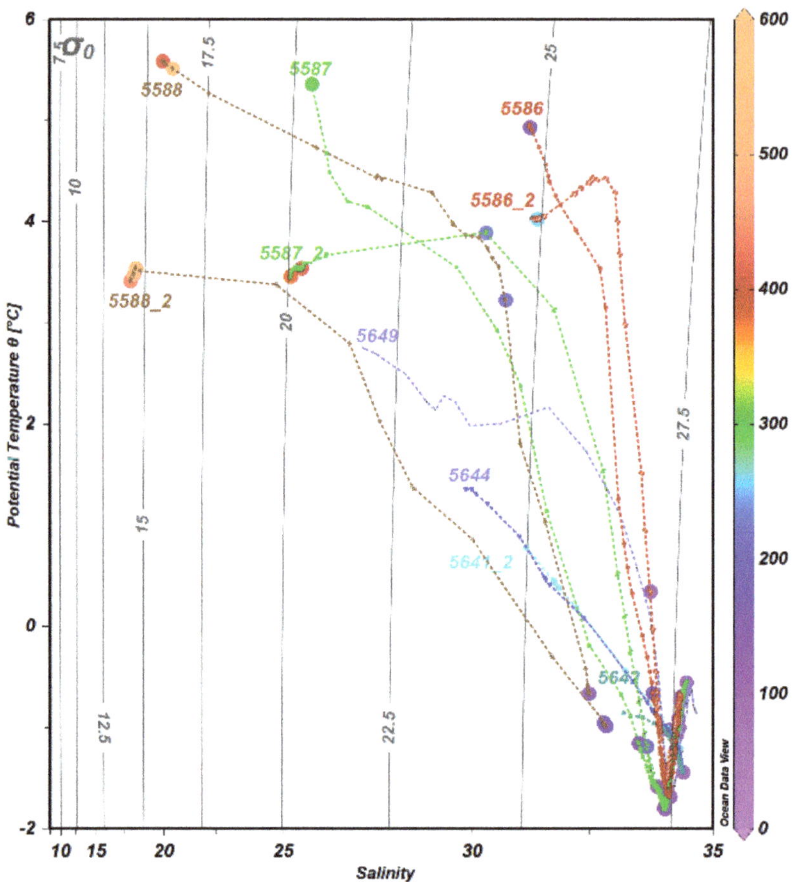

Figure 3. Temperature—salinity diagram for CTD profiles observed in the Kara Sea in Autumn, 2017. Color coding indicates DOC concentration (μM).

The core of CIL was located at depths of about 50 m. This agrees well with the data reported by Pavlov and Pfirman [42], who states that summer warming extends to depths of 20–60 m. The upper heated layer thickness in different areas varied from 5 m at station 5641 to 25 m at the Ob-Yenisei coast (5588), and was absent completely in the inner part of Blagopoluchiya Bay (5642). It was reported previously that the depth of the surface layer in the Kara Sea was 6–8 m in shallow parts of the sea, ranging up to 20–30 m over deeper regions [42]. Under the CIL core, the temperature profiles were quite different. The increase of temperature with depth in the south (5586) is much sharper than the one in the northern part of the Novaya Zemlya basin (5587, 5649). In the Novaya Zemlya Trough at depths above 120 m, we suggest the active admixing of waters, which flow from the Barents Sea along the St. Anna Trough. A month later (stations 5586_2, 5587_2, and 5588_2), the surface water temperature dropped by 1–2 °C. The remaining profiles have changed insignificantly (Figure 3). Salinity in the studied areas of the Kara Sea varied from 20, measured for the Ob-Yenisei coast surface waters, to 34.6 (deep waters). Below 50 m, all profiles are similar, except for the station 5587, which is characterized by fresher deep waters. High salinity >31 was measured in the Bay and at the southern station 5586, where the waters of the Ob and Yenisei did not extend. The influence of Ob and/or Yenisei riverine waters was reported in 2015 in the Oga and Tsivol'ki Bays of the Severny Island of Novaya Zemlya [35,43], when the western distribution of low-salinity waters [44] was observed. The temperature in the

CIL core at station 5587 was −1.8 °C. This roughly corresponds to the freezing point of water with a salinity of 32.8. Taking into account that the formation of CIL is related to vertical mixing during the winter period, we suggest that the salinity in the upper layer of the Kara Sea in winter is quite high—above 32. The waters with lower salinities distributed in the Kara Sea in the summer and fall periods is the result of current-year freshening caused by the Ob and Yenisei rivers, the Baydaratskaya Bay, and melt waters. In the study of Olsson and Anderson [45], such a threshold salinity value in case of the Siberian shelf seas was reported to be 24.

3.2.2. Laptev Sea

Along the transects in the Lena Delta region, water was warmed up to 10–15 m (Figure 4). Closer to the coastline, the temperature of the low-salinity water layer was 5 °C and decreased to 1–2 °C away from the Lena Delta region. The bottom water layer was characterized by temperatures below zero. The mixing of riverine and sea waters occurred throughout the entire water column of 20 m depth in the southern part of the sections. Compared to 2015, the spreading of freshened waters along the transect was weaker. In 2015, the salinity value of 25 was observed 200 km farther to the north. The waters in the area of the shallow shelf were freshened up to the 20 m depth. A frontal zone was observed above the sloping shelf brow at about 20 m depth, were the waters with salinity of 15 or less were distributed. In opposite, salinity of 15 was measured 50 km southerly in 2015 (Figure 4). We assume that a more active horizontal mixing of sea and river waters took place in 2015. Consequently, low-salinity waters spread far to the north, but at the same time low salinity values were observed closer to the Lena Delta. The halocline was located at depths of 10–15 m along the entire transect, and isohaline concentration within the frontal zone increased above the brow of the slope. This result does not completely support a schematic diagram of the interactions occurring in the Laptev Sea continental shelf close to the Lena River delta region reported by Gonçalves-Araujo et al. [46], since it implied reducing the thickness of plume-influenced freshened water layer, but agrees well with the data on salinity and temperature vertical distribution along the 130° E transect discussed by Bauch et al. [47]. A small lens of fresher and warmer waters was observed at station 5595. A similar lens was noted in 2015, but it was larger and located further from the coast. It very likely was formed due to the "offshore" atmospheric forcing [47].

The vertical salinity structure and distribution of freshwater fraction at the transect from the Khatanga River estuary to the continental slope are discussed in detail by Osadchiev et al. [48]. Shortly, the Khatanga plume was weakly-stratified and occupied the whole water column in the shallow inner part of the estuary (stations 5627–2629) due to intense tidal mixing in the Khatanga Gulf. Tidal-induced dilution caused an increase of surface salinity and depth of the plume from 4–7 m (station 5628) to 17–25 m (station 5630) at 120 km along the transect. In the outer part of the estuary, the plume detached from sea bottom and its depth steadily decreased to 11 m, while surface salinity increased to 21 (station 5632). Penetration of marine waters (with salinity above 30) into the bay was observed up to Bolshoy Begichev Island. Closer to the mouth in the bottom layer, salinity gradually decreased from 25 to 7 (Figure 5). The impact of the Khatanga River fresh waters decreased sharply not far from the entrance to the open sea (station 5633) due to primarily moderate Khatanga River flow, which is about five times smaller than that of the Lena River [49]. The boundary of freshened water lies above the 25 m isobath. The surface water temperature varied from 3 °C in the Khatanga River estuary to −1 °C near the continental slope (station 5635). Deep waters (>30 m) were characterized by temperatures below zero (Figure 5).

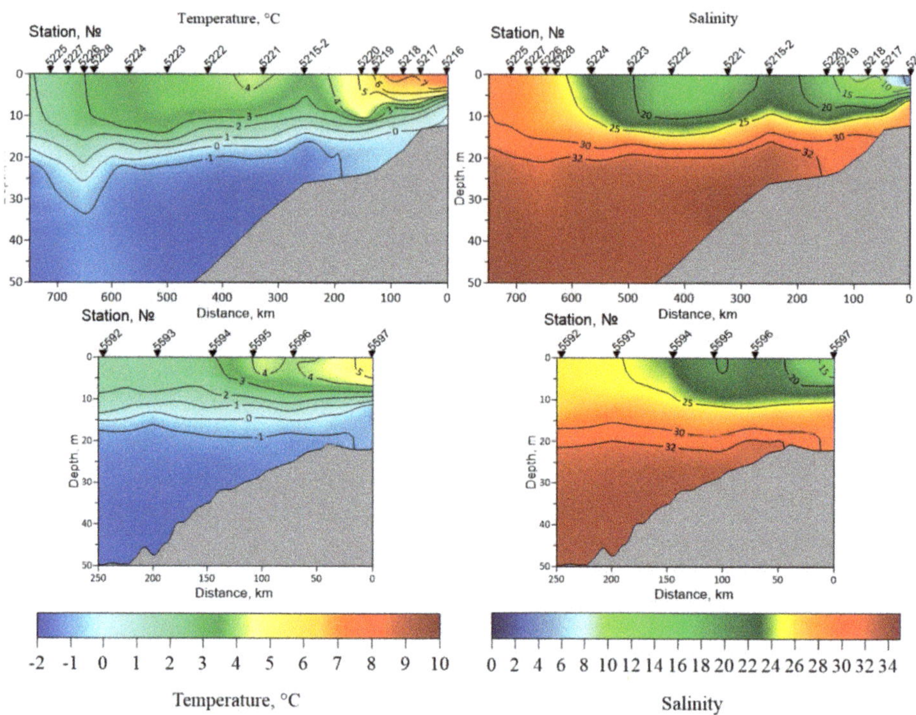

Figure 4. Temperature and salinity across the mixing zones of the Lena River in fall 2015 (**upper**) and 2017 (**lower**).

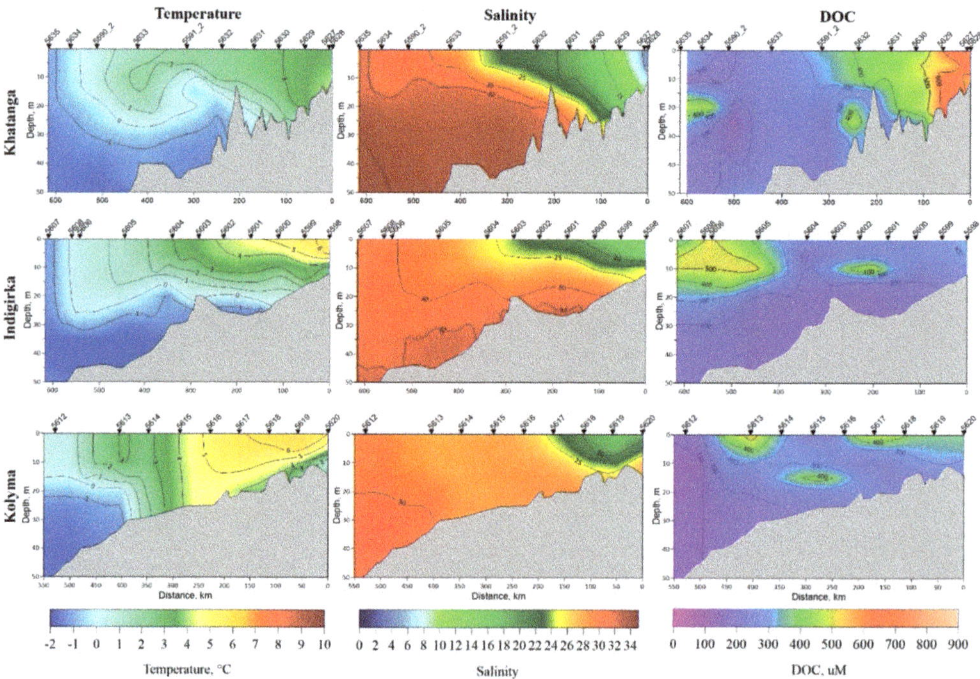

Figure 5. Distribution of temperature, salinity, and DOC along the Khatanga, Indigirka, and Kolyma transects.

3.2.3. East Siberian Sea

In the East Siberian Sea, surface water temperature decreased from south to north (Figure 5). At the section from the Indigirka River to the continental slope, the surface water temperature near the river mouth reached 6.2 °C and gradually decreased to −1.4 °C at station 5607 located near the ice edge. Negative temperatures of bottom water were observed on the inner shelf (station 5601) at the depth of 26 m. Low-salinity waters supplied by the river runoff extend to a depth of 10–15 m. At the transect from Kolyma River estuarine region to the continental slope, the surface water temperature on the inner shelf was 6.7 °C (stations 5619 and 5620) and decreased to 0.5 at the northern station 5612. In most of the Kolyma section, the total temperature difference in the water column did not exceed 1 degree. Salinity of surface waters increased from southwest to northeast (Figure 5). In the shelf area adjacent to the estuaries of the Kolyma and Indigirka, the minimum salinity was 17 and 15, respectively. The maximum salinity values were recorded at the northernmost stations 5607 (30 and 32.5 at the surface and bottom, respectively) and 5612 (29.2 and 31.2). The vertical distribution of salinity at the Indigirka section showed a pronounced freshened surface water layer, typical for the Arctic shelf under the influence of continental runoff. A unique feature of the eastern part of the East Siberian Sea (Kolyma transect) was a region of nearly 150 km of practically homogenous vertically mixed water column. This area may provide vertical convection down to the bottom during autumn water cooling.

3.3. *DOC*

The concentration of DOC in seawater of Arctic seas varied in a wide range between 82.8 and 886.7 µM. The complete DOC dataset is given in Supplementary Table S2. In the Kara and Laptev seas, the higher DOC concentrations were measured for the upper fresher water layer in the mixing zones formed under the influence of Ob, Yenisei, Lena, and Khatanga runoff (Figures 3 and 5, Tables 1 and 2). In the Kara Sea, the mean DOC content in the upper 25-m water layer was approximately two-fold higher compared to deep waters below 25 m. In the Blagopoluchiya Bay, the DOC was lower, compared with the data obtained recently for the Oga and Tsivolki Bays in the case of the western distribution of low-salinity riverine waters [35]. It confirms that terrestrial-derived DOM, supplied by Ob and Yenisei rivers, represents the main DOM source in the bays of Severny Island of Novaya Zemlya archipelago. At the transect from the Kolyma River mixing zone to the East Siberian Sea continental slope, DOC varied between 125.8 and 505.0 µM for the salinity rage 17.0–31.5. The Indigirka transect covered a larger salinity gradient from 15.2 to 33.4. The values of DOC varied there from 165.0 to 526.7 µM. The surface waters of the transect were characterized by moderate DOC concentrations of 236.7–393.3 µM with its local increase at the station 5606 up to 520.0 µM. In contrast to the Kara and Laptev seas, the linkage between DOC and hydrological parameters in the East Siberian Sea was not observed. DOC was distributed rather randomly, showing that significant DOC concentrations of 300–526.7 µM were typical for both Indigirka and Kolyma mixing zones and continental slope region. The DOC values measured in 2017 in the East Siberian Sea were higher than the ones reported by Alling et al. [50]. Thus, in 2008, the upper water layer of 15 m to the west of 160° E was characterized by mean DOC of 170 µM (mean salinity S = 22), and to the east of 160° E 93 µM (S = 28). In the present study, corresponding mean DOC concentrations were 285.1 µM (S = 26.0) and 168.1 µM (S = 29.7).

Table 1. Variation of DOC and optical characteristics of the Kara Sea waters.

	0–25 m	>25 m	Blagopoluchiya Bay
DOC (μM)	114.2–575.0	110.0–175.8	96.7–145.8
$a_{CDOM}(350)$ (m^{-1})	0.22–5.73	0.27–1.07	0.32–0.75
$a_{CDOM}(375)$ (m^{-1})	0.14–3.64	0.16–0.64	0.19–0.45
$S_{275–295}$ (μm^{-1})	17.78–26.46	22.79–29.32	16.65–18.63
S_R	0.98–3.08	0.92–1.08	n/a
SUVA (m^2gC^{-1})	0.48–2.30	0.39–0.79	0.65–0.76

Table 2. Variation of DOC and optical characteristics of the Laptev Sea and East Siberian Sea shelf waters (2017).

	0–10 m	>10 m
	Lena	
DOC (μM)	242.5–886.7	125.0–337.5
$a_{CDOM}(350)$ (m^{-1})	2.36–11.17	0.54–1.72
$a_{CDOM}(375)$ (m^{-1})	1.44–7.20	0.34–1.48
$S_{275–295}$ (μm^{-1})	16.38–20.35	19.53–23.35
S_R	0.91–1.15	1.12–2.08
SUVA (m^2 g C^{-1})	1.72–2.52	0.38–1.28
	Khatanga	
DOC (μM)	145.8–727.5	158.3–678.3
$a_{CDOM}(350)$ (m^{-1})	0.57–11.21	0.27–7.02
$a_{CDOM}(375)$ (m^{-1})	0.31–7.21	0.15–4.68
$S_{275–295}$ (μm^{-1})	15.81–24.84	14.14–25.24
S_R	0.92–1.39	0.97–2.39
SUVA (m^2 g C^{-1})	0.59–2.48	0.24–3.35
	Indigirka	
DOC (μM)	195.8–526.7	165.0–319.17
$a_{CDOM}(350)$ (m^{-1})	0.74–3.77	0.64–2.26
$a_{CDOM}(375)$ (m^{-1})	0.42–2.41	0.38–1.38
$S_{275–295}$ (μm^{-1})	18.73–24.45	19.17–23.81
S_R	1.00–1.38	0.92–1.97
SUVA (m^2 g C^{-1})	0.35–2.6	0.1–2.0
	Kolyma	
DOC (μM)	125.83–505	129.0–425.0
$a_{CDOM}(350)$ (m^{-1})	0.48–3.35	0.47–1.21
$a_{CDOM}(375)$ (m^{-1})	0.43–3.07	0.27–0.73
$S_{275–295}$ (μm^{-1})	19.36–26.21	21.39–25.84
S_R	0.94–2.07	0.91–2.20
SUVA (m^2 g C^{-1})	0.32–1.71	0.12–1.19

3.4. Optical indices

3.4.1. Kara Sea

CDOM concentrations, depicted as $a_{CDOM}(375)$ [51], followed similar trends to that observed for DOC. The surface of mostly freshened waters had the highest $a_{CDOM}(375)$ (up to 3.64 and the mean value 1.43 m^{-1}), while $a_{CDOM}(375)$ of Blagopoluchiya Bay and Kara Sea deep waters did not exceed 0.64 m^{-1} (Table 1).

Spectral slope ratio S_R varies between 0.9–3.4. It strongly correlates with $a_{CDOM}(375)$ showing exponential decrease to the values, typical for the Ob (~0.87) and Yenisei (~0.91) freshwaters reported by Stedmon et al. [26], see also Discussion Section. Similar dependence was obtained for the $S_{275–295}$ spectral slope, indicating the predominance of terrestrial material in a fewer number of samples ($S_{275–295} < 20$) and mixed or mostly autochthonous DOM character for the others. Interestingly, the data from Blagopoluchiya Bay are grouped separately in the $S_{275–295}$—$a_{CDOM}(375)$ plot and differ by the lower $S_{275–295}$ values (Figure 6). It apparently reflects local input of terrestrial-derived material from Novaya Zemlya island.

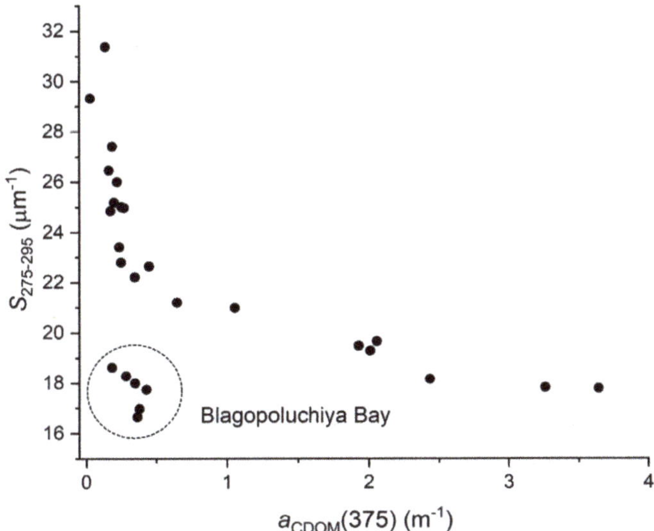

Figure 6. Spectral slope $S_{275-295}$ plotted against $a_{CDOM}(375)$ for the Kara Sea waters.

SUVA varied between 0.39 and 2.30 m^2 g C^{-1} with the highest values observed for the upper water layer. Since SUVA has been shown to be positively correlated to molecular weight [52] and also to be one of the most reliable parameters for the DOM source discrimination with regard to its changes during bio- and photodegradation [53], we suggest the upper water layer DOM to have higher molecular weight, which is explained by a larger fraction of humic acids [54] supplied by Ob and Yenisei rivers. SUVA values of the Kara Sea waters were found to be lower than the ones of the Ob (2.58–2.75 m^2 g C^{-1}) and Yenisei (1.95–2.97 m^2 g C^{-1}) end members [26].

3.4.2. Laptev Sea

CDOM optical properties suggest the dominance of terrigenous humic substances in surface and thermocline waters (~10 m layer) at the transect along the 130° E (see Table 2). Absorption coefficient $a_{CDOM}(375)$ varied between 0.34 and 7.20 m^{-1}. The lowest absorption coefficient $a_{CDOM}(350)$ was measured for the bottom waters from the station 5592 and amounted to 0.5 m^{-1}, which is close to the ones reported for the Polar Waters of the East Greenland Current [55,56]. In the study of Pugach et al. [57] a comparable spatial variability of $a_{CDOM}(350)$ was demonstrated. The maximal $a_{CDOM}(350)$ of 11.2 measured for the surface waters with salinity 6.6 (station 5596_2) was slightly lower than the one reported for a mid-flow regime of the Lena River of about 13.1 m^{-1} [58]. The values of $a_{CDOM}(350)$ were found to be lower compared to the data reported by Gonçalves-Araujo et al. for the Lena Delta region [46] ($0.9 < a_{CDOM}(350) < 15.7$ m^{-1}), which is likely explained by a smaller terrestrial CDOM contribution. Spectral slope $S_{275-295}$ varied between 16.38–20.35 μm^{-1}. Spectral slope ratios S_R obtained in the present study ($0.9 < S_R < 1.2$) are generally consistent with the results of Pugach et al. [57] and Gonçalves-Araujo et al. [46] (2015) (~0.87–1.00) for the Lena Delta—sea mixing zone. For comparison, S_R values of the Lena River water samples were reported to vary between seasons and estimated by Stedmon et al. [26] as 0.81–0.89. Higher S_R values were typical for deep waters at depths 15–44 m, see Table 2. Values of SUVA of the upper 10 m water layer fall in the range 1.72 < SUVA < 2.52 m^2 g C^{-1}, which is comparable with the results 1.33 < SUVA < 4.80 m^2 g C^{-1} reported by Gonçalves-Araujo et al. [46] for salinities of 0.90–32.63. For deep waters with salinities 30.1–33.9, SUVA varied from 0.38 to 1.28 m^2 g C^{-1}, indicating a lower impact of terrestrial-derived DOM.

The Khatanga discharge experienced intense estuarine tidal mixing and therefore was distributed from surface to the bottom in the inner estuary and over the 20–25 m deep water column in the outer estuary, see Section 3.2.2 and the study of Osadchiev et al. [48]. The difference in DOC concentration and CDOM absorption between the upper and deep waters was, therefore, not as pronounced as for the Kara Sea and the Lena Delta region Tables 1 and 2. The data for salinities above and below 25 are summarized in Supplementary Table S4 for convenience. At the beginning of the transect (stations 5627, 5629 and 5630), the entire water column was characterized by maximal along the transect values of absorption ($a_{CDOM}(375)$ was 2.29–7.10 m^{-1}) and specific absorbance SUVA (1.44–2.48 m^2 g C^{-1}), while $S_{275–295}$ (15.73–16.40 μm^{-1}) and S_R (0.92–1.09) were low. This indicates the predominance of terrigenous CDOM in this location [12,37]. Further north (stations 5631–5633), the contribution of terrigenous CDOM decreases and becomes significant in the upper 10–20 m water layer only (Figure 7). In contrast, deep waters had lower absorption and SUVA (0.32–0.93, mean 0.75 m^2 g C^{-1}). The increase of spectral slope $S_{275–295}$ was observed for the bottom waters. At the northernmost stations of the section, the waters of different optical characteristics were observed at depths 10–40 m. They were characterized as being higher compared to oceanic waters [12] absorption at 375 nm ($a_{CDOM}(375)$ was 3.0 m^{-1}) and $S_{275–295}$ values typical for estuarine and coastal waters with strong humic character (14.1–18.5 μm^{-1}). At the same time, high salinities and spectral slope ratio S_R varying from 2.23–2.39 clearly indicates the presence of oceanic and/or photodegraded terrestrial CDOM. We, therefore, assume that increased absorption at 375 nm is related to the recently produced CDOM [59,60].

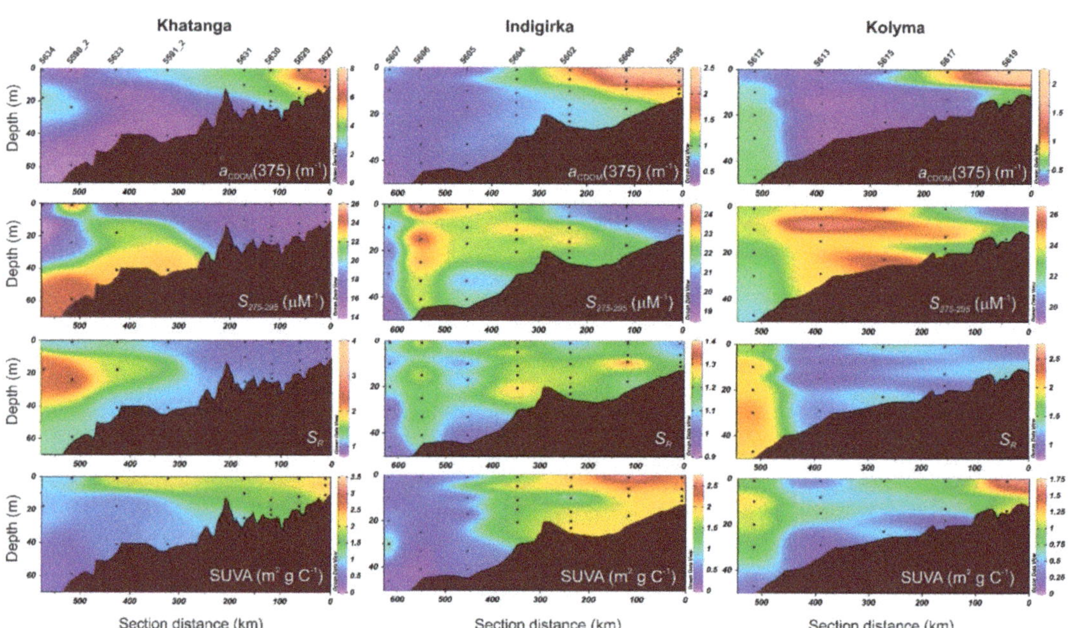

Figure 7. Distribution of $a_{CDOM}(375)$, $S_{275–295}$, S_R and SUVA along the Khatanga, Indigirka, and Kolyma transects.

3.4.3. East Siberian Sea

At the beginning of the transect from the Indigirka River mixing zone to the continental slope, the values of $a_{CDOM}(375)$ were typical for estuaries and coastal waters [12] and accounted for 0.7–2.4 m^{-1}. The waters farthest from the coast (stations 5605–5607) exhibited lower $a_{CDOM}(375)$ values of 0.38–0.56 m^{-1}, which is a characteristic of oceanic waters [59]. Spectral slope $S_{275–295}$ and S_R were distributed along the transect rather irregularly (Figure 7), resulting in similar mean values for the upper 10 m water layer and deep waters. Spectral slope ratio S_R varied between 0.9 and 1.4, indicating that CDOM had a weak humic character and intermediate or strong autochthonous component. The mean values of the specific UV absorbance were 1.45 m^2 g C^{-1} for the upper water layer and 1.0 m^2 g C^{-1} for the deep waters and testify the predominance of algae and bacterial CDOM. The surface waters of the inner plume (stations 5598–5602) exhibited SUVA up to 2.6 m^2 g C^{-1} caused by the presence of terrestrial-derived DOM supplied by the Indigirka River.

Pronounced plume was not seen from the Kolyma River. The impact of the Kolyma River waters was notable for over 150 km from the beginning of the transect (stations 2619 and 5617). Absorption at 375 nm was about 1.5 times lower compared to the Indigirka transect. The maximal $a_{CDOM}(375)$ values were measured for the inner plume (1–2.1 m^{-1}), as well as for the entire water column at the northernmost station 5612. Similar to the Khatanga transect, an increase in absorption at 375 nm, observed at salinities >29, was accompanied by a decrease in $S_{275–295}$ and by an increase in S_R, which indicates the recently produced CDOM. The values of $a_{CDOM}(350)$ and S_R obtained in the present study are consistent with the data reported by Pugach et al. [57]. The values of SUVA were below 1.8 m^2g C^{-1} for all the samples, showing little freshwater input.

4. Discussion

4.1. Conservative DOC Behavior in the Kara and Laptev Seas

The DOC versus salinity plot (Figure 8) for the Kara Sea showed that the data fell close to the DOC = 927.3 − 23.0 × *Salinity* regression line, characterized by the coefficient of determination R^2 = 0.91. No difference in conservative DOM distribution was found between surface (open circles) and depth profile (filled circles) samples. This agreed well with the observations reported by Stein et al. [7], suggesting similar vertical and horizontal mixing in the Ob and Yenisei estuaries. The conservative DOM behavior in the Kara Sea was demonstrated recently by several studies [19,35,61–63]. We summarized the coefficients a and b for the regression line DOC = a − b× *Salinity*, obtained during late summer and fall periods 1997–2017 in Supplementary Table S3.

In September 2015, the DOC at the transect along 130° E (Laptev Sea) was distributed conservatively. A negative correlation of DOC with respect to salinity is described by the following equation: DOC = 545.5 − 10.4× *Salinity*, R^2 = 0.74. In September 2017, four stations were examined for DOC (stations 5592 and 5596 on the way there and back, 5592_2 and 5596_2). Higher DOC values were observed and accounted to 887 µM at salinity 6.6. Assuming conservative DOM behavior, our estimates for the DOC in fresh water are 545.5 µM C and 1015.4 µM C in 2015 and 2017, respectively. These results are in a good agreement with the data reported previously (506–1252 µM C) [26,46,64–68]. The mean DOC concentrations for the upper and deep waters are consistent with the data given by Alling et al. [50].

At the transect from the Khatanga River to continental slope, a linear correlation DOC − salinity was described as DOC = 802.6 − 19.3 × *Salinity*, R^2 = 0.97, for salinities varying from 3.5–31.5 (Figure 8). Considering conservative DOC behavior within the Khatanga River mixing zone, our evaluation of DOC in the Khatanga River is 802 µM. It is almost twofold higher compared to the DOC value of 472 µM reported by Wheeler at al. [33]. This assessment was based on the mean TOC in the Khatanga River and typical for Russian rivers mean fraction of DOC/TOC = 0.9 [34]. According to the data on annual variation of DOC in six major Arctic rivers [26], DOC concentration in September can exceed the mean annual value no more than 10% (Ob and Kolyma rivers). For the Mackenzie, Yenisei, Yukon, and Lena rivers, the mean annual DOC was found to be even higher than DOC measured during the period from the end of August to the beginning of October. We, therefore, assume that the mean annual DOC concentration of 472 µM in the Khatanga River waters might be underestimated.

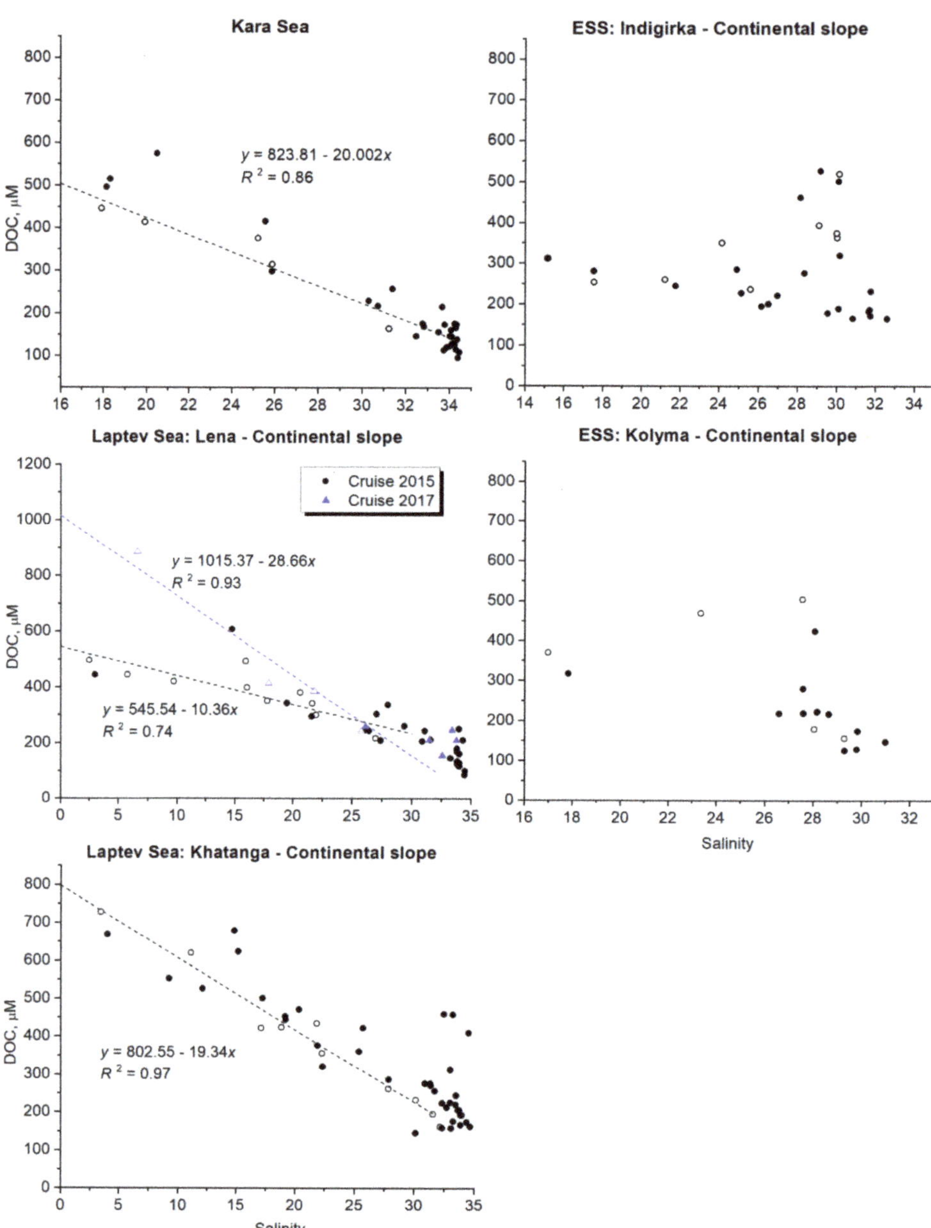

Figure 8. Distribution of DOC along the salinity gradient in the Kara, Laptev, and East Siberian seas. The data on surface and subsurface (1 m depth) waters are shown by open circles.

4.2. Nonconservative DOC Behavior in the East Siberian Sea

In the study of Alling et al. [50], nonconservative DOM behavior was revealed in the East Siberian Sea. It was related to the DOM removal, which explained the net DOC deficit. The field studies conducted in 2017 have also demonstrated nonconservative DOC distribution along the salinity gradient in the East Siberian Sea (Figure 8). While in the

Arctic region, the maximum DOC content is usually a characteristic of low-salinity waters of the mixing zones affected by river runoff, in this study the regions of the Arctic shelf remote from the estuaries and deltas showed DOC concentrations that were comparable with the ones observed at lower salinities. In order to identify the samples with a high DOC content, which was not caused by newly released terrestrial-derived material, we plotted DOC against absorption coefficient at 350 nm (Figure 9A). The substantial DOC concentrations of >300 μM and $a_{CDOM}(350) < 2$ m^{-1} were found at the farthest from the coast stations of the Indigirka section (5605–5607) as well as in the upper 15 m water layer throughout the entire Kolyma section (stations 5613, 5615 and 5617). The possible mechanisms of formation of high salinity waters exhibiting high DOC concentration is an autochthonous DOM production, which is one of the major DOM sources to the marine environment with limited continental influence [69,70]. This assumption is supported by the increase in the above areas the spectral slope $S_{275–295}$ (Figure 7). A similar increase in the DOC content, accompanied by an increase in absorption in the short-wavelength spectral range, was observed in the region of the continental slope in the Khatanga section (station 5634).

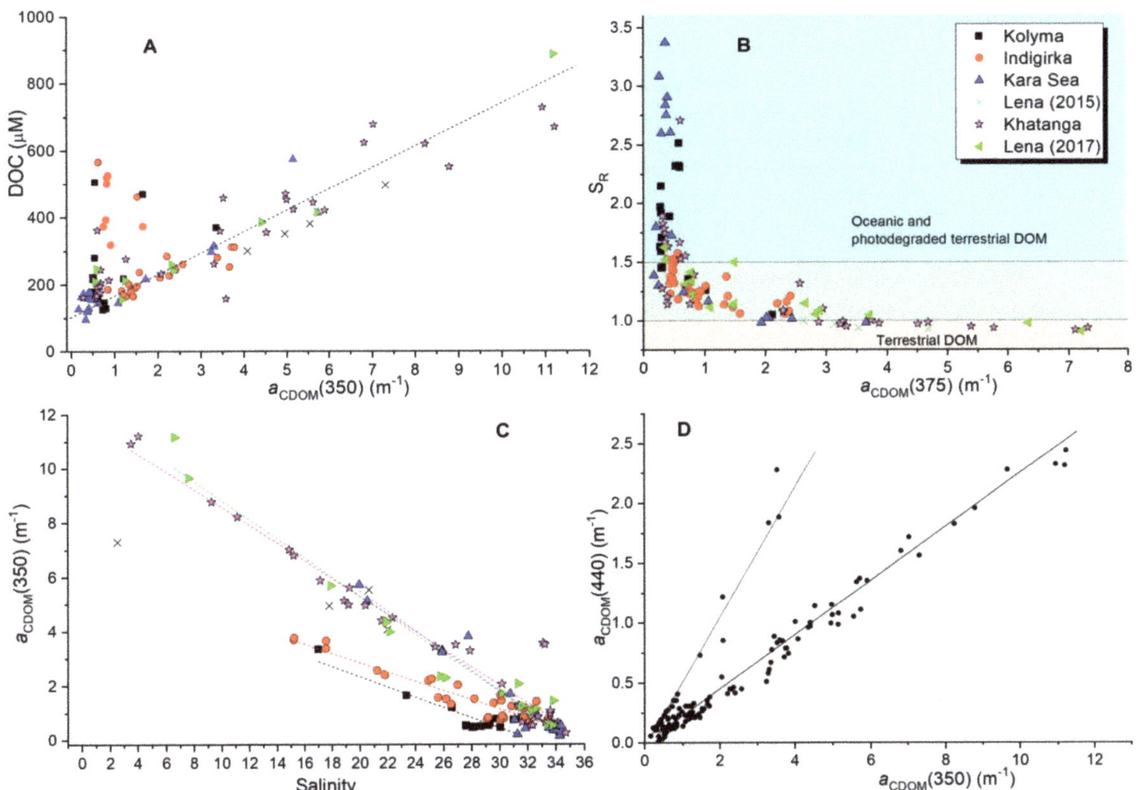

Figure 9. (**A**)—DOC against absorption coefficient at 350 nm for the studied samples; (**B**)—spectral slope ratio plotted against the absorption coefficient at 375 nm; (**C**)—$a_{CDOM}(350)$ against salinity for the individual mixing zones; (**D**)—$a_{CDOM}(440)$ plotted against $a_{CDOM}(350)$.

4.3. CDOM Sources

A criterion suggested by Helms et al. [37] for identifying the sources of CDOM shows that the dominant contribution of terrigenous OM ($S_R < 1$) is a characteristic of the mixing zones of the Khatanga and Lena rivers (salinities 3.5–21.5) (Figure 9B). S_R values varied between 0.91 and 1. For comparison, S_R values for river water samples were reported to vary between seasons and estimated by Stedmon et al. [26] as 0.82–0.92 (Kolyma), 0.81–0.89 (Lena), 0.83–0.92 (Ob), and 0.79–0.93 (Yenisei). CDOM of the Indigirka transect was of mixed autochthonous-allochthonous character, while the stations east of 160° E (Kolyma section) are distinguished by the presence of autochthonous CDOM in seawater. This is consistent with the results, published by Semiletov et al. [71], demonstrating that a significant component of freshwater from Siberian river inflows into the coastal East Siberian Sea, extending to approximately 160° E, where the long-term average position of the Pacific frontal zone is located.

The predominance of autochthonous CDOM was also demonstrated in the area of the Novaya Zemlya Trough, in the Blagopoluchiya Bay, and the northern part of the Khatanga transect. Expectedly, the decrease of the influence of the Lena, Khatanga, Indigirka, and Kolyma river runoff farther seaward along the studied transects was accompanied by an increase of salinity, $S_{275–295}$ and S_R, while the water temperature, CDOM absorption at 350 and 375 nm, and SUVA decreased. At the northern stations of the Khatanga and Kolyma sections, however, an increase of $a_{CDOM}(375)$ was observed and very likely was associated with the recently produced DOM [59].

4.4. CDOM Absorption at 350 nm and 440 nm

The absorption coefficient $a_{CDOM}(350)$ was repeatedly used earlier as a quantitative measure for CDOM concentrations, see for example [58], due to its ability to estimate lignin concentrations and inputs of terrestrial DOM to the Arctic Ocean [26,72]. While no correlation between DOC and salinity was found in the case of mixing zones of the Kolyma and Indigirka rivers, $a_{CDOM}(350)$ plotted against salinity (Figure 9C) showed good correlations described separately for each studied water area in Table 3. We suggest that Indigirka and Kolyma river waters were characterized by similar $a_{CDOM}(350)$ values of ~6.2 m^{-1}. This is about twofold smaller compared to the Khatanga, Lena, and Ob/Yenisei rivers (~12.3 m^{-1}). The obtained results are consistent with data on the Ob, Lena, and Kolyma rivers during mid-flow [58]. Lower $a_{CDOM}(350)$ values in the Kolyma River were explained by lower vascular plant inputs during freshet and its more extensive microbial degradation in the Kolyma watershed.

Table 3. Coefficients a and b of the $a_{CDOM}(350) = a + b \times Salinity$ regression line and corresponding coefficients of determination obtained for the Kara, Laptev, and East Siberian seas during August–September 2017.

	A	*B*	R^2
Kara Sea	12.9 ± 0.6	−0.365 ± 0.019	0.93
Lena	12.3 ± 0.5	−0.352 ± 0.017	0.96
Khatanga	11.8 ± 0.4	−0.327 ± 0.013	0.94
Indigirka	6.2 ± 0.3	−0.170 ± 0.011	0.87
Kolyma	6.1 ± 0.7	−0.188 ± 0.026	0.77

Another important optical characteristic of seawater is a CDOM absorption coefficient at 440 nm $a_{CDOM}(440)$. CDOM represents an essential constituent affecting ocean color. Thus, in the Arctic Ocean, the contribution of $a_{CDOM}(443)$ to the total non-water absorption can reach ~50% [73]. It was shown that systematic differences in chlorophyll retrievals resulting from different ocean color models are related to each model's ability to account for the absorption of light by CDOM [14]. In the present study, most of the $a_{CDOM}(440)$ data showed a good negative correlation with salinity, similar to the ones reported for the $a_{CDOM}(350)$ absorption coefficients (Figure 9C). Some of the water samples, however, were failed to be described by the linear dependence on salinity due to high absorption coefficients $a_{CDOM}(440)$; they were taken in the Blagopoluchiya Bay, at the station 5586 in the Kara Sea and northern parts of the Khatanga (stations 5633 surface waters, 5590_2

24 m, and 5634 surface waters and 18 m) and Kolyma (5612) transects. In the $a_{CDOM}(350)$ against $a_{CDOM}(440)$ plot (Figure 9D), such data points are grouped separately, they are characterized by $a_{CDOM}(350)$ below 4 m^{-1} and higher $a_{CDOM}(440)$ values. This group of points that are aligned on a regression line characterized by different slopes represents a group of CDOM absorption spectra characterized by a shallower slope coefficient. The increase of $a_{CDOM}(440)$ may be caused by autochthonous CDOM. Thus, in the study of the South Brazilian Bight, a strong correlation between Chl-a and $a_{CDOM}(440)$ was revealed and described with a regression line lying close to the one observed for the global pelagic oceans reported by Bricaud et al. [74]. The CDOM—Chl-a correlation allowed suggesting the presence of an autochthonous source of CDOM to the region driven mostly by the phytoplankton community over the shelf domain [75]. Our assumption is also supported by the study of phytoplankton of the Khatanga transect [76]. It was found that the area of the continental slope in the western Laptev Sea represents a specific local biotope. Phytoplankton in the area of the continental slope was characterized by high abundance and biomass, dominance of diatoms, and the formation of the deep maximum formed by actively growing algae. At the station 5633, maximum of phytoplankton biomass was observed in surface waters, while at station 5635 (about 40 km north from 5634), it was found at 45 m depth.

Obtained results on CDOM absorption can be valuable in remote sensing and modeling issues. For example, calculated from satellite data, CDOM absorption coefficients may be used as an effective indicator of the Kara Sea surface desalinated layer distribution and dynamics [31]. As the values of light absorption in this layer are significantly higher than in surrounding seawaters [56,77], its characteristics must be taken into account in heat budget models.

5. Conclusions

The complex field studies, conducted in fall 2015 and 2017, covered a large area of the eastern Arctic shelf of the Kara, Laptev and East Siberian seas. Analysis of DOC concentration and CDOM optical properties, supported by CTD data, allowed us to consider DOM distribution and its quality in the mixing zones of the Ob/Yenisei, Khatanga, Lena, Indigirka, and Kolyma rivers. It was demonstrated that the Kara and Laptev mixing zones were characterized by conservative DOC and $a_{CDOM}(350)$ behavior, while the East Siberian sea waters showed nonconservative DOC distribution. We provide the first estimates on DOC content, based on the high-temperature combustion technique, in the Khatanga River during mid-flow regime, it accounted for 802.6 μM (9.6 mg/L). Assuming conservative DOM behavior, our estimates for the DOC in fresh water are 545.5 μM C and 1015.4 μM C in 2015 and 2017, respectively, which is consistent with the results of previous studies. Despite the individual watershed characteristics of the rivers flowing into the eastern Arctic shelf seas, variation of the absorption at 350 nm along the salinity gradient was found to be similar for the Laptev and Kara seas. Absorption of the East Siberian Sea waters was found to be two-fold smaller, which is explained by lower CDOM content in the Indigirka and Kolyma rivers, as well as degradation of humic substances supplied by the Lena River during the transport to the East Siberian Sea through the Dmitry Laptev Strait.

Estuarine and delta regions were characterized by the predominance of terrestrial-derived DOM supplied by river runoff. The increase of DOC content was observed at the most distant from the shore stations in the area of the continental slope. It was frequently accompanied by growth of absorption at short-wave spectral range ($S_{275–295}$), S_R, and $a_{CDOM}(440)$, which indicates the production of autochthonous DOM by marine biota to be the dominant CDOM source at those locations. The literature overview also demonstrated the correlation between high DOC values and the increase of phyto- or zooplankton populations.

The OLCI ocean color scanners launched in February 2016 (Sentinel-3A) and in April 2018 (Sentinel-3B) should provide satellite data in the next decade. It was shown by Glukhovets et al. [31] that the standard OLCI algorithm for estimating the CDOM absorption coefficient ADG443_NN gives high errors in the Arctic seas. The dataset presented in this work may be used to improve existing standard and regional [78] algorithms and to create new ones.

Supplementary Materials: The following are available online at https://www.mdpi.com/2072-4292/13/6/1145/s1, Table S1: Sampling dates during the 63rd and 69th cruises of R/V *Akademik Mstislav Keldysh*, Table S2: Salinity and DOC concentration of water samples, Table S3: Coefficients *a* and *b* of the DOC = $a + b \times Salinity$ regression line and corresponding coefficients of determination obtained for the Kara Sea waters during August–September periods 1997–2017, Table S4: DOC and optical characteristics of the Khatanga transect waters (Laptev Sea).

Author Contributions: Conceptualization, A.N.D.; data curation, A.N.D., A.A.N., N.V.L. and S.A.S.; formal analysis, A.N.D.; funding acquisition, A.N.D.; investigation, A.N.D., A.A.N., N.V.L., S.V.P. and S.A.S.; supervision, A.N.D.; visualization, A.A.N. and A.N.D.; writing—original draft preparation, A.N.D. and A.A.N.; writing—review and editing, A.N.D. All authors have read and agreed to the published version of the manuscript.

Funding: The study of the Kara Sea waters was performed in the framework of the state assignment of IO RAS (theme 0128-2021-0016). The study of DOC along the Khatanga and Lena Delta transects was supported by the RFBR (project 18-05-60214). Investigation of CDOM of the East Siberian Sea was supported by the RSF Grant (project 18-77-00053).

Institutional Review Board Statement: Not applicable.

Informed Consent Statement: Not applicable.

Data Availability Statement: The data on DOC and salinity of are given in Supplementary Materials to this study. Spectroscopic indices and CDOM absorbance spectra are provided in [32].

Acknowledgments: The data are included as supporting information. The authors are grateful to Mikhail V. Flint for the opportunity to take part in the impressive complex expeditions to the Arctic shelf seas and the crew of the R/V *Akademik Mstislav Keldysh* for their contribution during field studies. We thank Dmitry I. Glukhovets, Alexander A. Osadchiev and Marina D. Kravchishina for fruitful discussion, as well as anonymous reviewers for their comments on an earlier version of this paper.

Conflicts of Interest: The authors declare no conflict of interest.

References

1. Hansell, D.A.; Carlson, C.A.; Repeta, D.J.; Schlitzer, R. Dissolved organic matter in the ocean: A controversy stimulates new insights. *Oceanography* **2009**, *22*, 202–211. [CrossRef]
2. Wagner, S.; Schubotz, F.; Kaiser, K.; Hallmann, C.; Waska, H.; Rossel, P.E.; Galy, V. Soothsaying DOM: A current perspective on the future of oceanic dissolved organic carbon. *Front. Mar. Sci.* **2020**, *7*, 341. [CrossRef]
3. Romankevich, E.A.; Vetrov, A.A.; Peresypkin, V.I. Organic matter of the World Ocean. *Russ. Geol. Geophys.* **2009**, *50*, 299–307. [CrossRef]
4. Stein, R.; Stein, R.; MacDonald, R.W. *The Organic Carbon Cycle in the Arctic Ocean*; Springer: Berlin, Germany, 2004.
5. Opsahl, S.; Benner, R.; Amon, R.M.W. Major flux of terrigenous dissolved organic matter through the Arctic Ocean. *Limnol. Oceanogr.* **1999**, *44*, 2017–2023. [CrossRef]
6. McClelland, J.W.; Holmes, R.M.; Dunton, K.H.; Macdonald, R.W. The Arctic Ocean Estuary. *Chesap. Sci.* **2012**, *35*, 353–368. [CrossRef]
7. Stein, R.; Fahl, K.; Fütterer, D.; Galimov, E.M.; Stepanets, O.V. *Siberian River Run-off in the Kara Sea: Characterisation, Quantification, Variability, and Environmental Significance*; Elsevier: Amsterdam, The Netherlands, 2003.
8. Yentsch, C.S. The influence of phytoplankton pigments on the colour of sea water. *Deep. Sea Res. (1953)* **1960**, *7*, 1–9. [CrossRef]
9. Coble, P.G. Characterization of marine and terrestrial DOM in seawater using excitation-emission matrix spectroscopy. *Mar. Chem.* **1996**, *51*, 325–346. [CrossRef]
10. Arrigo, K.; Brown, C. Impact of chromophoric dissolved organic matter on UV inhibition of primary productivity in the sea. *Mar. Ecol. Prog. Ser.* **1996**, *140*, 207–216. [CrossRef]
11. Siegel, D.A.; Maritorena, S.; Nelson, N.B.; Hansell, D.A.; Lorenzi-Kayser, M. Global distribution and dynamics of colored dissolved and detrital organic materials. *J. Geophys. Res. Space Phys.* **2002**, *107*, 21-1–21-14. [CrossRef]
12. Stedmon, C.A.; Nelson, N.B. The Optical Properties of DOM in the Ocean. In *Biogeochemistry of Marine Dissolved Organic Matter*; Elsevier BV: Amsterdam, The Netherlands, 2015; pp. 481–508.

13. Kowalczuk, P.; Olszewski, J.; Darecki, M.; Kaczmarek, S. Empirical relationships between coloured dissolved organic matter (CDOM) absorption and apparent optical properties in Baltic Sea waters. *Int. J. Remote. Sens.* **2005**, *26*, 345–370. [CrossRef]
14. Siegel, D.A.; Maritorena, S.; Nelson, N.B.; Behrenfeld, M.J.; McClain, C.R. Colored dissolved organic matter and its influence on the satellite-based characterization of the ocean biosphere. *Geophys. Res. Lett.* **2005**, *32*. [CrossRef]
15. D'Ortenzio, F.; Marullo, S.; Ragni, M.; d'Alcalà, M.R.; Santoleri, R. Validation of empirical SeaWiFS algorithms for chlorophyll-a retrieval in the Mediterranean Sea: A case study for oligotrophic seas. *Remote Sens. Environ.* **2002**, *82*, 79–94. [CrossRef]
16. Palmer, S.C.; Pelevin, V.V.; Goncharenko, I.; Kovács, A.W.; Zlinszky, A.; Présing, M.; Tóth, V.R. Ultraviolet fluorescence LiDAR (UFL) as a measurement tool for water quality parameters in turbid lake conditions. *Remote Sens.* **2013**, *5*, 4405–4422. [CrossRef]
17. Bailly, J.-S.; Montes-Hugo, M.; Pastol, Y.; Baghdadi, N. LiDAR Measurements and Applications in Coastal and Continental Waters. In *Land Surface Remote Sensing in Urban and Coastal Areas*; Elsevier BV: Amsterdam, The Netherlands, 2016; pp. 185–229.
18. Romankevich, E.A.; Vetrov, A.A. *Carbon Cycle in the Arctic Seas of Russia*; Nauka Publishers: Moscow, Russia, 2001.
19. Amon, R.M.W. The Role of Dissolved Organic Matter for the Organic Carbon Cycle in the Arctic Ocean. In *The Organic Carbon Cycle in the Arctic Ocean*; Springer International Publishing: New York, NY, USA, 2004; pp. 83–99.
20. Hansell Dennis, A.; Craig, A. *Carlson. Biogeochemistry of Marine Dissolved Organic Matter*; Academic Press: Cambridge, MA, USA, 2014.
21. Mann, P.J.; Spencer, R.G.M.; Hernes, P.J.; Esix, J.; Aiken, G.R.; Tank, S.E.; McClelland, J.W.; Butler, K.D.; Dyda, R.Y.; Holmes, R.M. Pan-Arctic Trends in Terrestrial Dissolved Organic Matter from Optical Measurements. *Front. Earth Sci.* **2016**, *4*, 25. [CrossRef]
22. Dai, M.; Yin, Z.; Meng, F.; Liu, Q.; Cai, W.-J. Spatial distribution of riverine DOC inputs to the ocean: An updated global synthesis. *Curr. Opin. Environ. Sustain.* **2012**, *4*, 170–178. [CrossRef]
23. Nihoul, J. *Influence of Climate Change on the Changing Arctic and Sub-Arctic Conditions*; Springer: Berlin/Heidelberg, Germany, 2009.
24. McGuire, A.D.; Anderson, L.G.; Christensen, T.R.; Dallimore, S.; Guo, L.; Hayes, D.J.; Heimann, M.; Lorenson, T.D.; Macdonald, R.W.; Roulet, N. Sensitivity of the carbon cycle in the Arctic to climate change. *Ecol. Monogr.* **2009**, *79*, 523–555. [CrossRef]
25. Meier, W.N.; Hovelsrud, G.K.; Van Oort, B.E.; Key, J.R.; Kovacs, K.M.; Michel, C.; Haas, C.; Granskog, M.A.; Gerland, S.; Perovich, D.K.; et al. Arctic sea ice in transformation: A review of recent observed changes and impacts on biology and human activity. *Rev. Geophys.* **2014**, *52*, 185–217. [CrossRef]
26. Stedmon, C.; Amon, R.; Rinehart, A.; Walker, S. The supply and characteristics of colored dissolved organic matter (CDOM) in the Arctic Ocean: Pan Arctic trends and differences. *Mar. Chem.* **2011**, *124*, 108–118. [CrossRef]
27. Jaffé, R.; McKnight, D.; Maie, N.; Cory, R.; McDowell, W.H.; Campbell, J.L. Spatial and temporal variations in DOM composition in ecosystems: The importance of long-term monitoring of optical properties. *J. Geophys. Res. Space Phys.* **2008**, *113*. [CrossRef]
28. Harrison, J.A.; Caraco, N.; Seitzinger, S.P. Global patterns and sources of dissolved organic matter export to the coastal zone: Results from a spatially explicit, global model. *Glob. Biogeochem. Cycles* **2005**, *19*, GB4S04. [CrossRef]
29. Griffin, C.; McClelland, J.; Frey, K.; Fiske, G.; Holmes, R. Quantifying CDOM and DOC in major Arctic rivers during ice-free conditions using Landsat TM and ETM+ data. *Remote. Sens. Environ.* **2018**, *209*, 395–409. [CrossRef]
30. Gonçalves-Araujo, R.; Rabe, B.; Peeken, I.; Bracher, A. High colored dissolved organic matter (CDOM) absorption in surface waters of the central-eastern Arctic Ocean: Implications for biogeochemistry and ocean color algorithms. *PLoS ONE* **2018**, *13*, e0190838. [CrossRef]
31. Glukhovets, D.; Kopelevich, O.; Yushmanova, A.; Vazyulya, S.; Sheberstov, S.; Karalli, P.; Sahling, I. Evaluation of the CDOM Absorption Coefficient in the Arctic Seas Based on Sentinel-3 OLCI Data. *Remote. Sens.* **2020**, *12*, 3210. [CrossRef]
32. Drozdova, A.N.; Puiman, M.S.; Krylov, I.N.; Patsaeva, S.V.; Shatravin, A.V. Dataset on optical characteristics and spectroscopic indices of dissolved organic matter of the Kara, Laptev, and East Siberian seas in August–September. *Data Brief* **2019**, *26*, 104562. [CrossRef]
33. Wheeler, P.; Watkins, J.; Hansing, R. Nutrients, organic carbon and organic nitrogen in the upper water column of the Arctic Ocean: Implications for the sources of dissolved organic carbon. *Deep. Sea Res. Part II Top. Stud. Oceanogr.* **1997**, *44*, 1571–1592. [CrossRef]
34. Gordeev, V.V.; Martin, J.M.; Sidorov, I.S.; Sidorova, M.V. A reassessment of the Eurasian river input of water, sediment, major elements, and nutrients to the Arctic Ocean. *Am. J. Sci.* **1996**, *296*, 664–691. [CrossRef]
35. Drozdova, A.N.; Patsaeva, S.V.; Khundzhua, D.A. Fluorescence of dissolved organic matter as a marker for distribution of desalinated waters in the Kara Sea and bays of Novaya Zemlya archipelago. *Oceanology* **2017**, *57*, 41–47. [CrossRef]
36. Amante, C.; Eakins, B.W. ETOPO1 1 arc-minute global relief model: Procedures, data sources and analysis. NOAA Technical Memorandum NESDIS NGDC-24. *Natl. Geophys. Data Cent.* **2009**, *10*, V5C8276M.
37. Helms, J.R.; Stubbins, A.; Ritchie, J.D.; Minor, E.C.; Kieber, D.J.; Mopper, K. Absorption spectral slopes and slope ratios as indicators of molecular weight, source, and photobleaching of chromophoric dissolved organic matter. *Limnol. Oceanogr.* **2008**, *53*, 955–969. [CrossRef]
38. Bricaud, A.; Morel, A.; Prieur, L. Absorption by dissolved organic matter of the sea (yellow substance) in the UV and visible domains1. *Limnol. Oceanogr.* **1981**, *26*, 43–53. [CrossRef]
39. Derrien, M.; Yang, L.; Hur, J. Lipid biomarkers and spectroscopic indices for identifying organic matter sources in aquatic environments: A review. *Water Res.* **2017**, *112*, 58–71. [CrossRef]

40. Weishaar, J.L.; Aiken, G.R.; Bergamaschi, B.A.; Fram, M.S.; Fujii, R.; Mopper, K. Evaluation of Specific Ultraviolet Absorbance as an Indicator of the Chemical Composition and Reactivity of Dissolved Organic Carbon. *Environ. Sci. Technol.* **2003**, *37*, 4702–4708. [CrossRef] [PubMed]
41. Drozdova, A.N.; Kravchishina, M.D.; Khundzhua, D.A.; Freidkin, M.P.; Patsaeva, S.V. Fluorescence quantum yield of CDOM in coastal zones of the Arctic seas. *Int. J. Remote Sens.* **2018**, *39*, 9356–9379. [CrossRef]
42. Pavlov, V.K.; Pfirman, S.L. Hydrographic structure and variability of the Kara Sea: Implications for pollutant distribution. *Deep Sea Res. Part Ii Top. Stud. Oceanogr.* **1995**, *42*, 1369–1390. [CrossRef]
43. Glukhovets, D.I.; Goldin, Y.A. Surface layer desalination of the bays on the east coast of Novaya Zemlya identified by shipboard and satellite data. *Oceanology* **2019**, *61*, 68–77. [CrossRef]
44. Zatsepin, A.G.; Zavialov, P.O.; Kremenetskiy, V.V.; Poyarkov, S.G.; Soloviev, D.M. The upper desalinated layer in the Kara Sea. *Oceanology* **2010**, *50*, 657–667. [CrossRef]
45. Olsson, K.; Anderson, L.G. Input and biogeochemical transformation of dissolved carbon in the Siberian shelf seas. *Cont. Shelf Res.* **1997**, *17*, 819–833. [CrossRef]
46. Gonçalves-Araujo, R.; Stedmon, C.A.; Heim, B.; Dubinenkov, I.; Kraberg, A.; Moiseev, D.; Bracher, A. From Fresh to Marine Waters: Characterization and Fate of Dissolved Organic Matter in the Lena River Delta Region, Siberia. *Front. Mar. Sci.* **2015**, *2*, 108. [CrossRef]
47. Bauch, D.; Dmitrenko, I.A.; Wegner, C.; Hölemann, J.; Kirillov, S.A.; Timokhov, L.A.; Kassens, H. Exchange of Laptev Sea and Arctic Ocean halocline waters in response to atmospheric forcing. *J. Geophys. Res. Space Phys.* **2009**, *114*, C05008. [CrossRef]
48. Osadchiev, A.; Medvedev, I.; Shchuka, S.; Kulikov, M.; Spivak, E.; Pisareva, M.; Semiletov, I. Influence of estuarine tidal mixing on structure and spatial scales of large river plumes. *Ocean Sci.* **2020**, *16*, 781–798. [CrossRef]
49. Vetrov, A.; Romankevich, E.; Romankevich, E.A. *Carbon Cycle in the Russian Arctic Seas*; Springer Science & Business Media: Berlin, Germany, 2004.
50. Alling, V.; Sanchez-Garcia, L.; Porcelli, D.; Pugach, S.; Vonk, J.E.; Van Dongen, B.; Mörth, C.-M.; Anderson, L.G.; Sokolov, A.; Gustafsson, Ö.; et al. Nonconservative behavior of dissolved organic carbon across the Laptev and East Siberian seas. *Glob. Biogeochem. Cycles* **2010**, *24*. [CrossRef]
51. Stedmon, C.; Markager, S.; Kaas, H. Optical Properties and Signatures of Chromophoric Dissolved Organic Matter (CDOM) in Danish Coastal Waters. *Estuar. Coast. Shelf Sci.* **2000**, *51*, 267–278. [CrossRef]
52. Chin, Y.-P.; Aiken, G.; O'Loughlin, E. Molecular Weight, Polydispersity, and Spectroscopic Properties of Aquatic Humic Substances. *Environ. Sci. Technol.* **1994**, *28*, 1853–1858. [CrossRef]
53. Lee, M.-H.; Osburn, C.L.; Shin, K.-H.; Hur, J. New insight into the applicability of spectroscopic indices for dissolved organic matter (DOM) source discrimination in aquatic systems affected by biogeochemical processes. *Water Res.* **2018**, *147*, 164–176. [CrossRef] [PubMed]
54. Allard, B.; Borén, H.; Pettersson, C.; Zhang, G. Degradation of humic substances by UV irradiation. *Environ. Int.* **1994**, *20*, 97–101. [CrossRef]
55. Pavlov, A.K.; Granskog, M.A.; Stedmon, C.A.; Ivanov, B.V.; Hudson, S.R.; Falk-Petersen, S. Contrasting optical properties of surface waters across the Fram Strait and its potential biological implications. *J. Mar. Syst.* **2015**, *143*, 62–72. [CrossRef]
56. Granskog, M.A.; Pavlov, A.K.; Sagan, S.; Kowalczuk, P.; Raczkowska, A.; Stedmon, C.A. Effect of sea-ice melt on inherent optical properties and vertical distribution of solar radiant heating in Arctic surface waters. *J. Geophys. Res. Ocean.* **2015**, *120*, 7028–7039. [CrossRef]
57. Pugach, S.P.; Pipko, I.I.; Shakhova, N.E.; Shirshin, E.A.; Perminova, I.V.; Gustafsson, Ö.; Bondur, V.G.; Ruban, A.S.; Semiletov, I.P. Dissolved organic matter and its optical characteristics in the Laptev and East Siberian seas: Spatial distribution and interannual variability (2003–2011). *Ocean Sci.* **2018**, *14*, 87–103. [CrossRef]
58. Walker, S.A.; Amon, R.M.W.; Stedmon, C.A. Variations in high-latitude riverine fluorescent dissolved organic matter: A comparison of large Arctic rivers. *J. Geophys. Res. Biogeosci.* **2013**, *118*, 1689–1702. [CrossRef]
59. Stedmon, C.A.; Markager, S. The optics of chromophoric dissolved organic matter (CDOM) in the Greenland Sea: An algorithm for differentiation between marine and terrestrially derived organic matter. *Limnol. Oceanogr.* **2001**, *46*, 2087–2093. [CrossRef]
60. Kitidis, V.; Stubbins, A.P.; Uher, G.; Goddard, R.C.U.; Law, C.S.; Woodward, E.M.S. Variability of chromophoric organic matter in surface waters of the Atlantic Ocean. *Deep. Sea Res. Part II Top. Stud. Oceanogr.* **2006**, *53*, 1666–1684. [CrossRef]
61. Belyaev, N.A.; Peresypkin, V.I.; Ponyaev, M.S. The organic carbon in the water, the particulate matter, and the upper layer of the bottom sediments of the west Kara Sea. *Oceanology* **2010**, *50*, 706–715. [CrossRef]
62. Belyaev, N.A.; Ponyaev, M.S.; Kiriutin, A.M. Organic carbon in water, particulate matter, and upper layer of bottom sediments of the central part of the Kara Sea. *Oceanology* **2015**, *55*, 508–520. [CrossRef]
63. Meon, B.; Amon, R. Heterotrophic bacterial activity and fluxes of dissolved free amino acids and glucose in the Arctic rivers Ob, Yenisei and the adjacent Kara Sea. *Aquat. Microb. Ecol.* **2004**, *37*, 121–135. [CrossRef]
64. Cauwet, G.; Sidorov, I. The biogeochemistry of Lena River: Organic carbon and nutrients distribution. *Mar. Chem.* **1996**, *53*, 211–227. [CrossRef]
65. Fouest, V.L.; Babin, M.; Tremblay, J.É. The fate of riverine nutrients on Arctic shelves. *Biogeosciences* **2013**, *10*, 3661–3677. [CrossRef]
66. Lara, R.J.; Rachold, V.; Kattner, G.; Hubberten, H.W.; Guggenberger, G.; Skoog, A.; Thomas, D.N. Dissolved organic matter and nutrients in the Lena River, Siberian Arctic: Characteristics and distribution. *Mar. Chem.* **1998**, *59*, 301–309. [CrossRef]

67. Lobbes, J.M.; Fitznar, H.P.; Kattner, G. Biogeochemical characteristics of dissolved and particulate organic matter in Russian rivers entering the Arctic Ocean. *Geochim. Et Cosmochim. Acta* **2000**, *64*, 2973–2983. [CrossRef]
68. Juhls, B.; Stedmon, C.A.; Morgenstern, A.; Meyer, H.; Hölemann, J.; Heim, B.; Povazhnyi, V.; Overduin, P.P. Identifying Drivers of Seasonality in Lena River Biogeochemistry and Dissolved Organic Matter Fluxes. *Front. Environ. Sci.* **2020**, *8*, 53. [CrossRef]
69. Organelli, E.; Claustre, H. Small Phytoplankton Shapes Colored Dissolved Organic Matter Dynamics in the North Atlantic Subtropical Gyre. *Geophys. Res. Lett.* **2019**, *46*, 12183–12191. [CrossRef] [PubMed]
70. Stedmon, C.A.; Markager, S. Tracing the production and degradation of autochthonous fractions of dissolved organic matter by fluorescence analysis. *Limnol. Oceanogr.* **2005**, *50*, 1415–1426. [CrossRef]
71. Semiletov, I.; Dudarev, O.; Luchin, V.; Charkin, A.; Shin, K.H.; Tanaka, N. The East Siberian Sea as a transition zone between Pacific-derived waters and Arctic shelf waters. *Geophys. Res. Lett.* **2005**, *32*, L10614. [CrossRef]
72. Spencer, R.G.; Aiken, G.R.; Butler, K.D.; Dornblaser, M.M.; Striegl, R.G.; Hernes, P.J. Utilizing chromophoric dissolved organic matter measurements to derive export and reactivity of dissolved organic carbon exported to the Arctic Ocean: A case study of the Yukon River, Alaska. *Geophys. Res. Lett.* **2009**, *36*, L06401. [CrossRef]
73. Kowalczuk, P.; Sagan, S.; Makarewicz, A.; Meler, J.; Borzycka, K.; Zabłocka, M.; Pavlov, A.K. Bio-optical properties of surface waters in the Atlantic Water inflow region off Spitsbergen (Arctic Ocean). *J. Geophys. Res. Ocean.* **2019**, *124*, 1964–1987. [CrossRef]
74. Bricaud, A.; Morel, A.; Babin, M.; Allali, K.; Claustre, H. Variations of light absorption by suspended particles with chlorophyll a concentration in oceanic (case 1) waters: Analysis and implications for bio-optical models. *J. Geophys. Res. Ocean.* **1998**, *103*, 31033–31044. [CrossRef]
75. Gonçalves-Araujo, R.; Röttgers, R.; Haraguchi, L.; Brandini, F.P. Hydrography-driven variability of optically active constituents of water in the South Brazilian Bight: Biogeochemical implications. *Front. Mar. Sci.* **2019**, *6*, 716. [CrossRef]
76. Sukhanova, I.N.; Flint, M.V.; Fedorov, A.V.; Sakharova, E.G.; Artemyev, V.A.; Makkaveev, P.N.; Nedospasov, A.A. Phytoplankton of the Khatanga Bay, Shelf and Continental Slope of the Western Laptev Sea. *Oceanology* **2019**, *59*, 648–657. [CrossRef]
77. Burenkov, V.I.; Goldin, Y.A.; Artem'Ev, V.A.; Sheberstov, S.V. Optical characteristics of the Kara Sea derived from shipborne and satellite data. *Oceanology* **2010**, *50*, 675–687. [CrossRef]
78. Vazyulya, S.V.; Kopelevich, O.V.; Sheberstov, S.V.; Artemiev, V.A. Satellite estimation of the coefficients of CDOM absorption and diffuse attenuation in the White and Kara seas. *Curr. Probl. Remote Sens. Earth Space* **2014**, *11*, 31–41.

Letter

Low-cost Fiberoptic Probe for Ammonia Early Detection in Fish Farms

Arnaldo G. Leal-Junior [1,*], Anselmo Frizera [1] and Carlos Marques [2]

1 Telecommunications Laboratory (LABTEL), Graduate Program in Electrical Engineering Federal University of Espírito Santo, Fernando Ferrari avenue, Vitória-ES 29075-910, Brazil; frizera@ieee.org
2 Physics Department & I3N, Universidade de Aveiro, Campus Universitário de Santiago, 3810-193 Aveiro, Portugal; carlos.marques@ua.pt
* Correspondence: leal-junior.arnaldo@ieee.org

Received: 15 April 2020; Accepted: 29 April 2020; Published: 1 May 2020

Abstract: Recirculating aquaculture systems (RAS) are complex systems in which there is an interaction between the fish biomass and water chemistry, where small variations in the environment can lead to major effects in the production. Ammonia is one of the key limiting factors in RAS and its early detection in small concentrations prevents fish mortality and improves the production quality. Aiming at this background, this paper presents a low-cost fiberoptic probe for the early detection of ammonia. The sensor was based on the chemical interaction between the Oxazine 170 perchlorate layer, deposited in an uncladed polymer optical fiber (POF), and the ammonia dissolved in water. In addition, a thin metallic layer (composed by gold and palladium) was deposited in the fiber end facet and acted as a reflector for the optical signals, enabling the use of the proposed sensor in reflection mode. Different configurations of the sensor were tested, where the effects of polydimethylsiloxane (PDMS) protective layer, thermal treatments, and the use on reflection or transmission modes were compared in the assessment of ammonia concentrations in the range of 100 ppb to 900 ppb. Results showed a better performance (as a function of the sensor sensitivity and linearity) of the sensor with the annealing thermal treatment and without the PDMS layer. Then, the proposed fiberoptic probe was applied on the ammonia detection in high-salinity water, where ammonia concentrations as low as 100 ppb were detected.

Keywords: polymer optical fibers; ammonia detection; optical fiber coating; aquaculture

---

1. Introduction

Aquaculture is the fastest growing food production sector in the world [1] and is therefore responding to the demands of ensuring sustainable global food security. Global aquaculture production has grown by nearly 7% per year. These demands are expected to continue to rise in response to projections of increased global population and the associated need for new protein sources. The marine fish aquaculture industry in Europe is dominated by Atlantic salmon (*Salmo salar*) in the north and by gilthead seabream (*Sparus aurata*) and European seabass (*Dicentrarchus labrax*) in the south.

Since the publication of the European Commission's 2013 Strategic Guidelines on the sustainable development of European aquaculture [2], the Commission has worked closely with Member States to address the barriers hampering the development of the sector while launching several campaigns to promote sustainable and competitive fish farming. A large part of this expansion and species diversification is expected to take place in land-based aquaculture facilities or intensive systems where water is reused in some form, such as recirculating aquaculture systems (RAS).

RAS are currently operating for many different species and have recently been developed to the large industrial scale. The highly technological RAS allows a very high degree of water reuse, normally in the range of 95–99% when compared to the water consumption in a flow-through system. Such facilities need to be designed for safe operation in the long-term, involving continuous and close monitoring of several critical parameters to safeguard fish welfare and profitability. The benefits of RAS are the reduced fresh and saltwater usage, reduced land requirement due to the high stocking density, reduced wastewater effluent volume, increased biosecurity by effectively treating disease outbreaks, and independence from weather and variable environmental conditions. Indeed, RAS satisfy the objectives of European Union for sustainable aquaculture by producing food while sustaining natural resources with a minimum ecological impact. However, RAS are complex systems where fish biomass and water chemistry/quality interact and where small variations may result in suboptimal conditions. These variations may cause stress and reduced feed intake and may ultimately result in reduced growth performance or even mortalities [3].

Ammonia is a limiting factor for producing fish in RAS [4,5], but the tolerance affecting mortality and growth varies [6]. Different water quality parameters affect the kinetics of the RAS biological filters (BF) [7], causing variation in ammonia removal. This gives rise to the dilemma of how to control and limit ammonia levels in order to optimize production under different conditions. Tolerance to ammonia varies between fish species and within different life stages and physiological status (starved, stresses, and health status) of the same species [8,9]. Furthermore, the type of rearing conditions may play a part. Other water quality parameters like temperature, carbon dioxide, oxygen, and salinity may also interact and influence the fish response to water quality and stress [8], which again can vary between different fish farms, and even throughout a production cycle. In addition, ammonia tolerance depends on how the fish acclimate to ammonia [10] and to changing water quality [11]. Not least, the different efficiency in which different RAS system convert ammonia to less toxic N-compounds varies depending on the RAS-design [12]. Acute levels of ammonia lead to acute mortality, while chronic levels of ammonia lead to suboptimal conditions for the fish. These suboptimal conditions are difficult to monitor but may result in reduced appetite and growth and eventually increased mortality [13]. Even in well-managed RAS, this is probably a common issue due to the many and intricate interactions between the fish and the other water quality parameters mentioned. The Food and Agriculture Organization (FAO) of the United Nations has advised that the ammonia levels should cover a range of at least 1-100 ng/mL [14,15].

In this way, water quality and toxicity levels in recirculation systems are critical, and ammonia is an important parameter which is closely linked to the fish excretion and efficacy of the BF. Growth and stocking density (biomass production and loading) are the main factors affecting the ammonia fluctuations and load in RAS [7]. The existing sensors for ammonia are generally used in sewage treatment plants and do not give consistent values in saltwater systems [9]. Moreover, there is a lack of available technology to add sensing elements before and after the BF and degassing column of a RAS.

In this respect, suitable multipoint monitoring elements for ammonia are essential. The use of optical fiber sensor-based devices offers many well-known and desirable features for label-free methods when compared with conventional electrical transducers. Examples are size, immunity to electromagnetic interference, cost, light path control, remote sensor deployment, high transmission rate, multiple sensors just on a single fiber and the use of biocompatible and biofunctionalized materials, and intrinsic safety and inert nature, which enables them to have minimal impact on the environment [16]. These advantages motivate the use of optical fiber sensors in medicine [17], environmental monitoring [18], and antibody detection [19]. Optical fiber sensors are selected for such applications due to their operational safety in aqueous environments and ease of introduction into the tanks, which eliminates the need to take samples to be tested in external instruments. Additionally, sensing can be performed with either a handheld probe or as a set of remotely operated devices along an optical fiber cable.

The optical fiber sensors approach for ammonia assessment is generally related to the use of nanoparticles or dyers that interact with the dissolved ammonia, resulting in intensity [20] or wavelength [21] variations in the optical fiber probe. An optical fiber interferometer was proposed by the authors of [21] for the ammonia detection. Using a nanocomposite coating, the sensor was able of detecting ammonia concentrations as low as 14 ppm, which is higher than the lower bound of ammonia detection in RAS [14]. In addition, wavelength-based sensors need an optical spectrum analyzer, which is generally a high-cost equipment. In order to obtain a low-cost sensor system, intensity variation-based systems have been proposed, such as the one presented by the authors of [22], where the luminescence from the nanoparticles upconversion was used on the sensor development. In this case, the sensor was tested in a range of 100-10000 ppm, i.e., far greater than the ammonia levels recommended in water. Similar techniques employed dyers in the lateral section of an optical fiber to evaluate the differences in the fluorescence as function of the ammonia levels [23]. The authors of [24] proposed a porous waveguide with a combination of copolymers as pH indicator, where ammonia concentrations as low as 10 ppm were measured. Then, with further improvements in the Oxazine 170 perchlorate deposition in polymer optical fibers (POFs), the authors of [20] proposed a highly sensitive ammonia sensor with detection limit of 1.4 ppm, the sensor was tested in stagnated and dynamic water. However, the approach was based on the transmission mode, which, for field applications, can lead to practical disadvantages on the sensor positioning. Moreover, intensity variation-based sensors are also sensitive to light source power deviations, and self-compensation techniques for these deviations need to be addressed for a higher resolution of the sensors.

Considering this background, this paper presents a low-cost POF sensor to cover the ammonia levels required by RAS and suitable to add in critical locations of a RAS, in which the proposed sensor was based on the reaction between the POF cladding with Oxazine 170 perchlorate and the water-dissolved ammonia. Such reaction led to variations in the evanescent field, resulting in optical power variations. In addition, a reflector was manufactured in the optical fiber tip in order to operate in the reflection mode, which brought important advantages regarding to the sensor positioning inside the water tank. Another advantageous feature was the possibility of using unclad POFs recycled from the LCD monitors (as depicted by the authors of [25]) for the sensor manufacturing, where the cladding was based on the Oxazine 170 perchlorate solution.

2. Sensors Development and Experimental Setup

An uncladed multimode POF with polymethyl methacrylate (PMMA) core was used on the fabrication of the optical fiber probe with a 980-μm core (ESKA Mitsubishi, Japan). The fiber diameter was about 1 mm and the fabrication of the proposed ammonia sensor occurred in a few steps, as summarized in Figure 1. The first step involved the gold-palladium (Au/Pd) thin layer deposition using a sputter coater. The sputter target was composed of 20% palladium and 80% gold. The POF end facet was positioned on the sputter target and the Au/Pd thin layer was created, where this layer acted as a mirror and reflects the optical signals, which enabled the use of the proposed sensor in the reflection mode.

Then, in the second step, the Oxazine 170 perchlorate was deposited on the fiber uncladed region by dip coating. The Oxazine 170 perchlorate deposition followed the methods discussed by the authors of [20], where a dye solution was prepared using 1 mg of Oxazine dissolved in 100 mL of distillated water though a temperature-controlled magnetic stirrer. The uncladed region of the POF was dipped in the Oxazine solution (without dipping the fiber end facet, which had the Au/Pd thin layer) for about 2 hours at 70 °C, as shown in Step 2 of Figure 1, where an u-bend was made on the fiber to position it without dipping its end facet. Thereafter, the fiber was removed, and the temperature of the solution was decreased to 20 °C. After the solution reached the input temperature (20°C), the fiber was dipped again for 10 hours. The Oxazine 170 perchlorate dipping step was the one in which the sensor element was fabricated. The operation principle of the proposed sensor was based on the chemical reactions between the dissolved ammonia and the Oxazine layer on the uncladed POF. Such reactions led to differences in the evanescent field, which resulted in the optical power variation in the POF. The operation principle of the proposed sensor is depicted in Figure 2a, whereas a photograph of the sensor (Sample 3) is shown in Figure 2b.

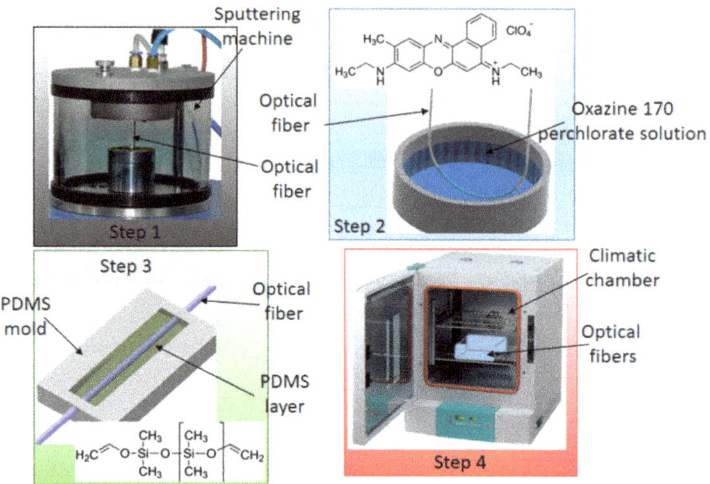

Figure 1. Schematic representation of the steps for the samples preparation.

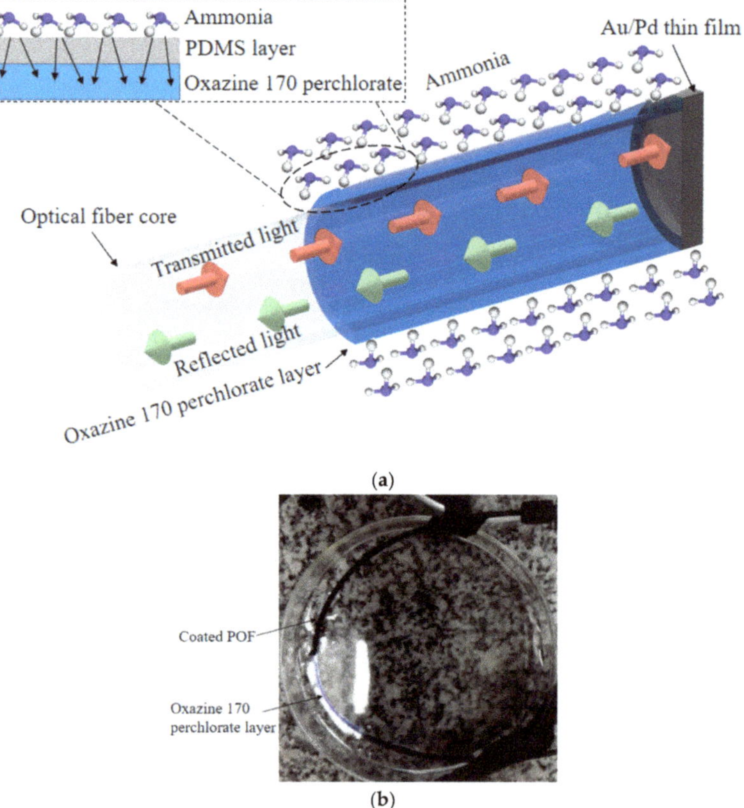

Figure 2. (**a**) Operation principle of the proposed sensors. The arrows are only a schematic representation, since the polymer optical fiber (POF) used had multiple propagating modes instead of just the ones presented in the figure. Figure inset shows ammonia diffusion in the sensing layer. (**b**) Photograph of the proposed sensor (Sample 3).

The chemical reaction between the Oxazine and the ammonia occurs at aqueous solution or in the presence of moisture, which is in accordance with the proposed application, i.e., ammonia sensing on water tanks. However, the direct contact between the sensor and the water can result in a lower lifespan of the proposed sensor if the water can remove the Oxazine layer. For this reason, in the third step (after the Oxazine 170 perchlorate deposition), a polydimethylsiloxane (PDMS) layer was applied. The PDMS layer was also applied by dip coating the optical fiber in the region where the Oxazine was deposited. As proposed by the authors of [20], the optical fiber was dipped in a water container before the application of the PDMS layer in order to create a region with entrapped water, which was needed for the chemical reactions between the Oxazine and ammonia. Then, the fiber was dipped in the PDMS curing agent solution with 24 hours for the layer curing. In this case, there was an additional mechanism in the sensor operation principle, which was the ammonia diffusion in the PDMS layer due to the pressure difference between the layer and the solution containing ammonia [20]. A schematic representation of the ammonia diffusion is presented in Figure 2a inset, where the pressure gradients in conjunction with the ammonia concentration resulted in higher ammonia diffusion into the PDMS layer. This additional mechanism can lead to higher response times for the sensor when compared with the one without the PDMS layer. However, as presented in previous works [20], the ammonia diffusion was fast (of a few seconds), and it was expected that the sensors with and without the PDMS layer would present similar response times. Nevertheless, the ammonia diffusion was also related with its concentration. For this reason, in the tests with variation of ammonia concentration, the sensor with the PDMS layer presented lower linearity than the one without this layer, which can be attributed to the concentration dependency of the ammonia diffusion.

In the last step, a thermal annealing was performed in the fiberoptic probe, where the sensor was positioned inside a climatic chamber at about 50 °C for 6 hours in order to reduce residual stresses created in the fiber during its manufacturing or during the previous steps of the sensor fabrication. In addition, the thermal annealing also accelerated the PDMS curing and provided a better adhesion of the Oxazine coating layer on the uncladed POF.

For the fiberoptic probe interrogation unit, an optical fiber coupler 2x1 IF 562 (Industrial Fiber Optics, USA) with 50:50 coupling ratio was used. The light source, comprising of a red laser (Phywe, Germany) centered at 650 nm (optical power of 4 mW) and a photodetector IF-D92 (Industrial Fiber Optics, USA), was positioned in the two input ports, whereas the other port was positioned in the fiberoptic probe. The photodetector was connected to a microcontroller with a 16-bit analog-to-digital converter FRDM KL25Z (NXP, Netherlands). The proposed probe was positioned inside a container filled with distillated water in which different concentrations of ammonium hydroxide solutions were injected. In this case, the ammonia concentration in the solutions ranged from 100 ppb to 900 ppb. In addition, the sensors were tested in sodium chloride solutions to verify its capability of detecting ammonia in high-salinity conditions, which are commonly found in fish farms.

In order to evaluate each of the steps for the fiberoptic probe fabrication, different set of samples (with three samples) were fabricated using different combinations of the steps. Figure 3 shows the experimental setup and the indication of which steps were used in each set of samples. It is worth mentioning that all samples were subjected to Step 2, in which the sensing layer was created, which resulted in similar Oxazine 170 perchlorate thickness of 35±5 μm for all samples. The comparison between samples was based on the sensitivity and linearity as a function of the ammonia concentration. In the first analysis, the sensors operating at reflection and transmission modes (with and without Step 1, respectively) were compared (Samples 2 and 4). Thereafter, the influence of the PDMS layer on the sensor responses was analyzed (Samples 1 and 2). Finally, annealed and nonannealed samples were compared, i.e., Samples 2 and 3. In the last test, the differences between the proposed fiberoptic probe spectral responses at high-salinity water were obtained with a spectrometer at the wavelength range of 400 nm to 900 nm USB4000 (Ocean Optics, USA) in order to verify if there were any variations in the spectral responses due to the water salinity.

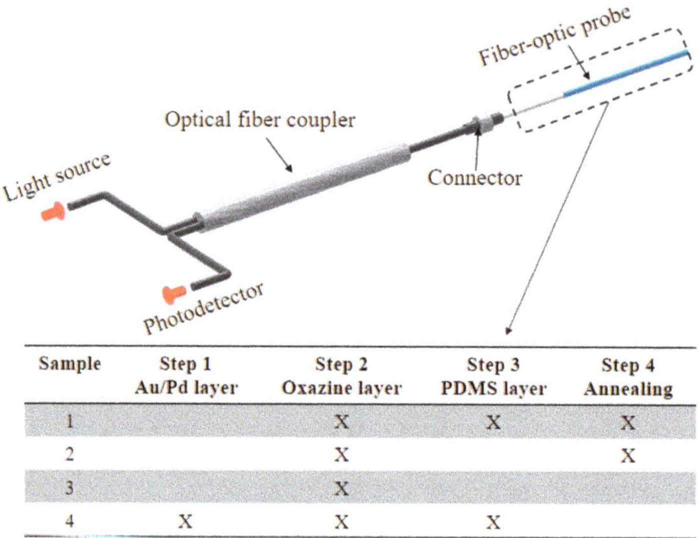

Sample	Step 1 Au/Pd layer	Step 2 Oxazine layer	Step 3 PDMS layer	Step 4 Annealing
1		X	X	X
2		X		X
3		X		
4	X	X	X	

Figure 3. Experimental setup for the sensor's response acquisition. The description of which steps were followed for each set of samples is also shown.

3. Results and Discussion

First, a statistical analysis was performed for each sample, where 15 consecutive measurements of each ammonia concentration were analyzed. In this case, one-way ANOVA was applied for each sample result to verify if there were significant statistical differences between measurements for a specific sample set. The one-way ANOVA tests perform for each sample and ammonia concentrations showed a p-value lower than 0.05 in all cases. These results indicate that there were no significant statistical differences in 15 measurement for each sample in different ammonia concentrations.

Figure 4a shows the mean and standard deviation results of the sensors with and without the PDMS layer for six ammonia concentrations in distillated water, from 0 ppb to 500 ppb in 100-ppb steps. It is worth mentioning that the responses were not in the same plot due to the larger differences in the sensors' sensitivities. In addition, the linearity of the sensors was presented as the determination coefficient (R^2) between the sensors' responses and a linear regression. The results in Figure 4a show that the PDMS layer led to lower linearity of the sensor when compared with the samples without the PDMS layer. Although some differences in the sensors' sensitivities were expected due to some minor differences in the Oxazine layer thickness, there was a high difference in the sensitivity of the sensors with and without the PDMS layer. The sensors without the additional PDMS layer had a sensitivity of 0.025 ppm^{-1}, which was seven-times higher than the one obtained from the sensor with the PDMS layer (0.0036 ppm^{-1}). Therefore, concerning the sensors' performance, the sensor without the PDMS layer was preferable for the proposed application. It is worth noting that the fiberoptic probe without the PDMS layer can have lower lifespan than the one with the PDMS protection. This feature can be verified in Figure 4a by the lower standard deviation of the probe with PDMS layer. However, the sensors were fabricated with low-cost methods (a few U$ per sensor) and with easy assembly that enabled a scalable production and the replacement of probes in field applications.

In order to verify the sensor repeatability and reusability of the sensor without the PDMS layer (since it presented a better performance when compared with the one with PDMS layer), Figure 4b shows the results of the sensor as a function of time for two consecutive cycles, one in 300-ppb concentration, where the sensor was kept in this concentration for about 10 seconds, and the other in 600-ppb concentration. It is possible to observe in both cases that the sensor presented a response reversibility, since the normalized optical power returned to its initial value when the sensor was not immersed in ammonia solutions, where response times close to 5 seconds were found for both increasing and decreasing concentrations of ammonia. As another performance analysis of the sensor, its hysteresis was analyzed in cycles of ammonia concentration. As shown in Figure 4c, the sensor presented low hysteresis of 2.48%, which indicates the suitability of the sensor in detecting ammonia concentrations in dynamic cycles with low errors.

Regarding to the thermal treatment effects on the sensors' performance, Samples 2 and 3 were tested in a larger range of ammonia concentrations (from 0 ppb to 900 ppb) (see Figure 5). The results show that the nonannealed samples presented a nonlinear behavior that could be accurately represented by an exponential regression. Nevertheless, in concentrations below 60 ppb, the nonannealed samples showed lower sensitivity than the annealed ones. The annealed samples showed a high linearity, with R^2 as high as 0.98. However, the sensitivity of nonannealed samples was higher than the annealed one for ammonia concentrations higher than 600 ppb. For early ammonia detection, a sensor with higher linearity in low concentrations of ammonia is desirable. For this reason, the annealed samples were preferable for the proposed application, but it is worth noting that in environments with high ammonia concentrations, the nonannealed samples may lead to a sensor system with higher resolution in the ammonia detection.

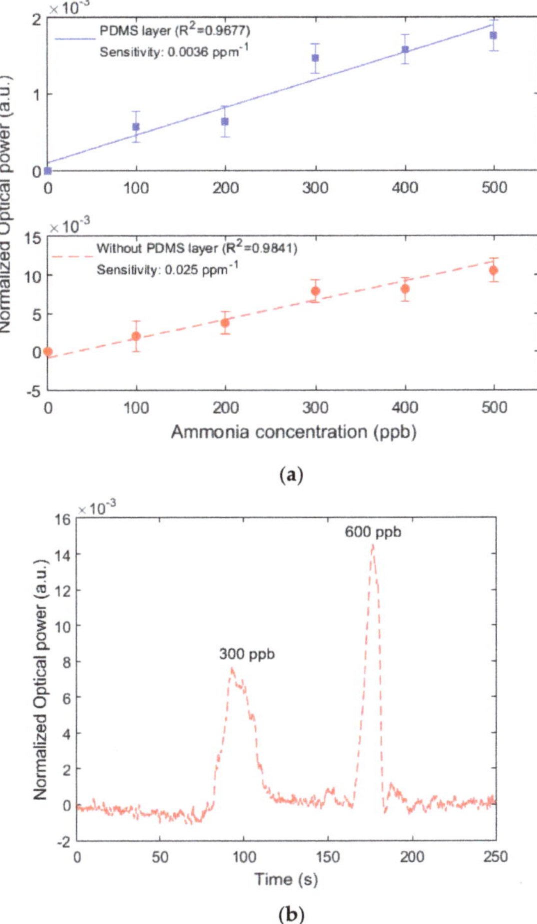

Figure 4. *Cont.*

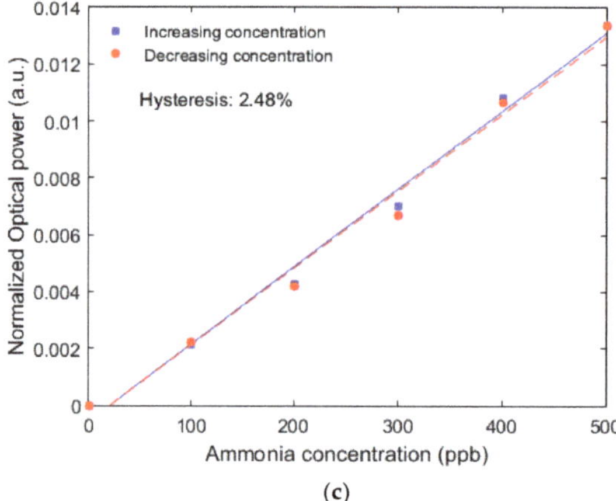

(c)

Figure 4. (**a**) Comparison between the fiberoptic probe responses with and without the PDMS layer as a function of the ammonia concentration. (**b**) Fiberoptic probe responses without PDMS layer as a function of time for two consecutive ammonia concentration cycles. (**c**) Hysteresis analysis of the proposed fiberoptic probe for ammonia detection.

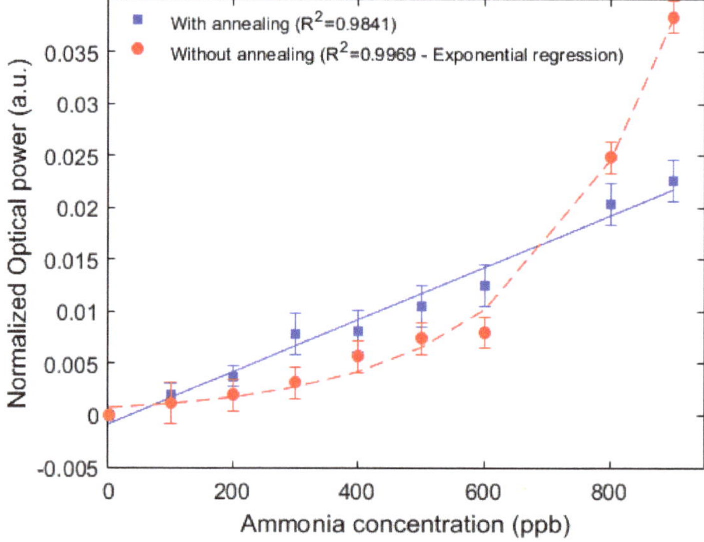

Figure 5. Comparison between the fiberoptic probe responses with and without annealing as a function of the ammonia concentration.

The tests performed in the fiberoptic probes operating at transmission and reflection modes are presented in Figure 6, where both sensors presented similar sensitivities, i.e., 0.035 ppm^{-1} and 0.041 ppm^{-1} for transmission and reflection fiberoptic probes, respectively. Regarding to the sensors' linearity, the transmission mode presented higher linearity than the one of the reflection mode. However, the R^2 values of both sensors were close to each other, which indicates compatibility between both operation modes. Therefore, due to the necessity of regular substitution of the probes and to facilitate their installation, the probe in the reflection mode was the preferable one, since both modes had similar performance.

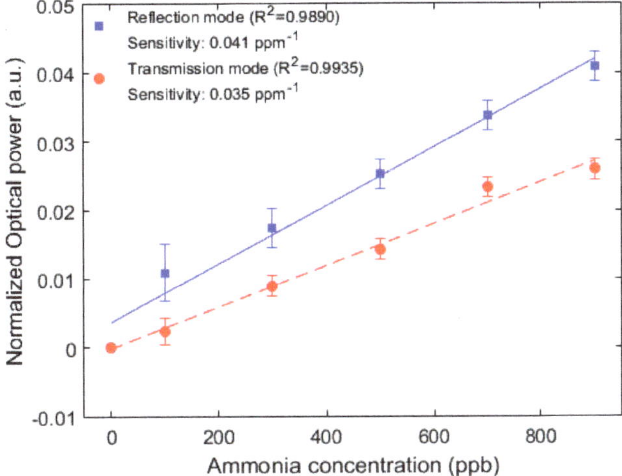

Figure 6. Comparison between the fiberoptic probe responses in the transmission and reflection mode as a function of the ammonia concentration.

Then, the fiberoptic probe in the reflection mode with annealing and without the PDMS layer (Samples 1 and 2 in Figure 3) was tested in a sodium chloride solution to simulate an environment with high-salinity water as observed in RAS. Figure 7a presents the spectra of the sensors at each ammonia concentration to verify if the water salinity leads to other variations in the reflected spectrum (such as shift in the wavelength or spectral width) besides the intensity variation. However, the results in Figure 7a show only an amplitude (or intensity) variation on the reflected spectra, which indicate a similar performance of the sensor when compared with the results in distillated water. The Figure 7a inset shows the magnified view of the peaks, where the differences in the peak intensities can be observed. In addition, the proposed sensor was able to detect ammonia concentrations as low as 100 ppb with a straightforward and low-cost sensor approach, which can be used in many RAS industries for early detection of ammonia. Figure 7b shows the comparison between the sensor responses at distillated water and saltwater, where it is possible to observe that the sensor presented similar responses with a slight sensitivity reduction in saltwater. However, considering the standard deviations between measurements, the upper limit of the sensitivity in saltwater was 0.041 ppm^{-1}, whereas the lower limit of the sensitivity in distilled water was 0.040 ppm^{-1}. This analysis indicates that there were no significant sensitivity differences in sensors responses in distilled water and saltwater.

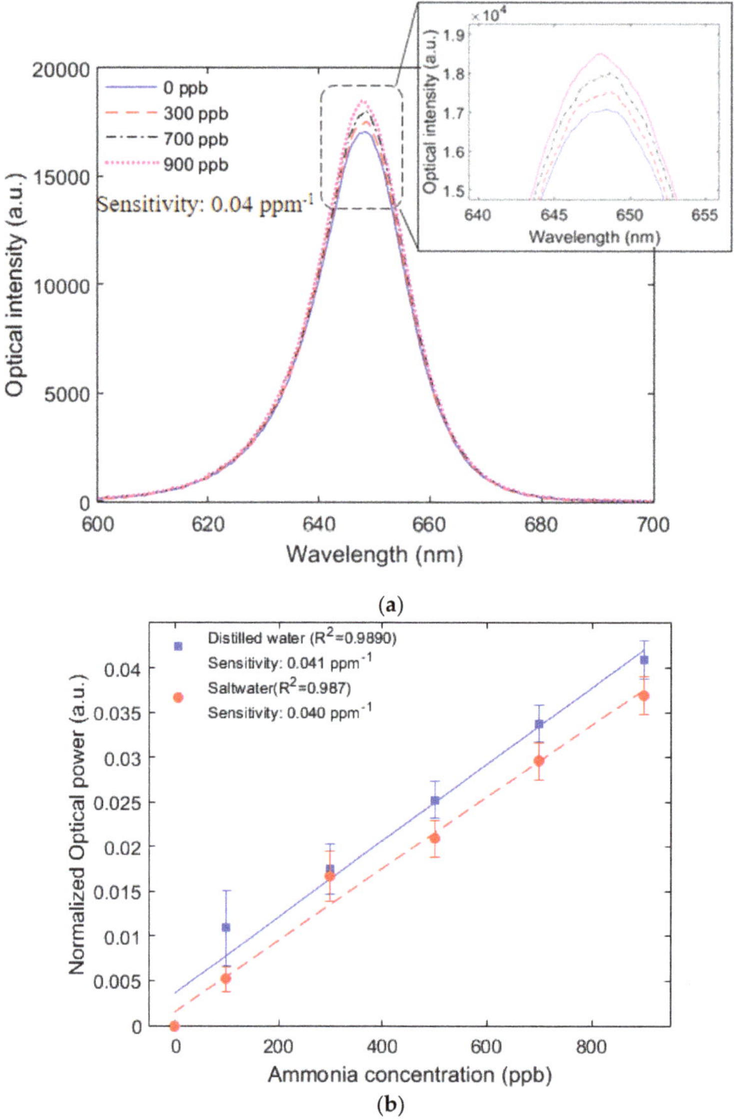

Figure 7. (**a**) Fiberoptic probe reflected spectra for different ammonia concentrations in high-salinity water. (**b**) Comparison between the sensors responses as a function of the ammonia concentration in distilled water and saltwater.

4. Conclusions

This paper presented the design and validation of a fiberoptic probe for ammonia detection. The sensor was based on the chemical reaction between Oxazine 170 perchlorate coating applied in an uncladed POF. Different design options were investigated, including the use of a PDMS layer in order to increase the sensor lifespan as well as thermal treatments (e.g., annealing) and their influence on the sensor performance parameters. Namely, the sensitivity and linearity were investigated. The transmission and reflection operation modes of the sensor were also investigated, where a thin layer of Au/Pd was deposited in the POF end facet in order to create a reflector for the probe operation in reflection mode. The sensor was tested not only in distillated water, but also in high-salinity water to emulate different operation conditions. Table 1 presents a comparison between different optical sensors for ammonia detection.

Table 1. Comparison between lowest measured concentration and sensitivities of different ammonia sensors.

Sensor	Lowest Measured Concentration	Sensitivity
Upconverting nanoparticles [22]	2.5 mM (about 42.5 ppm)	Not reported (estimated 0.0011 ppm^{-1})
Intensity variation-transmission [20]	1.4 ppm	Not reported
Etched-tapered interferometer [21]	14.2 ppm	300 a.u./% (0.03 a.u./ppm)
PDMS composite [26]	2.0 ppm	0.0382 V/ppm
Proposed sensor-reflection	0.1 ppm	0.041 ppm^{-1}

The proposed fiberoptic probe shows its feasibility on early ammonia detection, where concentrations as low as 100 ppb can be estimated with the proposed design. Compared with previously reported sensors in Table 1, the proposed sensor presented the lowest detection limit and highest sensitivity. Therefore, the proposed probe can be applied in different fish farms, including the ones with RAS, as a low-cost method for water quality assessment. In addition, with the development of recycled uncladed POFs [25], the proposed approach can be applied in such fibers for an even higher reduction of the system cost. In future works, the integration of this system with a multiplexing technique for the development of multiple probes and its application in a RAS system will be investigated.

Author Contributions: Conceptualization, A.G.L.-J. and C.M.; methodology, A.G.L.-J. and C.M.; formal analysis, A.G.L.-J., A.F. and C.M; investigation, A.G.L.-J. and C.M.; resources, A.F and C.M.; writing—original draft preparation, A.G.L.-J.; writing—review and editing, A.G.L.-J., A.F. and C.M; funding acquisition, A.F. and C.M. All authors have read and agreed to the published version of the manuscript.

Funding: This research is financed by FAPES (85426300 and 84336650), CNPq (304049/2019-0 and 427054/2018-4) and Petrobras (2017/00702-6). C. Marques acknowledges Fundação para a Ciência e a Tecnologia (FCT) through the CEECIND/00034/2018 (iFish project) and this work was developed within the scope of the project i3N, UIDB/50025/2020 & UIDP/50025/2020, financed by national funds through the FCT/MEC. This work is also funded by national funds (OE), through FCT, I.P., in the scope of the framework contract foreseen in the numbers 4, 5 and 6 of the article 23, of the Decree-Law 57/2016, of August 29, changed by Law 57/2017, of July 19.

Acknowledgments: The authors would like to thank Tiago Silva, from Department of Materials and Ceramic Engineering (DEMaC), University of Aveiro, for his valuable support during the Au/Pd thin layer deposition.

Conflicts of Interest: The authors declare no conflict of interest.

References

1. FAO Aquaculture Development. Available online: http://www.fao.org/aquaculture/en/ (accessed on 20 March 2020).
2. European Environment Agency Aquaculture Production in Europe. Available online: Eea.europa.eu/data-and-maps/indicators/aquaculture-production-4 (accessed on 20 March 2020).
3. Ogawa, K.; Ito, F.; Nagae, M.; Nishimura, T.; Yamaguchi, M.; Ishimatsu, A. Effects of acid stress on reproductive functions in immature carp, Cyprinus Carpio. *Water Air Soil Pollut.* **2001**, *130*, 887–892. [CrossRef]
4. Besson, M.; Komen, H.; Aubin, J.; De Boer, I.J.M.; Poelman, M.; Quillet, E.; Vancoillie, C.; Vandeputte, M.; Van Arendonk, J.A. Economic values of growth and feed efficiency for fish farming in recirculating aquaculture system with density and nitrogen output limitations: A case study with African catfish (clarias gariepinus). *J. Anim. Sci.* **2014**, *92*, 5394–54052. [CrossRef] [PubMed]
5. Fivelstad, S. Waterflow requirements for salmonids in single-pass and semi-closed land-based seawater and freshwater systems. *Aquac. Eng.* **1988**, *7*, 183–200. [CrossRef]
6. Ruyet, J.P.L.; Chartois, H.; Quemener, L. Comparative acute ammonia toxicity in marine fish and plasma ammonia response. *Aquaculture* **1995**, *136*, 181–194. [CrossRef]
7. Chen, S.; Ling, J.; Blancheton, J.P. Nitrification kinetics of biofilm as affected by water quality factors. *Aquac. Eng.* **2006**, *34*, 179–197. [CrossRef]
8. Conte, F.S. Stress and the welfare of cultured fish. *Appl. Anim. Behav. Sci.* **2004**, *86*, 205–223. [CrossRef]
9. Randall, D.; Tsui, T.K. Ammonia toxicity in fish. *Mar. Pollut. Bull.* **2002**, *45*, 17–23. [CrossRef]

10. Wright, P.A.; Wood, C.M. Seven things fish know about ammonia and we don't. *Respir. Physiol. Neurobiol.* **2012**, *184*, 231–240. [CrossRef]
11. Foss, A.; Kristensen, T.; Åtland, Å.; Hustveit, H.; Hovland, H.; Øfsti, A.; Imsland, A.K. Effects of water reuse and stocking density on water quality, blood physiology and growth rate of juvenile cod (Gadus morhua). *Aquaculture* **2006**, *256*, 255–263. [CrossRef]
12. Badiola, M.; Basurko, O.C.; Piedrahita, R.; Hundley, P.; Mendiola, D. Energy use in recirculating aquaculture systems (RAS): A review. *Aquac. Eng.* **2018**, *81*, 57–70. [CrossRef]
13. Rodrigues, R.V.; Schwarz, M.H.; Delbos, B.C.; Sampaio, L.A. Acute toxicity and sublethal effects of ammonia and nitrite for juvenile cobia Rachycentron canadum. *Aquaculture* **2007**, *271*, 553–557. [CrossRef]
14. Kolarevic, J.; Takle, H.; Felip, O.; Ytteborg, E.; Selset, R.; Good, C.M.; Baeverfjord, G.; Åsgård, T.; Terjesen, B.F. Molecular and physiological responses to long-term sublethal ammonia exposure in Atlantic salmon (Salmo salar). *Aquat. Toxicol.* **2012**, *124–125*, 48–57. [CrossRef] [PubMed]
15. Bregnballe, J. *A Guide to Recirculation Aquaculture*; Food and Agriculture Organization of the United Nations (FAO), EUROFISH International Organisation: Rome, Italy, 2015; p. 100.
16. Alwis, L.; Sun, T.; Grattan, K.T.V. Developments in optical fibre sensors for industrial applications. *Opt. Laser Technol.* **2016**, *78*, 62–66. [CrossRef]
17. Majchrowicz, D.; Kosowska, M.; Sankaran, K.J.; Sobaszek, M.; Haenen, K. Nitrogen-doped diamond film for optical investigation of hemoglobin concentration. *Materials* **2018**, *11*, 109. [CrossRef]
18. Zhang, Y.; Peng, H.; Qian, X.; Zhang, Y.; An, G.; Zhao, Y. Recent advancements in optical fiber hydrogen sensors. *Sens. Actuators B Chem.* **2017**, *244*, 393–416. [CrossRef]
19. Emiliyanov, G.; Høiby, P.E.; Pedersen, L.H.; Bang, O. Selective serial multi-antibody biosensing with TOPAS microstructured polymer optical fibers. *Sensors* **2013**, *13*, 3242–3251. [CrossRef]
20. Jalal, A.H.; Yu, J.; Agwu Nnanna, A.G. Fabrication and calibration of Oxazine-based optic fiber sensor for detection of ammonia in water. *Appl. Opt.* **2012**, *51*, 3768. [CrossRef]
21. Mohammed, H.A.; Rashid, S.A.; Abu Bakar, M.H.; Ahmad Anas, S.B.; Mahdi, M.A.; Yaacob, M.H. Fabrication and characterizations of a novel etched-tapered single mode optical fiber ammonia sensors integrating PANI/GNF nanocomposite. *Sens. Actuators B Chem.* **2019**, *287*, 71–77. [CrossRef]
22. Mader, H.S.; Wolfbeis, O.S. Optical ammonia sensor based on upconverting luminescent nanoparticles. *Anal. Chem.* **2010**, *82*, 5002–5004. [CrossRef]
23. Chu, C.-S.; Chen, Y.-F. Development of ratiometric optical fiber sensor for ammonia gas detection. In Proceedings of the 2017 25th Optical Fiber Sensors Conference (OFS), Jeju, South Korea, 24–28 April 2017.
24. Zhou, Q.; Kritz, D.; Bonnell, L.; Sigel, G.H. Porous plastic optical fiber sensor for ammonia measurement. *Appl. Opt.* **1989**, *28*, 2022. [CrossRef]
25. Prado, A.R.; Leal-Junior, A.G.; Marques, C.; Leite, S.; de Sena, G.L.; Machado, L.C.; Frizera, A.; Ribeiro, M.R.N.; Pontes, M.J. Polymethyl methacrylate (PMMA) recycling for the production of optical fiber sensor systems. *Opt. Express* **2017**, *25*, 30051. [CrossRef] [PubMed]
26. Ozhikandathil, J.; Badilescu, S.; Packirisamy, M. Polymer composite optically integrated lab on chip for the detection of ammonia. *J. Electrochem. Soc.* **2018**, *165*, B3078–B3083. [CrossRef]

 remote sensing

MDPI

Review

Applications of Unmanned Aerial Systems (UASs) in Hydrology: A Review

Mercedes Vélez-Nicolás [1,*], Santiago García-López [1], Luis Barbero [1], Verónica Ruiz-Ortiz [2] and Ángel Sánchez-Bellón [1]

[1] Department of Earth Sciences, Faculty of Marine and Environmental Sciences, University of Cádiz, 11510 Puerto Real, Spain; santiago.garcia@uca.es (S.G.-L.); luis.barbero@uca.es (L.B.); angel.sanchez@uca.es (Á.S.-B.)

[2] Department of Industrial and Civil Engineering, Higher Polytechnic School of Algeciras, University of Cádiz, 11202 Algeciras, Spain; veronica.ruiz@uca.es

* Correspondence: mercedes.velez@uca.es

Abstract: In less than two decades, UASs (unmanned aerial systems) have revolutionized the field of hydrology, bridging the gap between traditional satellite observations and ground-based measurements and allowing the limitations of manned aircraft to be overcome. With unparalleled spatial and temporal resolutions and product-tailoring possibilities, UAS are contributing to the acquisition of large volumes of data on water bodies, submerged parameters and their interactions in different hydrological contexts and in inaccessible or hazardous locations. This paper provides a comprehensive review of 122 works on the applications of UASs in surface water and groundwater research with a purpose-oriented approach. Concretely, the review addresses: (i) the current applications of UAS in surface and groundwater studies, (ii) the type of platforms and sensors mainly used in these tasks, (iii) types of products generated from UAS-borne data, (iv) the associated advantages and limitations, and (v) knowledge gaps and future prospects of UASs application in hydrology. The first aim of this review is to serve as a reference or introductory document for all researchers and water managers who are interested in embracing this novel technology. The second aim is to unify in a single document all the possibilities, potential approaches and results obtained by different authors through the implementation of UASs.

Keywords: drone applications; surface water; groundwater; photogrammetry; optical sensing; thermal infrared

Citation: Vélez-Nicolás, M.; García-López, S.; Barbero, L.; Ruiz-Ortiz, V.; Sánchez-Bellón, Á. Applications of Unmanned Aerial Systems (UASs) in Hydrology: A Review. *Remote Sens.* **2021**, *13*, 1359. https://doi.org/10.3390/rs13071359

Academic Editor: Giacomo De Carolis

Received: 25 February 2021
Accepted: 31 March 2021
Published: 1 April 2021

1. Introduction

Sustainable water management has become a major concern over the past decades; as water demand increases with socioeconomic development and population growth, the availability of freshwater resources shrinks due to climate change, aquatic ecosystem degradation and anthropogenic impacts. According to UNESCO [1], water use has been increasing 1% annually worldwide since the 1980s and the global demand for water is expected to keep a similar trend until 2050. This would account for an increase of 20% to 30% above the current level of water use. Under these circumstances, ensuring water in adequate quantity and quality to meet food security, environmental targets, public health requirements and the production of energy, services and other goods remains one of the greatest challenges for water managers in coming years. This is especially critical in regions such as sub-Saharan Africa, central and southwest Asia, which are affected by persistent multi-year droughts, or the Mediterranean basin, where water resources are unevenly distributed and present severe deficiencies in its southern and eastern parts. On the other side of the coin, extreme precipitation events and alterations in flood frequency and duration are affecting an increasing number of countries globally, causing loss of lives, health-related issues and multiple social and economic damages. These extreme hydrological phenomena are likely to be exacerbated as a result of climate change, with increases in their frequency according to the Fifth Assessment Report of the IPCC [2]. Likewise, water quality issues are becoming a major concern. Numerous aquatic systems have undergone severe pollution processes linked to agricultural, domestic and industrial activities and waste disposal, situation that compromises the supply of clean potable water and have deleterious effects on the ecosystem functioning and biota.

This scenario demands effective management and intervention in catchments, which necessarily involves gaining further knowledge of hydrological systems and filling current information gaps. In this regard, unmanned aerial systems (UAS) also known as unmanned aerial vehicles (UAVs), remotely piloted aircraft systems (RPAS) or drones, have recently emerged as new allies in environmental monitoring and management. UAS have not only made bird's eye observation a reality, but also offer a vast range of applications that is continually growing as technology advances. Although initially devised to support military operations, the civilian and scientific applications of UAS have attracted increasing attention in recent years, experiencing an exponential growth in their commercial, governmental and amateur use. The advances in fabrication, remote control capabilities and power systems along with the improvements in sensing technologies installed onboard, have led to the development of a wide range of UAS that can be used to obtain valuable information in different contexts. Numerous advantages of UASs over other systems can be cited; they are portable and enable the retrieval of data with very-high spatial resolutions and unprecedent temporal coverage. They also enable engagement with areas that would otherwise be inaccessible or cost-prohibitive, especially if compared with methods such as airborne campaigns. Moreover, these platforms are easy to deploy, can be flown in small enclosed areas [3] and their imagery might constitute systematic and permanent data that can be used by other individuals and organizations [4]. Regarding the latter aspect, it should be noted that there is still a long way to go in terms of standardization of the methods and available information. To date, UASs have been implemented in a wide range of fields, such as wildlife research and monitoring [5,6] forestry [7,8], precision agriculture [9], architecture, engineering and construction [10,11], disaster management [12] and social research just to name a few.

The need for monitoring the elements of the hydrological cycle is a widely recognized issue. Although in the last decades the field of hydrology has witnessed tremendous advances that have resulted in an increasing number of papers on the subject, as McCabe et al. [13] pointed out, there are still significant gaps in our hydrological knowledge and analysis capabilities such as the estimation of water and mass transport processes across aquatic–terrestrial interfaces, groundwater depth and storage, deep soil moisture, evaporation or snow water equivalent, among others. In this regard, the emergence of UAS is leading to improvements in the understanding of hydrological processes and the management of water resources [14].

This article presents a review focused on the latest applications and advances of UAS technology in the field of hydrology, concretely in water resource research, comprising surface water and groundwater. Although there are other reviews in this field, some have a general scope, addressing the application of UAS to a wide range of civilian and environmental purposes [15] while others focus on very specific aspects of hydrology. For instance, Carrivick and Smith [16] reviewed the application of the structure from motion (SfM) algorithms in photogrammetry for aquatic and fluvial environments, Tomsett and Leyland [17] the current applications of remote sensing in river corridors and Rhee et al. [18] the application of UAS remote sensing to fluvial environments exclusively. DeBell et al. [14] focused on pragmatic concepts of UAS technology and sensors adequate for water resource management. Several authors reviewed the applications of UAS in harmful algal bloom studies [19,20]. Here, nevertheless, we address the vast range of UASs applications in inland water bodies with a purpose-oriented approach, classifying them according to the aim of the research conducted. This paper addresses (i) the current applications of UAS in surface and groundwater research, (ii) the type of UASs and sensors mainly used in these tasks, (iii) the type of products generated from UAS-borne data, (iv) the advantages and limitations associated with their use and (v) knowledge gaps and future prospects of UASs application in freshwater research and management. Studies on oceanic-coastal applications, hydric erosion, atmospheric water cycle, glaciology and aquatic biodiversity/ecology were excluded from this review.

2. Methodology

An exhaustive search of papers published in academic journals was carried out from August to November 2020. Papers without direct relevance to the use of UAS in freshwater research were discarded. The academic databases and search engines used were Scopus, Google Scholar, Science Direct and Web of Science. Search results were restricted to English language. Table 1 depicts the keywords used in the bibliographic search, including drone-related and hydrology-related terms.

Table 1. Search terms.

UAS and Sensor-Related Terms		Hydrology-Related Terms	
Drone LiDAR Remote sensing RPAS UAS UAV Unmanned Aerial System Unmanned Aerial Vehicle Structure from Motion RGB	Photogrammetry Multispectral Hyperspectral Thermal infrared	Algal bloom Aquifer Bathymetry Flooding Freshwater Groundwater Hydrology Morphology River River discharge Riverine plastic Runoff	Submarine groundwater discharge Subsidence Surface velocity Surface-Groundwater interaction Thermal plume Thermal structure Water budget Water level Water pollution Water sampling Water storage Water table Wetland

For each application covered in this paper, search was conducted using different combinations of keywords, such as: "UAS hydrology"; "unmanned vehicle bathymetry"; "river bathymetry UAV"; "UAV river discharge"; "water surface velocity UAV", "algal bloom UAV", "UAS submarine groundwater discharge", etc.

Only those papers published over the last two decades (2000–2020) were considered in order to report the most updated information and ensure the inclusion of the most relevant works, given the recent development of this technology.

The study focuses on the applications of UASs in hydrology, concretely in freshwater research, thus the literature search process was restricted to those experiences in which UAS were used to study rivers, lakes, reservoirs, wetlands, groundwater bodies and the interactions among them. Figure 1 shows the steps followed in the literature search and classification process, from the initial screening and selection of eligible articles to their final inclusion. Research articles were manually screened by reading the title and abstract, and to ensure relevance, were subsequently analyzed in detail. After discarding duplicates and non-relevant works, a total of 122 research papers were selected and subsequently categorized according to the focus of the research work (Surface-groundwater interactions, water levels, contamination studies, system dynamics, etc.). It should be noted however, that some of the UAS applications presented here could fit into more than one category. For this reason, the papers have been classified according to the main objective of the study.

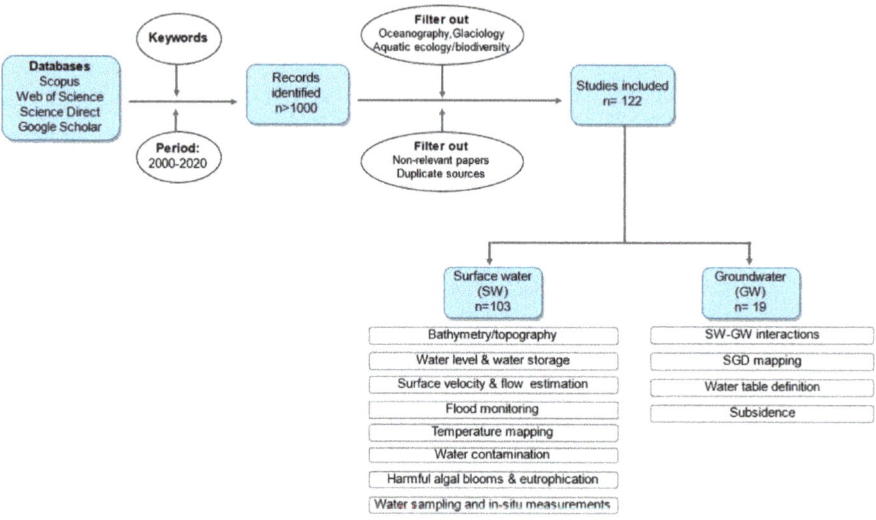

Figure 1. Process of literature search and classification.

3. Types of UAS, Sensors and Other Payloads

3.1. Platforms

These sections do not intend to make an exhaustive overview on the type of UAS, sensor, power systems, etc., but to provide the necessary background for the comprehension of the review as a whole. The vehicle itself is just one part of the complete unmanned aerial system as recognized by other authors [21]. Components such as the control and command element, the payload, take off and recovery system, data link and transmission, and the most important part: the human component, are essential for the system to perform its task.

An unmanned aerial system consists of a flying device that includes: (i) A platform with the structural, mechanical and electronic elements necessary for the flight, its control and stability, (ii) a set of sensors and devices for the acquisition of information from the environment and (iii) a ground control station. UAS characterization can be made based on different variables. Normally, they are classified based on weight [22,23]. Terms such as "nano drone", "micro drone", "mini drone", "pico-drone", etc., are the most frequently used. Apart from weight, other authors include in the classification the typology of aerial vehicle itself or the take-off and landing mode: horizontal take-off and landing (HTOL) versus vertical take-off and landing (VTOL) [24]. Unconventional UAS such as bio-drones, smart dust, air-land-water hybrid drones are being considered in recent times [24]. In any case, most of the applications discussed in this work refer to "lightweight UAS", which due to their reduced maximum take-off weight (MTOW) (less than 7 kg), affordable price, easy deployment and high performance are widely used in scientific research. The most affordable models range between 500 and 1500 €, while the most sophisticated can reach prices up to 120,000 € including the sensing equipment. Although some authors such as Watts et al. [25] propose a classification for civil applications that ranges from very small UAS to large remotely controlled vehicles that can reach thousands of meters in height and many hours of autonomy, legal restrictions in most countries justify a more restrictive classification. DeBell et al. [14], classified the lightweight UASs and analyzed their main characteristics and the available sensors with application to the management of water resources. More recently, Johnston [26] established a new classification and provided a current overview of platforms and sensors that, although it is oriented at marine research and conservation, it is also of interest for research on water resources.

In general, for most hydrological applications these platforms can be categorized according to their airframe configuration, propulsion method, and flight characteristics into four main types: multi-rotor, fixed wing, transitional and others (balloons, kites and blimps). Multi-rotor UASs have multiple engines and propellers that allow the lift, movement and orientation of the platform (pitch, roll, and yaw). Multi-rotors, also known as multicopters, typically employ four (quadcopter), six (hexacopter), or eight (octocopter) motors and propellers; by increasing or decreasing the output of individual motors it is possible to control the movement of such vehicles. A greater number of motors, especially in larger UASs, reduces the risk of falling due to mechanical failures, even though this redundancy is often accompanied by a reduction in efficiency. In fact, the main limitation of multirotor systems is their flight time, mostly due to higher energy demands, battery weight and energy storage constraints. Finally, these types of UASs are manufactured from lightweight and resistant materials such as aluminum, composites (carbon fiber) and plastic.

Fixed-wing UASs have one or two wings that provide maneuverability and lift; they can have several wings configurations (gliders, delta wings, canards, among others). Fixed-wing UASs are generally driven by one or two motors and propellers in a push or pull configuration. These types of UASs are manufactured from very lightweight materials, including various forms of expanded foam (polyolefin, polypropylene or polystyrene) and composite. The flight efficiency of these aircraft is usually higher than in the multi-rotors UASs, due to their lower relative weight and the fact that the lift is of the passive type, through large airfoil surfaces.

Transitional UASs combine aspects of multi-rotor and fixed-wing devices to provide greater flexibility in their use. They intend to incorporate the advantages of the vertical take-off and landing of the former with the greater autonomy and range of action of the latter. Vertical flight is typically driven through three or four motors facing upward, and horizontal flight is driven by a motor and propeller in a push configuration while lift is produced by the wings.

Finally, the category "other" includes devices that either do not have autonomous means of propulsion (kites and balloons) or, if they do, the lift is given by gas lighter than air (blimps). In recent years, these devices have been gradually replaced in most applications by the previous types, due to technological development which have notably improved their performance. Table 2 shows the main characteristics of the platform classes and Figure 2 illustrates different types of platforms and sensors.

Albeit different power supply and energy management systems can be used (see a review on this topic in [27]) most UASs used for research purposes are battery powered. Lithium polymer batteries provide significant energy density and can discharge enough electrical current to meet the variable electrical current demands of the engine and payload. Some systems are hybrid, employing combustion engines together with battery-powered electric motors.

The flight control system, sometimes named autopilot, is the core of the vehicle. It normally includes a flight control processing unit composed of an inertial measurement unit (IMU), electronic speed control (ESC) units, barometer and a global navigation satellite system (GNSS). The IMU consists of a set of accelerometers (normally 3), gyroscopes and magnetometers whose main purpose is determining the orientation, linear and angular accelerations to estimate the movement direction with respect to the earth magnetic field. A data storage system and communication modules (telemetry) are also additional parts of the flight control system. UASs rely on two main navigation technologies, Inertial Navigation Systems (INS) and Global Navigation Satellite Systems (GNSS), so that they can work in a GNSS-only or INS/GNSS coupling mode. The cost reduction and miniaturization of high-performance GNSS units, which allow connection to a nearby reference station and reception of correctional data in real time (kinematic correction), can provide accuracies of 3–5 cm in the dynamic position of the platform, with the consequent improvement in the observations made using sensors and devices.

Table 2. Classification of lightweight unmanned aerial systems (UAS) platforms and their main characteristics.

Main Features	Multi-Rotor	Fixed-Wing	Transitional	Other (Balloons, Kites and Blimp)
Diameter or wing-span	35–150 cm	100–200 cm	100–200 cm	Up to 5 m
Flight time	15–50 min	25–75 min	25–90 min	Hours
Payload Capability	1–2.5 kg	1–2 kg	1–2 kg	>2.5 kg
Maneuverability	High	Medium	Medium-high	Low
Wind resistivity	10–15 m/s	8–20m/s	12 m/s	Highly variable
Ability to fly under windy conditions	Medium	Medium-high	Medium-high	Low
Spatial coverage in a single flight	20–40 ha	80–320 ha	80–320 ha	Variable
Expertise required	Low	Medium	High	Low
Take-off and landing capability	Vertical take-off and landing (VTOL)	Launch line, catapult, or hand launch; open area for landing or parachute	Vertical take-off and landing (VTOL)	Vertical take-off and landing (VTOL)
Velocity or thrust failure	Crash	Glide capability, controlled crash, or parachute	Glide capability, controlled crash, or parachute	Variable
Ability to carry sampling devices	Yes	No	No	No

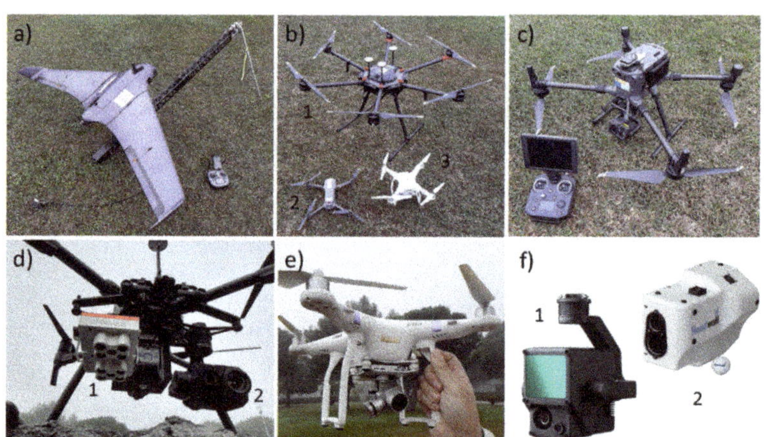

Figure 2. (**a**) Fixed wing UAS Atyges-FV1 on catapult; (**b**) 1. Hexacopter DJI Matrice 600 pro, 2. Micro UAS Quadcopter DJI Mavic Pro, 3. Micro UAS Quadcopter DJI Phantom 3 professional; (**c**) Quadcopter DJI Matrice 300 RTK and radiocontrol station; (**d**) 1. Multispectral camera Micasense dual (10 bands) and 2. visible and thermal camera Zenmuse XT2 onboard a DJI Matrice 210 RTKv2; (**e**) RGB camera Sony EXMOS onboard a DJI Phantom 3 professional; and (**f**) 1. LiDAR DJI Zenmuse L1 and 2. hyperspectral camera Headwall VNIR-SWIR. Courtesy of the Drone Service of the University of Cádiz.

3.2. Sensors and Other Payloads

The application of UASs in hydrology fundamentally encompasses the field of low altitude aerial Photogrammetry and Remote Sensing (PaRS), although other non-remote applications have also been successfully developed. In any case, an important limitation of these systems is the size and weight of the payload, especially in lightweight UASs, which rarely exceed 2.5 kg. This implies that the sensors and devices included in the UAS must meet special requirements in terms of lightness and packaging. Fortunately, in recent years a substantial number of sensor systems for PaRS applications that meet these conditions have been specifically developed by manufacturers to be mounted in lightweight UAS. Most of the sensors used in hydrological UAS-based mission are essentially digital cameras, which are optical-electronic sensors. These sensors consist of an optical system (lens) that captures and projects electromagnetic radiation onto a CCD (Charge Coupled Device) or alternatively onto a CMOS (Complementary Metal-Oxide-Semiconductor) device, so that the light generates in each element of a matrix an electrical current that is registered and encoded. Currently, there is a wide range of passive optical-electronic sensors available for UAS, which covers a wide spectrum range (400–14,000 nm) and includes RGB, multispectral, hyperspectral and thermal infrared (TIR) cameras.

The usefulness of RGB cameras, which record radiation in the visible spectrum, lies in obtaining aerial videos and imagery that can be processed to orthorectified image mosaics and digital surface models (DSM) using photogrammetry Although some models already exist, usually RGB cameras onboard UAS are non-metric, which implies that the interior orientation elements (focal length, principal point location, distortion parameters, etc.) have not been determined through a calibration process, in addition to not meeting certain requirements for lens quality and dimensional stability of its components. This aspect, which constitutes a serious drawback for stereoscopic photogrammetry, has been solved using new digital photogrammetry methods such as Structure from Motion (S*f*M). By using multiple overlapping images, S*f*M reconstructs the 3D scene structure, camera positions and orientations from a set of feature correspondences. The incorporation of bundle adjustment techniques, which consist of a non-linear refinement of camera and point parameters, has enabled to optimize S*f*M 3D reconstructions and to minimize reprojection errors. [28]. Many UAS are endowed with consumer-grade RGB cameras that provide excellent image quality in a relatively small and light package along with flexibility in imaging through their interchangeable lenses. As well, some RGB cameras have been created specifically for UAS applications.

Multi- and hyperspectral cameras are typically used for material identification and mapping purposes based on the study of the spectral signature of the different materials (vegetation, soils, water constituents, waste, etc.) rather than defining metric properties. These sensors sample multiple bands of the electromagnetic spectrum, from blue (400 nm) to near infrared (840 nm), short wave-length infrared (1800 nm) and mid-wavelength infrared (4000 nm). While multispectral cameras detect radiation in a small number of bands with relatively large bandwidth (20–40 nm), hyperspectral cameras can sample a broad range with a much higher spectral resolution, even higher than 200, with narrower bandwidths (2–5 nm). The reduced need for correction of the atmospheric effect thanks to the very low flight height of these platforms facilitates calibration using standards and field radiometry. Nevertheless, several environmental factors such as light angle, shades, water vapor content or technological constraints like the relatively high signal-to-noise ratio makes it necessary to apply certain corrections during image processing. Adao et al. [7] or He and Weng [29] among others, reviewed a range of hyperspectral sensors used with UASs.

TIR sensors are aimed at detecting long-wave infrared energy (8000–14,000 nm) emitted by objects in the camera's field of view. Since the emitted energy is directly proportional to the fourth power of the thermodynamic temperature of the surface, TIR sensors can be used to generate surface temperature maps but need to account for emissivity effects such as angle of observation or material. Another aspect to consider is the time of observation; while during day the recorded energy is a mixture of purely emitted and reflected radiation, in the absence of incoming radiation (e.g., during night) the energy and temperature maps show the truly emitted radiation of the surface. TIR sensors use microbolometers as detectors; devices that change their electrical resistance as a function of temperature.

This resistance change is measured and transformed into temperatures which can be used to create an image. Currently, there are two main types of commercial TIR systems; (i) sensors with cooled microbolometer arrays, which measure the short-medium wave IR, offer greater resolution and thermal sensitivity but are bulkier and more expensive and (ii) sensors with uncooled microbolometers, which work in the LWIR spectrum and are less susceptible to solar reflection. As cooled microbolometers are heavier, they are normally used in airplanes or satellites. Uncooled microbolometer on the other hand are lighter and cheaper, what makes them suitable to be mounted on UAS, but they can experience temperature drift issues [30]. This temperature drift problem is an inherent consequence of the set-up of the sensor; uncooled microbolometers are infrared sensors distributed in an array where each unit function as single sensor elements. Temperature drift in the system will occur due to temperature changes inside the camera or due to heating of the lenses or focal plane array. The most common approach to tackle with this issue is the application of non-uniformity corrections to remove noise, what leads to a more homogenized response signal along the microbolometer array [31,32]. In this regard, ref. [32] proposes a workflow for data acquisition and preprocessing along with best practices to reduce camera uncertainty in long and short-term noise. In recent years LiDAR (Light Detection and Ranging) sensors are being incorporated into lightweight UAS platforms. These sensor packages, which are active type, use lasers to scan the environment to produce 3D point clouds of the terrain, bottom of surface water bodies and vegetation.

Finally, in addition to the devices that remotely acquire information from objects avoiding any physical contact, UAS platforms can carry instruments such as multiparameter probes (temperature, electrical conductivity, pH, Eh, dissolved oxygen, oxidation-reduction potential, etc.) to be dipped into the water column, different types of samplers, Geiger counters or magnetometers or multi-gas monitors among others. In all cases, UAS constitute a platform from which conventional measurements can be carried out with great flexibility in terms of location and movement, reducing campaign costs and minimizing risks [33]. A general overview of the different types of sensors and their advantages, limitations and potential applications is provided in Table 3.

Table 3. Overview and some examples of sensors commonly used with UAS in hydrological research.

Types of Sensor	Spectral Ranges (nm)	Camera Examples from Cited Works	Hydrological Applications	Main Advantages and Disadvantages
RGB	~400–700	Canon Powershot G5/Canon EOS, [34] Zenmuse X3 FC350, [35] RGB OLYMPUS EP-2, [36] Nikon D500/D5100, [37]	Visual analysis, bathymetry, DEM photogrammetry, water stages, flood monitoring, particle velocymetry, HAB [1] monitoring, mapping and classification of surfaces.	Advantages: (1) Wide range of prices, resolutions and weights available depending on the model. (2) Video capture. (3) Easy integration in different platforms. Disadvantages: (1) Lower spectral resolution that makes them unsuitable for many tasks. (2) Very sensitive to environmental and ilumination conditions. (3) Some lack of geometric and radiometric calibration.
Multispectral	~400–1000	Rededge Micasense [38,39]	Bathymetry, HAB monitoring, river/lake trophic status, flood monitoring, SW–GW [2] interactions, wetland/river mapping, surface/material identification.	Advantages: (1) Wider range of applications compared to RGB sensors, allowing to discriminate/identify a variety of materials. (2) Some present means of radiometric calibration. (3) Allow geometric reconstruction. (4) Allow sub-decimetric mapping. Disadvantages: (1) Relatively high prices. (2) Detect radiation in a small number of broad wavelength bands, limiting their applications. (3) Currently, sensors are not optimised for aquatic applications. (4) Limited compatibility to UASs.

Table 3. *Cont.*

Types of Sensor	Spectral Ranges (nm)	Camera Examples from Cited Works	Hydrological Applications	Main Advantages and Disadvantages
Hyperspectral	~500–2500	Rikola 2D, [36]	Bathymetry, HAB monitoring, river/lake trophic status, flood monitoring, water quality monitoring, wetland/river mapping, surface/material identification.	Advantages: (1) High spectral resolution data, with many narrow contiguous spectral bands that improve their ability to discriminate/identify materials.
		Nano Hyperspec, [40]		Disadvantages: (1) High costs. (2) Size of sensors. (3) Need for spesialised software. (4) Low signal-to-noise ratio.
Thermal infrared	~8000–14,000	ThermoMAP, [41]	River/lake temperature mapping, SW- GWD identification, thermal plumes identification, river discharge.	Advantages: (1) Validation is not required if only relativetemperatures are needed. (2) Relatively low cost. (3) Wide range of models and resolutions.
		DJI Zenmuse XT, [42]		
		ICI Mirage 640, [43]		Disadvantages: (1) Temperature drift issues. (2) Radiation emitted from near-bank objects may impact the sensor, resulting in erroneous image interpretation. (3) Need for radiometric corrections. (4) TIR imagery interpretation can be complex and requires expertise. (5) Highly sensitive to <30° observation angles and changes in surface roughness.
		FLIR TAU2 640 [44]		
UAS LiDAR	~500–900	Phoenix Scout SL1 [45]	Bathymetry, 3-D mapping, water stages, flood monitoring.	Advantages: (1) Less susceptible to environmental conditions. (2) provides direct geometric measurements. (3) Possibility of discriminating the effect of vegetation.
		ASTRALiTE edge [43]		Disadvantages: (1) Limited by water clarity and bottom reflectivity. (2) High costs. (3) Few models compatible with UAS. (4) Need for groundfiltering corrections. (5) Strong dependence on accurate dynamic positioning systems.

[1] Harmful algal bloom; [2] surface water–groundwater.

4. UAS Applications in Surface Water Research

In the last two decades, UAS have become a widely applied tool in fluvial environments and specially in the study of river morphology. This has resulted in a large body of available literature reporting the use of UAS for the study of channel evolution and bedform migration, bank erosion, bed grain size and fluvial topography/bathymetry mapping among others [46–49]. Likewise, there is a growing number of papers addressing other hydrological aspects such as river stage fluctuations, water budgets, river discharge, surface velocity or flooding and temperature mapping. In addition to remote retrieval of data, in recent years, UAS are being used as platforms to deploy a variety of sensors, as water samplers or as key elements of water monitoring systems. This section presents a collection of studies focused on hydrological issues, the advances they represent and the role that UAS have played in their execution.

4.1. Bathymetry and Submerged Topography

Knowledge of river and stream bathymetry is fundamental to study fluvial processes and provides essential inputs to hydrodynamic models. Bathymetric measurements were traditionally carried out manually or using single/multibeam echo sounders mounted on vessels. These methods, which are not only time-consuming but also difficult to implement in deep streams and rivers with strong currents, have been progressively complemented with high resolution systems (remote sensing techniques, satellite imagery, aerial photography and laser scanning) and particularly, with UAS. In the context of remote sensing, bathymetric mapping can be tackled with three approaches; (i) spectral methods based on the attenuation of the electromagnetic wave in the water column and the reflection from the bottom of the water body, (ii) photogrammetry, which uses sets of images recorded with imaging sensors to identify coordinates of points, boundaries and features in the images, and (iii) bathymetric LiDAR, which measures the time from pulse emission and echo reception, scattered back from the river/lake bottom within the instantaneous field of view. These remote sensing tools can significantly improve the accuracy and reliability of river geomorphology mapping and modeling specially when high topographical resolution is required.

Optical methods such as UAS photogrammetry were traditionally restricted to above-water studies, but nowadays their scope of application has been extended to below-water areas due to the development of refraction-correction algorithms. This has made possible to capture water surface and bottom features and water column characteristics as long as the system is not masked by vegetation or other obstacles and water is clear. One of earliest examples of this type of UAS application is provided by Lejot et al. [34], who launched a paramotor-paraglider equipped with a conventional RGB camera to acquire high resolution imagery and study channel water depth and gravel bar geometry at the Ain and Drome rivers (France). A simple empirical model that related radiometric signal with water depth was used to characterize aquatic zones of different size, yielding spatial resolutions of 5–7 cm. The water depth surveyed ranged from 0 to 5 m, and the results showed good accuracy until 3 m. The gravel bar was characterized through classical photogrammetry with a DEM generated using stereoscopic pairs, achieving 5 cm resolution. Although heterogeneous, the quality of both products was sufficient for bathymetric and channel microtopography mapping. In a pilot study Zinke and Flener [50] generated bathymetric maps at 2 sites of a gravel riverbed in Norway using drone-borne photography and the Lyzenga's deep water correction algorithm. Although the results of this study offered variable levels of accuracy and precision, the authors demonstrated the usefulness of UAS-based optical remote sensing methods to improve the effectiveness of bathymetric surveys. Flener et al. [37] created a seamless DEM of a subarctic river channel and plains combining mobile laser scanning and UAS-photography based bathymetric modeling. Albeit water depth in the study area did not exceed 1.5 m, the authors found quality issues in the UAS imagery that were mostly related to reflection (sunglint) and illumination changes during the flight. Accuracy varied depending on the combination of methods applied. Despite these problems, the UAS-photography-bathymetric model gave depth accuracies below 10 cm, which is suitable for applications such as hydraulic modeling or habitat studies, but insufficient for more detailed studies.

On the other hand, structure from motion (S*f*M), has been widely used in fluvial bathymetry in recent years. For instance, Woodget et al. [51] quantified the exposed and submerged topography of two river reaches (depths ranging from 0.14 to 0.70 m) in UK through S*f*M photogrammetry. An UAS equipped with a RGB consumer-grade camera was used to produce DEMs with hyperspatial resolutions of 2–15 cm. Mean errors in submerged areas ranged between 1.6–8.9 cm but were significantly reduced after the application of a simple refraction correction, reaching values between −2.9 to 5.3 cm. Dietrich [52] retrieved fluvial bathymetric measurements from UAS imagery and S*f*M and proposed a multicamera-based refraction correction method for off-nadir S*f*M datasets. In this case, the surveyed water depth varied between 0 and 1.5m. The bathymetric datasets obtained after the correction showed precisions of ~0.1% of the flying altitude (40 and 60 m above ground level), which makes this method suitable for a range of fluvial applications. Entwistle and Heritage [53] generated a S*f*M-derived DEM with imagery collected from a small UAS to predict water depth and map bathymetric surfaces. The results from the S*f*M survey were compared with others from a theodolite survey, showing similar accuracy and precision (0.85 R^2 value after using a 1.02 multiplier on the regression line up to depths of 1 m), thus demonstrating that the UAS approach proposed could achieve good depth estimations. Carrivick and Smith [16] reviewed the applications of S*f*M photogrammetry and UAS technology in aquatic environments, highlighting the need of automated procedures to correct refraction. The issue of refraction correction has been addressed in several works; Woodget et al. [35] quantified above and below-water geomorphic changes in a river through SfM photogrammetry and analyzed the implications of refraction, water surface elevation and the spatial variability of topographic errors. They demonstrated that it is possible to quantify submerged geomorphic changes with levels of accuracy of less than 4 cm similar to that from exposed areas without the need of calibration data and that, using nadir imagery, the results obtained after different refraction corrections are practically the same. Partama et al. [54] presented a novel technique based on co-registered image sequences or video frames to reduce the effects of water-surface reflection on UAS-based photogrammetry. This promising method, applied in a river reach 0–1.5 m deep, achieved accuracies of 5–15 cm, clarified the reflected signal from the bottom bed and enabled to reduce moving light patterns. The issue of refraction correction has been addressed by several authors, although most works focus on the marine environment, for instance Skarlatos and Agrafiotis [55] proposed an iterative algorithm that, applied at the photo level, could reduce the effect of refraction to 2 times the ground pixel size within the photogrammetric workflow. Also in marine environments, Agrafiotis et al. [56,57] provided a deep-learning framework to automatically correct water refraction errors. The method consisted in a support vector regression model based on known depth observations from bathymetric LiDAR surveys, which enabled to estimate the real depth of point clouds obtained from SfM and multi-view-stereo techniques. The model demonstrated great potential in terms of depth accuracy in shallow waters and can be used when no LiDAR data are available.

River depth estimations have also been successfully retrieved from UAS hyperspectral data; Gentile et al. [36] used very high spatial resolution hyperspectral images and an empirical model to map the bathymetry of a shallow river, achieving a good fit in the spectral range from 700–800 nm and average errors of even less than 13 cm at depths between 9–101 cm.

On the other hand, LiDAR sensors were rarely implemented in UAS owing to their high costs and usually excessive weight, however, these issues are being overcome leading to the democratization of these systems and making their use increasingly frequent. For these reasons, until recently, a widely applied alternative to optimize photogrammetric DEMs and produce more accurate products was the complementary use of satellite, airborne or land-based LiDAR data and UAS optical images. The performance of multispectral satellite imagery, hyperspectral data from manned aircraft and UAS, and water penetrating green LiDAR to map fluvial bathymetry was assessed by [40] in a short reach of Sacramento River with a mean depth of 1.8 m. LiDAR systems offered greater accuracies, while UAS-borne products tended to underpredict depths, reaching R^2 values of 0.95 and 0.88 respectively. However, the maximum detectable depth for the LiDAR sensor was an important constraint due to the lack of bottom returns at depths greater than 2 m. Given the constraints of remote sensing of river bathymetry, the authors advocate for a hybrid field and airborne approach but recommend the use of hyperspectral or multispectral UAS imagery for small channels and fine-scale information owing to their versatility, high spatial resolution and the possibility of collecting targeted and task-specific data. A major leap forward for river mapping is the methodology presented Mandlburger et al. [58] who developed a novel compact topo-bathymetric laser scanner designed for integration on UASs and other types of aircrafts. The sensor comprised an IR laser that emitted pulses at a wavelength of 532 nm with a duration about 1.5 ns pulse and a pulse repetition rate of 50–200 kHz. The laser had a 10–30 cm footprint on the ground and yielded a point density of 20–50 points/m^2, which made it particularly well suited for capturing river bathymetry. The system was tested in freshwater ponds with an overall depth of 5–6 m and turbid bottoms. Although the system presented a maximum vertical deviation of 7.8 cm and a systematic depth-dependent error, the sensor showed great potential for hydrology and fluvial morphology applications owing to its good depth performance, accuracy and high spatial resolution.

Bathymetric analysis has also been addressed with multiplatform and multisensor approaches. Kasvi et al. [59] addressed the accuracy of shallow water (0–1.5 m) bathymetric mapping through the implementation of three remote sensing techniques: optical and bathymetric S*f*M modeling, both from UAS data, and a remote controlled acoustic doppler current profiler (ADCP) coupled with an echo-sounder. Their results (Figure 3) emphasized how the characteristics of the study site such as river size, depth and water quality determine the results and suitability of each method.

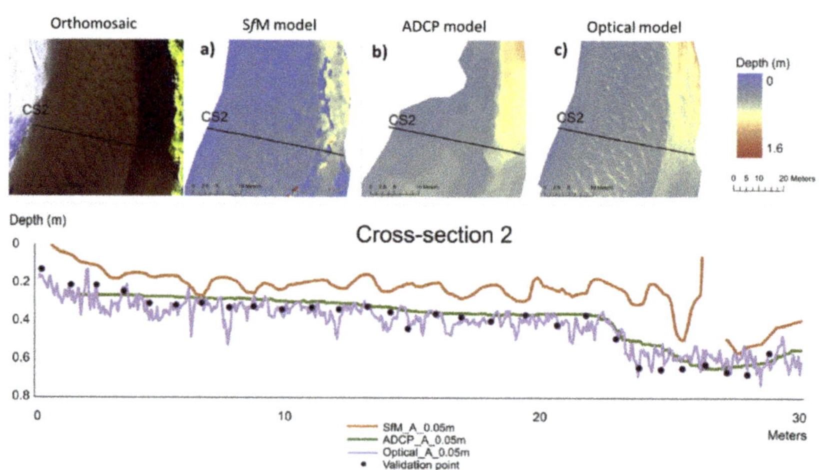

Figure 3. Orthomosaic and bathymetric models of 5 cm resolution obtained from (**a**) structure from motion (SfM), (**b**) acoustic doppler profiler and (**c**) optical modeling. CS2: cross-section of the river studied. Modified from [59].

Although the results of the echo-sounding were less detailed and restricted to depths over 0.2 m, they were the most accurate. On the other hand, the bathymetric S*f*M was highly sensitive to flow turbidity, color and therefore depth, with mean errors that increased notably at depths >0.8 m and notably changed between autumn (−50 cm mean error) to spring (−120 cm). Optical modeling, was very sensitive to substrate variability, color and depth. This evidences that bathymetric S*f*M is more suitable for clear waters and structured and visible riverbeds. Erena et al. [60] proposed a method to create updated capacity curves for 21 reservoirs of the Segura River Basin (Spain). Their study integrated bathymetric measurements acquired using an Unmanned Surface Vehicle (USV) and an Unmanned Underwater Vehicle (UUV) with a DSM generated from UAS imagery and airborne LiDAR data. Apart from evidencing a substantial silting in the reservoirs, this multiplatform methodology demonstrated being more efficient than the traditional methods.

UAS have also been used to deploy bathymetric measurement instruments such as sonars. This is the case of the study by Bandini et al. [61], who measured accurate water depths in lakes and rivers of Denmark using a tethered floating sonar controlled by an UAS. This method, which enables to monitor dangerous or inaccessible areas, yielded significantly accurate results (~2.1%) although some observational bias (attributed to the dependence of the sound wave speed on temperature, salinity, and pressure) was detected. In a U.S. reservoir, Álvarez et al. [62] implemented a novel bathymetric survey method by merging two UAS-based sampling techniques; (i) a small UAS that propelled a mini boat equipped with a single-beam echosounder for surveying submerged topography in deeper waters and (ii) S*f*M photogrammetry to cover shallower areas visible from the UAS but not detected by the echosounder. The UAS-S*f*M data successfully complemented the echosounder survey and minimum depths from 0 to 5 cm were detected.

Finally, it should be noted that many studies on bathymetric reconstruction of shallow streams and reservoirs proved that models derived from high resolution UAS data are more robust than previous models based on conventional methods, reducing inaccuracies and biases, and enabling better hydraulic modeling performance [63–66].

4.2. Water Level Measurement and Water Storage

Albeit the temporal and spatial fluctuations of water stored in rivers, lakes, reservoirs and wetlands are among the most important hydrologic observations, the current knowledge of these variables is still considerably improvable, limiting our water management and forecasting capacity. Water level measurements have been traditionally made with limnimetric scales and limnigraphs, which despite being useful for determining water stages, only provide information of very specific points in the water body. The translation of water levels into stored volumes can be complicated and give imprecise results. In this regard, achieving a better understanding of these processes and reliable hydrologic predictions require accurate observations of several variables, such as surface water area, water level elevation, slope and temporal changes [67]. In this context, UAS are playing a key role in quantitatively characterizing and monitoring water level and their usefulness has been demonstrated in recent works. For instance, UAS allow to retrieve data from inaccessible environments and to overcome the disadvantages of traditional static sensors, which can be deployed in limited numbers and imply higher personnel and maintenance costs. UAS imagery has been used to obtain rapid observations of water level fluctuations in large-scale hydrological systems such as reservoirs. Ridolfi and Manciola [68] proposed a sensing platform that allowed to measure the water level and collect hydraulic information during adverse operation conditions such as floods. Water level in a dam was retrieved from the combination of UAS-based RGB imagery and optical methods based on ground control points and edge detectors. Their results were promising and proved the suitability of the platform for hydraulic measurement acquisition, however, the pixelization of the images and the perspective constituted an important source of uncertainty. Gao et al. [69] devised an innovative methodology that integrated UAS photogrammetry and image recognition to define water level in a complex hydrological environment (plunge pool downstream a hydropower station) in China. Water surface fluctuations were captured by an UAS and the regions of interest were extracted from imagery. The relationship between image pixel scale and actual spatial scale, the real water surface elevation and fluctuations, as well as the maximum, average and minimum water levels were obtained applying specific calibration coefficients. The method was successfully implemented and offered reliable results that back its potential to monitor water level.

The water level in rivers and lakes was estimated in [70] using three types of sensors (radar, sonar and a camera-based laser distance sensor) capable of measuring the range to water surface onboard lightweight UAS. Water surface level was calculated by subtracting the range measured by the sensors from the vertical position retrieved by the onboard GNSS receiver. The radar offered the best results in terms of accuracy (0.5% of the range) and longest maximum range (60 m), whereas the sonar was more appropriate for stable and low flight altitude. The laser system gave less accurate results but proved to be useful for narrow fields of view. This system proved that water level measurements acquired with ranging sensors and GNSS receivers can provide more accuracy than spaceborne or airborne altimetry. In a more recent work, ref. [71] retrieved water surface elevation (WSE) measurements in very small vegetated streams (1–2 m wide) using a drone-based radar altimetry solution with full waveform analysis. The water surface elevations measurements provided by this system were one order of magnitude better than those obtained from LiDAR or photogrammetry. Following a similar approach, Jiang et al. [72] deployed a UAS-based radar altimetry system to map spatially distributed WSE in a small vegetated stream and subsequently used the dataset obtained to calibrate and validate roughness parameters in hydrodynamic models. UAS altimetry delivered WSE observations with 3 cm accuracy and 0.5 m resolution and could identify significant variations of the Manning–Strickler coefficients over time. The study demonstrated that UAS-borne WSE constitutes a useful tool to understand the variations of hydraulic roughness. In order to assess the usefulness of various UAS geospatial products in water body recognition, Tymkow et al. [73] combined data from UAS-based laser scanner, thermal infrared and RGB imagery to estimate the water extent of several rivers and lakes. According to their results, the most suitable product in water body detection was four bands RGB + TIR (Thermal InfraRed) orthomosaic, achieving an average kappa coefficient above 0.9.

On the other hand, *Sf*M is currently offering promising results in the reconstruction of free water surfaces. The potential innovative applications of this technique have been highlighted in several recent works. Niedzielski et al. [74] observed the water stages in the Scinawka river (SW Poland) by combining UAS RGB imagery and *Sf*M algorithms to generate multitemporal orthophotomaps without using ground control points. The authors detected statistically significant increments in water surface areas between orthophotomaps using asymptotic and bootstrapped versions of the Student's t-test. This approach proved to be effective in detecting statistically significant transitions produced by all characteristic water levels, from low and mean, to high stages, and can be used to verify hydrodynamic models, monitor the water level in ungauged basins and predict high flows. A methodology to estimate the water storage in seasonal ponds is provided by García-López et al. [75] (Figure 4). The authors defined the geometry of a coastal wetland's basin in pre-flood conditions via UAS photogrammetry and *Sf*M techniques. The DEM generated allowed a spatial resolution of 6.9 cm and a mean squared error in Z of 5.9 cm with a practically nil bias and resulted an order of magnitude better than the pre-existing official LiDAR cartographic products. Although the effect of vegetation on the DEM was the main obstacle to the application of this method and it had to be minimized through spectral and morphological techniques, as a whole, this approach proved to be useful for calculating the water balance in this type of systems.

Figure 4. Prediction of the flooded area according to different water levels ((**a**) 1.80, (**b**) 1.90, (**c**) 2.00 and (**d**) 2.20 m with respect to the mean sea level) from UAS-derived DEM. Modified from [75].

To identify the river stages along the Kilim river (Malaysia), Mohamad et al. [76] generated DSM and orthomosaics for different tidal phases through the combination of UAS-based photogrammetry, GNSS vertical data and the use of the *Sf*M algorithm. Although the study achieved a limited vertical precision, it may be useful for future works on water level measurement in estuaries and systems influenced by tides. *Sf*M techniques have also been employed in the mining field by Yucel et al. [77] to analyze the areal changes and evolution of lakes originated by acid mine drainage. More recently, Chen et al. [45] developed a ground-filtering method to remove the effect of vegetation in wetland´s water budget estimations, thus improving the DEM obtained from UAS photogrammetry. Their approach comprised the combination of airborne LiDAR products with UAS LiDAR, UAS photogrammetric surveys and the application of a linear-weighted-average filter to generate a filtered DEM where the vegetation interference in the bathymetric data was reduced. The ground-filtered DEM showed a 90% correlation with in situ measurements, diminished the errors in water balance and reduced the uncertainties in the results.

4.3. Surface Velocity and Flow Estimations

Runoff and river discharge constitute a fundamental element of the water balance and is of critical importance for the sustainable management of water resources, freshwater ecosystems integrity and flood forecasting, among others. Although river discharge is one of the most accurately measured components of the hydrological cycle, the access to this information remains limited. The reasons are that large extensions of river basins around the globe are still poorly gauged or even ungauged [78], the decline in hydrometric stations, the gaps in datasets, the need for improved methods for the storage, retrieval and processing of hydrometric information [79].

Traditionally, discharge measurements have been carried out through direct and indirect methods. The first consist of the definition of the river section and flow velocity with mechanical current-meters, ADCP, tracers, etc. Indirect methods are based on the measurement of water surface level at gauging stations, which provide flow estimations through rating curves. Remote sensing and satellite altimetry on the other hand, has been widely used to successfully calculate river discharge in well-known large rivers of the world [80,81]. However, this methodology is less suitable for medium/small rivers and presents limitations related to the measurement of the average flow velocity and slope, the need of calibration with in situ measurements, poor spatial and temporal resolution and other issues. In this line, UAS have been increasingly used in the estimation of river discharge and flow as they overcome the handicaps of satellite remote sensing and present some advantages over it in terms of data accuracy.

Some authors propose the combined use of UAS with traditional hydrological formulas to calculate river discharge in ungauged areas. Yang et al. [82] calculated the river discharge in ten sections in the Tibetan Plateau and Province of Xinjiang by (i) retrieving the parameters of the slope-area method using an UAS to generate a digital orthophoto map and a DSM and (ii) applying three different formulas (Manning–Strickler, Saint-Venant and Darcy–Weisbach). In Yang et al. [83], UAS remote sensing, high altitude remote sensing and in situ measurements are combined to estimate river flow in medium/small wide shallow rivers in arid areas. The attenuation coefficient and discharge of the ungauged Hotan river (Central Asia) was calculated by applying energy equations and a trapezoidal cross-section discharge equation. Similarly, UAS-borne data were used to verify the river section morphology through orthophotomaps and a DSM, reaching a centimeter level accuracy. Both studies proved the usefulness of UAS for carrying out rapid river discharge assessments and provide relevant hydrological data in poorly gauged or ungauged systems. Lou et al. [84] integrated UAS and satellite remote sensing (Gaofen-2, SPOT-5, and Sentinel-2) with the application of hydrological formulas (Manning–Strickler) to obtain accurate discharge values from 24 ungauged rivers in the Tibetan Plateau on a long term scale, extending the previously mentioned works.

In addition, the monitoring of surface velocity has been addressed through UAS-based videos and the application of particle tracking velocimetry methods using natural and artificial tracers. Koutalakis et al. [85] compared the results obtained from the analysis of UAS imagery with 3 specific software based on different velocimetry methods: large-scale particle image velocimetry (LSPIV), large scale particle tracking velocimetry (LSPTV) and space-time image velocimetry (STIV). Although the three methods displayed similar trends, the surface velocity values differed slightly in some vegetated parts of the river reach and the authors point out the need for verification. Similarly, ref. [86] used UAS-borne imagery to evaluate the sensitivity of 5 image velocimetry algorithms under low flow conditions and concluded that, for surface velocities of approximately 0.12 m/s, image velocimetry techniques can provide results comparable to traditional techniques such as ADCPs. Tauro et al. [87] presented a custom-built UAS platform equipped with a ground-facing orthogonal camera that did not require image orthorectification, to monitor the surface flow through LSPIV in a natural stream. This low-cost system expands the use of traditional fixed cameras employed in particle velocimetry by enabling access to otherwise inaccessible areas. Tauro et al. [88] assessed the suitability of a low cost UAS for creating accurate surface flow maps of small-scale streams. After calibration, flow velocity maps were extracted from the motion of stream floaters at real-time by applying the LSPIV high-speed cross-correlation algorithm. Similarly, ref. [89] integrated LSPIV and UAS imagery and compared the discharge values obtained with that from other methods, such as a fixed camera, ADCP, stream gauges and a propeller meter. The authors also report the minimum steps required to produce accurate LSPIV measurements using UAS.

Despite its many advantages, the accuracy of the image velocimetry method is conditioned by several factors such as seeding density, the environmental and experimental setting conditions or the presence of floating materials on the stream surface. In this regard, ref. [90] used high-definition UAS footage to test the precision of LSPIV and particle tracking velocimetry methods under different seeding and hydrological conditions. The authors demonstrated that seeding density, dispersion index and the spatial variance of tracer dimension are statistically relevant metrics in the estimation of velocity and that a prior knowledge of seeding conditions can be useful for selecting the processing intervals, image velocimetry techniques to apply and for controlling the accuracy of results.

Time series of temperature images have also been used for estimating surface flow velocities. Kinzel and Legeiter [43] proposed a methodology to estimate river discharge through the combination of thermal velocimetry and scanning polarizing LiDAR, both UAS-based. The velocity of surface features, expressed as small differences in temperature, was tracked with a particle image velocimetry algorithm. The system was able to detect the movement of flow features expressed as subtle differences in temperature, with velocity measurements that agreed closely with that made in situ. However, depths inferred from the LiDAR showed less agreement with in situ measurements in the deeper sections of the river. Another innovative approach was proposed by Thumser et al. [91] who measured real-time river velocity using floating, infrared light-emitting particles and a color vision sensor onboard a UAS to track the position of objects. The system, termed by the authors as RAPTOR-UAV, enabled to capture large-scale, nonstationary regions of the velocity field where the flow fields are persistent and chaotic. Recently, remote sensing of surface velocity and river discharge has been addressed in [92] through UAS-based doppler radars yielding results comparable to conventional streamgaging if cross-sectional area is available.

4.4. Flood Monitoring

Floods are among the most important driving forces in fluvial environments. These periodic events not only reshape floodplain landscape and riverbeds, but also contribute to basin connectivity facilitating the exchange of water, sediments, organic matter, and nutrients between rivers, floodplains, and riparian wetlands. However, floods can also have dramatic impacts on human societies, causing economic damage, health issues, and loss of human life, infrastructures and productive land. For this reason, gathering information on flooding conditions and dynamics and determining the extent of floods at different scales is crucial for predicting and mitigating the impact of this natural hazard. Over several decades, flood hazard assessment and inundation mapping has been addressed through the use of satellite remote sensing. Nowadays, UAS provide very high-resolution imagery on-demand and are being increasingly used in geohazard studies and monitoring.

UAS-based photography has been extensively used to acquire data in inaccessible locations and create high resolution DEMs, map flood-prone areas and to enhance hydraulic modeling [93–97]. In particular, S*f*M techniques have seen a strong development as an alternative approach to classical digital photogrammetry in the study of floods, leading to an increasing body of literature in the last years; Özcan and Özcan [98] extracted a high-resolution DEM of a flood-vulnerable river in Turkey applying UAS-based SfM. Landform alterations due to erosion and deposition after flood events were analyzed through the application of DEM of difference algorithm (DoD) and geomorphic change detection (GCD) techniques. After a 2-year monitoring period, the authors verified that UAS S*f*M and DoD were useful tools for geomorphological dynamic mapping and flood event monitoring. Diakakis et al. [99] investigated one of the most devastating floods occurred in Greece in 40 years through the combination of systematic ground measurements and UAS observations during and after the event. Highly detailed DSM with 2.7 cm resolution derived from *Sf*M were used to obtain the flood extent, depth maps and the peak discharge, as well as a comprehensive description of the physical characteristics of floodwaters across the inundated area (Figure 5). An analysis of the flood impacts on geomorphology, vegetation, infrastructures and human population was also carried out.

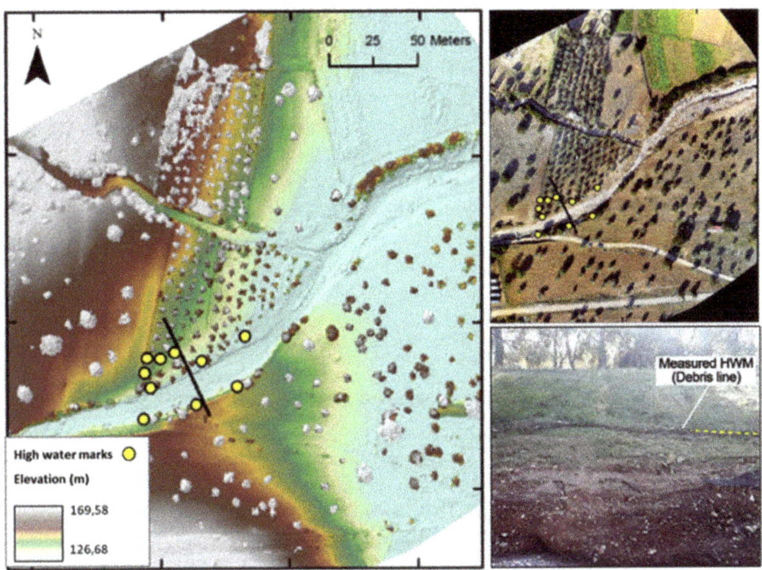

Figure 5. Digital surface models (DSM) derived from SfM and high-water marks identified upstream and downstream a cross section (indicated by a black line) of the river after a flood event. Modified from [99]. Copyright 2021 Elsevier.

Moreover, UAS *Sf*M photogrammetry is being increasingly applied to flood research as an alternative to LiDAR techniques owing to its more flexible image acquisition, the lower costs and the availability of powerful processing tools that generate DEM and DSMs in a single workflow. As several authors have pointed out, the recent advances of UAS-*Sf*M provide accuracies comparable to that obtained by LiDAR. For instance, Leitão et al. [100] evaluated the suitability of several UAS-derived DEMs (obtained with different flight parameters) for urban overland flow and flood modeling in Switzerland and compared them with an aerial LiDAR DEM from the Federal Office of Topography (SwissTopo LiDAR). According to their results, the quality of both products were comparable and differences were not substantial, especially in building, vegetation and tree free areas. For instance, the minimum, maximum, mean, and standard deviation of the elevation differences between the UAS and LiDAR DEM was −0.468, 0.306, 0.06, and 0.119 m respectively, and differences in slope were always below 10%. However, the UAS DEM suffered from a major limitation; it did not cover the whole area of interest (e.g., areas behind buildings). Hashemi-Beni et al. [101] assessed the quality of UAS-derived DEMs for mapping a flood event and its extent during a hurricane in North Carolina (USA). The water surface extracted from the UAS-derived DEM was compared with LiDAR and stage gage data, demonstrating general agreement between models. More recently, ref. [102] used a fixed-wing UAS equipped with a conventional RGB camera to generate very high resolution bare-earth DEMs of a small river floodplain in Maryland, USA. The accuracy of the model was compared with the pre-exiting LiDAR models, obtaining a trivial bias of 1.6 cm and a root mean square derivation of 39 cm, which demonstrate the suitability of this method. Annis et al. [103] compared the suitability of UAS-derived DEMs, freely available large-scale DEM and LiDAR DEMs for flood modeling in small basins. The authors proved that DEMs generated from UAS imagery significantly outperformed the large-scale DEM, and that drone-derived topography constitutes an appropriate alternative to LiDAR DEMs.

Another recent development in DEM generation methods is the combination of S*f*M and multi-view stereo (MVS) 3D reconstruction techniques. In a typical S*f*M-based process, the application of MVS algorithms follows image matching and allows to generate dense 3D point clouds from large sets of images. Villanueva et al. [104] assessed the DEM accuracy achieved from a S*f*M-MVS processing chain using LiDAR-derived control points and compared the estimations of flood volume and area obtained from UAS and LiDAR data. The authors concluded that UAS based DEMs were as accurate as LiDAR DEMs.

Reconstruction of flood episodes is also crucial to better evaluate and prevent potential risks in the future, being thus necessary precise information on flood extents, flow peaks, depths as well as on factors that might worsen the outcomes of the event. High resolution pre- and post-flood topographic data and UAS-derived ortho-imagery (4–5 cm/pixel) were employed by [105] to analyze three-dimensional morphodynamic changes associated with an extreme flood event in the Elbow River (Canada). Subsequently, geomorphic changes were assessed using the DEMs and their relationship with flow conditions were analyzed with a 2D hydrodynamic model. The authors documented large elevation changes, widespread bank erosion and channel widening after the flood and highlighted potential relationships between flood forces and geomorphic change. Abdelkader et al. [106] employed an UAS as a platform for Lagrangian flood sensing. The aircraft dropped small disposable buoyant wireless sensors that were carried away by the flood and transmitted real-time data that were used to map the extent of the inundated area. This method enabled accurate estimations of local flood parameters and short-term forecast of flood propagation.

Computer vision techniques and machine learning have also been used to classify remotely sensed imagery of flood events, e.g., [107] proposed a novel method for the detection and segmentation of flooded areas based on texture and color analysis in a deep neural network. The system, supported by ground control stations and a fixed wing UAS for image acquisition, increased the accuracy of flooded area detection up to 99.12%. Gebrehiwot et al. [108] applied a deep learning approach based on convolutional neural networks (CNN) to extract the flooded areas from UAS-borne (fixed wing and multicopter) RGB imagery. This method, implemented immediately after two hurricane events over 3 flood-prone areas in the USA, yielded a highly accurate classification (97.5%) of the remotely sensed data. Jakovljevic et al. [109] generated flood risk maps from LiDAR and UAS-derived data through a novel methodology based on raw point cloud classification and ground point filtering using neural networks. Flood risk assessment was calculated at 12 different vertical water levels. Although the UAS-derived model was less accurate than LiDAR and tended to overestimate the elevation, the overall approach met the accuracy required for flood mapping according to European Flood Directive standards.

In addition to the aforementioned applications, UASs constitute the most efficient and fastest tool for situational awareness in disaster management, not only providing accurate and spatially detailed hydrologic information during flooding events, but also enabling to identify safe shelter points, detect stranded people and damaged properties and to define future action strategies [110,111].

Nevertheless, although UASs have demonstrated to be a very useful tool in flood monitoring and delineation, they cannot substitute ground-truth measurements in this field since they cannot provide data on flood depth and the quality of the topographic products is compromised when the terrain is masked by vegetation or anthropogenic features. This issue has been recently addressed by [112], who reconstructed the flash flood events of two ungauged ephemeral streams in Olympiada region (North Greece) through the combination of ground-based and UAS observations and hydraulic (HEC-RAS) with hydrological (SCS-CN) models. Although the comparison between the modeled and observed flow extents showed a good performance of the hydraulic model, the modeled flood depths displayed an overestimation attributable to the low resolution and quality of the DEM especially in the urbanized and vegetated areas of the floodplain.

4.5. Temperature Mapping

Temperature is one of the most important variables that control the chemical and biological processes within water bodies and determine the spatio-temporal distribution of habitat niches for many aquatic species. TIR remote sensing not only enables to detect groundwater discharge into inland and coastal systems, an aspect that will be discussed in Section 5, but also allows to measure longitudinal and transversal temperature heterogeneities in rivers, to study their temporal variability and to identify aquatic thermal refuges among other features.

UASs have emerged as alternative tools to traditional high-resolution TIR remote sensing or ground-based thermography, allowing to collect data at a spatial scale intermediate to both approaches and to conduct multiple surveys in a brief period of time. However, as TIR river mapping is based on the quantification of temperature from the amount of energy emitted by a water body, the quality of the measurements can be impacted by diffuse reflections from solar radiation if surveys are not conducted during the night or by factors such as surface roughness (e.g., riffles) that might alter the energy received by the sensor. Emissivity on the other hand can be altered by suspended sediment and dissolved minerals. Other sources of uncertainty are shadows cast by riverbank objects or the presence of foam, which can be misleading in the detection of discrete features such as cool water patches. In addition, the angle of observation is of the utmost importance. At oblique viewing angles above 30° water's emissivity is reduced as a function of increased specular reflection [113]. These problems can be ameliorated by careful selection of flight parameters (flight time, flight altitude and angle of observation) and type of sensors, and by supporting thermal with simultaneous optical imagery.

Early attempts of thermal image acquisition consisted of handheld TIR cameras adapted to be mounted on UASs [114,115]. Jensen et al. [114] combined a low-cost UAS-based visible, NIR and TIR imagery to generate surface temperature maps with 30 cm resolutions in order to identify the thermal patterns of a highly dynamic stream in northern Utah (USA). On the other hand, the thermal patterns of braided rivers in the French Alps are analyzed in [115] using an UAS equipped with an RGB and a TIR camera. In this case, a powered paraglider was employed to acquire the TIR images, which allowed to predict habitat diversity from temperature heterogeneity. More recently, Collas et al. [41] obtained UAS-borne high resolution imagery of a river side channel using a consumer-grade TIR camera to map sub-daily temperature heterogeneity and habitat suitability for native and alien fish species. The authors achieved an accuracy in water temperature estimates of 0.53° over all flights, thus demonstrating the usefulness of UASs TIR imagery for temperature mapping and habitat management. Dugdale et al. [42] assessed the results of UAS-based TIR surveys applied to the study of diffuse temperature and discrete thermal inputs (springs, culverts, tributaries) in rivers of USA and Scotland. The data obtained were strongly biased, which the authors attributed to sun glint (linked to flight characteristics), the impact of radiative warming on the camera and to internal heating of the sensor. This generated substantial differences between radiant and kinetic temperatures in the rivers studied, making it necessary to apply substantial corrections. Temperature drift, inherent to uncooled microbolometers, can be compensated through hardware and software solutions. This issue has been recently addressed by several authors such as Mesas-Carrascosa et al. [31] who provide an overview on the topic and present a correction methodology based on redundant information from multiple overlapping images, or Kelly et al. [116] who propose a set of best practices to minimize temperature drift of TIR cameras.

Another limitation is that TIR sensors can only measure the skin temperature of water bodies. This issue, however, has been recently overcome through temperature sensor-integrated-UASs, which enable the mapping of the thermal structure of shallow water bodies. Chung et al. [117] developed an UAS equipped with a temperature probe that was dipped into the water to record temperature throughout the water column. Albeit the temperatures measured from the UAS were higher than those obtained in situ even after calibration, this system have the potential to enable high resolution 3D temperature mapping if some minor improvements are applied. Similarly, Koparan et al. [118] designed an UAS-based temperature profiling system that measured water temperature and depth. A key factor in their study is that the UAS system could land on and take off from water surface, avoiding hoovering during the measurements, reducing battery use and enabling more precise depth measurements.

To geolocate thermal plumes and collect temperature depth profiles over relatively larger scales (100 m to kms), DeMario et al. [119] integrated an IR camera and an immersible temperature probe in an UAS (Figure 6). This approach is especially useful for monitoring thermal effluents from power plants and can potentially displace the current boat-based operations usually implemented to characterize such discharges.

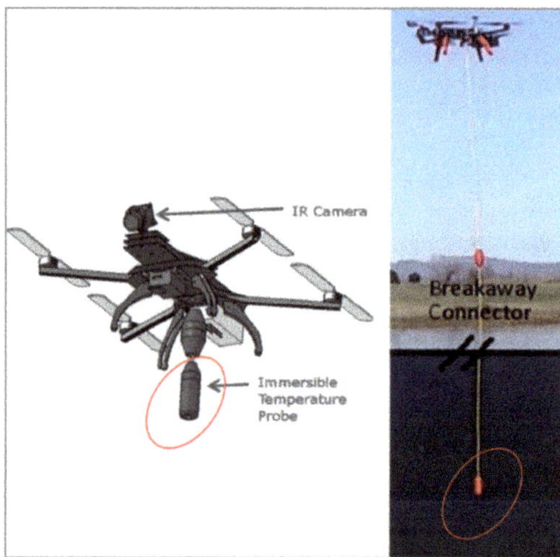

Figure 6. UAS platform designed by De Mario et al. (2017), equipped with an infrared (IR) camera for mapping water surface and a temperature probe to record the temperature depth profile. Modified from De Mario et al. [119].

Other applications of UAS include the monitoring and assessment of the effects of stormwater on the temperature of urban streams. Caldwell et al. [120] characterized the temperature response of a large urban stream in Syracuse (NY) to stormwater inputs by combining UASs-based TIR imagery with in situ sensors and pairing the observations with a deterministic stream temperature model. The authors quantified and characterized the creek´s temperature heterogeneities and the relative magnitude and extent of stormwater plumes. The study demonstrated the utility of airborne TIR imagery for simulating hydrologic processes, heat exchange and for revealing complex interactions between urban effluents and stream water temperature.

4.6. Water Contamination

Pollution of freshwater bodies is a ubiquitous problem that has drawn the attention of the scientific community, water managers and policy makers for a long time due to the serious public health and environmental risks that it poses. The pollutants that reach rivers, lakes and wetlands are extremely varied in nature, composition and source, being closely linked to anthropogenic activities and land use. The effective management of these substances and the prevention of their potentially harmful effects require the understanding and prediction of their transport, dispersion and behavior within the aquatic systems.

4.6.1. Dispersion Processes

UAS technology is enabling the rapid filling of knowledge gaps in the transport of pollutants and other hazardous agents while overcoming the difficulties inherent to in situ measurements. Powers et al. [121] coordinately used an UAS and an USV to detect and track dye released in a reservoir. The UAS was used to capture videos and images of the dye plume and to provide visual navigation of the tracer for the pilot of the USV. The dye concentration measurements collected by the USV were compared with that from UAS imagery, yielding similar results and demonstrating that processed images from UAS can be used to predict dye concentrations near the water surface. Baek et al. [122] estimated the concentration of fluorescent dye in an open channel using UAS RGB imagery. The authors applied an artificial neural network (ANN) to establish the empirical relationships between the digital numbers in the images and the spatio-temporal distribution of dye concentration (Figure 7). Although the ANN models required simultaneous in situ measurements, the authors demonstrated the feasibility of generating accurate high-resolution dye concentration maps with this novel and cost-effective approach.

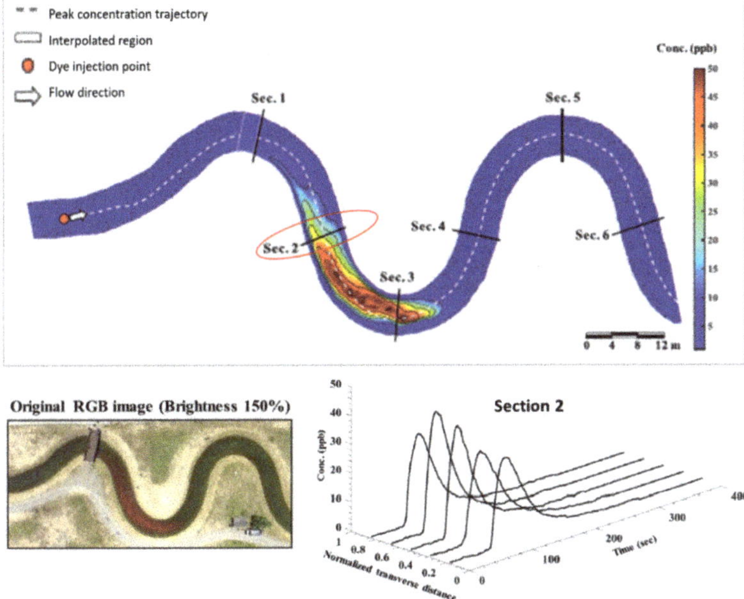

Figure 7. Transport of the tracer cloud at t = 90 s, original UAS RGB image and dye concentration distribution at Section 2. Modified from [122]. Copyright 2021 Advances in Water Resources.

Legleiter [123] assessed the suitability of various types of remotely sensed data (field spectra, hyperspectral images from manned and unmanned aircraft, and high resolution digital aerial photography) to estimate concentrations of a visible tracer. The results showed that a broad range of visible wavelengths were strongly correlated with dye concentration and therefore, tracer dispersion could be accurately mapped using RGB imagery.

4.6.2. Plastic Pollution

Similarly, the problem of plastic pollution in aquatic ecosystems has garnered widespread public and scientific attention. The rampant proliferation of plastics has led to explore the use of UAS in the evaluation and dynamics of plastics in the aquatic environment and to the development of numerous works on this issue. Several authors such as [124–126] have explored methodologies for the monitoring and automatic quantification of anthropogenic marine debris based on UAS imagery and machine learning techniques. It should be noted however, that most of these studies have been conducted in marine and coastal environments and to date, only a few works focus on fluvial systems. Geraeds et al. [127] presented a novel methodology to quantify the floating riverine plastic transport in the Klang River, Malaysia. In order to make detailed cross-sectional plastic transport profiles, the authors conducted an aerial survey with flight transects perpendicular to the river flow direction. The flight path was set up at three different altitudes relative to the instantaneous water level. Depending on the altitude at which images were taken, they served different purposes. Although sudden changes in weather conditions can lead to a loss of accuracy in flight altitude, the authors documented similar plastic densities and transport profiles using UAS aerial surveys, visual observations and plastic sampling. Jakovljevic et al. [128] utilized high resolution orthophotos obtained from UAS and *SfM* algorithms to map floating plastics in rivers and lakes in Bosnia Herzegovina. The authors developed an end-to end semantic segmentation algorithm based on U-Net architecture that enabled to accurately detect and classify 3 types of plastic (OPS, Nylon and PET) at different spatial resolutions with an underestimation of the plastic area of only 3.4%. More recently, Hengstmann and Fischer [129] identified sources for plastics and influences on their distribution at four sandy bank border segments of Lake Tollense in Germany. They compared field-based observations with the results of anthropogenic litter detection via UAS imagery and obtained good recovery rates when minimizing the flight height. In addition, the aerial images were used to test automatic and supervised image analysis to detect and classify plastics.

4.7. Harmful Algal Blooms (HABs) and Eutrophication

Harmful algae blooms (HABs) outbreaks are closely linked to anthropogenic eutrophication and have dramatically increased in the past three decades. HABs not only give rise to a rapid depletion of dissolved oxygen in water bodies and massive deaths of aquatic organisms but can also pose a serious threat to human health owing to the production of neurotoxins of some species [130]. As in other disciplines, HAB monitoring evolved from expensive and labor-intensive water sampling techniques to satellite monitoring. However, HABs are highly dynamic and their monitoring, even with high resolution satellite data, is not always suitable owing to their limited temporal resolution (large time intervals between image re-acquisition). The quality of satellite images can be affected by weather conditions, cloudiness and atmospheric absorption. Remote sensing from manned aircrafts, nevertheless, can provide high spatio-temporal resolutions but its implementation is costly and usually limited by safety, logistical and operational issues. In this context, UAS offer the flight flexibility, temporal and spatial resolution and rapid response needed to detect and monitor HAB dynamics with minimal environmental disturbance. Despite these advantages, since the development of UAS-based techniques and the commercialization of suitable sensors is very recent, the current number of studies focused on the study of HABs with UAS is limited. One of the applications given to UAS imagery has been the visual inspection of water bodies, e.g., [131] detected the beginning of eutrophication process and tracked the proliferation of ragweed in a small lake in Hungary using high resolution photography from two different models of fixed wing UAS and an hexacopter.

Other early works include the use of conventional cameras separately or in combination with NIR sensors to map water quality and quantify algae concentrations. Ngo et al. [132] studied the relationship between algae concentration and water color through the implementation of a regression model. A RGB camera mounted on an UAS was used to capture videos and retrieve individual frames to subsequently extract the color component values of green and blue pixels. Van der Merwe and Price [133] exploited the contrast between clear water and cyanobacteria in color-infrared imagery. The authors employed UAS equipped with modified cameras that captured NIR and blue wavelengths to quantify algae densities at water surface through the formulation of a parameter termed as "blue normalized difference vegetation index". Their work emphasizes the role of UAS remote sensing as a complement to traditional methods and the usefulness of drones when HABs need to be characterized and tracked with a high spatio-temporal precision and accuracy. Su and Chou [134] utilized UAS RGB and NIR imagery and established regression models to find the best correlations between water quality parameters and band ratio in order to map the trophic state of a small reservoir. The authors emphasized that, compared with traditional photogrammetry or satellite remote sensing techniques, UAS offer a better cost/profit ratio in terms of reservoir mapping. Jang et al. [135] monitored HABs in one of the largest rivers in South Korea and developed a modified Algal Bloom Detection Index that included the red band (625 nm) to better distinguish algal bloom from clear water. Their study combined the use of UASs equipped with a S110 RGB camera and a NIR camera with in situ measurements of spectral reflectance and water quality analysis. Aguirre-Gómez et al. [136] combined UAS high-definition imagery with in situ radiometric measurements and algae sampling to conduct spatio-temporal analysis of the extension and distribution of cyanobacteria in urban lakes of Mexico. Their approach included dark object subtraction techniques in order to correct sun/sky glint errors and proved to be an accurate, flexible and rapid method to detect and predict eutrophication and cyanobacterial blooms in reservoirs. Salarux and Kaewplang [137] employed UAS-derived RGB and NIR imagery to calculate vegetation indexes and assess their performance using mathematical models (Figure 8). This approach proved to be useful for estimating algal biomass.

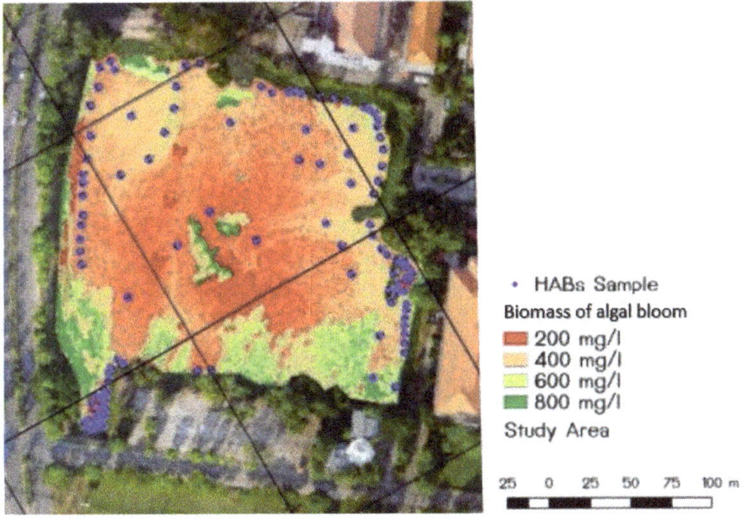

Figure 8. Map of algae biomass. Modified from [137].

Since the recent development of the first lightweight hyperspectral cameras adapted to be mounted on UASs, the applications of this type of imagery have increased in numerous research fields, including algal bloom monitoring. Although hyperspectral sensors are still expensive, these are emerging as alternatives to conventional RGB and multispectral cameras owing to their improved capacity of quantifying water constituents and identifying algal and phytoplankton groups species and genus by their spectral pigment absorption. Some recent experiences using drone-borne hyperspectral imagery can be cited; in an early work, Honkavaara et al. [138] presented a promising water quality monitoring technique that combined small-manned vehicles and UAS equipped with a Fabry-Perot interferometer (FPI) hyperspectral camera. Becker et al. [139] deployed two different low-cost UAS configurations in a lake and a river to monitor water quality and assess potential algal blooms in near-real time. UAS were equipped with hyperspectral radiometers to measure chlorophyll-a, cyanobacteria index, surface scums and suspended minerals. The UAS spectral data presented a quality similar to that from ground based spectroradiometers and enabled the construction of transect maps of derived cyanobacterial index products that showed the distribution of algae abundance in the systems. Penmetcha et al. [140] proposed a deep learning-based algae detector and a multi-robot system algae removal planner integrated by an UAS and a USV. More recently, Castro et al. [38] presented a multisensor tool to monitor water quality in a Spanish reservoir affected by eutrophication. Their approach combined satellite, in situ data and multispectral UAS imagery to retrieve chl-a concentrations, which indicates the trophic status of the system. The performance of the different sensors, empirical models and band indices was evaluated. The authors concluded that multi-platform and multi-sensor approaches have great potential for small reservoir eutrophication monitoring. Kwon et al. [141] retrieved hyperspectral data from portable and drone-based sensors to develop bio-optical algorithms, measure the vertical pigment-concentration profile and map the distribution of phycocyanin in a deep reservoir. The bio-optical algorithms implemented using UAS-based reflectance measurements performed slightly worse than those using in situ remote sensing. The authors point out that future work should aim to improve remote-sensing algorithms through the identification of the relationship between pigment cumulation depth and light attenuation. Zhang et al. [142] mapped the concentration of eutrophication-related parameters in an urban river through the development of a self-adapting ANN based on UAS hyperspectral imagery and using the modified spectral reflectance of water measured with a ground-based analytical spectral device.

4.8. Water Sampling and In Situ Parameter Measurement

In addition to acquiring images, UAS are being increasingly used to take water samples, as they eliminate sampling risks to humans and allow to reach otherwise inaccessible locations. These innovative sampling platforms reduce costs and efforts, increasing the speed and range at which samples were traditionally obtained and allow to obtain undisturbed samples when hoovered at an adequate height to avoid induced mixing due to downwash. Some of the first references on the use of these instruments date correspond to Ore et al. [143] and Detweiler et al. [144], who designed a mechanism for sampling water autonomously from an UAS. The system was able to collect 3 samples of 20 mL per mission and the properties of the water collected matched that from manual samples, proving the adequacy of this technique. Similar works on the use of UAS as water sampling tools can be found in [145,146].

More recently, water sampling from UAS has gone a step further by including a wide range of sensors to conduct in situ measurements of different physical-chemical parameters. Song et al. [147] compared the reliability and effectiveness of UAS-based in situ physico-chemical measurements and water sampling with traditional and sensor-based methods. Although UAS enhance the reliability of data, there are still barriers to their full integration in limnological studies; the drone´s maximum carrying capacity limits the number and volume of samples, making these less representative of water chemistry compared to manually collected samples. Elijah et al. [148] implemented a smart river monitoring system that included an UAS to collect water samples and was equipped with and array of sensors and probes. Koparan et al. [149] designed a lightweight hexacopter equipped with a multiprobe system to measure temperature, electrical conductivity (EC), dissolved oxygen (DO), and pH. To avoid hoovering, the platform had flotation elements to allow the UAS to land on water surface at the waypoints in the flight mission. Esakki et al. [150] designed a UAS-based platform equipped with a robotic arm, a water pump and several sensors to collect 500 mL water samples and conduct in situ measurements of water quality. The results gathered by this system were in close agreement with laboratory test, with a 98% accuracy. To take samples of DNA fragments from a reservoir, Doi et al. [151] developed a new method for water collection consisting of a UAS with an attachment for a 1-L water bottle. The performance and contamination risk of the system was compared with that of samples obtained by boat, yielding similar results. Banerjee et al. [152] developed an electromechanical and pneumatic system to collect water samples from UAS, facilitating the access to inaccessible water bodies or dangerous mine sites and enabling the analysis of several physical-chemical variables. Castendyk et al. [153] presented the first application of UASs for deep water sampling (up to 80 m) in pit lakes. Their system collected samples using Nisking sampler bottles and measured in situ the physical-chemical profiles of the water body. Later, Castendyk et al. [154] built on this work and on similar experiences to assess the state-of-the-art of drone water sampling of pit lakes in North America and to demonstrate how UAS-derived profiles can be used to select optimal sampling depths.

The use of UAS is not restricted to HAB monitoring and detection; in recent years UAS have also been employed to counteract them. Jung et al. [155] developed an autonomous robotic system to remove algal blooms that combined a catamaran-type USV instrumented with an electrocoagulation and flotation reactor with an UAS. The latter detected the HABs through an image-based algorithm and sent GPS coordinates to the USV for path planning. This method achieved a 98.3% of cyanobacteria removal.

Finally, the recent advances in technology have facilitated the creation of "multipurpose drones" that apart from completing standard tasks such as the acquisition of videos and photos, can monitor and take samples in a variety of environments and situations. An example of multipurpose UAS is provided by Agarwal et al. [156] who designed a platform capable of taking water samples, monitoring air and water quality in situ and that included video recording functions. Similarly, in order to access to an extreme environment such as the Lusi mud eruption in Indonesia, Di Stefano et al. [33] designed a multipurpose UAS. The platform was able to complete video surveys, high resolution photogrammetry, TIR mapping and was equipped with deployable thermometers and gas, mud and water samplers. An UAS equipped with a hydrophilic with a lipophilic balance (HLB), thin-film solid-phase microextraction (TF-SPME) sampler was developed by Grandy et al. [157] to remotely screen a wide range of pollutants present in water bodies. Finally, a recent review on drones use for water sampling can be found in [158].

5. UAS Applications in Hydrogeology

The number of papers documenting the application of UAS to groundwater (GW) and aquifer research is rather scarce, especially if compared with the extensive literature dealing with their use in riverine and oceanic environments. The reasons are the inherent limitations of remote sensing and the very early stage in which the technological developments in this field are currently. The major constraint in GW detection and mapping is related to the insufficient penetration capacity that remote sensing offers, only enabling the acquisition of data at the ground surface or within shallow subsurface layers a few meters deep. Apart from geophysical techniques, ground-penetrating radar (GPR) is the only technology that can survey depths of several meters. However, at present, owing to the mass, size and very high costs of these sensors, radars onboard UAS are still uncommon, and most experiences are restricted to applications such as landmine detection. Currently, some technological advances such as the Frequency Domain Electromagnetic (FDEM) method allow to perform geophysical surveys from UASs. These are aimed at the detection of underground properties such as magnetism and resistivity, which can be subsequently related with the presence of groundwater and its characteristics (e.g., salinity). Nevertheless, this field is still to be developed in scientific terms.

On the other hand, the thermal inertia of groundwater emergence in coastal areas (SGD) and fluvial systems has been recently exploited to generate temperature maps and monitor warm plumes by means of TIR-equipped UAS yielding promising results. Current available research on UASs applications in hydrogeology includes studies on surface-groundwater (SW–GW) exchange flow, submarine groundwater discharge (SGD), piezometric level delimitation and subsidence processes associated to groundwater withdrawals.

5.1. Surface Water–Groundwater (SW–GW) Interactions

Surface water–groundwater (SW–GW) interactions are critical to calculate water balances and sustainable levels of water extraction. These interactions have been traditionally studied through an array of methods (e.g., differential gauging, seepage meters, tracer injection experiments, monitoring of piezometric heads or satellite remote sensing) whose selection depend on the temporal and spatial scale, limitations and uncertainties inherent to each technique. Recently, UAS have been implemented in SW–GW research since they are the only tools that enable indirect observations of these processes with a sufficiently high spatial resolution.

As several experiences have demonstrated, UAS-derived TIR imagery can provide useful information on the size and extent of groundwater plumes emerging from springs and near streambanks, especially when the temperature contrast between SW and GW is marked. Abolt et al. [159] mapped thermal refugia associated with groundwater discharge in Devil´s river (Texas) using a small UAS equipped with an inexpensive uncooled microbolometer and proposed a method for stabilizing the resulting image mosaic and compensating the pixel bias. This low-cost platform produced more consistent results than those obtained from a more expensive TIR camera system and allowed to generate high quality maps of surface temperature in riparian ecosystems.

TIR imagery has also been supported by optical products; Harvey et al. [160] evaluated the suitability of UAS equipped with lightweight TIR sensors to identify groundwater discharges and validate thermal refugia goals in a hydrological restored peatland. TIR, visible, and DSM products were compared to evaluate the landscape forms and thermal signature of groundwater inflows. The authors detected substantial inflows of warm groundwater that were visible along the restored channel, along with seepage areas, discrete seeps and thermally stable pools of ecological importance (Figure 9). This work emphasizes the usefulness of UAS-based TIR for mapping groundwater seeps in wetlands with a spatial coverage that is unimpeded by site-access considerations. However, the approach presents several limitations: Firstly, while the method proved to be effective in a continental climate with cold winter conditions, it is less effective when applied in spring, autumn or in temperate to tropical climates as the temperature contrast between SW–GW is less evident and becomes quickly lost by surface mixing. Secondly, TIR sensors cannot penetrate vegetation, meaning that surface waters may not always be visible because of foliage. Thirdly, shallow groundwater temperature can be similar to the mean annual surface temperature, weakening the TIR images contrast. Casas-Mulet et al. [161] generated high-resolution TIR and RGB orthomosaics to characterize cold water patches associated with groundwater inputs over a 95 km-long river section using simultaneous UAS flights. This method allowed to identify riverscape spatial patterns of temperature and to detect and classify these patches. Finally, UAS-based TIR has also been applied in the field of mining by Rautio et al. [162]. The authors combined airborne and UAS TIR imagery to identify SW–GW interactions for the planning and siting of mining facilities in order to prevent potential acid mine drainage pollution.

Other studies focused on SW–GW interactions rely on the combination of ground-based measurements, UAS photogrammetry and hydrologic modeling. Pai et al. [39] quantified sinuosity-driven GW-SW exchange in a river employing a suite of UAS-derived products (water surface elevation, normalized difference vegetation index (NDVI) maps and vegetation-top elevation distribution along meanders) and distributed in situ temperature measurements. *Sf*M photogrammetry was used to generate a DSM that allowed to estimate the river gradient, the hydraulic gradient across the meander necks, river-reach topography, and vegetation-top elevations. Compared with the surveyed ground control points, the modeled surface presented a 3.8 cm mean absolute error (less than aerial LiDAR) and a precision of 2.5 cm. The NDVI maps obtained presented a resolution better than 10 cm, enabling to document even individual plants. The combination of topographic analysis with low-cost multispectral imaging enabled to identify GW shortcutting, which occurred through the necks of the meander bends, where hydraulic gradients were found to be larger. Bandini et al. [163] evaluated the potential of spatially-distributed UASs observations for improving hydrological models and in particular, estimates of GW-SW exchange flow. The authors simulated a river and its catchment using an integrated hydrological model that was calibrated through 2 methods; first, against river discharge retrieved by in situ stations and the piezometric head of the aquifers and second, against dense spatially distributed water level observations obtained with UASs. After calibration, the sharpness of the estimates of GW–SW time series improved by 50% using UASs and the root mean square error decreased by R75% compared with the values provided by the model calibrated against discharge only. Tang et al., [164] simulated a flood event with the groundwater model HydroGeosphere to study the spatial and temporal variability of riverbed topography and hydraulic conductivity on SW–GW exchange and groundwater heads. The authors combined several state-of-the-art techniques; UAS-based photogrammetry, physically based measurements and the ensemble Kalman Filter. This combination resulted in substantially improved hydrological predictions and enabled the estimation of river-aquifer fluxes. Briggs et al. [165] combined remote sensing with ground- and drone-based measurements to characterize the enhancement of SW–GW interactions and changes in water chemistry induced by beaver activity along two alluvial mountain streams in USA. Several UAS were deployed to map the river corridor and beaver ponds and SfM techniques were applied to generate time-specific digital elevation models of floodplain structure and channel geomorphology. The UAS information was complemented with historical imagery, TIR data from handheld cameras and measurements of water quality (metals) and seepage associated to beaver pond return flows. The authors reported a multi-seasonal enhancement of the floodplain hydrologic connectivity and an increase in groundwater storage associated with beaver activity. Furlan et al. [166] combined electrical resistivity tomography with UAS photogrammetry to map the topography, internal morphology, water storage and hydrologic flow paths in a savanna wetland in Brazil. The authors used a fixed-wing UAS with a RGB sensor to obtain very high-resolution images and create an orthomosaic and a DSM for the relief analysis. The wetland was compartmentalized and the area and volume of each sector calculated for the subsequent hydrological modeling. On the other hand, the geophysical surveys provided information about groundwater behavior and infiltration zone architecture and enabled to produce a 2D inversion and a pseudo-3D model to visualize the subsurface geologic structure and hydrologic flow paths. The combined application of very high resolution UAS images and electrical surveys allowed the authors to propose a broader hydrologic interpretation of the wetland functioning and to complete surface and subsurface imaging of the system.

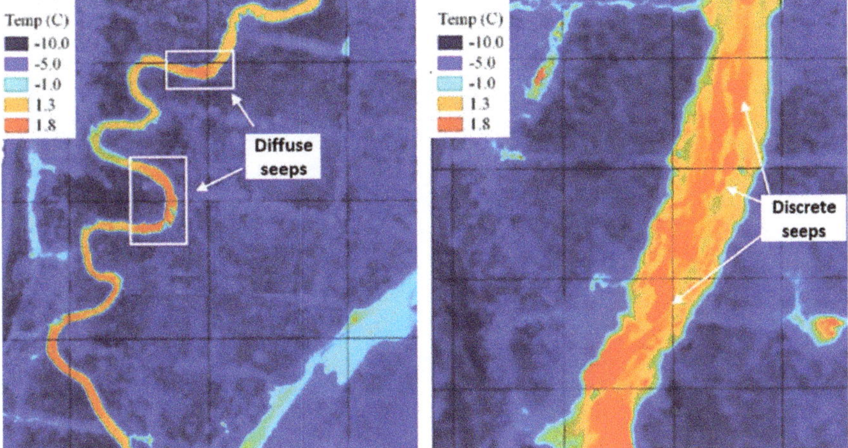

Figure 9. Thermal infrared (TIR) imagery showing the diffuse and discrete warm groundwater inputs into the stream. Modified from [160].

5.2. Submarine Groundwater Discharge (SGD) Mapping

Submarine groundwater discharge has a suggested impact on marine environment and geochemical cycles, playing a key role in the transport of nutrients, contaminants, and other chemical substances to coastal water. Although neglected for many years owing to the intrinsic difficulty of its quantification and the poor understanding of the process, in the last decades there has been a sharp increase in the number of publications on this topic. SGD has been traditionally studied using geochemical tracers, geophysical techniques, piezometric levels, water budgets and hydrological modeling. These classical methods, however, have large uncertainties associated, are labor-intensive and difficult to implement over large areas. Most SGD research is based on the detection of groundwater discharge from temperature anomalies and the seasonal contrast between ocean and groundwater temperature. TIR imagery from manned aircraft, although costly, has proved to be useful in mapping SGD, especially in areas where temperature contrasts between groundwater and seawater are marked. Conversely, satellite monitoring suffers either from an inadequate spatial resolution for detailed studies, or it is restricted by established schedules (revisit times) that may not be adequate for the needs of each study. In this context, since the implementation of UAS in SGD research is still relatively new, only a few recent studies are available, highlighting UAS´s suitability as affordable alternatives to the aforementioned methods and to overcome limitations in terms of spatial resolution.

Siebert [167] studied what they termed as "sub-lake groundwater discharges" in the Dead Sea; off-shore springs that drained the surrounding mountain freshwater aquifers. The authors integrated data obtained from ground-truth measurements, high-precision and high-resolution bathymetric campaigns, side-scan sonar imaging from an USV and imagery of sea surface temperature from a TIR sensor mounted on an UAS. Their approach proved to be suitable for precisely mapping SGD locations from remotely sensed thermal anomalies and even led the authors to hypothesize that the anomaly size reflects discharge quantities.

In a pioneer work, Lee et al. [168] successfully characterized and quantified SGD in an island of the Korean Peninsula through UAS-based TIR mapping supplemented by ground-based observations. Thermal signatures of SGD plumes and their tidal-derived fluctuations were captured with great detail (Figure 10).

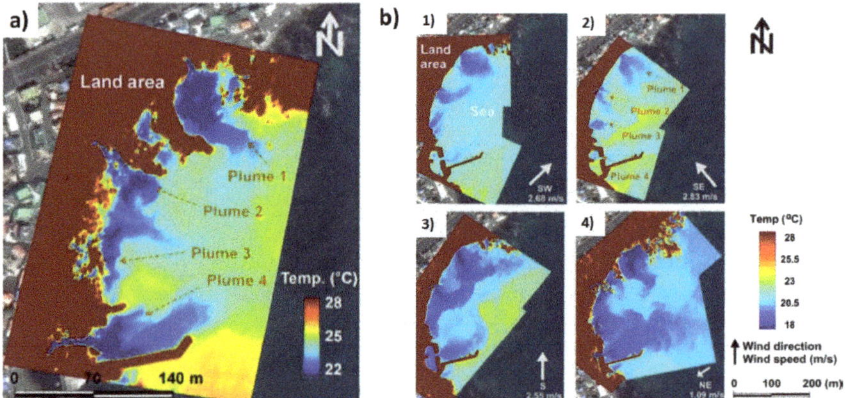

Figure 10. (**a**) Sea surface temperature maps and (**b1–4**) temperature fluctuations depending on the tidal stage (1. rising tide, 2. high tide, 3. outgoing tide and 4, low tide) captured by the UAS-based TIR. Extracted from [168].

This work evidenced that UASs offer better results in the study of SGD dynamics in small, localized target areas (0.01–1 km^2) than manned aircrafts, which are more adequate for regional characterizations. The drone-based methodology allows an easier control of the spatial resolution, more flexibility in field operations and lower costs compared to conventional aerial surveys, making it a powerful tool to study SGD and other coastal processes.

More recently, Mallast and Siebert [44] investigated variations in thermal radiation induced by focused and diffuse SGD in a sedimentary fan of the Dead Sea. The authors used a hoovering UAS equipped with a long-wave TIR camera and a radiometry module and assumed that thermally stable areas indicated focused SGD whereas highly variable areas indicated diffuse SGD. After applying specific subjective variance thresholds and spatio-temporal analysis, their results highlighted that the spatio-temporal behavior of a SGD-induced thermal radiation pattern can vary in size and over time by up to 155% in the case of focused SGDs and by up to 600% in the case of diffuse SGDs owing to underlying flow dynamics. The authors recommend the application of this approach prior any in situ sampling in order to identify adequate sampling locations and intervals.

5.3. Water Table Definition

The definition of the water table of an aquifer is usually carried out from observations at specific points in the territory with the subsequent application of the interpolation methods devised for this purpose (linear, polynomial, IDW, kriging, etc.). Direct observations are made by measuring water level depth with specific instruments (electrical probes, pressure transducers, radar, etc.) at observation points that reach the saturated zone (wells, boreholes, piezometers, pits, excavations). The piezometric level is inferred from the subtraction of water level depth from ground elevation with respect to datum. Photogrammetry carried out from UAS can provide high accuracy in the definition of the morphology of the terrain and is useful for levelling the observation points. Nevertheless, the determination of the level depth using UAS require certain exposure of the water sheet. Therefore, it is not possible to detect the piezometric level when the observation points are closed (installed boreholes or piezometers with covers) or when their geometric characteristics (diameter) do not favor inspection from air. That is why there are very few works in this field. The only applications described refer to peatlands and karst aquifers affected by large dissolution/collapse structures (cenotes).

Thus, Rahman et al. [169] proposed a methodology for mapping groundwater in a treed-bog peatland using orthophotography and photogrammetric point clouds acquired from an UAS. DSM and a DEM were used to obtain a canopy height model and open water objects were converted into a continuous surface. The elevations of the samples were interpolated to generate a water table surface, which was then subtracted from the DEM to obtain the depth to water surface. This method demonstrated great potential for measuring groundwater levels over large, complex and inaccessible areas, although its performance and effectiveness were hindered in densely vegetated areas. Bandini et al. [170] proposed a methodology based on the observation from UAS to define the water surface elevation (WSE) in cenotes in the calcareous aquifer of the Yucatán Peninsula (Mexico), which is useful to feed hydrological models and estimate hydraulic gradients and groundwater flow directions. WSE observations were retrieved using a radar and a global navigation satellite system on-board a multicopter platform. Moreover, water depth was measured using a tethered floating sonar controlled by the UAS. Later, these observations are transformed to orthometric water height above mean sea level and compared with the ground-truth observations retrieved by a GNSS rover station. The authors estimate an accuracy better than 5–7 cm in the WSE and they highlight that UASs allow monitoring of remote areas located in the jungle, which are difficult to access by human operators.

Other applications related to the detection of the saturated zone depth using UAS are focused on the identification of drainage systems in agricultural fields; Allred et al. [171] conducted UAS surveys to detect drainage pipes in a pilot agricultural field where documentation on the pipe network was available. The study was caried out using visible, NIR, and TIR images during the dry period. Under these field conditions, drainage pipes could not be detected with the VIS and NIR imagery. Conversely, the TIR image detected roughly 60% of the subsurface drainage infrastructure. Kratt et al. [172] followed a similar approach and obtained better results with visible images than with the TIR sensor, which did not detect the phenomenon owing to thermal differences lower than the sensitivity of the instrument, a consequence of non-optimal environmental conditions.

5.4. Subsidence

Subsidence is a slow and gradual movement caused by tension-induced changes over natural terrains or built surfaces. This geological hazard can affect all types of terrains and is produced by a range of natural factors and human activities. With regards to the latter, groundwater extraction plays a significant role in the occurrence of land subsidence; pumping can lead to substantial drawdowns and water table depressions in the neighboring areas of boreholes, resulting higher water extraction costs and, in the worst-case scenario, to material and human damage.

This phenomenon has been extensively documented worldwide using different techniques like interferometry synthetic aperture radar (InSAR) from satellites or numerical models [173–175]. Nowadays, the implementation of UASs as complements or enhancing tools for conventional surveying and mapping of subsidence is drawing increasing attention among the scientific community, however, this discipline is still in its earlier stages of development.

Although works on the application of UASs in the study of subsidence linked to groundwater extraction are scarce, some pioneering attempts are worth mentioning. One of them was conducted in Arizona (USAEE.UU), where the central and southeastern regions present subsidence problems driven by groundwater pumping, leading to fissures and cracks caused by tensile failure of sediment and soil. These fissures have resulted in property damages and increase the risk of groundwater contamination from surface pollutants. To address this issue, the Arizona Geological Survey (AZGS) founded the earth fissure program in 2007 to systematically identify, map, and monitor earth fissures. In 2018 the AZGS incorporated the use of UAS-S*f*M for fissure monitoring and terrain mapping [176]. As a result, they produced accurate DSM of the areas affected by fissures, highlighted the diffuse nature of ground surface disturbance around fissure openings and proposed to extend the use of UASs to other zones of Arizona due to the better precision of the method. On the other hand, the effect of subsidence on groundwater levels has been studied by [177,178]. The authors used RGB images from a rotatory-wing UAS to extract subsided cultivated land in high-groundwater level coal mines. After calculating several vegetation indexes and applying a hierarchical extraction method to define subsided farmland, the comparison of the results of the UAS-based images (Kappa coefficient of 0.96) with a traditional method (Kappa coefficient of 0.86) demonstrated the highest accuracy of the former. However, the methodology displayed some weaknesses; although the remote sensing images obtained by UAS at low altitude had high spatial resolution, the amount of data was relatively large, slowing down data processing. In addition, the high energy demands of the rotary-wing UAS only allowed to conduct shorter flights, what hindered the obtention of data over larger areas, and the resolution of the images was affected by meteorological conditions (cloudiness and fog).

6. Results and Conclusions

6.1. Current Applications of UASs in Water Resource Research

The advent of UASs as remote sensing platforms for hydrological monitoring constitutes a major breakthrough in water research, overcoming the shortcomings of traditional manned-aircraft and satellite observations, and opening up new opportunities to achieve the spatio-temporal resolutions required to understand small-scale hydrological processes and dynamics. In this regard, UASs are bridging the gaps between satellites, airborne imagery and ground-based measurements. In addition, the increasingly competitive prices of these platforms, together with the parallel advances in sensor development and software tools that automate data processing, have led to a growing interest of the scientific community on this technology.

In this paper a total of 122 research works is summarized, providing an overview on UASs recent advances and state-of-the-art methodologies implemented in the study of surface water and groundwater, including notions about the types of sensors, platforms and advantages and limitations of this technology that is in rapid and continuous development. The works reviewed were classified according to the phenomenon or process under study and the type of water body considered.

The number of studies focused on surface water notably exceeds those on groundwater investigations (Figure 11). In fact, UASs applications on surface water account for 84.4% of all the papers reviewed, of which river bathymetry and the study of floods are the prevalent studies. Conversely, groundwater research using aerial drones constitute only a 15.6% of the total, being most of these papers related to SW–GW interactions.

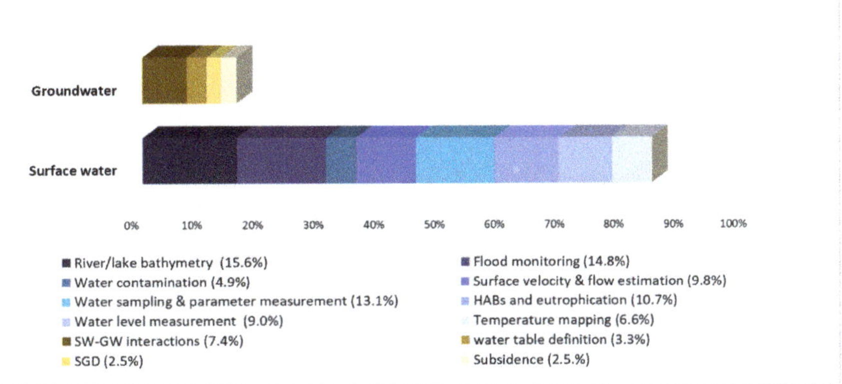

Figure 11. Percentage distribution and number of reviewed studies of UASs applications in hydrology.

As this review evidences, a wide number of approaches have been developed for observing surface and submerged features of rivers, lakes and other wetlands. For instance, UAS-based optical methods have been used to create bathymetric maps by stablishing correlations between pixel values or spectral properties of the images. The combination of UAS images and videos with optical velocimetry and particle tracking techniques have enabled to characterize surface flows and velocity fields of rivers and streams. UAS-derived data have been extensively used to produce high accuracy DEMs to identify the extent of inundations and improve flood models. IR sensors mounted on UASs have been also deployed to detect temperature patterns within rivers or lakes, and to identify groundwater inputs or warm effluents from power plants. Likewise, RGB, IR and more recently hyperspectral cameras are being deployed from UASs to early detect and monitor eutrophication and harmful algal blooms. Moreover, beyond being tools for the remotely acquisition of data, UAS have gone one step further, with devices that allow the collection of water samples and the measurement of parameters in situ, overcoming the many limitations of traditional manual sampling.

Nevertheless, the application of UAS technology to surface water research is subject to several limitations and challenges. Among the most obvious are the weather conditions, such as strong winds, rain, cloudiness and sunglint contamination that may hamper the flights and result in geometrically distorted images. These factors might degrade the quality of the images, making necessary the formulation of correction algorithms specific for the flight and environmental conditions. Likewise, the characteristics and distribution of vegetation can hinder the acquisition of data from the terrain in both aerial and underwater conditions and can even represent a serious risk for the movement and safety of unmanned aerial platforms. The accuracy in vertical and horizontal position (x,y,z axis) is crucial in many studies and rely heavily on the GCP or the availability/accuracy of the IMU/INS/RTK units. In this regard, ensuring a good geolocation and alignment of UAS imagery in highly dynamic environments (e.g., rivers), areas with scarce ground control points or surfaces rich in contrast, constitutes another problematic aspect. Fortunately, the recent advances and miniaturization of differential GNSS are yielding significant improvements in flight performance both in terms of accuracy and integrity. Other indirect problem associated with the large body of information generated by UASs, is the high data storage and processing capacity needed. Additional constraints related to low payload capacity, which limits the number of sensors that can be mounted on the platform, short flight times and battery duration also requires previous consideration.

On the one hand, the implementation of UAS remote sensing to the study of underground environments and processes, hidden from direct observations, is even more challenging since this technology was initially devised to detect features directly from the sky. In this context, the main limitation is the insufficient penetration capacity that most remote sensing techniques (except for long wave radar) present beyond the uppermost soil layer. If conditions are favorable and in the absence of vegetation, long wave radar imagery can detect groundwater levels situated a few meters deep. Although the application of UAS remote sensing to groundwater exploration is limited by technological development, several recent studies demonstrate the usefulness of this technique and represent a promising step forward in the knowledge of these water bodies. Most of these works are based on the detection of groundwater-induced thermal anomalies in coastal waters or riverine systems to characterize plumes and produce temperature maps. On the other hand, a few studies have taken advantage of the particular configuration of certain areas such as cenotes and peatlands to map water table. A research field currently under development is the use of geophysical instruments from UASs for the detection or characterization of groundwater (e.g., the aforementioned FDEM).

Regarding the types of UASs employed in hydrological research, as Tables 4 and 5 show, the prevalent type was rotary-wing, both in surface water and groundwater, although fixed wing and other types (paraglider) were also reported. The reason is that, even though fixed-wing UAS can cover larger areas and offer longer flight times, rotary wings allow greater maneuverability, allowing vertical movements and hoovering, as well as working in inaccessible areas, which are characteristic often needed in many data collection tasks. However, some authors have made a combined use of both types of platforms in order to complement their strengths and compensate for their weaknesses. Among the 122 works reviewed, 10 did not specify the type or model of UAS used and were excluded from tables.

Table 4. Number of reviewed works by UAS type in surface water research.

	Bathymetry	Water Level	Surface Velocity/Discharge	Flood Monitoring	TermperaTure Mapping	Water Contamination	HABs	Water Sampling	Sum
Fixed-wing	1	1	0	10	2	0	2	0	16
Rotary-wing	17	9	12	6	6	5	8	14	77
Both	0	0	0	2	0	0	2	0	4
Other	1	0	0	0	0	0	0	0	1

Table 5. Number of reviewed works by UAS type in groundwater research.

	SW–GW Interactions	SGD	Water Table Definition	Subsidence	Sum
Fixed-wing	1	0	1	0	2
Rotary-wing	4	2	3	3	12
Both	0	0	0	0	0

6.2. UAS Application in Freshwater Research: Future Prospects

Since their recent emergence in the civil sphere, UASs have been subject to continuous improvements and innovations developed in parallel with advances in sensors. Their incorporation into the research field has opened a new era in remote data acquisition, with a range of applications that is continuously growing.

UASs are providing researchers and water managers the opportunity of gathering exceptionally large data volumes in a safe, cost-effective and relatively easier manner than traditional methods. In fact, UASs have come to revolutionize measurement and sampling campaigns and, owing to their unprecedent spatio-temporal coverage and the recent advances in sensor technologies that allow to capture surface and submerged processes, it is possible to analyze aquatic systems from a holistic point of view more than ever. In a period of just two decades, UASs have deeply transformed the discipline of hydrology and its use alone or in combination with other instruments have contributed to the generation of highly accurate representations of fluvial morphology, to map surface properties and study system´s interactions.

In addition, UASs have rapidly evolved from early prototypes like balloons, paragliders and paramotors to sophisticated lightweight platforms able to operate in difficult to reach and hazardous environments preventing risks for operators. The decreasing prices of the electronic components and the many possibilities available in terms of platform customization and sensor combination have democratized their use and made them affordable even for small research groups and modest institutions.

In the light of the broad literature selection presented here, there are several opportunities and challenges that need to be considered for the full and efficient implementation of UAS in hydrological research and management:

Platform limitations: The use of non-metric cameras and the development of low-cost platforms have been a key in the rapid uptake of UASs for the obtention of high-resolution data on demand. However, UASs are subject to a number of limitations. Among the most important and evident ones, is UAS maximum payload, which have hindered their coupling with certain sensors and storage devices so far (e.g., to date there are only a few commercially available models of LiDAR sensors compatible with UAS). This limitation has prevented drones from carrying more integrated systems with multiple sensors, data processing equipment and communication hardware. Another hurdle of this technology is UAS´s high energy requirements (an inherent issue of scaling down the vehicle), which limit their autonomy, spatial coverage and makes necessary to interrupt the flight multiple times to replace batteries. In addition, most UASs cannot fly autonomously without GPS and may not be deployed under adverse meteorological conditions, situation that is recently changing with the development of new industrial ratings (e.g., IP44 and IP45) that ensure protection against debris and water. The widespread and more effective implementation of UASs in research calls for further miniaturization of components, the development of alternative power sources such as thin-film photovoltaic panels or tethered systems that extend the battery duration and provide a reliable wired communication, and improvements in vehicle autonomy, minimizing human intervention.

Methodological robustness: The many papers reviewed here reflect the necessity of increasing research efforts and carrying out more experiences on the application of UASs in hydrology, with a special focus on groundwater. While some applications based on visual imagery to extract river morphology, bathymetry or flood extents have gathered highly accurate results, other methods such as multispectral approaches to infer water quality, TIR imagery to study fluvial temperature patterns, video data to estimate stream flow or UAS-based approaches to define water table are still subject to important limitations. These methodologies will have to be refined as UAS and sensing technologies mature and optimized processing tools and algorithms are developed.

Information gaps: The availability of detailed hydrological information at global scale is still uneven in terms of spatial distribution and type of data. Thus, we believe that future research efforts should aim at increasing the current collection of UAS imagery over different types of basins, hydrological conditions and events, with especial focus on ungauged systems and information-scarce areas. On the other hand, as this review evidences, most UAS remote sensing techniques are focused on morphological aspects, while works on dynamic processes such water surface elevation, discharge estimations and surface-groundwater interactions are less common owing to technological, spatial and temporal limitations. In this regard, future advances in the field of UASs, together with multiplatform and multisensor approaches will narrow these information voids and will enable a deeper understanding of water bodies functioning and interactions.

Image processing and the role of machine learning: Manual extraction of regions of interest from the large volumes of information generated by UASs is a costly and time demanding task, even more in long-term monitoring programs involving repeated surveys. In this context, the emergence of machine learning algorithms provides great opportunities to automatize image processing workflows and address the challenges of hydrological research. To date, some works have successfully implemented machine learning approaches such as support vector machine or ANN for feature extraction and classification, or to solve non-linear problems in hydrological forecasting. Nevertheless, despite machine learning usually outperforms physically based models, their predictive accuracy is restricted to the range of input data used in the training stage. Thus, a long-term goal to improve model predictions at broader scales is to fill information gaps and acquire consistent spatial-temporal hydrological input data for their subsequent integration into artificial intelligence networks. As technology matures, UASs will be extremely useful tools for the acquisition of different types of training data.

Uneven legal framework: The rapid expansion of UASs has not been balanced by an exhaustive regulation and to date, there is not a single harmonized regulatory instrument at international scale. This situation stems from the shift in the use of UAS technology from the initial military purpose to professional/commercial and recreational use, leading to new legal implications. UASs must comply with principles of safety and privacy and every activity performed by a drone is subject to a convoluted set of specific sectorial rules such as data protection, privacy law, aviation law, telecommunication law, environmental law, product liability or criminal law, which hinders the potential and competitiveness of the drone market. The integration of civil UAS into all areas of the airspace while removing the technical barriers that limit their flight and addressing the inherent security, privacy and ethical issues still one of the main obstacles to overcome. Recently, the European Regulation 2019/947 [179] on the rules and procedures for the operation and registration of unmanned aircraft has constituted a leap forward in the unification and integration of civil drones into all areas of the airspace. The document sets the rules for conducting operational risk assessment and cross-border operations within the EU, the conditions under which drone operations can be allowed and establishes the requirements for the registration of UAS operators and certified UAS among others.

Finally, this paper presents a comprehensive literature review on the diverse applications of UAS in the fields of hydrology and hydrogeology, classifying them according to the aim of the study and covering a range of aspects such as the types of platforms and sensors used, their strengths, limitations and the new possibilities opened up by their combination with other measurement instruments. The compilation of experiences here summarized has demonstrated the great supporting role of UASs to ground-based and satellite-borne observations, covering the demand of near real-time and cost-effective data especially in the study of highly dynamic processes. Despite the rapid advances experienced and all the advantages UASs offer, their integration in water resource research (as in many other research fields) it is still in its infancy; there is room for improvement in many aspects such as image processing, standardization of the techniques and procedures adopted, or UASs integration with traditional instruments and analysis tools. In this regard, experience and data sharing among researchers can further improve hydrological research, help to fill knowledge gaps and lay the foundations for the design of methodologies and protocols appropriate to different hydrological contexts. This review aims at serving as a reference or introductory document for all the researchers and water managers who are interested in embracing this novel technology to meet their needs and to unify in a single document all the possibilities, potential approaches and the results obtained by different authors through the implementation of UASs.

Author Contributions: Conceptualization, M.V.-N., S.G.-L. and L.B.; methodology, M.V.-N., S.G.-L. and L.B.; investigation, M.V.-N., S.G.-L., V.R.-O. and Á.S.-B.; data curation, M.V.-N., S.G.-L., V.R.-O. and Á.S.-B.; writing—original draft preparation, M.V.-N. and S.G.-L.; writing—review and editing, L.B., V.R.-O. and Á.S.-B.; supervision, S.G.-L. All authors have read and agreed to the published version of the manuscript.

Funding: This work has received financial support from the research group RNM373 Geoscience-UCA of Junta de Andalucía.

Institutional Review Board Statement: Not applicable.

Informed Consent Statement: Not applicable.

Data Availability Statement: No new data were created or analyzed in this study. Data sharing is not applicable to this article.

Acknowledgments: The authors want to thank the support received from the plan "Programa de Fomento e Impulso de la Investigación y Transferencia" of the University of Cádiz. The authors also want to express their gratitude to the reviewers and in particular, to Ulf Mallast, whose encouraging and insightful comments have greatly contributed to improve this review. We also thank the Drone Service of the University of Cádiz, which kindly lent their equipment to illustrate this work and to the infrastructure project EQC2018-004446-P of the Ministry of Economy and Competitiveness of the Government of Spain.

Conflicts of Interest: The authors declare no conflict of interests.

References

1. WWAP (UNESCO World Water Assessment Programme). *The United Nations World Water Development Report 2019: Leaving No One Behind*; UNESCO: Paris, France, 2019.
2. IPCC. *Climate Change 2014: Synthesis Report. Contribution of Working Groups I, II and III to the Fifth Assessment Report of the Intergovernmental Panel on Climate Change*; Core Writing Team, Pachauri, R.K., Meyer, L.A., Eds.; IPCC: Geneva, Switzerland, 2014; 151p.
3. Chamoso, P.; González-Briones, A.; Rivas, A.; De Mata, F.B.; Corchado, J.M. The Use of Drones in Spain: Towards a Platform for Controlling UAVs in Urban Environments. *Sensors* **2018**, *18*, 1416. [CrossRef] [PubMed]
4. Linchant, J.; Lisein, J.; Semeki, J.; Lejeune, P.; Vermeulen, C. Are unmanned aircraft systems (UASs) the future of wildlife monitoring? A review of accomplishments and challenges. *Mammal Rev.* **2015**, *45*, 239–252. [CrossRef]
5. Chabot, D.; Bird, D.M. Wildlife research and management methods in the 21st century: Where do unmanned aircraft fit in? *J. Unmanned Veh. Syst.* **2015**, *3*, 137–155. [CrossRef]
6. Christie, K.S.; Gilbert, S.L.; Brown, C.L.; Hatfield, M.; Hanson, L. Unmanned aircraft systems in wildlife research: Current and future applications of a transformative technology. *Front. Ecol. Environ.* **2016**, *14*, 241–251. [CrossRef]
7. Adão, T.; Hruška, J.; Pádua, L.; Bessa, J.; Peres, E.; Morais, R.; Sousa, J.J. Hyperspectral Imaging: A Review on UAV-Based Sensors, Data Processing and Applications for Agriculture and Forestry. *Remote Sens.* **2017**, *9*, 1110. [CrossRef]
8. Wieser, M.; Mandlburger, G.; Hollaus, M.; Otepka, J.; Glira, P.; Pfeifer, N. A Case Study of UAS Borne Laser Scanning for Measurement of Tree Stem Diameter. *Remote Sens.* **2017**, *9*, 1154. [CrossRef]
9. Radoglou-Grammatikis, P.; Sarigiannidis, P.; Lagkas, T.; Moscholios, I. A compilation of UAV applications for precision agriculture. *Comput. Netw.* **2020**, *172*, 107148. [CrossRef]
10. Rakha, T.; Gorodetsky, A. Review of Unmanned Aerial System (UAS) applications in the built environment: Towards automated building inspection procedures using drones. *Autom. Constr.* **2018**, *93*, 252–264. [CrossRef]
11. Zhou, S.; Gheisari, M. Unmanned aerial system applications in construction: A systematic review. *Constr. Innov.* **2018**, *18*, 453–468. [CrossRef]
12. Restas, A. Drone Applications for Supporting Disaster Management. *World J. Eng. Technol.* **2015**, *3*, 316–321. [CrossRef]
13. McCabe, M.; Rodell, M.; Alsdorf, D.; Miralles, D.; Uijlenhoet, R.; Wagner, W.; Lucieer, A.; Houborg, R.; Verhoest, N.; Franz, T.; et al. The future of earth observation in Hydrology. *Air Space Eur.* **2017**, *2*, 42–44. [CrossRef]
14. Debell, L.; Anderson, K.; Brazier, R.; King, N.; Jones, L. Water resource management at catchment scales using lightweight UAVs: Current capabilities and future perspectives. *J. Unmanned Veh. Syst.* **2016**, *4*, 7–30. [CrossRef]

15. Colomina, I.; Molina, P. Unmanned aerial systems for photogrammetry and remote sensing: A review. *ISPRS J. Photogramm. Remote Sens.* **2014**, *92*, 79–97. [CrossRef]
16. Carrivick, J.L.; Smith, M.W. Fluvial and aquatic applications of Structure from Motion photogrammetry and unmanned aerial vehicle/drone technology. *Wiley Interdiscip. Rev. Water* **2019**, *6*, e1328. [CrossRef]
17. Tomsett, C.; Leyland, J. Remote sensing of river corridors: A review of current trends and future directions. *River Res. Appl.* **2019**, *35*, 779–803. [CrossRef]
18. Rhee, D.S.; Kim, Y.D.; Kang, B.; Kim, D. Applications of unmanned aerial vehicles in fluvial remote sensing: An overview of recent achievements. *KSCE J. Civ. Eng.* **2017**, *22*, 588–602. [CrossRef]
19. Kislik, C.; Dronova, I.; Kelly, M. UAVs in Support of Algal Bloom Research: A Review of Current Applications and Future Opportunities. *Drones* **2018**, *2*, 35. [CrossRef]
20. Wu, D.; Li, R.; Zhang, F.; Liu, J. A review on drone-based harmful algae blooms monitoring. *Environ. Monit. Assess.* **2019**, *191*, 211. [CrossRef] [PubMed]
21. Terwilliger, B.; Ison, D.C.; Robbins, J.; Vincenzi, D. *Small Unmanned Aircraft Systems Guide: Exploring Designs, Operations, Regulations, and Economics*; Aviation Supplies & Academics: Washington, DC, USA, 2017.
22. Arjomandi, M.; Agostino, S.; Mammone, M.; Nelson, M.; Zhou, T. *Classification of Unmanned Aerial Vehicles*; Mech. Eng.; University of Adelaide: Adelaide, Australia, 2006.
23. Weibel, R.E.; Hansman, R.J. Safety considerations for operation of different classes of UAVs in the NAS. In Proceedings of the Aiaa 4th Aviation Technology, Integration and Operations (Atio) Forum, Chicago, IL, USA, 20–22 September 2004; Volume 1, pp. 341–351. [CrossRef]
24. Hassanalian, M.; Abdelkefi, A. Classifications, applications, and design challenges of drones: A review. *Prog. Aerosp. Sci.* **2017**, *91*, 99–131. [CrossRef]
25. Watts, A.C.; Ambrosia, V.G.; Hinkley, E.A. Unmanned Aircraft Systems in Remote Sensing and Scientific Research: Classification and Considerations of Use. *Remote Sens.* **2012**, *4*, 1671–1692. [CrossRef]
26. Johnston, D.W. Unoccupied Aircraft Systems in Marine Science and Conservation. *Annu. Rev. Mar. Sci.* **2019**, *11*, 439–463. [CrossRef]
27. Boukoberine, M.N.; Zhou, Z.; Benbouzid, M. A critical review on unmanned aerial vehicles power supply and energy management: Solutions, strategies, and prospects. *Appl. Energy* **2019**, *255*, 113823. [CrossRef]
28. Snavely, N.; Seitz, S.M.; Szeliski, R. Modeling the World from Internet Photo Collections. *Int. J. Comput. Vis.* **2008**, *80*, 189–210. [CrossRef]
29. He, Y.; Weng, Q. *High Spatial Resolution Remote Sensing: Data, Analysis, and Applications*; CRC Press: Boca Raton, FL, USA, 2018; ISBN 9781498767682.
30. Dugdale, S.J. A practitioner's guide to thermal infrared remote sensing of rivers and streams: Recent advances, precautions and considerations. *Wiley Interdiscip. Rev. Water* **2016**, *3*, 251–268. [CrossRef]
31. Mesas-Carrascosa, F.-J.; Pérez-Porras, F.; De Larriva, J.E.M.; Frau, C.M.; Agüera-Vega, F.; Carvajal-Ramírez, F.; Martínez-Carricondo, P.; García-Ferrer, A. Drift Correction of Lightweight Microbolometer Thermal Sensors On-Board Unmanned Aerial Vehicles. *Remote Sens.* **2018**, *10*, 615. [CrossRef]
32. Döpper, V.; Gränzig, T.; Kleinschmit, B.; Förster, M. Challenges in UAS-Based TIR Imagery Processing: Image Alignment and Uncertainty Quantification. *Remote Sens.* **2020**, *12*, 1552. [CrossRef]
33. Di Stefano, G.; Romeo, G.; Mazzini, A.; Iarocci, A.; Hadi, S.; Pelphrey, S. The Lusi drone: A multidisciplinary tool to access extreme environments. *Mar. Pet. Geol.* **2018**, *90*, 26–37. [CrossRef]
34. Lejot, J.; Delacourt, C.; Piégay, H.; Fournier, T.; Trémélo, M.-L.; Allemand, P. Very high spatial resolution imagery for channel bathymetry and topography from an unmanned mapping controlled platform. *Earth Surf. Process. Landf.* **2007**, *32*, 1705–1725. [CrossRef]
35. Woodget, A.S.; Dietrich, J.T.; Wilson, R.T. Quantifying Below-Water Fluvial Geomorphic Change: The Implications of Refraction Correction, Water Surface Elevations, and Spatially Variable Error. *Remote Sens.* **2019**, *11*, 2415. [CrossRef]
36. Gentile, V.; Mróz, M.; Spitoni, M.; Lejot, J.; Piégay, H.; Demarchi, L. Bathymetric Mapping of Shallow Rivers with UAV Hyperspectral Data. In Proceedings of the Fifth International Conference on Telecommunications and Remote Sensing, Milan, Italy, 10–11 October 2016; pp. 43–49.
37. Flener, C.; Vaaja, M.; Jaakkola, A.; Krooks, A.; Kaartinen, H.; Kukko, A.; Kasvi, E.; Hyyppä, H.; Hyyppä, J.; Alho, P. Seamless Mapping of River Channels at High Resolution Using Mobile LiDAR and UAV-Photography. *Remote Sens.* **2013**, *5*, 6382–6407. [CrossRef]
38. Castro, C.C.; Gómez, J.A.D.; Martín, J.D.; Sánchez, B.A.H.; Arango, J.L.C.; Tuya, F.A.C.; Díaz-Varela, R. An UAV and Satellite Multispectral Data Approach to Monitor Water Quality in Small Reservoirs. *Remote Sens.* **2020**, *12*, 1514. [CrossRef]
39. Pai, H.; Malenda, H.F.; Briggs, M.A.; Singha, K.; González-Pinzón, R.; Gooseff, M.N.; Tyler, S.W. Potential for Small Unmanned Aircraft Systems Applications for Identifying Groundwater-Surface Water Exchange in a Meandering River Reach. *Geophys. Res. Lett.* **2017**, *44*, 11868–11877. [CrossRef]
40. Legleiter, C.J.; Harrison, L.R. Remote Sensing of River Bathymetry: Evaluating a Range of Sensors, Platforms, and Algorithms on the Upper Sacramento River, California, USA. *Water Resour. Res.* **2019**, *55*, 2142–2169. [CrossRef]

41. Collas, F.P.; Van Iersel, W.K.; Straatsma, M.W.; Buijse, A.D.; Leuven, R.S. Sub-Daily Temperature Heterogeneity in a Side Channel and the Influence on Habitat Suitability of Freshwater Fish. *Remote Sens.* **2019**, *11*, 2367. [CrossRef]
42. Dugdale, S.J.; Kelleher, C.A.; Malcolm, I.A.; Caldwell, S.; Hannah, D.M. Assessing the potential of drone-based thermal infrared imagery for quantifying river temperature heterogeneity. *Hydrol. Process.* **2019**, *33*, 1152–1163. [CrossRef]
43. Kinzel, P.J.; Legleiter, C.J. sUAS-Based Remote Sensing of River Discharge Using Thermal Particle Image Velocimetry and Bathymetric Lidar. *Remote Sens.* **2019**, *11*, 2317. [CrossRef]
44. Mallast, U.; Siebert, C. Combining continuous spatial and temporal scales for SGD investigations using UAV-based thermal infrared measurements. *Hydrol. Earth Syst. Sci.* **2019**, *23*, 1375–1392. [CrossRef]
45. Chen, S.; Johnson, F.; Drummond, C.; Glamore, W. A new method to improve the accuracy of remotely sensed data for wetland water balance estimates. *J. Hydrol. Reg. Stud.* **2020**, *29*, 100689. [CrossRef]
46. Vázquez-Tarrío, D.; Borgniet, L.; Liébault, F.; Recking, A. Using UAS optical imagery and SfM photogrammetry to characterize the surface grain size of gravel bars in a braided river (Vénéon River, French Alps). *Geomorphology* **2017**, *285*, 94–105. [CrossRef]
47. Li, H.; Chen, L.; Wang, Z.; Yu, Z. Mapping of River Terraces with Low-Cost UAS Based Structure-from-Motion Photogrammetry in a Complex Terrain Setting. *Remote Sens.* **2019**, *11*, 464. [CrossRef]
48. Hemmelder, S.; Marra, W.; Markies, H.; De Jong, S.M. Monitoring river morphology & bank erosion using UAV imagery—A case study of the river Buëch, Hautes-Alpes, France. *Int. J. Appl. Earth Obs. Geoinf.* **2018**, *73*, 428–437. [CrossRef]
49. Mazzoleni, M.; Paron, P.; Reali, A.; Juizo, D.; Manane, J.; Brandimarte, L. Testing UAV-derived topography for hydraulic modelling in a tropical environment. *Nat. Hazards* **2020**, *103*, 139–163. [CrossRef]
50. Zinke, P.; Flenner, C. 2013. Experiences from the use of Unmanned Aerial Vehicles (UAV) for River Bathymetry Modelling in Norway. *Vann* **2013**, *48*, 351–360.
51. Woodget, A.S.; Carbonneau, P.E.; Visser, F.; Maddock, I.P. Quantifying submerged fluvial topography using hyperspatial resolution UAS imagery and structure from motion photogrammetry. *Earth Surf. Process. Landf.* **2014**, *40*, 47–64. [CrossRef]
52. Dietrich, J.T. Bathymetric Structure-from-Motion: Extracting shallow stream bathymetry from multi-view stereo photogrammetry. *Earth Surf. Process. Landf.* **2017**, *42*, 355–364. [CrossRef]
53. Entwistle, N.S.; Heritage, G.L. Small unmanned aerial model accuracy for photogrammetrical fluvial bathymetric survey. *J. Appl. Remote Sens.* **2019**, *13*, 014523. [CrossRef]
54. Partama, I.Y.; Kanno, A.; Ueda, M.; Akamatsu, Y.; Inui, R.; Sekine, M.; Yamamoto, K.; Imai, T.; Higuchi, T. Removal of water-surface reflection effects with a temporal minimum filter for UAV-based shallow-water photogrammetry. *Earth Surf. Process. Landf.* **2018**, *43*, 2673–2682. [CrossRef]
55. Skarlatos, D.; Agrafiotis, P. A Novel Iterative Water Refraction Correction Algorithm for Use in Structure from Motion Photogrammetric Pipeline. *J. Mar. Sci. Eng.* **2018**, *6*, 77. [CrossRef]
56. Agrafiotis, P.; Karantzalos, K.; Georgopoulos, A.; Skarlatos, D. Correcting Image Refraction: Towards Accurate Aerial Image-Based Bathymetry Mapping in Shallow Waters. *Remote Sens.* **2020**, *12*, 322. [CrossRef]
57. Agrafiotis, P.; Skarlatos, D.; Georgopoulos, A.; Karantzalos, K. DepthLearn: Learning to Correct the Refraction on Point Clouds Derived from Aerial Imagery for Accurate Dense Shallow Water Bathymetry Based on SVMs-Fusion with LiDAR Point Clouds. *Remote Sens.* **2019**, *11*, 2225. [CrossRef]
58. Mandlburger, G.; Pfennigbauer, M.; Schwarz, R.; Flöry, S.; Nussbaumer, L. Concept and Performance Evaluation of a Novel UAV-Borne Topo-Bathymetric LiDAR Sensor. *Remote Sens.* **2020**, *12*, 986. [CrossRef]
59. Kasvi, E.; Salmela, J.; Lotsari, E.; Kumpula, T.; Lane, S. Comparison of remote sensing based approaches for mapping bathymetry of shallow, clear water rivers. *Geomorphology* **2019**, *333*, 180–197. [CrossRef]
60. Erena, M.; Atenza, J.F.; García-Galiano, S.; Domínguez, J.A.; Bernabé, J.M. Use of Drones for the Topo-Bathymetric Monitoring of the Reservoirs of the Segura River Basin. *Water* **2019**, *11*, 445. [CrossRef]
61. Bandini, F.; Olesen, D.; Jakobsen, J.; Kittel, C.M.M.; Wang, S.; Garcia, M.; Bauer-Gottwein, P. Technical note: Bathymetry observations of inland water bodies using a tethered single-beam sonar controlled by an unmanned aerial vehicle. *Hydrol. Earth Syst. Sci.* **2018**, *22*, 4165–4181. [CrossRef]
62. Alvarez, L.V.; Moreno, H.A.; Segales, A.R.; Pham, T.G.; Pillar-Little, E.A.; Chilson, P.B. Merging Unmanned Aerial Systems (UAS) Imagery and Echo Soundings with an Adaptive Sampling Technique for Bathymetric Surveys. *Remote Sens.* **2018**, *10*, 1362. [CrossRef]
63. Kim, J.S.; Baek, D.; Seo, I.W.; Shin, J. Retrieving shallow stream bathymetry from UAV-assisted RGB imagery using a geospatial regression method. *Geomorphology* **2019**, *341*, 102–114. [CrossRef]
64. Langhammer, J.; Janský, B.; Kocum, J.; Minařík, R. 3-D reconstruction of an abandoned montane reservoir using UAV photogrammetry, aerial LiDAR and field survey. *Appl. Geogr.* **2018**, *98*, 9–21. [CrossRef]
65. Tamminga, A.D.; Hugenholtz, C.H.; Eaton, B.C.; Lapointe, M. Hyperspatial Remote Sensing of Channel Reach Morphology and Hydraulic Fish Habitat Using an Unmanned Aerial Vehicle (UAV): A First Assessment in the Context of River Research and Management. *River Res. Appl.* **2015**, *31*, 379–391. [CrossRef]
66. Templin, T.; Popielarczyk, D.; Kosecki, R. Application of Low-Cost Fixed-Wing UAV for Inland Lakes Shoreline Investigation. *Pure Appl. Geophys.* **2018**, *175*, 3263–3283. [CrossRef]
67. Alsdorf, D.E.; Rodríguez, E.; Lettenmaier, D.P. Measuring surface water from space. *Rev. Geophys.* **2007**, *45*. [CrossRef]
68. Ridolfi, E.; Manciola, P. Water Level Measurements from Drones: A Pilot Case Study at a Dam Site. *Water* **2018**, *10*, 297. [CrossRef]

69. Gao, A.; Wu, S.; Wang, F.; Wu, X.; Xu, P.; Yu, L.; Zhu, S. A Newly Developed Unmanned Aerial Vehicle (UAV) Imagery Based Technology for Field Measurement of Water Level. *Water* **2019**, *11*, 124. [CrossRef]
70. Bandini, F.; Jakobsen, J.; Olesen, D.H.; Reyna-Gutierrez, J.A.; Bauer-Gottwein, P. Measuring water level in rivers and lakes from lightweight Unmanned Aerial Vehicles. *J. Hydrol.* **2017**, *548*, 237–250. [CrossRef]
71. Bandini, F.; Sunding, T.P.; Linde, J.; Smith, O.; Jensen, I.K.; Köppl, C.J.; Butts, M.; Bauer-Gottwein, P. Unmanned Aerial System (UAS) observations of water surface elevation in a small stream: Comparison of radar altimetry, LIDAR and photogrammetry techniques. *Remote Sens. Environ.* **2020**, *237*, 111487. [CrossRef]
72. Jiang, L.; Bandini, F.; Smith, O.; Jensen, I.K.; Bauer-Gottwein, P. The Value of Distributed High-Resolution UAV-Borne Observations of Water Surface Elevation for River Management and Hydrodynamic Modeling. *Remote Sens.* **2020**, *12*, 1171. [CrossRef]
73. Tymków, P.; Jóźków, G.; Walicka, A.; Karpina, M.; Borkowski, A. Identification of Water Body Extent Based on Remote Sensing Data Collected with Unmanned Aerial Vehicle. *Water* **2019**, *11*, 338. [CrossRef]
74. Niedzielski, T.; Witek, M.; Spallek, W. Observing river stages using unmanned aerial vehicles. *Hydrol. Earth Syst. Sci.* **2016**, *20*, 3193–3205. [CrossRef]
75. García-López, S.; Ruiz-Ortiz, V.; Barbero, L.; Sánchez-Bellón, Á. Contribution of the UAS to the determination of the water budget in a coastal wetland: A case study in the natural park of the Bay of Cádiz (SW Spain). *Eur. J. Remote Sens.* **2018**, *51*, 965–977. [CrossRef]
76. Mohamad, N.; Khanan, M.F.A.; Ahmad, A.; Din, A.H.M.; Shahabi, H. Evaluating Water Level Changes at Different Tidal Phases Using UAV Photogrammetry and GNSS Vertical Data. *Sensors* **2019**, *19*, 3778. [CrossRef]
77. Yucel, M.A.; Turan, R.Y. Areal Change Detection and 3D Modeling of Mine Lakes Using High-Resolution Unmanned Aerial Vehicle Images. *Arab. J. Sci. Eng.* **2016**, *41*, 4867–4878. [CrossRef]
78. McGlynn, B.L.; Blöschl, G.; Borga, M.; Bormann, H.; Hurkmans, R.; Komma, J.; Nandagiri, L.; Uijlenhoet, R.; Wagener, T. *A Data Acquisition Framework for Runoff Prediction in Ungauged Basins, Runoff Prediction in Ungauged Basins*; Cambridge University Press: Cambridge, UK, 2013. [CrossRef]
79. Mishra, A.K.; Coulibaly, P. Developments in hydrometric network design: A review. *Rev. Geophys.* **2009**, *47*. [CrossRef]
80. Sichangi, A.W.; Wang, L.; Yang, K.; Chen, D.; Wang, Z.; Li, X.; Zhou, J.; Liu, W.; Kuria, D. Estimating continental river basin discharges using multiple remote sensing data sets. *Remote Sens. Environ.* **2016**, *179*, 36–53. [CrossRef]
81. Sneeuw, N.; Lorenz, C.; Devaraju, B.; Tourian, M.J.; Riegger, J.; Kunstmann, H.; Bárdossy, A. Estimating Runoff Using Hydro-Geodetic Approaches. *Surv. Geophys.* **2014**, *35*, 1333–1359. [CrossRef]
82. Yang, S.; Wang, J.; Wang, P.; Gong, T.; Liu, H. Low Altitude Unmanned Aerial Vehicles (UAVs) and Satellite Remote Sensing Are Used to Calculated River Discharge Attenuation Coefficients of Ungauged Catchments in Arid Desert. *Water* **2019**, *11*, 2633. [CrossRef]
83. Yang, S.; Wang, P.; Lou, H.; Wang, J.; Zhao, C.; Gong, T. Estimating River Discharges in Ungauged Catchments Using the Slope–Area Method and Unmanned Aerial Vehicle. *Water* **2019**, *11*, 2361. [CrossRef]
84. Lou, H.; Wang, P.; Yang, S.; Hao, F.; Ren, X.; Wang, Y.; Shi, L.; Wang, J.; Gong, T. Combining and Comparing an Unmanned Aerial Vehicle and Multiple Remote Sensing Satellites to Calculate Long-Term River Discharge in an Ungauged Water Source Region on the Tibetan Plateau. *Remote Sens.* **2020**, *12*, 2155. [CrossRef]
85. Koutalakis, P.; Tzoraki, O.; Zaimes, G. UAVs for Hydrologic Scopes: Application of a Low-Cost UAV to Estimate Surface Water Velocity by Using Three Different Image-Based Methods. *Drones* **2019**, *3*, 14. [CrossRef]
86. Pearce, S.; Ljubičić, R.; Peña-Haro, S.; Perks, M.; Tauro, F.; Pizarro, A.; Sasso, S.F.D.; Strelnikova, D.; Grimaldi, S.; Maddock, I.; et al. An Evaluation of Image Velocimetry Techniques under Low Flow Conditions and High Seeding Densities Using Unmanned Aerial Systems. *Remote Sens.* **2020**, *12*, 232. [CrossRef]
87. Tauro, F.; Pagano, C.; Phamduy, P.; Grimaldi, S.; Porfiri, M. Large-Scale Particle Image Velocimetry From an Unmanned Aerial Vehicle. *IEEE/ASME Trans. Mechatronics* **2015**, *20*, 3269–3275. [CrossRef]
88. Tauro, F.; Porfiri, M.; Grimaldi, S. Surface flow measurements from drones. *J. Hydrol.* **2016**, *540*, 240–245. [CrossRef]
89. Lewis, Q.W.; Lindroth, E.M.; Rhoads, B.L. Integrating unmanned aerial systems and LSPIV for rapid, cost-effective stream gauging. *J. Hydrol.* **2018**, *560*, 230–246. [CrossRef]
90. Sasso, S.F.D.; Pizarro, A.; Manfreda, S. Metrics for the Quantification of Seeding Characteristics to Enhance Image Velocimetry Performance in Rivers. *Remote Sens.* **2020**, *12*, 1789. [CrossRef]
91. Thumser, P.; Haas, C.; Tuhtan, J.A.; Fuentes-Pérez, J.F.; Toming, G. RAPTOR-UAV: Real-time particle tracking in rivers using an unmanned aerial vehicle. *Earth Surf. Process. Landf.* **2017**, *42*, 2439–2446. [CrossRef]
92. Fulton, J.; Anderson, I.; Chiu, C.-L.; Sommer, W.; Adams, J.; Moramarco, T.; Bjerklie, D.; Fulford, J.; Sloan, J.; Best, H.; et al. QCam: sUAS-Based Doppler Radar for Measuring River Discharge. *Remote Sens.* **2020**, *12*, 3317. [CrossRef]
93. Dyer, J.L.; Moorhead, R.J.; Hathcock, L. Identification and Analysis of Microscale Hydrologic Flood Impacts Using Unmanned Aerial Systems. *Remote Sens.* **2020**, *12*, 1549. [CrossRef]
94. Mourato, S.; Fernandez, P.; Pereira, L.; Moreira, M. Improving a DSM Obtained by Unmanned Aerial Vehicles for Flood Modelling. *IOP Conf. Series: Earth Environ. Sci.* **2017**, *95*, 022014. [CrossRef]
95. Muthusamy, M.; Casado, M.R.; Salmoral, G.; Irvine, T.; Leinster, P. A Remote Sensing Based Integrated Approach to Quantify the Impact of Fluvial and Pluvial Flooding in an Urban Catchment. *Remote Sens.* **2019**, *11*, 577. [CrossRef]

96. Şerban, G.; Rus, I.; Vele, D.; Breţcan, P.; Alexe, M.; Petrea, D. Flood-prone area delimitation using UAV technology, in the areas hard-to-reach for classic aircrafts: Case study in the north-east of Apuseni Mountains, Transylvania. *Nat. Hazards* **2016**, *82*, 1817–1832. [CrossRef]
97. Yalcin, E. Two-dimensional hydrodynamic modelling for urban flood risk assessment using unmanned aerial vehicle imagery: A case study of Kirsehir, Turkey. *J. Flood Risk Manag.* **2018**, *12*, e12499. [CrossRef]
98. Özcan, O. Multi-temporal UAV based repeat monitoring of rivers sensitive to flood. *J. Maps* **2020**, 1–8. [CrossRef]
99. Diakakis, M.; Andreadakis, E.; Nikolopoulos, E.; Spyrou, N.; Gogou, M.; Deligiannakis, G.; Katsetsiadou, N.; Antoniadis, Z.; Melaki, M.; Georgakopoulos, A.; et al. An integrated approach of ground and aerial observations in flash flood disaster investigations. The case of the 2017 Mandra flash flood in Greece. *Int. J. Disaster Risk Reduct.* **2019**, *33*, 290–309. [CrossRef]
100. Leitão, J.P.; De Vitry, M.M.; Scheidegger, A.; Rieckermann, J. Assessing the quality of digital elevation models obtained from mini unmanned aerial vehicles for overland flow modelling in urban areas. *Hydrol. Earth Syst. Sci.* **2016**, *20*, 1637–1653. [CrossRef]
101. Hashemi-Beni, L.; Jones, J.; Thompson, G.; Johnson, C.; Gebrehiwot, A. Challenges and Opportunities for UAV-Based Digital Elevation Model Generation for Flood-Risk Management: A Case of Princeville, North Carolina. *Sensors* **2018**, *18*, 3843. [CrossRef]
102. Schumann, G.J.P.; Muhlhausen, J.; Andreadis, K.M. Rapid mapping of small-scale river-floodplain environments using UAV SfM supports classical theory. *Remote Sens.* **2019**, *11*, 982. [CrossRef]
103. Annis, A.; Nardi, F.; Petroselli, A.; Apollonio, C.; Arcangeletti, E.; Tauro, F.; Belli, C.; Bianconi, R.; Grimaldi, S. UAV-DEMs for Small-Scale Flood Hazard Mapping. *Water* **2020**, *12*, 1717. [CrossRef]
104. Villanueva, J.R.E.; Martínez, L.I.; Montiel, J.I.P. DEM Generation from Fixed-Wing UAV Imaging and LiDAR-Derived Ground Control Points for Flood Estimations. *Sensors* **2019**, *19*, 3205. [CrossRef] [PubMed]
105. Tamminga, A.D.; Eaton, B.C.; Hugenholtz, C.H. UAS-based remote sensing of fluvial change following an extreme flood event. *Earth Surf. Process. Landf.* **2015**, *40*, 1464–1476. [CrossRef]
106. Abdelkader, M.; Shaqura, M.; Claudel, C.G.; Gueaieb, W. A UAV based system for real time flash flood monitoring in desert environments using Lagrangian microsensors. In Proceedings of the 2013 International Conference on Unmanned Aircraft Systems (ICUAS), Atlanta, GA, USA, 28–31 May 2013; pp. 25–34.
107. Popescu, D.; Ichim, L.; Stoican, F. Unmanned Aerial Vehicle Systems for Remote Estimation of Flooded Areas Based on Complex Image Processing. *Sensors* **2017**, *17*, 446. [CrossRef]
108. Gebrehiwot, A.; Hashemi-Beni, L.; Thompson, G.; Kordjamshidi, P.; Langan, T.E. Deep Convolutional Neural Network for Flood Extent Mapping Using Unmanned Aerial Vehicles Data. *Sensors* **2019**, *19*, 1486. [CrossRef]
109. Jakovljevic, G.; Govedarica, M.; Alvarez-Taboada, F.; Pajic, V. Accuracy Assessment of Deep Learning Based Classification of LiDAR and UAV Points Clouds for DTM Creation and Flood Risk Mapping. *Geosciences* **2019**, *9*, 323. [CrossRef]
110. Erdelj, M.; Natalizio, E.; Chowdhury, K.R.; Akyildiz, I.F. Help from the Sky: Leveraging UAVs for Disaster Management. *IEEE Pervasive Comput.* **2017**, *16*, 24–32. [CrossRef]
111. Salmoral, G.; Casado, M.R.; Muthusamy, M.; Butler, D.; Menon, P.P.; Leinster, P. Guidelines for the Use of Unmanned Aerial Systems in Flood Emergency Response. *Water* **2020**, *12*, 521. [CrossRef]
112. Kastridis, A.; Kirkenidis, C.; Sapountzis, M. An integrated approach of flash flood analysis in ungauged Mediterranean watersheds using post-flood surveys and unmanned aerial vehicles. *Hydrol. Process.* **2020**, *34*, 4920–4939. [CrossRef]
113. Handcock, R.N.; Torgersen, C.E.; Cherkauer, K.A.; Gillespie, A.R.; Tockner, K.; Faux, R.N.; Tan, J. Thermal Infrared Remote Sensing of Water Temperature in Riverine Landscapes. *Fluv. Remote Sens. Sci. Manag.* **2012**, *1*, 85–113. [CrossRef]
114. Jensen, A.M.; Neilson, B.T.; McKee, M.; Chen, Y. Thermal remote sensing with an autonomous unmanned aerial remote sensing platform for surface stream temperatures. *2012 IEEE Int. Geosci. Remote Sens. Symp.* **2012**, 5049–5052. [CrossRef]
115. Wawrzyniak, V.; Piégay, H.; Allemand, P.; Vaudor, L.; Grandjean, P. Prediction of water temperature heterogeneity of braided rivers using very high resolution thermal infrared (TIR) images. *Int. J. Remote Sens.* **2013**, *34*, 4812–4831. [CrossRef]
116. Kelly, J.; Kljun, N.; Olsson, P.-O.; Mihai, L.; Liljeblad, B.; Weslien, P.; Klemedtsson, L.; Eklundh, L. Challenges and Best Practices for Deriving Temperature Data from an Uncalibrated UAV Thermal Infrared Camera. *Remote Sens.* **2019**, *11*, 567. [CrossRef]
117. Chung, M.; Detweiler, C.; Hamilton, M.; Higgins, J.; Ore, J.-P.; Thompson, S. Obtaining the Thermal Structure of Lakes from the Air. *Water* **2015**, *7*, 6467–6482. [CrossRef]
118. Koparan, C.; Koc, A.B.; Sawyer, C.; Privette, C. Temperature Profiling of Waterbodies with a UAV-Integrated Sensor Subsystem. *Drones* **2020**, *4*, 35. [CrossRef]
119. Demario, A.; Lopez, P.; Plewka, E.; Wix, R.; Xia, H.; Zamora, E.; Gessler, D.; Yalin, A.P. Water Plume Temperature Measurements by an Unmanned Aerial System (UAS). *Sensors* **2017**, *17*, 306. [CrossRef] [PubMed]
120. Caldwell, S.; Kelleher, C.; Baker, E.; Lautz, L. Relative information from thermal infrared imagery via unoccupied aerial vehicle informs simulations and spatially-distributed assessments of stream temperature. *Sci. Total. Environ.* **2019**, *661*, 364–374. [CrossRef]
121. Powers, C.; Hanlon, R.; Schmale, D. Tracking of a Fluorescent Dye in a Freshwater Lake with an Unmanned Surface Vehicle and an Unmanned Aircraft System. *Remote Sens.* **2018**, *10*, 81. [CrossRef]
122. Baek, D.; Seo, I.W.; Kim, J.S.; Nelson, J.M. UAV-based measurements of spatio-temporal concentration distributions of fluorescent tracers in open channel flows. *Adv. Water Resour.* **2019**, *127*, 76–88. [CrossRef]
123. Legleiter, C.J.; McDonald, R.R.; Nelson, J.M.; Kinzel, P.J.; Perroy, R.L.; Baek, D.; Seo, I.W. Remote sensing of tracer dye concentrations to support dispersion studies in river channels. *J. Ecohydraulics* **2019**, *4*, 131–146. [CrossRef]

124. Martin, C.; Parkes, S.; Zhang, Q.; Zhang, X.; McCabe, M.F.; Duarte, C.M. Use of unmanned aerial vehicles for efficient beach litter monitoring. *Mar. Pollut. Bull.* **2018**, *131*, 662–673. [CrossRef] [PubMed]
125. Fallati, L.; Polidori, A.; Salvatore, C.; Saponari, L.; Savini, A.; Galli, P. Anthropogenic Marine Debris assessment with Unmanned Aerial Vehicle imagery and deep learning: A case study along the beaches of the Republic of Maldives. *Sci. Total Environ.* **2019**, *693*, 133581. [CrossRef] [PubMed]
126. Kako, S.; Morita, S.; Taneda, T. Estimation of plastic marine debris volumes on beaches using unmanned aerial vehicles and image processing based on deep learning. *Mar. Pollut. Bull.* **2020**, *155*, 111127. [CrossRef]
127. Geraeds, M.; Van Emmerik, T.; De Vries, R.; Bin Ab Razak, M.S. Riverine Plastic Litter Monitoring Using Unmanned Aerial Vehicles (UAVs). *Remote Sens.* **2019**, *11*, 2045. [CrossRef]
128. Jakovljevic, G.; Govedarica, M.; Alvarez-Taboada, F. A Deep Learning Model for Automatic Plastic Mapping Using Unmanned Aerial Vehicle (UAV) Data. *Remote Sens.* **2020**, *12*, 1515. [CrossRef]
129. Hengstmann, E.; Fischer, E.K. Anthropogenic litter in freshwater environments—Study on lake beaches evaluating marine guidelines and aerial imaging. *Environ. Res.* **2020**, *189*, 109945. [CrossRef]
130. Wurtsbaugh, W.A.; Paerl, H.W.; Dodds, W.K. Nutrients, eutrophication and harmful algal blooms along the freshwater to marine continuum. *Wiley Interdiscip. Rev. Water* **2019**, *6*. [CrossRef]
131. Fráter, T.; Juzsakova, T.; Lauer, J.; Dióssy, L.; Rédey, Á. Unmanned Aerial Vehicles in Environmental Monitoring—An Efficient Way for Remote Sensing. *J. Environ. Sci. Eng. A* **2015**, *4*, 85–91. [CrossRef]
132. Ngo, A.S.K.; Desingco, J.D.B.; Ii, M.O.C.; Uy, R.L.; Ong, P.M.B.; Punzalan, E.R.; Ilao, J.P. Determining the Correlation between Concentration Levels and the Visual Determining the Correlation Between Concentration Levels and the Visual Features of Algae in Water Surfaces. 2015. Available online: https://www.researchgate.net/publication/283086583_Determining_the_Correlation_Between_Concentration_Levels_and_the_Visual_Features_of_Algae_in_Water_Surfaces (accessed on 30 November 2020).
133. Van Der Merwe, D.; Price, K.P. Harmful Algal Bloom Characterization at Ultra-High Spatial and Temporal Resolution Using Small Unmanned Aircraft Systems. *Toxins* **2015**, *7*, 1065–1078. [CrossRef] [PubMed]
134. Su, T.-C.; Chou, H.-T. Application of Multispectral Sensors Carried on Unmanned Aerial Vehicle (UAV) to Trophic State Mapping of Small Reservoirs: A Case Study of Tain-Pu Reservoir in Kinmen, Taiwan. *Remote Sens.* **2015**, *7*, 10078–10097. [CrossRef]
135. Jang, S.W.; Yoon, H.J.; Kwak, S.N.; Sohn, B.Y.; Kim, S.G.; Kim, D.H. Algal Bloom Monitoring using UAVs Imagery. *Next Gener. Comput. Inf. Technol.* **2016**, *138*, 30–33. [CrossRef]
136. Aguirre-Gómez, R.; Salmerón-García, O.; Gómez-Rodríguez, G.; Peralta-Higuera, A. Use of unmanned aerial vehicles and remote sensors in urban lakes studies in Mexico. *Int. J. Remote Sens.* **2016**, *38*, 2771–2779. [CrossRef]
137. Salarux, C.; Kaewplang, S. Estimation of Algal Bloom Biomass Using UAV-Based Remote Sensing with NDVI and GRVI. *Mahasarakham Int. J. Eng. Technol.* **2020**, *6*, 1–6.
138. Honkavaara, E.; Hakala, T.; Kirjasniemi, J.; Lindfors, A.; Mäkynen, J.; Nurminen, K.; Ruokokoski, P.; Saari, H.; Markelin, L. New light-weight stereosopic spectrometric airborne imaging technology for high-resolution environmental remote sensing; case studies in water quality mapping. *ISPRS Int. Arch. Photogramm. Remote Sens. Spat. Inf. Sci.* **2013**, *XL-1/W1*, 139–144. [CrossRef]
139. Becker, R.H.; Sayers, M.; Dehm, D.; Shuchman, R.; Quintero, K.; Bosse, K.; Sawtell, R. Unmanned aerial system based spectroradiometer for monitoring harmful algal blooms: A new paradigm in water quality monitoring. *J. Great Lakes Res.* **2019**, *45*, 444–453. [CrossRef]
140. Penmetcha, M.; Luo, S.; Samantaray, A.; DIetz, J.E.; Yang, B.; Min, B.C. Computer vision-based algae removal planner for multi-robot teams. In Proceedings of the 2019 IEEE International Conference on Systems, Man and Cybernetics (SMC), Bari, Italy, 6–9 October 2019; pp. 1575–1581. [CrossRef]
141. Kwon, Y.S.; Pyo, J.; Duan, H.; Cho, K.H.; Park, Y. Drone-based hyperspectral remote sensing of cyanobacteria using vertical cumulative pigment concentration in a deep reservoir. *Remote Sens. Environ.* **2020**, *236*, 111517. [CrossRef]
142. Zhang, Y.; Wu, L.; Ren, H.; Liu, Y.; Zheng, Y.; Liu, Y.; Dong, J. Mapping Water Quality Parameters in Urban Rivers from Hyperspectral Images Using a New Self-Adapting Selection of Multiple Artificial Neural Networks. *Remote Sens.* **2020**, *12*, 336. [CrossRef]
143. Ore, J.P.; Elbaum, S.; Burgin, A.; Detweiler, C. Autonomous aerial water sampling. In *Field and Service Robotics*; Mejias, L., Corke, P., Roberts, J., Eds.; Springer International Publishing: Cham, Switzerland, 2015; pp. 137–151.
144. Detweiler, C.; Ore, J.-P.; Anthony, D.; Elbaum, S.; Burgin, A.J.; Lorenz, A. Environmental Reviews and Case Studies: Bringing Unmanned Aerial Systems Closer to the Environment. *Environ. Pract.* **2015**, *17*, 188–200. [CrossRef]
145. Schwarzbach, M.; Laiacker, M.; Mulero-Pazmany, M.; Kondak, K. Remote water sampling using flying robots. In Proceedings of the 2014 International conference on unmanned aircraft systems (ICUAS), Orlando, FL, USA, 27–30 May 2014; pp. 72–76. [CrossRef]
146. Terada, A.; Morita, Y.; Hashimoto, T.; Mori, T.; Ohba, T.; Yaguchi, M.; Kanda, W. Water sampling using a drone at Yugama crater lake, Kusatsu-Shirane volcano, Japan. *Earth Planets Space* **2018**, *70*, 64. [CrossRef]
147. Song, K.; Brewer, A.; Ahmadian, S.; Shankar, A.; Detweiler, C.; Burgin, A.J. Using unmanned aerial vehicles to sample aquatic ecosystems. *Limnol. Oceanogr. Methods* **2017**, *15*, 1021–1030. [CrossRef]
148. Elijah, O.; Rahman, T.A.; Leow, C.Y.; Yeen, H.C.; Sarijari, M.A.; Aris, A.; Salleh, J.; Chua, T.H. A concept paper on smart river monitoring system for sustainability in river. *Int. J. Integr. Eng.* **2018**, *10*, 130–139. [CrossRef]

149. Koparan, C.; Koc, A.B.; Privette, C.V.; Sawyer, C.B. In Situ Water Quality Measurements Using an Unmanned Aerial Vehicle (UAV) System. *Water* **2018**, *10*, 264. [CrossRef]
150. Esakki, B.; Ganesan, S.; Mathiyazhagan, S.; Ramasubramanian, K.; Gnanasekaran, B.; Son, B.; Park, S.W.; Choi, J.S. Design of Amphibious Vehicle for Unmanned Mission in Water Quality Monitoring Using Internet of Things. *Sensors* **2018**, *18*, 3318. [CrossRef]
151. Doi, H.; Akamatsu, Y.; Watanabe, Y.; Goto, M.; Inui, R.; Katano, I.; Nagano, M.; Takahara, T.; Minamoto, T. Water sampling for environmental DNA surveys by using an unmanned aerial vehicle. *Limnol. Oceanogr. Methods* **2017**, *15*, 939–944. [CrossRef]
152. Banerjee, B.; Raval, S.; Maslin, T.J.; Timms, W. Development of a UAV-mounted system for remotely collecting mine water samples. *Int. J. Mining Reclam. Environ.* **2018**, *34*, 385–396. [CrossRef]
153. Castendyk, D.; Straight, B.; Fllatreault, P.; Thlbeault, S.; Cameron, L. Aerial drones used to sample pit lake water reduce costs and improve safety. *Miner. Eng.* **2017**, *69*, 20–28.
154. Castendyk, D.; Straight, B.; Voorhis, J.; Somogyi, M.; Jepson, W.; Kucera, B. Using aerial drones to select sample depths in pit lakes. In Proceedings of the 13th International Conference on Mine Closure, Perth, Australia, 3–5 September 2019; pp. 1113–1126. [CrossRef]
155. Jung, S.; Cho, H.; Kim, N.; Kim, K.; Han, J.-I.; Myung, H. Development of Algal Bloom Removal System Using Unmanned Aerial Vehicle and Surface Vehicle. *IEEE Access* **2017**, *5*, 22166–22176. [CrossRef]
156. Agarwal, P.; Singh, M.K. A multipurpose drone for water sampling video surveillance. In Proceedings of the 2019 Second International Conference on Advanced Computational and Communication Paradigms (ICACCP), Gangtok, India, 25–28 February 2019; pp. 1–5. [CrossRef]
157. Grandy, J.J.; Galpin, V.; Singh, V.; Pawliszyn, J. Development of a Drone-Based Thin-Film Solid-Phase Microextraction Water Sampler to Facilitate On-Site Screening of Environmental Pollutants. *Anal. Chem.* **2020**, *92*, 12917–12924. [CrossRef] [PubMed]
158. Lally, H.; O'Connor, I.; Jensen, O.; Graham, C. Can drones be used to conduct water sampling in aquatic environments? A review. *Sci. Total Environ.* **2019**, *670*, 569–575. [CrossRef] [PubMed]
159. Abolt, C.; Caldwell, T.; Wolaver, B.; Pai, H. Unmanned aerial vehicle-based monitoring of groundwater inputs to surface waters using an economical thermal infrared camera. *Opt. Eng.* **2018**, *57*, 053113. [CrossRef]
160. Harvey, M.C.; Hare, D.K.; Hackman, A.; Davenport, G.; Haynes, A.B.; Helton, A.; Lane, J.J.W.; Briggs, M.A. Evaluation of Stream and Wetland Restoration Using UAS-Based Thermal Infrared Mapping. *Water* **2019**, *11*, 1568. [CrossRef]
161. Casas-Mulet, R.; Pander, J.; Ryu, D.; Stewardson, M.J.; Geist, J. Unmanned Aerial Vehicle (UAV)-Based Thermal Infra-Red (TIR) and Optical Imagery Reveals Multi-Spatial Scale Controls of Cold-Water Areas Over a Groundwater-Dominated Riverscape. *Front. Environ. Sci.* **2020**, *8*, 1–16. [CrossRef]
162. Rautio, A.; Korkka-Niemi, K.; Salonen, V.-P. *Thermal Infrared Remote Sensing in Assessing Ground/Surface Water Resources Related to the Hannukainen Mining Development Site*; Mine Water and Circular Economy IMWA: Lappeenranta, Finland, 2017.
163. Bandini, F.; Butts, M.; Jacobsen, T.V.; Bauer-Gottwein, P. Water level observations from unmanned aerial vehicles for improving estimates of surface water-groundwater interaction. *Hydrol. Process.* **2017**, *31*, 4371–4383. [CrossRef]
164. Tang, Q.; Schilling, O.S.; Kurtz, W.; Brunner, P.; Vereecken, H.; Franssen, H.-J.H. Simulating Flood-Induced Riverbed Transience Using Unmanned Aerial Vehicles, Physically Based Hydrological Modeling, and the Ensemble Kalman Filter. *Water Resour. Res.* **2018**, *54*, 9342–9363. [CrossRef]
165. Briggs, M.A.; Wang, C.; Day-Lewis, F.D.; Williams, K.H.; Dong, W.; Lane, J.W. Return flows from beaver ponds enhance floodplain-to-river metals exchange in alluvial mountain catchments. *Sci. Total Environ.* **2019**, *685*, 357–369. [CrossRef]
166. Furlan, L.M.; Rosolen, V.; Salles, J.; Moreira, C.A.; Ferreira, M.E.; Bueno, G.T.; Coelho, C.V.D.S.; Mounier, S. Natural superficial water storage and aquifer recharge assessment in Brazilian savanna wetland using unmanned aerial vehicle and geophysical survey. *J. Unmanned Veh. Syst.* **2020**, *8*, 224–244. [CrossRef]
167. Siebert, C.; Mallast, U.; Rödiger, T.; Ionescu, D.; Schwonke, F.; Hall, J.K.; Sade, A.R.; Pohl, T.; Merkel, B. Multiple sensor tracking of submarine groundwater discharge: Concept study along the Dead Sea. In Proceedings of the EGU General Assembly Conference Abstracts, Vienna, Austria, 24 April–2 May 2014; p. 11217.
168. Lee, E.; Yoon, H.; Hyun, S.P.; Burnett, W.C.; Koh, D.; Ha, K.; Kim, D.; Kim, Y.; Kang, K. Unmanned aerial vehicles (UAVs)-based thermal infrared (TIR) mapping, a novel approach to assess groundwater discharge into the coastal zone. *Limnol. Oceanogr. Methods* **2016**, *14*, 725–735. [CrossRef]
169. Rahman, M.M.; McDermid, G.J.; Strack, M.; Lovitt, J. A New Method to Map Groundwater Table in Peatlands Using Unmanned Aerial Vehicles. *Remote Sens.* **2017**, *9*, 1057. [CrossRef]
170. Bandini, F.; Lopez-Tamayo, A.; Merediz-Alonso, G.; Olesen, D.; Jakobsen, J.; Wang, S.; Garcia, M.; Bauer-Gottwein, P. Unmanned aerial vehicle observations of water surface elevation and bathymetry in the cenotes and lagoons of the Yucatan Peninsula, Mexico. *Hydrogeol. J.* **2018**, *26*, 2213–2228. [CrossRef]
171. Allred, B.; Eash, N.; Freeland, R.; Martinez, L.; Wishart, D. Effective and efficient agricultural drainage pipe mapping with UAS thermal infrared imagery: A case study. *Agric. Water Manag.* **2018**, *197*, 132–137. [CrossRef]
172. Kratt, C.; Woo, D.; Johnson, K.; Haagsma, M.; Kumar, P.; Selker, J.; Tyler, S. Field trials to detect drainage pipe networks using thermal and RGB data from unmanned aircraft. *Agric. Water Manag.* **2020**, *229*, 105895. [CrossRef]
173. Levy, M.C.; Neely, W.R.; Borsa, A.A.; Burney, J.A. Fine-scale spatiotemporal variation in subsidence across California's San Joaquin Valley explained by groundwater demand. *Environ. Res. Lett.* **2020**, *15*, 104083. [CrossRef]

174. Nayyeri, M.; Hosseini, S.A.; Javadi, S.; Sharafati, A. Spatial Differentiation Characteristics of Groundwater Stress Index and its Relation to Land Use and Subsidence in the Varamin Plain, Iran. *Nat. Resour. Res.* **2021**, *30*, 339–357. [CrossRef]
175. Yu, H.; Gong, H.; Chen, B.; Liu, K.; Gao, M. Analysis of the influence of groundwater on land subsidence in Beijing based on the geographical weighted regression (GWR) model. *Sci. Total Environ.* **2020**, *738*, 139405. [CrossRef]
176. Carlson, G.; Carnes, L.K.; Cook, J.P. Exploring Arizona earth fissures: An anthropogenic geologic hazard. *GSA Annu. Meet. Phoenix Ariz.* **2019**, *55*, 115–126. [CrossRef]
177. Hu, X.; Li, X. Information extraction of subsided cultivated land in high-groundwater-level coal mines based on unmanned aerial vehicle visible bands. *Environ. Earth Sci.* **2019**, *78*. [CrossRef]
178. Hu, X.; Li, X.; Min, X.; Niu, B. Optimal scale extraction of farmland in coal mining areas with high groundwater levels based on visible light images from an unmanned aerial vehicle (UAV). *Earth Sci. Inform.* **2020**, *13*, 1151–1162. [CrossRef]
179. Commission Implementing Regulation (EU) 2019/947 of 24 May 2019 on the Rules and Procedures for the Operation of Unmanned Aircraft. Available online: http://data.europa.eu/eli/reg_impl/2019/947/oj (accessed on 30 November 2020).

MDPI
St. Alban-Anlage 66
4052 Basel
Switzerland
Tel. +41 61 683 77 34
Fax +41 61 302 89 18
www.mdpi.com

Remote Sensing Editorial Office
E-mail: remotesensing@mdpi.com
www.mdpi.com/journal/remotesensing

www.ingramcontent.com/pod-product-compliance
Lightning Source LLC
LaVergne TN
LVHW070203170726
843515LV00007B/1990

* 9 7 8 3 0 3 6 5 1 6 6 6 0 *